Prentice-Hall, Inc., Englewood Cliffs, N.J. 07632

John Tobey, Jr.
North Shore Community College
Beverly, Massachusetts

BEGINNING ALGEBRA: A Direct Approach

Library of Congress Cataloging in Publication Data

Tobey, John,

Beginning algebra.

Includes index.

1. Algebra. I. Title.

QA152.2.T6 1984 512.9 83-13751

ISBN 0-13-071647-2

I dedicate this book to my parents, John and Lexie,
who have been for me model teachers for four decades

Editorial/production supervision: Karen J. Clemments
Interior design and cover design: Judith A. Matz
Manufacturing buyer: John Hall

Cover: Earth rising over the lunar landscape taken by Apollo astronauts in lunar orbit (courtesy NASA).

Placing a person on the moon and the recent advances in space technology have been among the outstanding achievements of humanity in recent history. Such technological advances are possible only with the use of both elementary algebra and other more advanced mathematical systems. It can be safely said that there is no part of the space exploration program (from the original design of the space vehicle to the final return to earth of the astronauts and their ship) that could have been done without the use of the procedures of elementary algebra.

Printed in the United States of America

10 9 8 7 6 5

ISBN 0-13-071647-2

Prentice-Hall International, Inc., *London*
Prentice-Hall of Australia Pty. Limited, *Sydney*
Editora Prentice-Hall do Brasil, Ltda., *Rio de Janeiro*
Prentice-Hall Canada Inc., *Toronto*
Prentice-Hall of India Private Limited, *New Delhi*
Prentice-Hall of Japan, Inc., *Tokyo*
Prentice-Hall of Southeast Asia Pte. Ltd., *Singapore*
Whitehall Books Limited, *Wellington, New Zealand*

Contents

Preface

A Special Note to the Student

Why This Book Was Made

This book was published to make the learning of mathematics both interesting and understandable. For many of you, it will actually become enjoyable if you make a sincere effort at learning from this book.

The book was designed to help you succeed in learning both what algebra is and how to use it. When you have completed this book you will better understand how mathematics is used in our technical society and you will be prepared to take more advanced mathematics courses.

What This Book Does Not Expect of You

This book has been created to help you learn the basic content area of algebra. It does not assume that you like mathematics. It does not assume that you did well in mathematics prior to this course. It does not assume you have had algebra before. If you have previously taken an algebra course, this book does not assume you remember all the main parts of the course.

How Can I Use This Book to Help Me?

As you encounter new definitions and procedures, take time to learn them and be able to explain them in your own words. In each section look over the sample examples carefully. In each case all the steps to solving a problem are shown and also explained in words. When you think you understand the sample examples in a given section try to work out the practice problems at the end of the section. Then turn over one or two pages and compare your answer to the completely worked out solution for each practice problem. If you made an error find out what mistake you made and how the step should be correctly done. Then you will be ready to do the homework exercise.

What Is The Secret to Understanding Algebra?

1. Beginning algebra is a course that introduces new *ideas* and *definitions*. You must take time to learn the definitions and be able to explain them in your own words. You will need to take the time to reflect on these new concepts and ideas

until they make sense to you. If you forget what a specific mathematical term means, turn to the glossary and review the definition.

2. Beginning algebra is a course that requires an *understanding* of generalized *concepts*. It is not merely a course where you "do a problem and get the right answer." It is important to ask what procedure is used and why that procedure works. You will need to take time to look over the problems. You need to see how one type of problem differs from another and to generalize what procedure should be used in each case.
3. Beginning algebra is a *skill* course. Like learning to type, swim, skateboard, or sky dive, it will take some practice until you really feel that you know what you are doing. So be sure to work out at least half the homework exercises in each set before you go on to the next section.

What Do I Do When I Finish a Chapter and I'm Nervous and Anxious About How Well I Will Do on The Test?

When you come to the end of a chapter review carefully each item in the chapter summary. Be sure you understand each concept and can use the concept in doing a problem. If you forget, go back to the numbered section where that concept was explained. When it is clear, try the problems in the practice quizzes and the practice test at the end of the chapter. When you have done these and checked your answers, you should be ready for any test your instructor may give on the content of the chapter.

A Special Note to the Instructor

Content and Philosophy

This book was written to cover all of the topics of an introductory algebra course in a clear and direct manner. It has been class tested for four semesters with students ranging in age from tenth graders in high school to sixty-five year old "senior students." It has been revised based on a wealth of suggestions from both students and faculty who used the book in manuscript form. The final printed result is designed to be intuitive and understandable and yet mathematically accurate; to be simple and direct and yet intellectually challenging.

Flexibility

This textbook was designed to be used successfully in any of these and other modes of instruction:

1. *Traditional Lecture*. The format of the book is appropriate for a lecture class. The book is not a cluttered collection of gimmicks, frames, and boxes. It presents material clearly, section by section, supported by a wealth of sample examples.
2. *Combined Lecture and Student Activity Approach*. In this approach the instructor stops the lecture at one or more points in the presentation and asks students to work one or more problems. In this text, practice problems are provided for each section of the book.
3. *Self-Paced Instruction*. In this approach students may be learning the material in a mathematics laboratory, a learning center, a no-lecture math course, or may be studying the material totally on their own. The inclusion of diagnostic chapter pretests, student learning objectives, practice problems, chapter summaries, practice chapter quizzes and tests are all helpful learning features that support the use of this book in a self-paced mode.

The content coverage of this book is very complete yet flexible, allowing an instructor to omit several sections or chapters not required at a particular school. (Specific suggestions are included in the Instructors Resource Manual.) The arithmetic review (Chapter Zero) can be included as part of the course or used as a supplementary resource for those students in the class that need it. To allow maximum flexibility the topics of ratio and proportion, the use of metric measurement, the conversion of units, the concept of negative integer exponents, and the use of scientific notation are covered in Appendixes B and C. They may be covered any time after Chapter One.

Features

1. A detailed review of fractions, decimals, and percents is presented in Chapter Zero and then the use of these concepts is carefully integrated throughout the remainder of the book.
2. Chapter Five is a detailed chapter that patiently guides the student to the solution of verbal problems. The student is taught how to reason and think through applied problem situations (without boxes, tricks, or gimmicks). All chapters after Chapter Five have a separate section devoted to application problems so that the student's skill in solving verbal problems is continually reinforced.
3. Effective use of second color highlights key procedures and emphasizes correct operations in specific areas where students are likely to make errors.
4. Clear, direct definitions of mathematical terms are emphasized. The student who really understands what is meant by the key mathematical terms of beginning algebra is the one who is likely to succeed in the course. A complete glossary of mathematical terms used in the book is provided.
5. A wealth of homework exercises are included *with* answers. The answers to all diagnostic pretests, exercises, practice quizzes, and practice tests are provided in the answer key in the back of the book. Each even-numbered problem is very similar to the preceding odd-numbered problem. Thus, a student who solves all even- or odd-numbered problems in a given section will have adequate exposure to each type of problem. If further drill or practice is necessary the student may do all the problems in a given section. Ample workspace is provided to enable the student to solve all the exercises directly in the book.
6. An ample supply of problem exercises appear *without* answers. The answers to the challenge problems for each section and the answers to the supplementary problems for each chapter do not have answers provided in the answer key. The answers are provided in the Instructor's Resource Manual. The supplementary problem sets may be used for graded homework assignments or quizzes, since students do not have access to these answers.
7. Chapter Six is a complete coverage of graphing techniques and a thorough review of common geometric formulas. This book provides a well-rounded chapter that is mathematically complete. (It is unlike many current elementary algebra texts that shortchange this important area).
8. Most instructors have heard the comment "I understood your lecture but when I went home to do the problems, I was not able to get the correct answer." To help students with this problem, there are three or four well-chosen practice problems at the conclusion of each section that illustrate the range of problems in the homework exercises. The student attempts to solve the practice problem, then turns to the answer and the worked out step-by-step solution. The student who makes an error is able to find quickly his or her mistake and correct it. Thus prepared by practice, the student is more likely to have success when completing the homework exercises.

9. Each chapter begins with a diagnostic pretest and ends with two or three practice quizzes and a practice test to assist the student in determining if he or she has mastered the content of the chapter. At the completion of the book a practice cumulative examination covering Chapters One to Nine is provided.

Opportunities to Foster Intellectual Curiosity and Academic Growth

1. Some elementary algebra texts are so simplistic and skill oriented that the better students are not challenged. This text provides a few challenge problems in each section to provide the better-prepared student with a meaningful challenge.
2. At the end of each chapter is a nontrivial application of the content of the chapter. This helps the student to broaden his or her intellectual horizons and to appreciate the wide range of applications of algebra.
3. In many places the introduction of a new concept is followed by a boxed question mark and a probing question. This feature is designed to help the student think through answers to such questions as "Why?," "Where did that formula come from?," "What is the purpose in using this method?," "Why is it necessary to perform that last step?"

Instructors Resource Manual

To aid in your program of instruction using *Beginning Algebra: A Direct Approach* a resource manual has been prepared. It includes suggested plans for adapting the topical coverage and amounts of coverage of the material in the text to the course at your school. It also includes suggestions for the use of the text in self-paced instruction situations. Quizzes, tests, and final examinations, ready for reproduction, are a part of the manual. The quizzes come in two versions, each covering approximately one half of a chapter for every part of the entire text. A total of 21 quiz units with two forms each are included. Chapter tests are available in five different equivalent versions for each of the nine chapters of the book. There are two forms of a cumulative final examination covering the content of Chapters One to Nine. Finally, the manual has answers to the challenge problems in each section and the supplementary problems for each chapter.

Acknowledgments

I would like to thank my editors at Prentice Hall: Robert Sickels, Dennis Hogan, Marcia Horton, Nancy Ranno, and Karen Clemments for their support, encouragement, and suggestions in bringing this project to completion.

I am deeply grateful to Thomas M. Rourke and F. Russell Sullivan who taught from the manuscript version of this book and provided helpful suggestions for its improvement. I would like to thank Cheryl Liacos-Halstead and Peter Leblanc for working out the problems in the text and checking the solutions. Joan Peabody patiently typed and retyped the various versions of the manuscript and her dedication and cooperation in this task have been invaluable. My colleagues and fellow authors, Jeffrey Slater and Hank Harmeling, encouraged me onward in the task of writing.

I am particularly grateful to Joseph Albree, Auburn University at Montgomery, Thomas M. Rourke, North Shore Community College, and Ralph D. Lewis, Salem State College for their detailed reviews and analyses of every aspect of the final text and their helpful suggestions. I would also like to thank Ann Bretscher, The University of Georgia, Athens, Georgia; Steven Brown, Auburn University, Auburn, Alabama; J. Curtis Chipman, Oakland University, Rochester, Michigan; Pat Deamer, Skyline College, San Bruno, California; Curtis Gooden, Cuyahoga Community College–

Metropolitan Campus, Cleveland, Ohio; James Jerkofsky, Benedictine College, Atchison, Kansas; Martin W. Johnson, Montgomery County Community College, Blue Bell, Pennsylvania; Robert Maynard, Rockingham Community College, Wentworth, North Carolina; William Neal, Fresno City College, Fresno, California; Richard Negley, North Georgia College, Dahlonega, Georgia; James Newsom, Tidewater Community College, Virginia Beach, Virginia; E. James Peake, Iowa State University, Ames, Iowa; Ray Pluta, Castleton State College, Castleton, Vermont; Nellie Tindoll, University of Mississippi, University, Mississippi; Thomas L. Van Wingen, Grand Rapids Junior College, Grand Rapids, Michigan; and Ben Zirkle, Virginia Western Community College, Roanoke, Virginia, for reviewing various parts of the text and providing constructive comments.

I would like to thank my wife, Nancy, and my children, Johnny, Melissa, and Marcia, who always had an encouraging word for me as I was writing this book. Finally, I thank God for the strength and energy to write and the opportunity to help others learn through this textbook.

J. T.

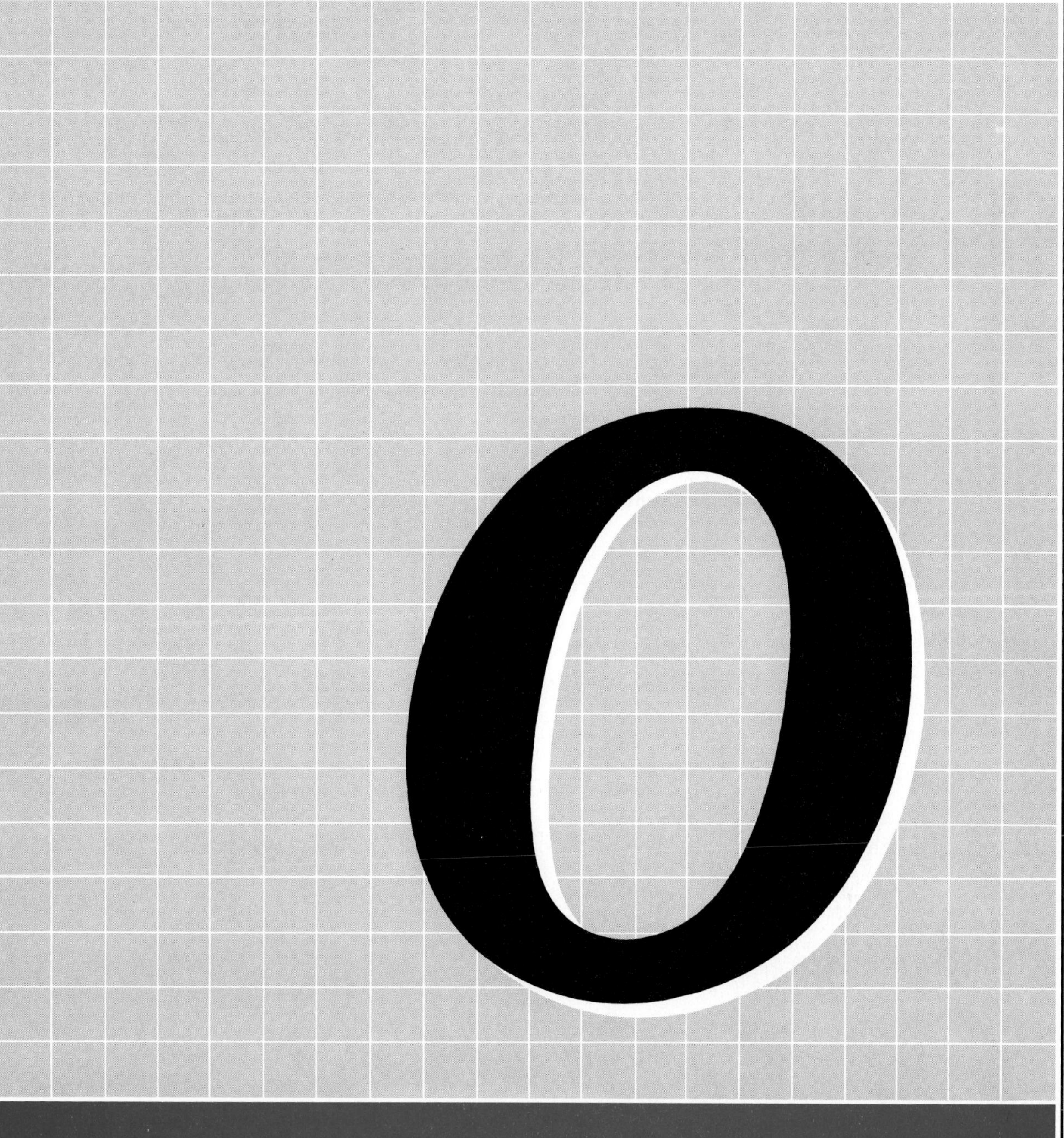

A Brief Review of Arithmetic Skills

Recent studies have indicated that college students taking a beginning algebra course often have difficulty because of a weakness in certain arithmetic skills using fractions, decimals, and percents. It is therefore important for you to determine if you have some deficiencies in this area before you begin the study of algebra in Chapter 1.

Please take the Diagnostic Pretest for Chapter 0 on the next two pages and compare your answers to the solutions for the pretest listed in the Answer Key at the end of the book. If you obtain the correct answer for 12 or more problems, you are ready to begin Chapter 1. If you did not get at least 12 correct answers, you should review the necessary sections of Chapter 0 so that you can improve your arithmetic skills. If you *take the time to master these arithmetic skills now*, you will definitely find Chapter 1 and remaining chapters of this book easier to complete.

Diagnostic Pretest for Chapter 0

Be sure to reduce all answers.

Section 0.1

1. Reduce this fraction:

$$\frac{21}{27} =$$

2. Write as a mixed fraction:

$$\frac{15}{4} =$$

3. Change the given fraction to a new form of the fraction with the specified denominator:

$$\frac{3}{7} = \frac{?}{42}$$

Section 0.2

Combine the fractions:

4. $\frac{8}{9} + \frac{3}{21} =$

5. $3\frac{1}{5} - 1\frac{3}{8} =$

Section 0.3

6. Multiply these fractions:

$$\frac{25}{7} \times \frac{14}{45} =$$

7. Divide these fractions:

$$2\frac{1}{3} \div 3\frac{1}{4} =$$

Section 0.4

Perform the following calculations.

8. $5.23 + 3.6 + 1.821 =$

9. $3.28 \times 0.63 =$

10. $3.015 \div 6.7 =$

11. Change to decimal form:

$$\frac{5}{8} =$$

Section 0.5

12. Convert to decimal form:

$32\% =$

13. What is 24% of 1630?

14. What percent of 500 is 36?

15. 88% of what number is 792?

0.1 Reducing and Transforming Fractions

Basic definitions

Whole numbers are the set of numbers 0, 1, 2, 3, 4, 5, 6, 7, ... They are used to describe whole objects, or entire quantities.

Fractions are a set of numbers that are used to describe parts of whole quantities. In the object shown in Figure 0.1 there are four equal parts. The three parts that are shaded are represented by the fraction $\frac{3}{4}$.

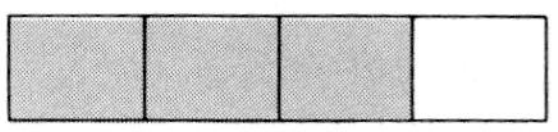

FIGURE 0.1

In the fraction $\frac{3}{4}$ the number 3 is called the *numerator* and the number 4, the *denominator*.

$$\frac{3}{4}\begin{array}{l}\longleftarrow \textit{Numerator} \text{ is on the top}\\ \longleftarrow \textit{Denominator} \text{ is on the bottom}\end{array}$$

The *denominator* of a fraction shows the number of parts in the whole and the *numerator* shows the number of these parts being talked about or being used.

Numerals are symbols we use to name numbers. There are many different numerals that can be used to describe the same number. We know that $\frac{1}{2} = \frac{2}{4}$. The fractional numerals $\frac{1}{2}$ and $\frac{2}{4}$ both describe the same number.

Usually, we find it more useful to use fractions that are reduced or simplified. A fraction is considered to be *reduced* when the numerator (top) and the denominator (bottom) can both be divided exactly by no number other than 1.

> $\frac{1}{2}$ is a *reduced* fraction.
>
> $\frac{2}{4}$ is not a *reduced* fraction since the numerator and the denominator can both be divided by 2.

Whole numbers can also be named with fractional notation. 3, $\frac{9}{3}$, $\frac{6}{2}$, and $\frac{3}{1}$ are ways of expressing the number *three*. Therefore,

$$3 = \frac{9}{3} = \frac{6}{2} = \frac{3}{1}$$

Similarly, the number *one* can be expressed as 1, $\frac{1}{1}$, $\frac{2}{2}$, $\frac{6}{6}$, $\frac{8}{8}$, etc., since

$$1 = \frac{1}{1} = \frac{2}{2} = \frac{6}{6} = \frac{8}{8}$$

0.1.1 Reducing Fractions

Natural numbers or *counting numbers* are the set of whole numbers excluding 0. Thus the natural numbers are the numbers 1, 2, 3, 4, 5, 6, ...

When two or more numbers are multiplied, each number that is multiplied is called a *factor*.

When we write $3 \times 7 \times 5$, each of the numbers 3, 7, and 5 is called a factor.

Prime numbers are all natural numbers greater than 1 whose only natural number factors are 1 and itself.

The number 5 is prime. The only natural number factors of 5 are 5 and 1.

$$5 = 5 \times 1$$

The number 6 is not prime. The natural number factors of 6 are 3 and 2 or 6 and 1.

$$6 = 3 \times 2 \qquad 6 = 6 \times 1$$

The first 15 prime numbers are

2, 3, 5, 7, 11, 13, 17, 19, 23, 29, 31, 37, 41, 43, 47

Any natural number greater than 1 is either prime or can be written as the product of prime numbers.

EXAMPLE A Indicate if the following numbers are prime. If they are not prime, write them as the product of prime numbers. **1.** 12 **2.** 30 **3.** 14 **4.** 19 **5.** 29

Solution We write as follows.

1. $12 = 2 \times 2 \times 3$

2. $30 = 2 \times 3 \times 5$

3. $14 = 2 \times 7$

4. 19 is a prime number.

5. 29 is a prime number.

Fractions can be reduced by writing each part of the fraction (numerator and denominator) as a product of prime numbers. Cross out any *factor* that appears in the *numerator* (top) and *denominator* (bottom) of the fraction.

EXAMPLE B Reduce each fraction. **1.** $\frac{14}{21}$ **2.** $\frac{15}{35}$ **3.** $\frac{20}{70}$

Solution **1.** $\frac{14}{21} = \frac{\cancel{7} \times 2}{\cancel{7} \times 3} = \frac{2}{3}$ **2.** $\frac{15}{35} = \frac{\cancel{5} \times 3}{\cancel{5} \times 7} = \frac{3}{7}$ **3.** $\frac{20}{70} = \frac{2 \times \cancel{2} \times \cancel{5}}{7 \times \cancel{2} \times \cancel{5}} = \frac{2}{7}$

If all the prime factors in the top (numerator) cancel out, we also leave a 1 there; otherwise, we would not realize that the answer was a fraction. Any number can be written as the product of itself and 1.

EXAMPLE C Reduce each fraction. **1.** $\frac{7}{21}$ **2.** $\frac{22}{66}$ **3.** $\frac{15}{105}$

Solution **1.** $\frac{7}{21} = \frac{\cancel{7} \times 1}{\cancel{7} \times 3} = \frac{1}{3}$ **2.** $\frac{22}{66} = \frac{\cancel{2} \times \cancel{11} \times 1}{3 \times \cancel{2} \times \cancel{11}} = \frac{1}{3}$ **3.** $\frac{15}{105} = \frac{\cancel{5} \times \cancel{3} \times 1}{7 \times \cancel{5} \times \cancel{3}} = \frac{1}{7}$

This is of course not necessary if all the prime numbers in the bottom (denominator) cancel out since we do not need to express the answer as a fraction. The answer is then a whole number and is not usually expressed in fractional form.

EXAMPLE D Reduce each fraction. **1.** $\frac{35}{7}$ **2.** $\frac{70}{10}$ **3.** $\frac{30}{6}$

Solution **1.** $\frac{35}{7} = \frac{5 \times \cancel{7}}{\cancel{7}} = 5$ **2.** $\frac{70}{10} = \frac{7 \times \cancel{5} \times \cancel{2}}{\cancel{5} \times \cancel{2}} = 7$ **3.** $\frac{30}{6} = \frac{5 \times \cancel{3} \times \cancel{2}}{\cancel{3} \times \cancel{2}} = 5$

?

Why does this method of reducing fractions work?
What is really going on when we reduce fractions?

We observe that $1 \times 5 = 5$, $1 \times 7 = 7$, $1 \times 108 = 108$, etc. When we reduce fractions, we are actually using the fact that 1 can be multiplied by any number without changing the value of that number. (Mathematicians usually call this the multiplicative identity.) Let's look again at one of the problems we reduced above.

$$\frac{14}{21} = \frac{7 \times 2}{7 \times 3}$$ Since $7 \times 2 = 14$ and since $7 \times 3 = 21$.

$$= \frac{7}{7} \times \frac{2}{3}$$ By the definition of multiplication of fractions (which will be reviewed in Section 0.3.1).

$$= 1 \times \frac{2}{3}$$ Since $\frac{7}{7}$ is another name for 1.

$$= \frac{2}{3}$$ Since 1 can be multiplied by any number without changing the value of that number.

So we see that

$$\frac{14}{21} = \frac{2}{3}$$

When we reduce fractions we are using this property of multiplying by 1.

There is a second concept that will help you understand reducing fractions and why the steps of reducing fractions really work. That second concept is the concept of *division*. The structure of *fractional notation* is based on the idea of *division of one whole number by another*. (In fact, fractions are sometimes defined as the quotient of two whole numbers where the divisor is not zero.) Let's look again at the fraction $\frac{35}{7}$, which we reduced earlier.

$$\frac{35}{7} = \frac{5 \times \not{7}}{\not{7}} = 5$$

Since fractions can be considered as the division of two whole numbers we can write

$$\frac{35}{7} = 35 \div 7 = 5$$

which is exactly the result we obtained previously. We can see that when we reduce fractions and obtain a whole number for an answer it is equivalent to dividing the numerator by the denominator.

Review the examples in this section and then try these practice problems.

Practice Problems

Reduce.

(a) $\frac{4}{12} =$ **(b)** $\frac{10}{16} =$ **(c)** $\frac{24}{36} =$

(d) $\frac{25}{125} =$ **(e)** $\frac{18}{6} =$ **(f)** $\frac{42}{36} =$

Solutions to Practice Problems

(a) $\frac{1}{3}$ **(b)** $\frac{5}{8}$ **(c)** $\frac{2}{3}$

(d) $\frac{1}{5}$ **(e)** 3 **(f)** $\frac{7}{6}$

0.1.2 Improper Fractions and Mixed Fractions

If the numerator is less than the denominator the fraction is a *proper fraction.* A proper fraction is used to describe a quantity smaller than a whole.

Fractions can also be used to describe quantities larger than a whole. Figure 0.2 shows two boxes that are equal in size. Each box is divided into 5 equal pieces. The first box is shaded in completely. The second box has 2 of the 5 pieces shaded in.

FIGURE 0.2

The shaded-in region can be represented by $\frac{7}{5}$ since 7 of the pieces (each of which is $\frac{1}{5}$ of a whole box) are shaded. The fraction $\frac{7}{5}$ is called an improper fraction. An *improper fraction* is one in which the numerator is larger than or equal to the denominator.

The shaded-in region can also be represented by 1 whole added to $\frac{2}{5}$ of a whole, or $1 + \frac{2}{5}$. This is written as $1\frac{2}{5}$. The fraction $1\frac{2}{5}$ is called a mixed fraction. A *mixed fraction* consists of a whole number added to a fraction whose numerator is smaller than its denominator. The addition is understood but not written. When we write $1\frac{2}{5}$ it represents $1 + \frac{2}{5}$. The numbers $1\frac{7}{8}$, $2\frac{3}{4}$, $8\frac{1}{3}$, and $126\frac{1}{10}$ are all mixed fractions.

From Figure 0.2 it seems clear that $\frac{7}{5} = 1\frac{2}{5}$. This suggests that we can change from one form to the other without changing the value of the fraction.

From a picture it is easy to see how to *change improper fractions to mixed fractions.* For example, if we start with the fraction $\frac{11}{3}$ and represent it by Figure 0.3 (where 11 of the pieces that are $\frac{1}{3}$ of a box are shaded), we see that $\frac{11}{3} = 3\frac{2}{3}$ since 3 whole boxes and $\frac{2}{3}$ of a box are shaded.

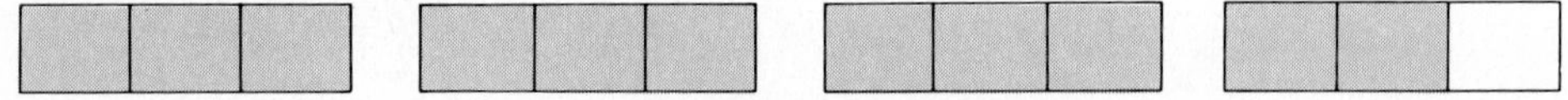

FIGURE 0.3

You can do the same procedure without a picture. For example, to change $\frac{15}{7}$ to a mixed fraction we can do the following:

$\frac{15}{7} = \frac{7}{7} + \frac{7}{7} + \frac{1}{7}$	Property of addition of fractions (will be discussed in detail in Section 0.2).
$= 1 + 1 + \frac{1}{7}$	Since $\frac{7}{7} = 1$.
$= 2 + \frac{1}{7}$	Since $1 + 1 = 2$.
$= 2\frac{1}{7}$	Notation used to write mixed fractions.

Now that you know how to perform the change and why it works, here is a shorter method.

To change an improper fraction to a mixed fraction, divide the denominator into the numerator. The result is the whole-number part of the mixed fraction.

The remainder from the division will be the numerator of the fraction. The denominator of the fraction remains unchanged.

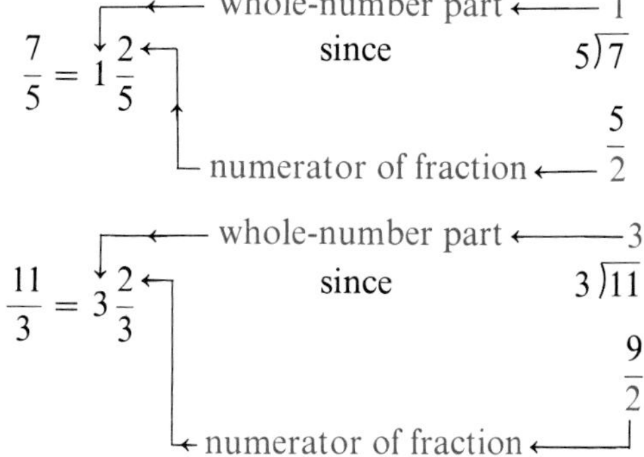

In the same way you can show that

$$\frac{9}{7} = 1\frac{2}{7} \qquad \frac{17}{8} = 2\frac{1}{8} \qquad \frac{19}{4} = 4\frac{3}{4}$$

It is not difficult to see how to change mixed fractions to improper fractions. Suppose that you wanted to write $2\frac{2}{3}$ as an improper fraction:

$$2\frac{2}{3} = 2 + \frac{2}{3} \qquad \text{The meaning of mixed fraction notation.}$$

$$= 1 + 1 + \frac{2}{3} \qquad \text{Since } 1 + 1 = 2.$$

$$= \frac{3}{3} + \frac{3}{3} + \frac{2}{3} \qquad \text{Since } 1 = \frac{3}{3}.$$

When we visualize this, we have Figure 0.4.

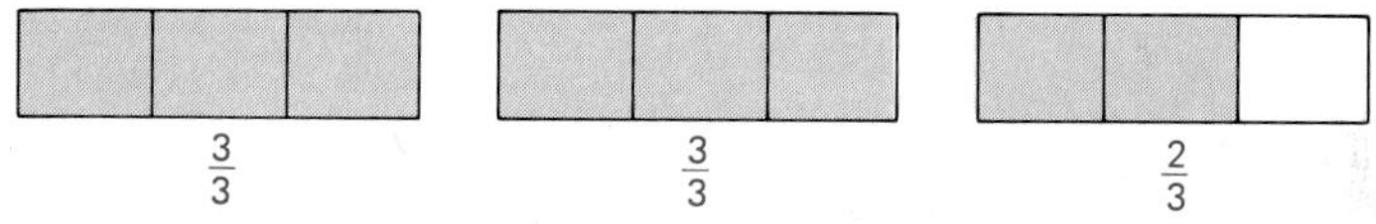

FIGURE 0.4

Thus if we count the squares (see Figure 0.5) we see that

$$\frac{3}{3} + \frac{3}{3} + \frac{2}{3} = \frac{8}{3}$$

FIGURE 0.5

Thus

$$2\frac{2}{3} = \frac{8}{3}$$

Now that you can see how this change can be done and why it works, here is a shorter method.

To change a mixed fraction to an improper fraction, multiply the whole number by the denominator and add this to the numerator. The result is the new numerator. The denominator does not change.

EXAMPLE E Change $2\frac{2}{3}$ to an improper fraction.

Solution

$$2\frac{2}{3} = \frac{(2 \times 3) + 2}{3} = \frac{6 + 2}{3} = \frac{8}{3}$$

EXAMPLE F Change to an improper fraction. **1.** $5\frac{1}{2}$ **2.** $3\frac{1}{7}$ **3.** $5\frac{4}{5}$ **4.** $3\frac{3}{4}$

Solution

1. $5\frac{1}{2} = \frac{(5 \times 2) + 1}{2} = \frac{10 + 1}{2} = \frac{11}{2}$

2. $3\frac{1}{7} = \frac{(3 \times 7) + 1}{7} = \frac{21 + 1}{7} = \frac{22}{7}$

3. $5\frac{4}{5} = \frac{(5 \times 5) + 4}{5} = \frac{25 + 4}{5} = \frac{29}{5}$

4. $3\frac{3}{4} = \frac{(3 \times 4) + 3}{4} = \frac{12 + 3}{4} = \frac{15}{4}$

Review the procedures and examples for changing mixed fractions to improper fractions and improper fractions to mixed fractions. Then try the following practice problems.

Practice Problems

Change to mixed fractions.

(a) $\frac{7}{4}$ **(b)** $\frac{8}{3}$ **(c)** $\frac{12}{7}$

(d) $\frac{14}{5}$

Change to improper fractions.

(e) $3\frac{2}{5}$ **(f)** $1\frac{3}{7}$ **(g)** $2\frac{6}{11}$

(h) $4\frac{2}{3}$ **(i)** $2\frac{5}{6}$

Solutions to Practice Problems

(a) $1\frac{3}{4}$ **(b)** $2\frac{2}{3}$ **(c)** $1\frac{5}{7}$

(d) $2\frac{4}{5}$ **(e)** $\frac{17}{5}$ **(f)** $\frac{10}{7}$

(g) $\frac{28}{11}$ **(h)** $\frac{14}{3}$ **(i)** $\frac{17}{6}$

0.1.3 Changing to Different Denominators

Fractions can be changed to an equivalent fraction with a different denominator by multiplying both numerator and denominator by the same number.

$$\frac{5}{6} = \frac{5 \times 2}{6 \times 2} = \frac{10}{12} \qquad \frac{3}{7} = \frac{3 \times 3}{7 \times 3} = \frac{9}{21}$$

We often want to get a *particular denominator.*

EXAMPLE G Find the missing number **1.** $\frac{3}{5} = \frac{?}{25}$ **2.** $\frac{3}{7} = \frac{?}{21}$ **3.** $\frac{2}{9} = \frac{?}{36}$

Solution

1. $\frac{3}{5} = \frac{?}{25}$ Here we observe that we need to multiply the denominator by 5 to obtain 25. We multiply the 3 by 5 also. Thus

$\frac{3 \times 5}{5 \times 5} = \frac{15}{25}$ The desired numerator is 15.

2. $\frac{3}{7} = \frac{?}{21}$ Observe that $7 \times 3 = 21$.

$\frac{3 \times 3}{7 \times 3} = \frac{9}{21}$ The desired numerator is 9.

3. $\frac{2}{9} = \frac{?}{36}$ Observe that $9 \times 4 = 36$.

$\frac{2 \times 4}{9 \times 4} = \frac{8}{36}$ The desired numerator is 8.

Study these examples and then try the following practice problems.

Practice Problems

(a) $\frac{3}{8} = \frac{?}{24}$ **(b)** $\frac{5}{6} = \frac{?}{30}$ **(c)** $\frac{12}{13} = \frac{?}{26}$

(d) $\frac{2}{7} = \frac{?}{56}$ **(e)** $\frac{5}{9} = \frac{?}{27}$ **(f)** $\frac{3}{10} = \frac{?}{60}$

(g) $\frac{3}{4} = \frac{?}{28}$ **(h)** $\frac{8}{11} = \frac{?}{55}$

Solutions to Practice Problems

(a) 9 **(b)** 25 **(c)** 24

(d) 16 **(e)** 15 **(f)** 18

(g) 21 **(h)** 40

Exercise 0.1

Reduce as much as possible.

1. $\frac{18}{24} =$ **2.** $\frac{35}{49} =$ **3.** $\frac{16}{40} =$

Change to an *improper fraction.*

4. $3\frac{1}{5} =$ **5.** $2\frac{6}{7} =$ **6.** $5\frac{2}{3} =$

Change to a *mixed fraction.*

7. $\frac{13}{8} =$ **8.** $\frac{39}{7} =$ **9.** $\frac{21}{9} =$

Find the missing numerator when changing each of the following to a new fraction with the specified denominator.

10. $\frac{3}{11} = \frac{?}{44}$ **11.** $\frac{5}{7} = \frac{?}{28}$ **12.** $\frac{3}{5} = \frac{?}{35}$

13. $\frac{2}{7} = \frac{?}{21}$ **14.** $\frac{9}{11} = \frac{?}{55}$ **15.** $\frac{13}{17} = \frac{?}{51}$

Solve these word problems.

16. A baseball player obtained a hit 342 times out of 620 times at bat. Express as a fraction in reduced form how much of the time she obtains a hit.

17. Juan got 7 out of 8 problems correct on a math quiz. The professor stated that there would be 136 problems all semester on tests and quizzes. How many *more* problems will Juan need to work correctly on future tests and quizzes to maintain the same fractional ratio of correct problems?

0.2 Addition and Subtraction of Fractions

0.2.1 Addition and Subtraction of Fractions with the Same Denominator

If fractions have the same denominator the numerators may be added or subtracted. For example,

EXAMPLE A Add the fractions. **1.** $\frac{2}{3} + \frac{1}{3}$ **2.** $\frac{5}{7} + \frac{1}{7}$ **3.** $\frac{1}{8} + \frac{1}{8} + \frac{3}{8} + \frac{2}{8}$

Solution **1.** $\frac{2}{3} + \frac{1}{3} = \frac{3}{3} = 1$ **2.** $\frac{5}{7} + \frac{1}{7} = \frac{6}{7}$ **3.** $\frac{1}{8} + \frac{1}{8} + \frac{3}{8} + \frac{2}{8} = \frac{7}{8}$

EXAMPLE B Subtract the fractions. **1.** $\frac{9}{11} - \frac{2}{11}$ **2.** $\frac{6}{7} - \frac{4}{7}$ **3.** $\frac{5}{13} - \frac{3}{13}$

Solution **1.** $\frac{9}{11} - \frac{2}{11} = \frac{7}{11}$ **2.** $\frac{6}{7} - \frac{4}{7} = \frac{2}{7}$ **3.** $\frac{5}{13} - \frac{3}{13} = \frac{2}{13}$

Study these examples and then try the following practice problems.

Practice Problems

(a) $\frac{3}{6}+\frac{2}{6}=$ **(b)** $\frac{3}{9}+\frac{2}{9}=$ **(c)** $\frac{3}{11}+\frac{2}{11}+\frac{6}{11}=$

(d) $\frac{7}{9}-\frac{3}{9}=$ **(e)** $\frac{5}{8}+\frac{2}{8}+\frac{4}{8}=$ **(f)** $\frac{11}{13}-\frac{6}{13}=$

Solutions to Practice Problems

(a) $\frac{5}{6}$ **(b)** $\frac{5}{9}$ **(c)** $\frac{11}{11}=1$

(d) $\frac{4}{9}$ **(e)** $\frac{11}{8}$ **(f)** $\frac{5}{13}$

0.2.2 Addition and Subtraction of Fractions that Do Not Have the Same Denominator

If fractions do not have the same denominator, we must change them to a common denominator before adding or subtracting. To save work we want to select the smallest possible common denominator. This is usually called the *least common denominator* or LCD. (Some people refer to it as the *lowest common denominator*.)

The LCD of two or more fractions is the smallest whole number that is exactly divisible by each denominator of the fractions.

EXAMPLE C Find the LCD:

$$\frac{2}{3} \text{ and } \frac{1}{4}$$

Solution The numbers are small enough to find the LCD by inspection. It is 12, since it is exactly divisible by 4 and by 3. There is no smaller number that is exactly divisible by 4 and 3.

EXAMPLE D Perform the following subtraction:

$$\frac{2}{3}-\frac{1}{4}$$

Solution We change each fraction to an equivalent fraction with a common denominator of 12, the LCD.

$$\frac{2}{3}=\frac{?}{12} \quad \frac{2}{3}\times\frac{4}{4}=\frac{8}{12} \quad \text{so} \quad \frac{2}{3}=\frac{8}{12}$$

$$\frac{1}{4}=\frac{?}{12} \quad \frac{1}{4}\times\frac{3}{3}=\frac{3}{12} \quad \text{so} \quad \frac{1}{4}=\frac{3}{12}$$

Then we rewrite the problem with common denominators:

$$\frac{2}{3}-\frac{1}{4}=\frac{8}{12}-\frac{3}{12}=\frac{8-3}{12}=\frac{5}{12}$$

Sometimes one of the denominators is the LCD. In such cases the fraction that has the LCD for the denominator will not need to be changed. If every other denominator divides into the largest denominator, then the largest denominator is the LCD.

EXAMPLE E Find the LCD:

$$\frac{3}{5} \text{ and } \frac{7}{20} \text{ and } \frac{1}{2}$$

Solution We can see by inspection that both 5 and 2 divide exactly into 20. Thus 20 is the LCD.

EXAMPLE F Add

$$\frac{3}{5}+\frac{7}{20}+\frac{1}{2}$$

Solution We change $\frac{3}{5}$ and $\frac{1}{2}$ to equivalent fractions with a common denominator of 20, the LCD.

$$\frac{3}{5}=\frac{?}{20} \quad \frac{3}{5}\times\frac{4}{4}=\frac{12}{20} \quad \text{so} \quad \frac{3}{5}=\frac{12}{20}$$

$$\frac{1}{2}=\frac{?}{20} \quad \frac{1}{2}\times\frac{10}{10}=\frac{10}{20} \quad \text{so} \quad \frac{1}{2}=\frac{10}{20}$$

Then we rewrite the problem with common denominators:

$$\frac{3}{5}+\frac{7}{20}+\frac{1}{2}=\frac{12}{20}+\frac{7}{20}+\frac{10}{20}=\frac{12+7+10}{20}=\frac{29}{20} \text{ or } 1\frac{9}{20}$$

In some cases the LCD cannot easily be determined by inspection. In such cases we can use a procedure that depends on the use of prime numbers. In this procedure we express numbers as prime factors. For example, if we write 30 as a product of prime factors, we would have $30 = 2 \cdot 3 \cdot 5$. The individual factors here are 2, 3, and 5. In algebra it is common to use a $(\cdot)$ to indicate multiplication. Thus $2 \cdot 3 \cdot 5 = 2 \times 3 \times 5$. You are encouraged to use the following procedure when you cannot readily determine the LCD by inspection.

Procedure to Find the LCD

1. **Write each denominator as the product of prime factors.**
2. **The LCD is the product of each different factor.**
3. **If a factor occurs more than once in any one denominator, the LCD will contain that factor repeated the greatest number of times that it occurs in any one denominator.**

EXAMPLE G Find the LCD

$$\frac{5}{6} \text{ and } \frac{1}{15}$$

by this new procedure:

Solution

$6 = 3 \cdot 2$

$15 = 5 \cdot 3$

We write each of the denominators as the product of prime factors.

$\text{LCD} = 2 \cdot 3 \cdot 5 = 30$

The LCD is the product of each different prime factor. The different factors are 2, 3, and 5.

EXAMPLE H Find the LCD:

$$\frac{3}{14} \text{ and } \frac{8}{35} \text{ and } \frac{6}{15}$$

Solution

$14 = 7 \cdot 2$

$35 = 7 \cdot 5$

$15 = 5 \cdot 3$

We write each of the denominators as the product of prime factors.

$\text{LCD} = 2 \cdot 3 \cdot 5 \cdot 7 = 210$

The LCD is the product of each different prime factor. The different factors are 2, 3, 5, 7.

Great care should be used to determine the LCD in the case of repeated factors.

EXAMPLE I Find the LCD:

$$\frac{5}{12} \text{ and } \frac{1}{15} \text{ and } \frac{7}{30}$$

Solution

$12 = 4 \cdot 3 = 2 \cdot 2 \cdot 3$

$15 = 5 \cdot 3$

We write each of the denominators as the product of prime factors.

$30 = 5 \cdot 6 = 5 \cdot 3 \cdot 2$

Note: The repeated factor is 2, which *occurs twice* in the factorization of 12.

$\text{LCD} = 2 \cdot 2 \cdot 3 \cdot 5 = 60$

The LCD is the product of each different factor with the factor 2 repeated twice since it occurred twice in one denominator.

EXAMPLE J **1.** Find the LCD:

$$\frac{4}{27} \text{ and } \frac{5}{18}$$

2. Now add the two fractions.

Solution **1.** $27 = 9 \cdot 3 = 3 \cdot 3 \cdot 3$

$18 = 9 \cdot 2 = 3 \cdot 3 \cdot 2$

We write each of the denominators as the product of prime factors. We observe that the factor 3 occurs three times in the factorization of 27.

$\text{LCD} = 3 \cdot 3 \cdot 3 \cdot 2 = 54$

The LCD is the product of each different factor. The factor 3 *occurred most* in the factorization of 27, where it was multiplied *three* times. Thus the LCD will be the product of *three* 3's and *one* 2.

2. Now we add:

$$\frac{4}{27} + \frac{5}{18}$$

$\frac{4}{27} = \frac{?}{54} \quad \frac{4}{27} \times \frac{2}{2} = \frac{8}{54}$

We write $\frac{4}{27}$ as an equivalent fraction $\frac{8}{54}$.

$\frac{5}{18} = \frac{?}{54} \quad \frac{5}{18} \times \frac{3}{3} = \frac{15}{54}$

We write $\frac{5}{18}$ as an equivalent fraction $\frac{15}{54}$.

(*Note*: You may need to divide 54 by 18 to discover that you need to multiply numerator and denominator by 3.)

$\frac{4}{27} + \frac{5}{18} = \frac{8}{54} + \frac{15}{54}$

Replace each fraction by equivalent fraction with the LCD for a denominator.

$= \frac{23}{54}$

Add the numerators.

Review Examples A to J and then try these practice problems.

Practice Problems

(a) $\frac{1}{8} + \frac{5}{12} =$ **(b)** $\frac{3}{5} + \frac{4}{25} =$ **(c)** $\frac{5}{6} - \frac{2}{15} =$

(d) $\frac{1}{12} - \frac{1}{30} =$ **(e)** $\frac{3}{14} + \frac{1}{21} =$ **(f)** $\frac{2}{27} + \frac{1}{18} =$

Solutions to Practice Problems

(a) $\frac{13}{24}$ **(b)** $\frac{19}{25}$ **(c)** $\frac{21}{30} = \frac{7}{10}$

(d) $\frac{3}{60} = \frac{1}{20}$ **(e)** $\frac{11}{42}$ **(f)** $\frac{7}{54}$

0.2.3 Addition and Subtraction of Mixed Fractions

If the problem you are adding or subtracting has *mixed fractions*, change them to *improper fractions* first, then perform the addition or subtraction. Often the final answer is converted to a mixed fraction. A good rule to follow is: If the original problem contains mixed fractions, express the result as a mixed fraction rather than an improper fraction.

EXAMPLE K Combine **1.** $5\frac{1}{2} + 2\frac{1}{3}$ **2.** $2\frac{1}{5} - 1\frac{3}{4}$

Solution **1.** $5\frac{1}{2} + 2\frac{1}{3} = \frac{11}{2} + \frac{7}{3} = \frac{33}{6} + \frac{14}{6} = \frac{47}{6} = 7\frac{5}{6}$ **2.** $2\frac{1}{5} - 1\frac{3}{4} = \frac{11}{5} - \frac{7}{4} = \frac{44}{20} - \frac{35}{20} = \frac{9}{20}$

Try this approach on the following practice problems.

Practice Problems

(a) $1\frac{2}{3} + 2\frac{1}{5} =$ **(b)** $3\frac{1}{4} - 2\frac{1}{3} =$ **(c)** $\frac{7}{8} - \frac{5}{6} =$

(d) $2\frac{1}{7} + \frac{5}{21} =$ **(e)** $\frac{5}{14} + \frac{1}{12} =$

Solutions to Practice Problems

(a) $3\frac{13}{15}$ **(b)** $\frac{11}{12}$ **(c)** $\frac{1}{24}$

(d) $2\frac{8}{21}$ **(e)** $\frac{37}{84}$

Take a few minutes to review the examples and the practice problems you have just completed in this section.

Exercise 0.2

Combine.

1. $\frac{3}{11}+\frac{2}{11}+\frac{4}{11}=$

2. $\frac{8}{17}-\frac{5}{17}=$

Combine. Be sure to reduce your answer whenever possible.

3. $\frac{5}{6}+\frac{8}{15}=$

4. $\frac{5}{14}-\frac{1}{4}=$

5. $\frac{2}{9}+\frac{5}{6}=$

6. $\frac{4}{5}-\frac{3}{8}=$

7. $\frac{9}{8}+\frac{7}{12}=$

8. $\frac{15}{16}-\frac{2}{3}=$

Combine. Be sure to reduce your answer whenever possible.

9. $3\frac{1}{5}+2\frac{2}{3}=$

10. $1\frac{5}{24}+\frac{5}{18}=$

11. $7\frac{1}{6}-2\frac{1}{4}=$

12. $9\frac{4}{5}-3\frac{1}{2}=$

13. $2\frac{1}{8}+3\frac{2}{3}=$

14. $12\frac{1}{8}-9\frac{3}{8}=$

Solve these word problems.

15. Saul jogged $3\frac{1}{8}$ miles on Monday, $2\frac{2}{3}$ miles on Tuesday, and $4\frac{1}{2}$ miles on Wednesday. What was his total jogging distance for the three days?

16. Alice has $8\frac{1}{2}$ hours to study this weekend. She estimates that it will take $2\frac{2}{3}$ hours to type a paper and $1\frac{3}{4}$ hours to read a book for literature class. How much study time will she have left?

0.3 Multiplication and Division of Fractions

0.3.1 *Multiplying Fractions*

To multiply two fractions, we multiply the two numerators and multiply the two denominators. Consider these examples.

EXAMPLE A Multiply. **1.** $\frac{3}{5} \times \frac{2}{7}$ **2.** $\frac{1}{3} \times \frac{5}{4}$ **3.** $\frac{7}{3} \times \frac{1}{5}$

Solution **1.** $\frac{3}{5} \times \frac{2}{7} = \frac{6}{35}$ **2.** $\frac{1}{3} \times \frac{5}{4} = \frac{5}{12}$ **3.** $\frac{7}{3} \times \frac{1}{5} = \frac{7}{15}$

It is possible to avoid having to reduce a fraction at the last step. In many cases we can cancel (cross out) a value that appears in both a numerator and a denominator. Often it is helpful to factor a number (write it as a product of prime numbers) in order to do this.

EXAMPLE B Multiply. **1.** $\frac{3}{5} \times \frac{5}{7}$ **2.** $\frac{4}{11} \times \frac{5}{2}$ **3.** $\frac{15}{8} \times \frac{10}{27}$ **4.** $\frac{8}{7} \times \frac{5}{12}$

Solution

1. $\frac{3}{\cancel{5}} \times \frac{\cancel{5}}{7} = \frac{3}{7}$

2. $\frac{4}{11} \times \frac{5}{2} = \frac{2 \times \cancel{2}}{11} \times \frac{5}{\cancel{2}} = \frac{10}{11}$

3. $\frac{15}{8} \times \frac{10}{27} = \frac{\cancel{3} \times 5}{2 \times 2 \times \cancel{2}} \times \frac{5 \times \cancel{2}}{\cancel{3} \times 3 \times 3} = \frac{25}{36}$

4. $\frac{8}{7} \times \frac{5}{12} = \frac{2 \times \cancel{2} \times \cancel{2}}{7} \times \frac{5}{\cancel{2} \times \cancel{2} \times 3} = \frac{10}{21}$

Why does this method of canceling a value that appears in both numerator and denominator work? What is really going on when we perform that step?

It really depends on the previously used property that any number can be multiplied by 1 without changing the value of the number. Let's reexamine one of the examples we have solved previously.

$$\frac{3}{5} \times \frac{5}{7} = \frac{3}{\cancel{5}} \times \frac{\cancel{5}}{7} = \frac{3}{7}$$

Consider the following steps and reasons.

$\frac{3}{5} \times \frac{5}{7} = \frac{3 \cdot 5}{5 \cdot 7}$ Definition of multiplication of fractions.

$= \frac{5 \cdot 3}{5 \cdot 7}$ Since $3 \cdot 5 = 5 \cdot 3$. This is called the commutative property of multiplication.

$= \frac{5}{5} \cdot \frac{3}{7}$ Definition of multiplication of fractions.

$= 1 \cdot \frac{3}{7}$ Since $\frac{5}{5}$ is another name for 1.

$= \frac{3}{7}$ Since any number can be multiplied by 1 without changing the value of the number.

Multiplication of a fraction by a whole number

As we mentioned in Section 0.1, any whole number can be written as a fraction whose denominator is 1. Therefore, when we multiply a fraction by a whole number we merely express the whole number as a fraction whose denominator is 1 and follow the multiplication rule. (In actual practice, we usually omit this step once we understand the operation.)

EXAMPLE C Multiply. **1.** $7 \times \frac{3}{5}$ **2.** $\frac{5}{23} \times 4$

Solution

1. $7 \times \frac{3}{5} = \frac{7}{1} \times \frac{3}{5} = \frac{21}{5} = 4\frac{1}{5}$

2. $\frac{5}{23} \times 4 = \frac{5}{23} \times \frac{4}{1} = \frac{20}{23}$

Multiplication of mixed fractions

When multiplying *mixed fractions*, we first change them to *improper fractions* and then follow the multiplication rule.

EXAMPLE D Multiply. **1.** $3\frac{1}{3} \times 2\frac{1}{2}$ **2.** $1\frac{2}{5} \times 2\frac{1}{3}$

Solution

1. $3\frac{1}{3} \times 2\frac{1}{2} = \frac{10}{3} \times \frac{5}{2} = \frac{\not{2} \cdot 5}{3} \times \frac{5}{\not{2}} = \frac{25}{3} = 8\frac{1}{3}$

2. $1\frac{2}{5} \times 2\frac{1}{3} = \frac{7}{5} \times \frac{7}{3} = \frac{49}{15} = 3\frac{4}{15}$

Review the examples in this section and then try these practice problems. Be sure to reduce your answers where possible.

Practice Problems

(a) $\frac{2}{7} \times \frac{5}{11} =$ **(b)** $\frac{3}{5} \times \frac{4}{3} =$ **(c)** $4 \times \frac{2}{7} =$

(d) $2\frac{1}{5} \times \frac{3}{7} =$ **(e)** $3\frac{1}{3} \times 1\frac{2}{5}$ **(f)** $\frac{5}{14} \times \frac{7}{25} =$

Solutions to Practice Problems

(a) $\frac{10}{77}$ **(b)** $\frac{4}{5}$ **(c)** $1\frac{1}{7}$

(d) $\frac{33}{35}$ **(e)** $4\frac{2}{3}$ **(f)** $\frac{1}{10}$

0.3.2 Division of Fractions

To divide two fractions, we invert the second fraction, and then multiply the two fractions.

EXAMPLE E Divide. **1.** $\frac{1}{3} \div \frac{1}{2}$ **2.** $\frac{2}{5} \div \frac{3}{10}$ **3.** $\frac{2}{3} \div \frac{7}{5}$

Solution

1. $\frac{1}{3} \div \frac{1}{2} = \frac{1}{3} \times \frac{2}{1} = \frac{2}{3}$

2. $\frac{2}{5} \div \frac{3}{10} = \frac{2}{5} \times \frac{10}{3} = \frac{2}{\not{5}} \times \frac{\not{5} \cdot 2}{3} = \frac{4}{3} = 1\frac{1}{3}$

3. $\frac{2}{3} \div \frac{7}{5} = \frac{2}{3} \times \frac{5}{7} = \frac{10}{21}$

The process of inverting the second fraction and then multiplying the two fractions should be done very carefully when one value is written as an integer.

EXAMPLE F Divide. **1.** $\frac{1}{3} \div 2$ **2.** $5 \div \frac{1}{3}$

Solution **1.** $\frac{1}{3} \div 2 = \frac{1}{3} \div \frac{2}{1} = \frac{1}{3} \times \frac{1}{2} = \frac{1}{6}$ **2.** $5 \div \frac{1}{3} = \frac{5}{1} \div \frac{1}{3} = \frac{5}{1} \times \frac{3}{1} = \frac{15}{1} = 15$

Sometimes division is written in the form of a complex fraction with one fraction in the numerator and one fraction in the denominator. It is best to write this in standard division notation before completing the problem.

EXAMPLE G Divide. **1.** $\frac{\frac{3}{7}}{\frac{3}{5}}$ **2.** $\frac{\frac{2}{9}}{\frac{5}{7}}$

Solution **1.** $\frac{\frac{3}{7}}{\frac{3}{5}} = \frac{3}{7} \div \frac{3}{5} = \frac{\cancel{3}}{7} \times \frac{5}{\cancel{3}} = \frac{5}{7}$ **2.** $\frac{\frac{2}{9}}{\frac{5}{7}} = \frac{2}{9} \div \frac{7}{7} = \frac{2}{9} \times \frac{7}{5} = \frac{14}{45}$

Why does this method of "invert and multiply" work? Why is it the way to perform division of fractions?

Again it really depends on the property that any number can be multiplied by 1 without changing the value of the number. Let's look carefully at an example of division of fractions.

$$\frac{2}{5} \div \frac{3}{7} = \frac{2}{5} \times \frac{7}{3} = \frac{14}{15}$$

$$\frac{2}{5} \div \frac{3}{7}$$ Suppose that we write the original problem using a complex fraction.

$$= \frac{\frac{2}{5}}{\frac{3}{7}}$$ This is the same division problem written using complex fraction notation.

$$= \frac{\frac{2}{5}}{\frac{3}{7}} \times 1$$ Since any number can be multiplied by 1 without changing the value of the number.

$$= \frac{\frac{2}{5}}{\frac{3}{7}} \times \frac{\frac{7}{3}}{\frac{7}{3}}$$ Since any nonzero number divided by itself equals 1, we know that

$$\frac{\frac{7}{3}}{\frac{7}{3}} = 1$$ We will choose this value as a multiplier to simplify the denominator.

$$= \frac{\frac{2}{5} \times \frac{7}{3}}{\frac{3}{7} \times \frac{7}{3}}$$ Definition of multiplication of fractions.

$$= \frac{\frac{2}{5} \times \frac{7}{3}}{1}$$

Since the product in the denominator equals 1.

$$= \frac{2}{5} \times \frac{7}{3}$$

Any number can be written as a fraction with a denominator of 1 without changing its value.

Thus we have shown that $\frac{2}{5} \div \frac{3}{7}$ is equivalent to $\frac{2}{5} \times \frac{7}{3}$ and have shown some justification for the "invert and multiply rule."

This method for division of fractions can be used with mixed fractions. However, we first would want to change them to improper fractions, then invert the second fraction, and finally multiply the two fractions.

EXAMPLE H Divide. **1.** $2\frac{1}{3} \div 3\frac{2}{3}$ **2.** $1\frac{5}{7} \div 3\frac{1}{2}$ **3.** $\frac{2}{3\frac{1}{2}}$

Solution **1.** $2\frac{1}{3} \div 3\frac{2}{3} = \frac{7}{3} \div \frac{11}{3} = \frac{7}{3} \times \frac{3}{11} = \frac{7}{11}$ **2.** $1\frac{5}{7} \div 3\frac{1}{2} = \frac{12}{7} \div \frac{7}{2} = \frac{12}{7} \times \frac{2}{7} = \frac{24}{49}$

3. $\frac{2}{3\frac{1}{2}} = 2 \div 3\frac{1}{2} = \frac{2}{1} \div \frac{7}{2} = \frac{2}{1} \times \frac{2}{7} = \frac{4}{7}$

Review the examples in this section and then try these practice problems. Be sure to reduce answers whenever possible.

Practice Problems

(a) $\frac{2}{5} \div \frac{1}{3} =$ **(b)** $\frac{3}{7} \div 6 =$ **(c)** $1\frac{2}{5} \div 2\frac{1}{3} =$

(d) $5\frac{2}{3} \div 7 =$ **(e)** $\frac{12}{13} \div \frac{4}{3} =$ **(f)** $\frac{1\frac{1}{5}}{1\frac{2}{7}} =$

Solutions to Practice Problems

(a) $1\frac{1}{5}$ **(b)** $\frac{1}{14}$ **(c)** $\frac{3}{5}$

(d) $\frac{17}{21}$ **(e)** $\frac{9}{13}$ **(f)** $\frac{14}{15}$

Exercise 0.3

Multiply. Reduce all answers.

1. $\frac{3}{5} \times \frac{2}{11} =$ **2.** $\frac{25}{3} \times \frac{6}{5} =$ **3.** $1\frac{1}{3} \times 2\frac{1}{2} =$

4. $3\frac{1}{3} \times \frac{1}{2} =$

5. $8 \times \frac{3}{7} =$

6. $\frac{3}{49} \times \frac{14}{6} =$

Divide. Reduce all answers.

7. $\frac{5}{3} \div \frac{5}{2} =$

8. $\frac{5\frac{1}{2}}{3\frac{1}{3}} =$

9. $\frac{3}{7} \div \frac{5}{3} =$

10. $\frac{12}{5} \div \frac{18}{25} =$

11. $1\frac{3}{7} \div \frac{1}{14} =$

12. $\frac{2}{1\frac{1}{5}} =$

Solve these word problems.

13. A certain size of man's shirt requires $2\frac{3}{4}$ yards of material. How many yards would be needed to make 26 shirts?

14. Martha jogged for $4\frac{1}{5}$ miles on Monday. Two-thirds of the distance was over a cross-country trail. How long is the cross-country trail?

15. The weight of the water in a container is 450 pounds. One cubic foot of water weighs $62\frac{1}{2}$ pounds. How many cubic feet of water are in the container?

0.4 Use of Decimals

0.4.1 Decimals and Fractions

When we want to express a portion of a whole quantity we can write the number by using a fraction or a decimal. Consider these equivalent forms:

$$\frac{3}{10} = 0.3 \qquad \frac{5}{100} = 0.05 \qquad \frac{172}{1000} = 0.172 \qquad \frac{58}{10{,}000} = 0.0058$$

A *decimal* is another way of writing a fraction whose denominator is 10, 100, 1000, 10,000, 100,000, etc.

The period in decimal notation is known as the *decimal point.* The number of digits in a number to the right of the decimal point are known as the number of *decimal places* of the number. The place value of decimals are shown below.

Hundred-thousands	Ten-thousands	Thousands	Hundreds	Tens	Ones	. ← Decimal point	Tenths	Hundredths	Thousandths	Ten-thousandths	Hundred-thousandths
100,000	10,000	1000	100	10	1	.	$\frac{1}{10}$	$\frac{1}{100}$	$\frac{1}{1000}$	$\frac{1}{10{,}000}$	$\frac{1}{100{,}000}$

EXAMPLE A Write the following decimals as a fraction. Give the number of decimal places, and write out how it would be spoken.

1. 0.6 **2.** 0.29 **3.** 0.527

4. 1.38 **5.** 256.3725 **6.** 0.00007

Solution

	Decimal Form	*Fraction Form*	*Number of Decimal Places*	*The Words Used to Describe the Fraction*
1.	0.6	$\frac{6}{10}$	one	six tenths
2.	0.29	$\frac{29}{100}$	two	twenty-nine hundredths
3.	0.527	$\frac{527}{1000}$	three	five hundred twenty-seven thousandths
4.	1.38	$1\frac{38}{100}$	two	one and thirty-eight hundredths
5.	256.3725	$256\frac{3725}{10,000}$	four	two hundred fifty-six and three thousand seven hundred twenty-five ten-thousandths
6.	0.00007	$\frac{7}{100,000}$	five	seven hundred-thousandths

You have seen that a given fraction can be written in several different but equivalent ways. This is true of the decimal form of fractions. The decimal 0.18 can be written in the following equivalent ways:

$$\text{Fractional form:} \quad \frac{18}{100} = \frac{180}{1000} = \frac{1800}{10,000} = \frac{18,000}{100,000}$$

$$\text{Decimal form:} \quad 0.18 = 0.180 = 0.1800 = 0.18000$$

Thus we see that any number of *zeros may be added onto the right-hand side* of a decimal without changing its value.

$$0.13 = 0.1300 \qquad 0.162 = 0.162000$$

Similarly, any number of *zeros may be removed* from the right-hand side of a decimal without changing its value.

$$0.2370 = 0.237 \qquad 0.001600 = 0.0016$$

0.4.2 Changing Fractional Form to Decimal Form

A fraction can be written in decimal form by dividing the denominator into the numerator.

EXAMPLE B Write each of the following fractions as a decimal. **1.** $\frac{3}{4}$ **2.** $\frac{21}{20}$ **3.** $\frac{1}{8}$ **4.** $\frac{3}{200}$

Solution **1.** $\frac{3}{4} = 0.75$ since

$$\begin{array}{r} 0.75 \\ 4\overline{)3.00} \\ 28 \\ \hline 20 \\ 20 \\ \hline \end{array}$$

2. $\frac{21}{20} = 1.05$ since

$$\begin{array}{r} 1.05 \\ 20\overline{)21.00} \\ 20 \\ \hline 100 \\ 100 \\ \hline \end{array}$$

3. $\frac{1}{8} = 0.125$ since $8\overline{)1.000}$ = 0.125 (8; 20; 16; 40; 40)

$$\begin{array}{r} 0.125 \\ 8\overline{)1.000} \\ \underline{8} \\ 20 \\ \underline{16} \\ 40 \\ \underline{40} \end{array}$$

4. $\frac{3}{200} = 0.015$ since

$$\begin{array}{r} 0.015 \\ 200\overline{)3.000} \\ \underline{200} \\ 1000 \\ \underline{1000} \end{array}$$

We are often required to carry out the division until there is no remainder.

Sometimes the division yields an infinite repeating decimal. We use three dots to indicate that the pattern continues forever. For example:

$$\frac{1}{3} = 0.3333\ldots \qquad \begin{array}{r} 0.333 \\ 3\overline{)1.000} \\ \underline{9} \\ 10 \\ \underline{9} \\ 10 \\ \underline{9} \\ 1 \end{array}$$

Note: An alternative notation is to place a bar over the repeating digits:

$$0.3333\ldots = 0.\overline{3}$$

$$0.575757\ldots = 0.\overline{57}$$

$$0.6342342342\ldots = 0.6\overline{342}$$

EXAMPLE C Change to decimal notation. **1.** $\frac{2}{11}$ **2.** $\frac{5}{6}$

Solution **1.** $\frac{2}{11} = 0.181818\ldots$

$$\begin{array}{r} 0.1818 \\ 11\overline{)2.0000} \\ \underline{11} \\ 90 \\ \underline{88} \\ 20 \\ \underline{11} \\ 90 \\ \underline{88} \\ 2 \end{array}$$

2. $\frac{5}{6} = 0.8333\ldots$

$$\begin{array}{r} 0.8333 \\ 6\overline{)5.0000} \\ \underline{48} \\ 20 \\ \underline{18} \\ 20 \\ \underline{18} \\ 20 \\ \underline{18} \\ 2 \end{array}$$

Note that the 8 does not repeat. Only the digit 3 is repeating.

Sometimes the division must be carried out to many places in order to observe the repeating pattern. This is true in the following example:

$\frac{2}{7} = 0.285714285714285714\ldots$ This can also be written as $\frac{2}{7} = 0.\overline{285714}$.

since all that is required is to add zeros to the right-hand side of the decimal (and we usually do not even write out that step).

As long as we line up the decimal points, we can add or subtract any decimal fractions.

> ### *Multiplication of Decimals*
>
> **To multiply decimals, you first multiply the appropriate digits. To determine the position of the decimal point, you count the total number of decimal places in the two numbers being multiplied. This will determine the number of decimal places that should appear in the answer.**

EXAMPLE F Multiply 0.8 × 0.4.

Solution

$$\begin{array}{rl} 0.8 & \text{(one decimal place)} \\ \times\ \ 0.4 & \text{(one decimal place)} \\ \hline 0.32 & \text{(two decimal places)} \end{array}$$

Note that you will often have to add zeros to the left of the digits obtained in the product so that you obtain the necessary number of decimal places.

EXAMPLE G Multiply 0.123 × 0.5.

Solution

$$\begin{array}{rl} 0.123 & \text{(three decimal places)} \\ \times\ \ \ \ 0.5 & \text{(one decimal place)} \\ \hline 0.0615 & \text{(four decimal places)} \end{array}$$

Why do we count the number of decimal places?

The rule really comes from the properties of fractions. If we write the problem above in fraction form, we have

$$(0.123) \times (0.5) = \frac{123}{1000} \times \frac{5}{10} = \frac{615}{10{,}000} = 0.0615$$

Here are some more examples that involve more decimal places.

EXAMPLE H Multiply. **1.** 2.56 × 0.003 **2.** 0.0036 × 0.008

Solution

1.

$$\begin{array}{rl} 2.56 & \text{(two decimal places)} \\ \times\ \ 0.003 & \text{(three decimal places)} \\ \hline 0.00768 & \text{(five decimal places)} \end{array}$$

2.

$$\begin{array}{rl} 0.0036 & \text{(four decimal places)} \\ \times\ \ \ \ 0.008 & \text{(three decimal places)} \\ \hline 0.0000288 & \text{(seven decimal places)} \end{array}$$

Since multiplication of fractions does not require a common denominator we do not need the same number of decimal places when we multiply decimals.

Practice Problems

Review the examples in this section and then try these practice problems.

Add or subtract.

(a) 3.12 + 5.08 + 1.42 = **(b)** 152.003 − 136.118 =

(c) 1.1 + 3.16 + 5.123 = **(d)** 1.0052 − 0.1234 =

Multiply.

(e) $0.5 \times 0.3 =$ (f) $0.12 \times 0.4 =$

(g) $1.23 \times 0.005 =$ (h) $4.1003 \times 0.2 =$

(i) $0.003 \times 0.00002 =$ (j) $3.61 \times 0.12 =$

(k) $18.3 \times 0.123 =$

Solutions to Practice Problems

(a) 9.62 (b) 15.885 (c) 9.383

(d) 0.8818 (e) 0.15 (f) 0.048

(g) 0.00615 (h) 0.82006 (i) 0.00000006

(j) 0.4332 (k) 2.2509

0.4.5 Division of Decimals

A quick review of terms used in division

When discussing division of decimals we frequently refer to the three primary parts of a division problem: the divisor, the dividend, and the quotient. Be sure you know the meaning of each term.

The *divisor* is the number you divide into another.

The *dividend* is the number to be divided.

The *quotient* is the result from division of one number by another.

In the problem $6 \div 2 = 3$ we represent each of these terms as follows:

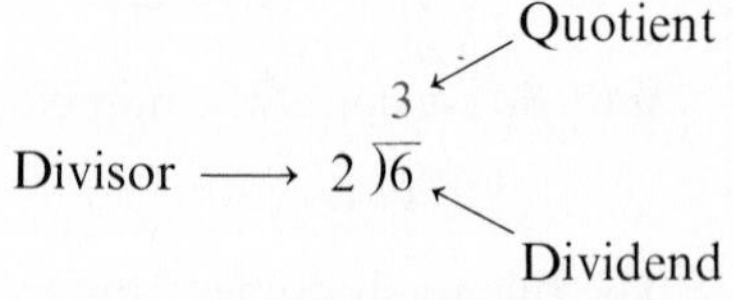

and in general

$$\text{Divisor}\overline{)\text{dividend}}\quad \text{quotient}$$

Division of decimals

When dividing two decimals, count the *number of decimal places* in the divisor. Then move the *decimal point to the right* that *same number of places* in both *the divisor* and *the dividend.* Mark that position with a caret (‸). Finally, perform the division. Be sure to line up the decimal point in the quotient with the position indicated by the caret in the dividend.

EXAMPLE 1 Divide: $32.68 \div 4$.

Solution

$$4\overline{)32.68}\quad 8.17$$

Since there are no decimal places in the divisor we do not need to move the decimal point.

We must be careful, however, to place the decimal point in the quotient directly above the decimal point in the dividend.

Thus $32.68 \div 4 = 8.17$.

EXAMPLE J Divide: 5.75 ÷ 0.5.

Solution

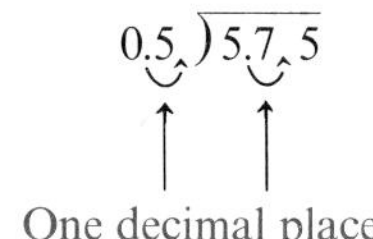

There is *one* decimal place in the divisor, so we move the decimal point *one place to the right* in the *divisor* and *dividend* and we mark that new position by a caret (‸).

```
      11.5
0.5‸)5.7‸5
     5
     7
     5
     25
     25
```

Now we perform the division as with whole numbers. The decimal point in the answer is directly above the caret in the dividend.

Thus 5.75 ÷ 0.5 = 11.5.

EXAMPLE K Divide: 0.1116 ÷ 0.18.

Solution

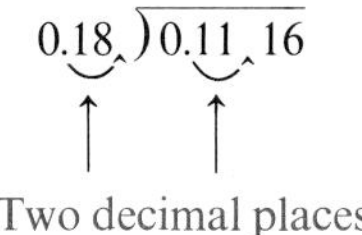

There are *two* decimal places in the divisor, so we move the decimal point *two places to the right* in the *divisor* and *dividend* and mark that new position by a caret.

```
          0. 62
0.18‸)0.11‸16
        10 8
           36
           36
```

Now we perform the division as with whole numbers. The decimal point in the answer is directly above the caret in the dividend.

Thus 0.1116 ÷ 0.18 = 0.62.

Note that sometimes we will need to place extra zeros in the dividend in order to move the decimal point the required number of places.

EXAMPLE L Divide: 16.2 ÷ 0.027.

Solution

```
0.027‸)16.200‸
```

Three decimal places

There are *three* decimal places in the divisor, so we move the decimal point *three places to the right* in the *divisor* and *dividend* and mark the new position by a caret. *Note* that we must add two zeros to 16.2 in order to do this.

```
              600.
0.027‸)16.200‸
        162
         000
```

Now perform the division as with whole numbers. The decimal point in the answer is directly above the caret in the dividend.

Thus 16.2 ÷ 0.027 = 600.

EXAMPLE M Divide: 0.04288 ÷ 3.2.

Solution

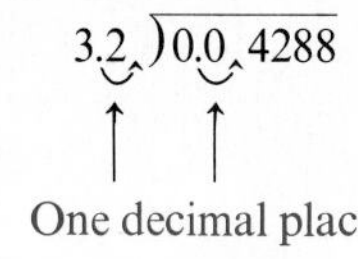

There is *one* decimal place in the divisor, so we move the decimal point *one place to the right* in the *divisor* and *dividend* and mark the new position by a caret.

```
         .0134
3.2.)0.0.4288
       32
       108
        96
        128
        128
          0
```

Now perform the division as with whole numbers. The decimal point in the answer is directly above the caret in the dividend.

Note the need for the initial zero after the decimal point in the answer.

Thus 0.04288 ÷ 3.2 = 0.0134.

Why does this method of dividing decimals work? Why is it necessary to move the decimal in the first place?

Essentially, we are using the steps we used in Section 0.1.3 to change a fraction to an equivalent fraction by multiplying both the numerator and denominator by the same number.

Let's reexamine Example M.

$0.04288 \div 3.2$

$= \dfrac{0.04288}{3.2}$ — We write the indicated division using equivalent fraction notation.

$= \dfrac{0.04288 \times 10}{3.2 \quad \times 10}$ — We multiply the numerator and denominator by 10.

$= \dfrac{0.4288}{32}$ — Result of multiplication by 10.

$= 0.4288 \div 32$ — Rewrite the fraction as an equivalent problem with division notation.

Notice that we have obtained the same problem that we had in Example M when we moved the decimal one place to the right in the divisor and dividend and marked the new position of the decimal point with a caret. Therefore, we see that *the reason we can move the decimal point so many places to the right in divisor and dividend* is that the numerator and denominator of a fraction are both being multiplied by 10, 100, 1000, etc., to obtain an equivalent fraction.

Practice Problems

Review Examples I to M and then try these practice problems.

(a) 0.0016 ÷ 0.04 = **(b)** 0.09 ÷ 0.3 = **(c)** 2.5 ÷ 0.005 =
(d) 0.01764 ÷ 4.9 = **(e)** 1800 ÷ 0.06 = **(f)** 0.065 ÷ 10.4 =

Solutions to Practice Problems

(a) 0.04 **(b)** 0.3 **(c)** 500
(d) 0.0036 **(e)** 30,000 **(f)** 0.00625

0.4.6 Multiplication and Division by a Multiple of 10

When multiplying by 10, 100, 1000, etc., a simple rule may be used to obtain the answer. For every zero in the multiplier, move the decimal point one place to the right.

EXAMPLE N Multiply. **1.** 3.24×10 **2.** 15.6×100 **3.** 0.0026×1000

Solution

1. $3.24 \times 10 = 32.4$ One zero—move decimal point one place to the right.

2. $15.6 \times 100 = 1560$ Two zeros—move decimal point two places to the right.

3. $0.0026 \times 1000 = 2.6$ Three zeros—move decimal point three places to the right.

The reverse rule is true for division. When dividing by 10, 100, 1000, 10,000, etc., move the decimal point one place to the left for every zero in the divisor.

EXAMPLE O Divide. **1.** $52.6 \div 10$ **2.** $0.0038 \div 100$ **3.** $5936.2 \div 1000$ **4.** $31.26 \div 10{,}000$

Solution

1. $\dfrac{52.6}{10} = 5.26$
(one place to the left)

2. $\dfrac{0.0038}{100} = 0.000038$
(two places to the left)

3. $\dfrac{5936.2}{1000} = 5.9362$
(three places to the left)

4. $\dfrac{31.26}{10{,}000} = 0.003126$
(four places to the left)

Review the examples in this section and then try these practice problems.

Practice Problems

(a) $0.0016 \times 100 =$ **(b)** $2.34 \times 1000 =$ **(c)** $\dfrac{5.82}{10} =$

(d) $123.4 \div 1000 =$ **(e)** $136.2 \times 10{,}000 =$ **(f)** $\dfrac{0.00614}{10{,}000} =$

Solutions to Practice Problems

(a) 0.16 **(b)** 2340 **(c)** 0.582

(d) 0.1234 **(e)** 1,362,000 **(f)** 0.000000614

Exercise 0.4

Convert to decimal form.

1. $\dfrac{5}{8} =$ **2.** $\dfrac{3}{20} =$ **3.** $\dfrac{16}{500} =$

Write each fraction in the following cases as a reduced fraction rather than a decimal.

4. $0.12 =$ **5.** $1.125 =$ **6.** $0.58 =$

Add or subtract.

7. $3.6 + 1.28 + 4.5 =$

8. $0.00381 - 0.00228 =$

9. $156. + 0.03 + 3.621 =$

10. $158.23 - 39.67 =$

Multiply.

11. $2.18 \times 1.3 =$

12. $5.26 \times 0.0015 =$

13. $0.0062 \times 0.018 =$

14. $368{,}000 \times 0.00021 =$

Divide.

15. $0.0455 \div 0.13 =$

16. $0.7208 \div 2.12 =$

17. $0.5230 \div 0.002 =$

18. $186.16 \div 5.2 =$

Do the following problems mentally.

19. $1.36 \times 1000 =$

20. $0.00243 \times 100{,}000 =$

21. $3.52 \div 1000 =$

22. $164{,}320 \div 10{,}000 =$

Solve the following problems.

23. Fred's car usually gets 28.5 miles per gallon when driven on the highway. The gas tank holds 23.4 gallons. What is the driving range of his car when driven on the highway?

24. Alice drove 478.6 miles last week. She used 26.8 gallons of gas. Find how many miles per gallon she averaged in her car last week. (Round your answer to the nearest tenth.)

25. Heather works a 40-hour week and is paid \$5.20 per hour. For every hour beyond a 40-hour week she is paid time and a half (1.5 times the normal hourly rate). Last week she worked 52 hours. What was her total pay last week?

0.5 Use of Percent

A *percent is a fraction that has a denominator of 100.* When you say "sixty-seven percent" or write "67%" you are just using another way of expressing the fraction $\frac{67}{100}$. The word "percent" is a shortened form of the Latin words *per centum,* which means "by the hundred." In everyday use, *percent means per one hundred.*

It is important to see that 49% means 49 out of 100 parts. It can also be written as a fraction $\frac{49}{100}$ or as a decimal 0.49. Understanding the meaning of the notation allows you to change from one notation to another.

Thus

$$49\% = 49 \text{ out of } 100 \text{ parts} = \frac{49}{100} = 0.49$$

EXAMPLE A Express as a percent and as a decimal: $\frac{11}{100}$

Solution

$\frac{11}{100}$ means 11 parts out of 100, or 11%

$\frac{11}{100}$ as a decimal is 0.11

Now that we understand the concept we can use some quick procedures to convert from decimals to percent, and vice versa.

0.5.1 Decimal/Percentage Conversion

Converting a decimal to percent

When changing a decimal to a percent, merely move the decimal point two places to the right and add the % symbol.

0.23 = 23%　　0.461 = 46.1%　　0.4 = 40%

0.0364 = 3.64%　　0.0026 = 0.26%　　0.0008 = 0.08%

1.48 = 148%　　1.08 = 108%

Converting a percent to a decimal

This process is the reverse of the one described above. Move the decimal point two places to the *left* and remove the % symbol.

16% = 0.16　　143% = 1.43　　4% = 0.04

3.2% = 0.032　　0.6% = 0.006　　0.09% = 0.0009

25.2% = 0.252

Study these examples and then try the following practice problems.

Practice Problems

Convert to percent.

(a) 0.92 = **(b)** 0.418 = **(c)** 1.02 =

(d) 0.0164 = **(e)** 0.0031 = **(f)** 0.00005 =

Convert to decimal form.

(g) 36% = **(h)** 154% = **(i)** 2.8% =

(j) 0.8% = **(k)** 0.08% = **(l)** 12.68% =

Solutions to Practice Problems

(a) 92% **(b)** 41.8% **(c)** 102%

(d) 1.64% **(e)** 0.31% **(f)** 0.005%

(g) 0.36 **(h)** 1.54 **(i)** 0.028

(j) 0.008 **(k)** 0.0008 **(l)** 0.1268

0.5.2 Applied Percent Problems

TYPE I Percent problem: finding the percent of a number

To find the percentage of a number, merely convert the percent to a decimal and multiply by the decimal.

EXAMPLE B Find **1.** 10% of 36 **2.** 2% of 350 **3.** 82% of 12 **4.** 0.3% of 42

Solution **1.** 10% of 36 = 0.10 × 36 = 3.6 **2.** 2% of 350 = 0.02 × 350 = 7

3. 82% of 12 = 0.82 × 12 = 9.84 **4.** 0.3% of 42 = 0.003 × 42 = 0.126

Study these examples and then try the following practice problems.

Practice Problems

(a) 18% of 50 = **(b)** 4% of 64 = **(c)** 156% of 35 =

(d) 0.8% of 60 = **(e)** 1.3% of 82 = **(f)** 0.002% of 564 =

Solutions to Practice Problems

(a) 9 **(b)** 2.56 **(c)** 54.6

(d) 0.48 **(e)** 1.066 **(f)** 0.01128

TYPE II Percent problem: converting a fraction to percent

To find what percent one number is of a second number: 1. Write a fraction with the first number as the numerator and the second number as a denominator. 2. Reduce the fraction (if possible). 3. Convert the fraction to a decimal. 4. Express the decimal as a percent.

A very common problem is the following: What percent of 24 is 15? This can be quickly solved as follows:

$\dfrac{15}{24}$	Expressing the relationship as a fraction.
$= \dfrac{5}{8}$	Reducing the fraction (when possible).
$= 0.625$	Converting the fraction to a decimal.
$= 62.5\%$	Converting the decimal to percent.

Some additional problems.

EXAMPLE C **1.** What percent of 200 is 82? **2.** What percent of 16 is 3.8? **3.** What percent of \$120 is \$150?

Solution **1.** What percent of 200 is 82?

$$\frac{82}{200} = \frac{41}{100} = 0.41 = 41\%$$

2. What percent of 16 is 3.8?

$$\frac{3.8}{16} = 0.2375 = 23.75\%$$

(*Note*: You need to divide 3.8 ÷ 16 by long division.)

3. What percent of \$120 is \$150?

$$\frac{150}{120} = \frac{5}{4} = 1.25 = 125\%$$

Study these examples and then try the following practice problems.

Practice Problems

(a) What percent of 48 is 3?

(b) What percent of 120 is 210?

(c) What percent of 25 is 4?

(d) What percent of 320 is 2?

Solutions to Practice Problems

(a) 6.25%

(b) 175%

(c) 16%

(d) 0.625%

TYPE III Percent problem: problems involving a percent of an unknown number

One last type of percent problem is somewhat more difficult. Consider the following: A salesman makes a sale to 40% of his potential customers. He made 150 sales last week. How many potential customers did he talk to?

The basic mathematical question here is: 40% of what number is 150? How can we solve this problem when we do not know the number that we take 40% of?

Let's see if we can compare it to a simple arithmetic problem. Suppose that someone said: "3 times a number will give you a result of 57, what is the number?" You would probably reason like this:

$3 \times ? = 57$ — Three times a number gives me 57. What is the number?

$? = \frac{57}{3}$ — Since 3 times the number gives 57, if I divide 57 by 3, I can find the number.

$? = 19$ — The number is 19.

From arithmetic we can reason as follows for the percent problem we introduced above. We know that 40% of an unknown number is 150.

$40\% \text{ of } ? = 150$ — Abbreviating the problem.

$0.40 \times ? = 150$ — Writing the problem as an equation, with the percent changed to decimal form.

$? = \frac{150}{0.40}$ — If 0.40 times an unknown number is 150, then 150 divided by 0.40 would yield that number.

$? = 375$ — The unknown number is 375.

Is that really correct? Let's check it!

Is 40% of 375 really 150?

$0.40 \times 375 = 150$ — We multiply the two values.

$150 = 150$ — Yes, it checks.

From this discussion let's form a general procedure to follow from this type of percent problem.

> ***TYPE III Percent Problem: Problems Involving a Percent of an Unknown Number.***
>
> **To solve a percent problem where we have been given that a percent of an unknown number is a certain amount we should:**
>
> **1. Change the percentage rate to decimal form.**
>
> **2. Divide the amount by the decimal.**

EXAMPLE D We find that 35% of a number is 54.6. What is the number?

Solution

$0.35 \times ? = 54.6$ — Change the percent to a decimal.

$? = \frac{54.6}{0.35}$ — Divide the amount by the decimal.

$? = 156$ — After performing the steps to divide this out, we obtain 156. The desired number is 156.

Check Does this seem correct?

Does 35% of 156 equal 54.6?

Does $0.35 \times 156 = 54.6$? — When we multiply the two numbers we see that it does check.

$54.6 = 54.6$

EXAMPLE E 125% of what number is 300?

Solution

$1.25 \times ? = 300$ — Change the percent to a decimal.

$? = \frac{300}{1.25}$ — Divide the amount by the decimal.

$? = 240$ — We divide this out to obtain 240, the unknown number.

Check Does this seem correct?

Does 125% of 240 equal 300?

Does $1.25 \times 240 = 300$? — When we multiply the two numbers we see that it does check.

$300 = 300$

EXAMPLE F A team has won 70% of their games. They have won 42 games. What is the total number of games that the team has played?

Solution

$70\% \text{ of } ? \text{ games} = 42$

$0.70 \times ? = 42$

$? = \frac{42}{0.70}$

$\frac{42}{0.70} = 60$ — Therefore, the team played *60 games.*

Review Examples D–F and then try these practice problems.

Practice Problems

(a) 50% of what number is 20?

(b) 40% of what number is 18?

(c) 60% of what number is 12?

(d) 55% of what number is 440?

(e) On a test a student answered 78% of the questions correctly. He answered 195 questions correctly. How many problems were on the test?

(f) It is known that 35% of the people in a given city have blood type B. There are 17,500 people in the city with blood type B. What is the population of the city?

Solutions to Practice Problems

(a) 40 **(b)** 45 **(c)** 20

(d) 800 **(e)** 250 **(f)** 50,000

Note. Students often confuse the different types of percentage problems discussed in these paragraphs. Please review the sample examples carefully so that the different methods used in each case are clear. When you are finished reviewing, complete the following problems.

Exercise 0.5

Convert to percent.

1. 0.624 =

2. 0.003 =

3. 1.56 =

4. 0.0006 =

Convert to decimal form.

5. 4% =

6. 0.24% =

7. 250% =

8. 3.58% =

Find the following.

9. What is 35.8% of 1000?

10. What is 0.3% of 54?

11. What is 140% of 212?

12. What is 0.08% of 220?

13. What percent of 600 is 30?

14. What percent of 120 is 6?

15. What percent of 56 is 84?

16. What percent of 500 is 2?

17. 25% of what number is 555?

18. 30% of what number is 0.948?

19. 28% of what number is 43.68?

20. 65% of what number is 110.5?

21. In a town of 30,000 people, 47% are registered as Democrats. How many people are Democrats in this town?

22. Frank made a purchase of a pair of Nike sneakers. The sales tax in his state is 5%. He paid $1.80 in tax. How much did the sneakers cost before the tax?

23. Melinda took an exam with 80 questions. She obtained 52 correct answers. What percent of the questions did she answer correctly?

24. The profits of a small computer company increased 20% from last year to this year. This year the profit was $3,000,000. What was the profit last year?

25. In a local college survey it was found that 97 out of 180 students smoked. What percent of the students in the sample said that they smoked? (Round your answer to the nearest hundredth of a percent.)

26. Susan is earning $4.20 an hour. She will get a 8% pay raise this year. What will be her new hourly pay rate? (Round your answer to the nearest cent.)

Practice Test for Chapter 0

Reduce.

1. $\frac{36}{48}$

2. $\frac{15}{50}$

3. $\frac{24}{72}$

4. $\frac{18}{30}$

5. Write as an *improper fraction*:

$$2\frac{5}{7}$$

6. Write as a *mixed fraction*:

$$\frac{34}{5}$$

7. Write as a *mixed fraction*:

$$\frac{27}{4}$$

Change the given fraction to an equivalent fraction with the specified denominator.

8. $\frac{5}{8} = \frac{?}{24}$

9. $\frac{1}{3} = \frac{?}{45}$

10. $\frac{4}{7} = \frac{?}{21}$

11. $\frac{2}{5} = \frac{?}{55}$

Combine.

12. $\frac{2}{3} + \frac{1}{4} =$

13. $\frac{1}{12} + \frac{3}{8} =$

14. $\frac{7}{20} - \frac{1}{12} =$

15. $\frac{5}{12} - \frac{1}{16} =$

16. $3\frac{1}{6} + 2\frac{3}{5} =$

17. $1\frac{1}{4} + 2\frac{7}{10} =$

18. $2\frac{1}{5} - 1\frac{2}{3} =$

19. $3\frac{1}{15} - 1\frac{3}{20} =$

Multiply.

20. $\frac{2}{3} \times \frac{9}{10} =$

21. $\frac{3}{5} \times \frac{10}{21} =$

22. $6 \times \frac{5}{11} =$

23. $2\frac{1}{3} \times 4\frac{1}{2} =$

24. $1\frac{1}{8} \times 2\frac{1}{9} =$

25. $\frac{4}{7} \times 5 =$

Divide.

26. $\frac{1}{3} \div \frac{1}{6} =$

27. $\frac{1}{2} \div \frac{1}{8} =$

28. $\frac{2}{5} \div \frac{4}{3} =$

29. $\frac{5}{7} \div \frac{15}{14} =$

30. $\frac{15}{16} \div 6\frac{1}{4} =$

31. $2\frac{6}{7} \div \frac{10}{21} =$

Combine.

32. $1.324 + 2.008 + 1.130 =$

33. $24.831 - 17.094 =$

34. $14.037 - 2.61 =$

35. $1.6 + 3.21 + 0.004 =$

36. $1.3 + 1.8 + 2.6 + 7.2 + 0.8 =$

37. $100.01 + 10.001 + 1.1011 =$

Multiply.

38. $0.002 \times 4.31 =$

39. $362.341 \times 1000 =$

40. $2.6 \times 0.03 \times 1.02 =$

41. $1.08 \times 0.06 \times 160 =$

Divide.

42. $0.186 \div 100 =$

43. $71.32 \div 1000 =$

44. $0.186 \div 93 =$

45. $1.35 \div 0.015 =$

46. $0.147 \div 2.1 =$

47. $0.19 \div 0.38 =$

48. Write as a percent:

$$\frac{3}{8}$$

49. Write as a reduced fraction:

$$24\%$$

In problems 50 to 53, write the percentages in decimal form.

50. $1.4\% =$

51. $36.1\% =$

52. $0.02\% =$

53. $125.3\% =$

54. What is 65% of 400?

55. Find 250% of 36.

56. 8.5% of what number is 17?

57. What is 1.8% of 1000?

58. What percent of 120 is 15?

59. What percent of 1250 is 750?

60. 75% of what number is 30?

61. 78% of what number is 195?

62. 36% of Americans have blood type B. If there are 270,000,000 Americans, how many have blood type B?

63. In a given university 720 of the 960 freshmen had a math deficiency. What percentage of the class had a math deficiency?

64. In a certain city that has had several factory closings, there are 8400 people who are unemployed. The unemployment rate is 35%. How many people live in the city?

65. If Steve added a woodburning stove to his home and used it as a primary heat source he could expect to save $\frac{5}{8}$ of his present annual heating bill. Last year he spent $776 to heat his home. How much will it cost him this year if he adds the stove?

66. Abdul is going to build 5 shelves in his room to hold textbooks. Each shelf will be $4\frac{3}{4}$ feet long. He is buying 5 boards that are 6 feet long for shelving. He will cut $4\frac{3}{4}$ feet from each to obtain the correct length for each shelf. How many linear feet of boards will he have left over?

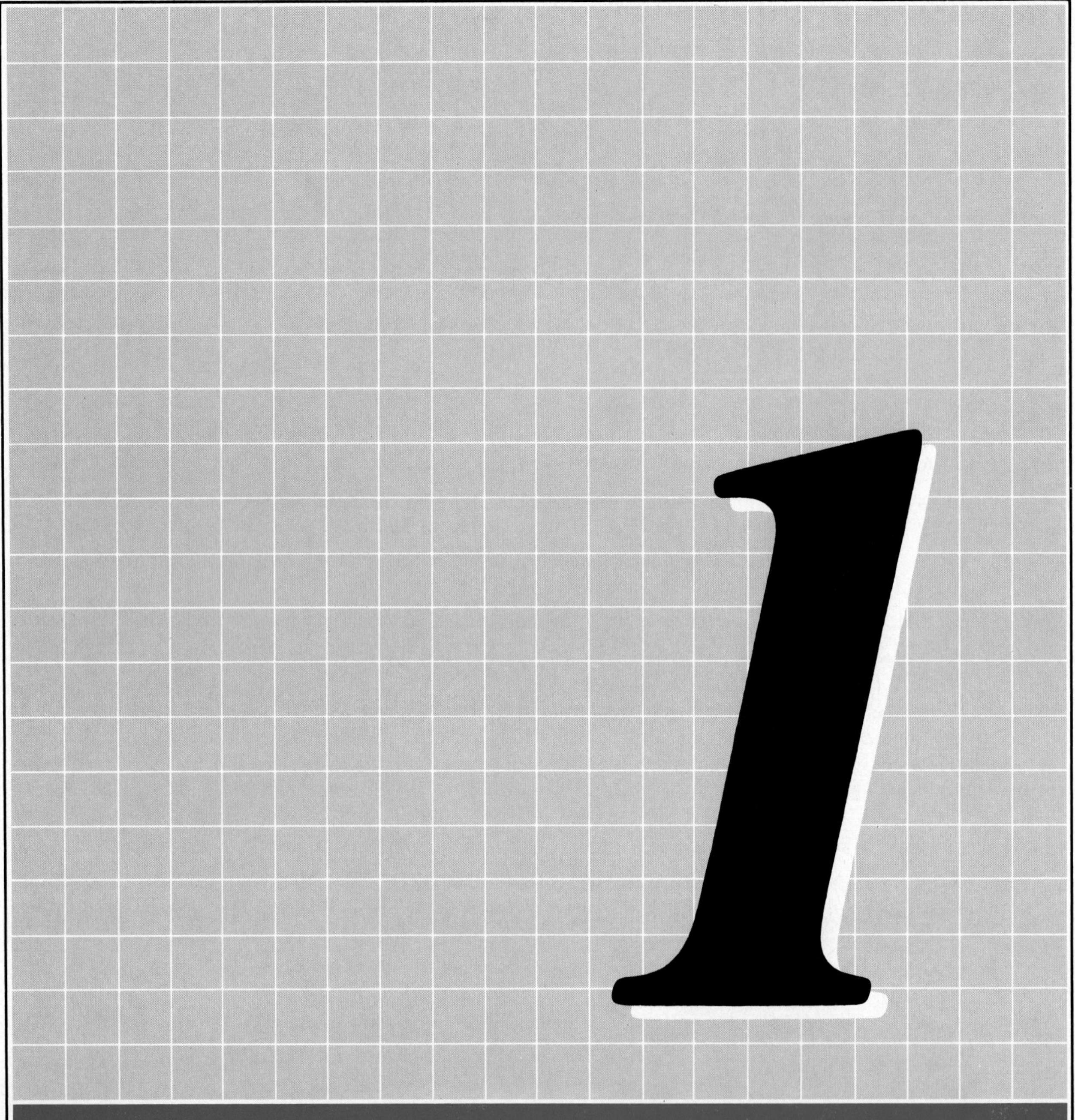

1

Basic Principles of Using Variables

Diagnostic Pretest for Chapter 1

This diagnostic pretest is intended for these students who have previously studied algebra. Work as many problems as you can. Then, check your answers against those listed at the back of the book. Where you have obtained an incorrect answer or have been unable to work the problem, you will need to study the corresponding sections of the chapter.

If you have not previously studied algebra, do not take this test now. You should study the chapter and work the problems and exercises. After finishing the chapter, you may use this test as a review or as a practice test.

Perform the necessary operation and simplify your answer.

Section 1.1

1. $(-3) - (-6) =$

2. $-5 + 6 - 3 - 2 =$

Section 1.2

3. $(-7)(-2)(+3)(-1) =$

4. $\dfrac{-\frac{2}{3}}{-\frac{1}{4}} =$

Section 1.3

5. Write in exponential notation:

$$(-3)(-3)(a)(a)(a)(b)(b)$$

6. Evaluate:

$$(-2)^6$$

7. Simplify:

$$(-3x^2y)(-2x^3y^4)$$

8. Simplify:

$$\frac{-12x^7y^3}{8x^9y}$$

9. Raise this expression to the indicated power:

$$\left(\frac{2x^3y}{z^2}\right)^4$$

Section 1.4

10. Combine like terms:

$$5x^2 - 3xy - 6x^2y - 8xy$$

Section 1.5

11. Multiply:

$$-2x(3x - 2xy + z)$$

Section 1.6

12. Evaluate:

$$3 - 2 \times 6 + (-2)^3$$

13. Evaluate:

$$3x^2 - 5x - 4 \text{ when } x = -2$$

Section 1.7

14. Simplify:

$$2x^2 - 3x[2x - (x + 2y)]$$

Student Learning Objectives for Chapter 1

After successfully learning the content of each section, you will be able to:

Section 1.1

A. Add and subtract signed numbers.

Section 1.2

B. Multiply and divide signed numbers.

Section 1.3

C. Write or evaluate numbers in exponential notation.

D. Use a variable to represent a number in exponential notation.

E. Simplify elementary expressions where variables in exponent form are multiplied or divided.

Section 1.4

F. Identify like terms and combine them.

Section 1.5

G. Use the distributive property to multiply a monomial by a polynomial.

Section 1.6

H. Learn and use the order of operations of numerical expressions involving addition, subtraction, multiplication, division, and raising to a power.

I. Evaluate variable expressions with a specified value.

Section 1.7

J. Simplify variable expressions with several grouping symbols.

1.1 Addition and Subtraction of Signed Numbers

In elementary algebra a student is required to master certain rules of signs. Extensive practice in this area will prove a great help in other areas of study. In elementary algebra we often refer to certain kinds of numbers.

Integers are numbers such as $\ldots, -3, -2, -1, 0, 1, 2, 3, \ldots$.

Rational numbers are numbers like $\frac{3}{2}, \frac{5}{7}, -\frac{3}{8}, -\frac{4}{13}$, etc. Rational numbers can be written as one integer divided by another (as long as the denominator is not zero!). Since integers can be written as fractions ($3 = \frac{3}{1}$, etc.) we can see why rational numbers include the integers. It is helpful to view the rational numbers displayed on a number line such as that follows.

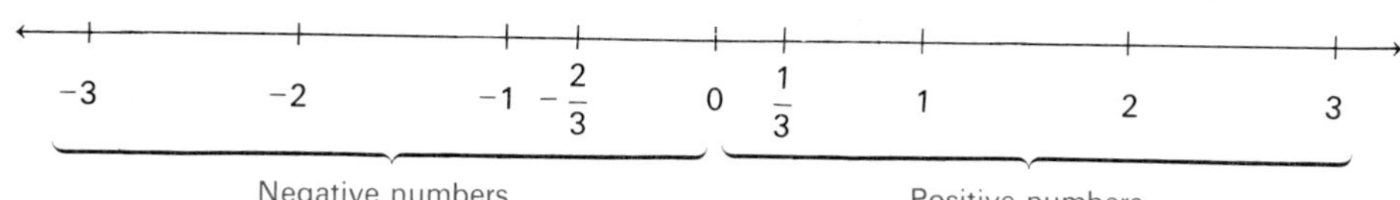

Positive numbers are to the right of 0 on the number line.

Negative numbers are to the left of 0 on the number line.

Irrational numbers are numbers that cannot be expressed as one integer divided by another. The numbers π, $\sqrt{2}$, and $\sqrt[3]{7}$ are irrational numbers. We will study irrational numbers extensively in Chapters 8 and 9. For the first few chapters of the text we will usually limit our discussion to rational numbers.

There are a number of examples in everyday life where people encounter practical examples of a number line that includes positive and negative rational numbers. For example, we can tell by reading the thermometer in Figure 1.1 that the temperature is 20° below 0. From the stock market report, we see that the stock opened at 36, closed at $34\frac{1}{2}$, and the net change for the day was $-1\frac{1}{2}$.

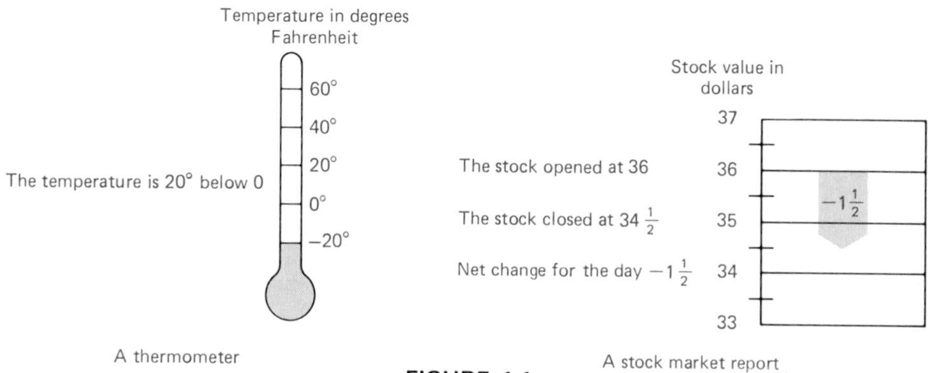

FIGURE 1.1

For a number of reasons we consider positive numbers to be the opposite of negative numbers. For example, let us look at Figure 1.2, which shows the yardage gained and lost during several plays in a football game. (The ball is initially placed at the point marked 0 in a football game.) A football team made a gain of 5 yards on the first down, but lost 8 yards on the second down. After two downs the ball was 3 yards behind the point marked 0 (the place where the ball was initially placed). The measured change in yardage was therefore -3.

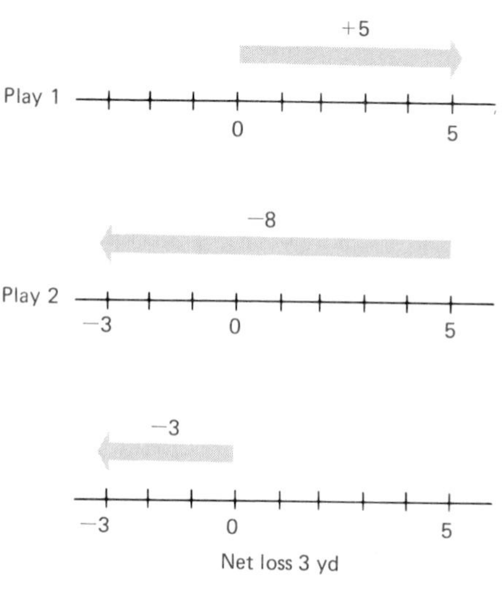

FIGURE 1.2

Opposites of numbers

We define positive numbers to be the *opposite* of negative numbers. Similarly, we define negative numbers to be the opposite of positive numbers. (We call -3 the opposite of 3, 2.16 the opposite of -2.16, etc.) We need to understand some basic rules that will allow us to combine signed numbers. We develop these rules in the next three sections.

1.1.1 Addition of Signed Numbers

In order to use signed numbers we need to be clear about the sign part of a signed number.

When we write the number positive three as $+3$ the sign indicates that it is a positive number. Frequently, the positive sign is omitted in writing positive numbers. If someone writes three (3) it is understood that it is a positive three $(+3)$. When we write a negative number such as negative three (-3) it is absolutely necessary to include the sign.

Sometimes we refer to a number without regard to sign. This property is sometimes referred to as the *magnitude* of a number or the *absolute value* of a number. We will study this in detail in Section 2.5. If, for example, we wanted to consider the number negative four (-4) without regard to sign, we would be considering the magnitude of the number 4.

Let's examine some properties of adding signed numbers by looking at a real-life illustration. Suppose that you are keeping track of your checking account at a local bank. You record deposits as positive numbers. A deposit of 5 dollars you record as $(+5)$. You record checks as debits to your account and therefore as negative numbers. When you write a check for 4 dollars you record it as (-4). Consider two situations.

Situation 1

You made a deposit of 20 dollars on one day and a deposit of 17 dollars the next day. You want to know the total value of your deposits.

Your record for situation 1

$(+20)$	$+$	$(+17)$	$=$	$(+37)$
The deposit of the first day	added to	the deposit of the second day	is	the total deposits made over the two days.

Situation 2

You write a check for 36 dollars to pay one bill and two days later write a check for 5 dollars. You want to know the total value of debits to your account for the two checks.

Your record for situation 2

(-36)	$+$	(-5)	$=$	(-41)
The value of the first check	added to	the value of the second check	is	the total debit to your account because of these two checks.

What do we observe from these illustrations? In each situation we found that we added the magnitude of each number. (That is, we added the numbers without regarding their sign.) The answer always contained the sign that was common to both numbers.

We will now state these results as a formal rule.

> ***Rule A***
>
> **To add two numbers with the same sign, we add the numbers without regard to sign and use the common sign in the answer.**

EXAMPLE A Add: $(+14) + (+16)$.

Solution

$$14 + 16 = 30$$ We add the numbers without regard to sign.

$$(+14) + (+16) = +30$$ We use the common sign in the answer.

EXAMPLE B Add: $(-8) + (-7)$.

Solution

$$8 + 7 = 15$$ We add the numbers without regard to sign.

$$(-8) + (-7) = -15$$ We use the common sign in the answer.

EXAMPLE C Add:

$$\left(+\frac{2}{3}\right) + \left(+\frac{1}{7}\right)$$

Solution

$$\left(+\frac{14}{21}\right) + \left(+\frac{3}{21}\right)$$ Change each fraction to an equivalent fraction with a common denominator of 21.

$$\frac{14}{21} + \frac{3}{21} = \frac{17}{21}$$ We add the numbers without regard to sign.

$$\left(+\frac{14}{21}\right) + \left(\frac{3}{21}\right) = +\frac{17}{21} \quad or \quad \frac{17}{21}$$ We use the common sign in the answer. Note that if no sign is written, it is understood to be a positive number.

EXAMPLE D Add: $(-4.2) + (-3.9)$.

Solution

$$4.2 + 3.9 = 8.1$$ We add the numbers without regard to sign.

$$(-4.2) + (-3.9) = -8.1$$ We use the common sign in the answer.

Rule A can be extended to adding more than two signed numbers. If we add more than two numbers with the same sign, the answer will have the common sign.

EXAMPLE E Add: $(-7) + (-2) + (-5)$.

Solution

$$\underline{(-7) + (-2)} + (-5)$$ We first add the first two numbers.

$$= (-9) + (-5)$$ We add $(-7) + (-2)$ and obtain (-9).

$$= -14$$ We add (-9) and (-5) to obtain -14.

What if the signs of the numbers are opposite when you add? Let's consider our illustrations of the checking account situation.

Situation 3

You made a deposit of 30 dollars on one day. On the next day you write a check for 25 dollars. You want to know the results of your two transactions.

Your record for situation 3

$(+30)$	$+$	(-25)	$=$	$(+5)$
The deposit on the first day is recorded as a positive number.	added to	The check on the next day is a debit and is recorded as a negative number.	is	The total result of the two transactions is a net increase of 5 dollars in your account.

Situation 4

You made a deposit of 20 dollars on one day. The next day you write a check for 40 dollars. You want to know the results of your two transactions.

Your record for situation 4

$(+20)$	$+$	(-40)	$=$	(-20)
The deposit on the first day is recorded as a positive number.	added to	The check on the next day is a debit and is recorded as a negative number.	is	The total result of the two transactions is a net decrease of 20 dollars in your account.

Since the check was larger than the deposit, the result is a negative twenty (-20). If you do not have at least 20 dollars in your account at the start of Situation 4, you have overdrawn your account.

What do we observe from the illustrations in Situations 3 and 4? In each case we found the difference of the two numbers without regard to sign. The sign of the result was always the sign of the larger number without regard to sign. In Situation 3 since 30 is larger than 25 (we are examining the numbers without regard to sign), the answer $(+5)$ had a positive sign since the sign of the larger number was positive. However, in Situation 4 the number 40 is larger than the number 20. Hence we had a negative number for an answer.

We will now state these results as a formal rule.

> ***Rule B***
>
> **To add two numbers with opposite signs, find the difference between the larger and the smaller without regard to sign. The answer will have the sign of the larger number.**

EXAMPLE F Add: $(+8) + (-7)$.

Solution

$(+8) + (-7)$ — Without regard to the signs of the numbers, 8 is larger than 7.

$8 - 7 = 1$ — We take the difference between the larger and the smaller.

$(+8) + (-7) = +1$ or 1 — The answer has the sign of the larger number, which is positive.

EXAMPLE G Add: $(-20) + (+13)$.

Solution

$(-20) + (+13)$ — Without regard to sign, 20 is larger than 13.

$20 - 13 = 7$ — We take the difference between the larger and the smaller.

$(-20) + (+13) = -7$ — The answer has the sign of the larger number, which is negative.

There are three properties of addition that you should understand and be able to use.

1. *Addition is commutative.* This property states that if two numbers are added, the result is the same if either number is written first.

$$(+3) + (+6) = (+6) + (+3) = 9$$

$$(-7) + (-8) = (-8) + (-7) = -15$$

$$(-15) + (+3) = (+3) + (-15) = -12$$

2. *Addition of zero to any given number will result in that given number again.*

$$0 + (+5) = 5$$

$$(-8) + (0) = -8$$

3. *Addition is associative.* This property states if three numbers are added, it does not matter which two numbers are added first.

$$3 + (5 + 7) = (3 + 5) + 7$$

$$3 + (12) = (8) + 7$$

$$15 = 15$$

First combine numbers inside parentheses, then combine the remaining numbers.

Rule B can be extended to addition of more than two numbers.

EXAMPLE H Add:

$$\left(+\frac{3}{17}\right) + \left(-\frac{8}{17}\right) + \left(+\frac{4}{17}\right)$$

Solution

$$\underbrace{\left(+\frac{3}{17}\right) + \left(-\frac{8}{17}\right)} + \left(+\frac{4}{17}\right)$$

We observe that all the fractions have a common denominator. We will add the first two numbers first.

$$= \left(-\frac{5}{17}\right) + \left(+\frac{4}{17}\right)$$

We add $\left(\frac{3}{17}\right) + \left(-\frac{8}{17}\right)$ and obtain $\left(-\frac{5}{17}\right)$. The answer is negative since the larger of the two numbers without regard to sign is negative.

$$= -\frac{1}{17}$$

We add $\left(-\frac{5}{17}\right)$ and $\left(+\frac{4}{17}\right)$ and obtain $\left(-\frac{1}{17}\right)$. The answer is negative since the larger of the two numbers without regard to sign is negative.

In a given problem of adding more than two numbers you will sometimes need to use both Rules A and B.

EXAMPLE I Add: $(-1.8) + (+1.4) + (-2.6)$.

Solution

$\underbrace{(-1.8) + (1.4)} + (-2.6)$	We first add the first two numbers, which are opposite in sign.
$= (-0.4) + (-2.6)$	We take the difference of 1.8 and 1.4 and use the sign of the larger number without regard to sign (by Rule B). Thus $(-1.8) + (1.4) = (-0.4)$.
$= -3.0$	Since the signs are the same we add the numbers without regard to sign and use the common sign (by Rule A). Thus $(-0.4) + (-2.6) = -3.0$.

If many signed numbers are added it is often easier to add numbers with like signs in a column format. Remember addition is commutative, therefore signed numbers can be *added in any order*. You do *not* need to combine the first two numbers as your first step.

EXAMPLE J Add: $(-8) + (+3) + (-5) + (-2) + (+6) + (+5)$.

Solution

$$\begin{array}{r} -8 \\ -5 \\ -2 \\ \hline -15 \end{array} \qquad\qquad \begin{array}{r} +3 \\ +6 \\ +5 \\ \hline +14 \end{array}$$

Add the three negative numbers by Rule A to obtain -15.

Add the three positive numbers by Rule A to obtain $+14$.

$(-15) + (+14) = -1$ — Add the two results using Rule B. The answer is negative since the larger number without regard to sign is negative.

In actual practice the positive sign and the parentheses are often omitted if it will not change the meaning of the mathematical expression. The only time we really need to show a sign of a number is when the number is negative (e.g., -3). The only time we need to show parentheses when we add signed numbers is when we have two different signs preceding a number [e.g., $-5 + (-6)$].

Complete Use of Parentheses and Positive Signs Used in Examples A to I		*Alternative Way of Writing Addition Examples, Minimizing the Use of Parentheses and Positive Signs*		*Answer to Each Example*
A. $(+14) + (+16)$	=	$14 + 16$	=	30
B. $(-8) + (-7)$	=	$-8 + (-7)$	=	-15
C. $\left(+\frac{2}{3}\right) + \left(+\frac{1}{7}\right)$	=	$\frac{2}{3} + \frac{1}{7}$	=	$\frac{17}{21}$
D. $(-4.2) + (-3.9)$	=	$-4.2 + (-3.9)$	=	-8.1
E. $(-7) + (-2) + (-5)$	=	$-7 + (-2) + (-5)$	=	-14
F. $(+8) + (-7)$	=	$8 + (-7)$	=	1
G. $(-20) + (+13)$	=	$-20 + 13$	=	-7
H. $\left(\frac{3}{17}\right) + \left(-\frac{8}{17}\right) + \left(+\frac{4}{17}\right)$	=	$\frac{3}{17} + \left(-\frac{8}{17}\right) + \frac{4}{17}$	=	$-\frac{1}{17}$
I. $(-1.8) + (1.4) + (-2.6)$	=	$-1.8 + 1.4 + (-2.6)$	=	-3.0

Study Examples A to J and then use Rules A and B to solve these practice problems.

Practice Problems for Exercise 1.1.1

Add.

1. $(-9) + (+3)$ **2.** $-6 + (-8) + 2$

3. $-6 + 5 + (-7) + (-2) + 5 + 3$

Now turn to the answers following Exercise 1.1.1 and check your solutions for these practice problems.

Exercise 1.1.1

Add.

1. $(-9) + (+5)$ **2.** $(-7) + (+2)$

3. $(-8) + (-5)$ **4.** $(-12) + (-6)$

5. $\left(-\frac{1}{3}\right) + \left(\frac{2}{3}\right)$ **6.** $\left(-\frac{1}{5}\right) + \left(-\frac{3}{5}\right)$

7. $0.6 + (-0.2)$ **8.** $-0.8 + 0.5$

9. $-12 + (-13)$ **10.** $-17 + (-21)$

11. $\frac{1}{3} + \left(-\frac{1}{4}\right)$ **12.** $-\frac{2}{7} + \left(-\frac{3}{14}\right)$

13. $2 + (-7) + (-6)$ **14.** $-5 + 3 + (-7)$

15. $-3 + 8 + 5 + (-7)$

16. $-2 + (-6) + 7 + 3$

17. $-2 + 8 + (-3) + (-5)$

18. $9 + 5 + (-3) + (-12)$

19. $31 + (-16) + 15 + (-17)$

20. $-22 + (-36) + 19 + 23$

Challenge Problems

21. $(-1.2) + (5.6) + (-2.8) + (1.3) + (5.7) + (-8.3) + (-2.4)$

22. $\left(-\frac{1}{3}\right) + \left(-\frac{1}{4}\right) + \left(\frac{1}{5}\right) + \left(-\frac{1}{2}\right) + \left(-\frac{1}{15}\right)$

Solutions to Practice Problems for Exercise 1.1.1

1. $(-9) + (+3) = -6$

We are adding two numbers with opposite signs, so we use Rule B. Without regarding the sign, the difference between 9 and 3 is 6. Since 9 is larger than 3, we use the sign of the 9, hence the answer is -6.

2. Notice that we can use the rules more than once. $(-6) + (-8) = -14$. This would use Rule A. Now we combine that result with the $+2$. $-14 + 2 = -12$, using Rule B. Thus

$$(-6) + (-8) + 2 = -12$$

We write

$$\begin{aligned} &-6 + (-8) + 2 \\ &= -14 + 2 \\ &= -12 \end{aligned}$$

3. In adding three or more signed numbers, most students add from left to right. However, sometimes it may be helpful to add all negative numbers and all positive numbers separately, especially if there are larger numbers. So here we would add negative numbers $-6 + (-7) + (-2) = -15$ and positive numbers $+5 + 5 + 3 = +13$, using Rule A in each case. Finally, we combine $-15 + 13$ to obtain -2, using Rule B. Thus

$$-6 + 5 + (-7) + (-2) + 5 + 3 = -2$$

We write these sums:

	Negative Numbers	*Positive Numbers*	*Final Sum*
	-6	$+5$	-15
	-7	$+5$	$+13$
	-2	$+3$	
=	-15	$+13$	-2

Once you understand how to obtain the correct answer in these three practice problems, do the problems in Exercise 1.1.1.

1.1.2 Subtraction of Signed Numbers

Now that you have mastered two rules to add signed numbers you can use the same rules in subtraction problems. What you do is change a subtraction problem to an addition problem that yields the same result.

Let's return to our illustration of a checking account to see an instance where we want to subtract signed numbers and would do it by changing to an addition problem.

Situation 5

You have a balance of 20 dollars in your checking account according to your bank statement. The bank calls you and says that a deposit of 5 dollars that belongs to another account was erroneously added to your account. They say they will correct the account balance to 15 dollars. The bank tells you that since they cannot take away the erroneous credit, they will add a new debit to your account. You want to keep track of what's happening to your account.

Your record for situation 5

$(+20)$	$-$	$(+5)$	$=$	15
The present balance on your statement is 20 dollars.	"take away" or subtract	erroneous deposit of 5 dollars credited to your account	will give a result	The new correct balance in your account will be 15 dollars.

This equation represents what the bank wants to do to your account. They want to subtract or "take away" the erroneous deposit of 5 dollars from the printed statement, which shows a balance of 20 dollars. The result will be a new correct account balance of 15 dollars.

The bank tells you that because the error happened in the past, they cannot "take it away." However, they can add to your account a debit (or debt) of 5 dollars.

$(+20)$	$+$	(-5)	$=$	15
The present balance on your statement is 20 dollars.	added to	A debit or debt of 5 dollars. This is represented by a negative number.	will give a result	The new balance in your account will be 15 dollars.

So we see that $(+20) - (+5)$ is equal to $(+20) + (-5)$. Subtracting a positive number seems to have the same effect as adding the *opposite* of that number. The *opposite* of a positive number is negative. The opposite of a negative number is positive.

In algebra, this type of change is the way subtraction is usually performed. We now state this as a formal rule.

> ***Rule C***
>
> **To subtract signed numbers, add the opposite of the second number to the first.**

EXAMPLE A Subtract: $(+6) - (-2)$.

Solution

$= (+6) + (+2)$ Replace the subtraction by addition. Replace the -2 by its opposite, which is $+2$.

$= 8$ Add the two signed numbers, using Rule A.

EXAMPLE B Subtract: $(-9)-(+3)$.

Solution $=(-9)+(-3)$ Replace the subtraction by addition. Replace the $+3$ by its opposite, which is -3.

$=-12$ Add the two signed numbers, using Rule A.

EXAMPLE C Subtract: $(-8)-(-6)$.

Solution $=(-8)+(+6)$ Replace the subtraction by addition. Replace the -6 by its opposite, which is $+6$.

$=-2$ Add the two signed numbers, using Rule B.

EXAMPLE D Subtract:

$$\left(+\frac{3}{7}\right)-\left(+\frac{6}{7}\right)$$

Solution $=\left(+\frac{3}{7}\right)+\left(-\frac{6}{7}\right)$ We note that the problem has two fractions with the same denominator. We change the subtraction problem to adding the opposite of the second number.

$=-\frac{3}{7}$ We add, using Rule B.

EXAMPLE E Subtract: $(-5.2)-(-5.2)$.

Solution $=(-5.2)+(+5.2)$ We change the subtraction problem to one of adding the opposite of the second number.

$=0$ We add, using Rule B.

This solution for Example E is sometimes called the *property of adding opposites.* When you add two real numbers that are opposites of each other you will obtain zero. Examples of this are the following:

$$(+5)+(-5)=0 \qquad (-186)+(186)=0 \qquad \left(-\frac{1}{8}\right)+\left(+\frac{1}{8}\right)=0$$

Extra parentheses and positive signs are often omitted in subtraction problems in a fashion similar to the problems we discussed in Section 1.1.1. We list below Examples A to E and show how each could be written minimizing the use of parentheses and positive signs.

	Complete Use of Parentheses and Positive Signs Used in Examples A to E		*Alternative Way of Writing Subtraction Examples, Minimizing the Use of Parentheses and Positive Signs*		*Equivalent Addition Problem*		*Answer to Each Example*
A.	$(+6)-(-2)$	$=$	$6-(-2)$	$=$	$6+2$	$=$	8
B.	$(-9)-(+3)$	$=$	$-9-3$	$=$	$-9+(-3)$	$=$	-12
C.	$(-8)-(-6)$	$=$	$-8-(-6)$	$=$	$-8+6$	$=$	-2
D.	$\left(+\frac{3}{7}\right)-\left(+\frac{6}{7}\right)$	$=$	$\frac{3}{7}-\frac{6}{7}$	$=$	$\frac{3}{7}+\left(-\frac{6}{7}\right)$	$=$	$-\frac{3}{7}$
E.	$(-5.2)-(-5.2)$	$=$	$-5.2-(-5.2)$	$=$	$-5.2+5.2$	$=$	0

In these examples you can see that we do not need to show a positive sign as the sign of a number, because it is understood. The only time you actually need to show parentheses in subtracting signed numbers is when you are subtracting a negative number.

This extra step to convert to addition may seem like more work at first, but after a little practice it comes naturally. See if you can obtain the correct answers for the following practice problems.

Practice Problems for Exercise 1.1.2

Subtract by first converting to an addition problem.

1. $(+9) - (+12)$

2. $(-6) - (-5)$

3. $-\frac{1}{5} - \left(+\frac{1}{4}\right)$

Now turn to the answers following Exercise 1.1.2 and check your solutions for these practice problems.

Exercise 1.1.2

Subtract by first converting to an addition problem.

1. $(+8) - (+5)$

2. $(+12) - (+6)$

3. $(+7) - (-3)$

4. $(-12) - (-8)$

5. $(-12) - (-4)$

6. $(-17) - (-6)$

7. $15 - 20$

8. $18 - 24$

9. $\frac{1}{3} - \left(-\frac{2}{5}\right)$

10. $\frac{3}{4} - \left(-\frac{3}{5}\right)$

11. $-0.6 - 0.3$

12. $-0.9 - 0.5$

13. $0 - (-7)$

14. $24 - (-24)$

15. $-17 - (-13)$

16. $-26 - (-7)$

17. $-\frac{2}{3} - \left(\frac{1}{4}\right)$

18. $-\frac{1}{5} - \left(\frac{3}{8}\right)$

19. $2.64 - (-1.83)$

20. $0.07 - (-0.09)$

21. $-34 - 7$

22. $16 - (-8)$

23. $\frac{1}{5} - 6$

24. $\frac{2}{7} - (-3)$

Challenge Problems

Combine the numbers inside the parentheses first. Then subtract.

25. $(3 - 7 + 2 - 4 + 3) - (2 - 6 - 4 + 3)$

26. $(3.1 - 6.2 - 4.3) - (0.8 - 2.7)$

Solutions to Practice Problems for Exercise 1.1.2

1. $9 - (+12)$ we rewrite as $9 + (-12)$ by Rule C. Now we add $9 + (-12) = -3$ by Rule B.
2. $-6 - (-5)$ we rewrite as $-6 + (+5)$ by Rule C. Now we add $-6 + 5 = -1$ by Rule B.
3. $-\frac{1}{5} - \left(+\frac{1}{4}\right)$ we rewrite as $-\frac{1}{5} + \left(-\frac{1}{4}\right)$ by Rule C. Now we change each fraction to an equivalent fraction with the common denominator of 20. Then we add

$$-\frac{4}{20} + \left(-\frac{5}{20}\right) = -\frac{9}{20}$$

by Rule A.

When you can obtain the correct answers for these practice problems, do the problems in Exercise 1.1.2.

1.2 Multiplication and Division of Signed Numbers

There are four properties of multiplication that you should understand and be able to use.

1. *Multiplication is commutative.* This property states that if two numbers are multiplied, the result is the same if either number is written first.

$$5 \times 7 = 7 \times 5 = 35 \qquad \frac{1}{3} \times \frac{2}{7} = \frac{2}{7} \times \frac{1}{3} = \frac{2}{21}$$

2. *Multiplication of any number by zero will result in zero.*

$$5 \times 0 = 0 \qquad 0 \times \frac{3}{8} = 0 \qquad 0 \times 0 = 0$$

3. *Multiplication of any number by 1 will result in that same number.*

$$5 \times 1 = 5 \qquad 1(-7) = -7 \qquad (1)\left(-\frac{5}{3}\right) = -\frac{5}{3}$$

4. *Multiplication is associative.* This property states that if three numbers are multiplied, the result is the same independent of which two numbers are multiplied first.

$$2 \times (3 \times 4) = (2 \times 3) \times 4$$

$$2 \times (12) = (6) \times 4$$

$$24 = 24$$

First multiply numbers in parentheses. Then multiply the remaining numbers.

Notation used to indicate multiplication

You should be aware that the multiplication of two numbers in algebra is commonly written in four different ways. You should be able to identify and use all of them. The following four expressions are all appropriate ways to indicate "three times five."

3×5	Using the multiplication symbol.
$(3)(5)$	Placing parentheses around each number.
$3(5)$	Placing parentheses around second number.
$3 \cdot 5$	Placing a raised dot between the two numbers.

Each of these forms of notation will be used in the exercises for this section. The most common notation will be with parentheses.

There are two properties of division that you should recall and be very confident that you understand.

1. *The result of dividing* 0 *by any nonzero number is* 0.

$$0 \div 5 = 0 \qquad 0 \div \frac{2}{3} = 0 \qquad \frac{0}{5.6} = 0 \qquad \frac{0}{1000} = 0$$

2. *Division of any number by* 0 *is not allowed.*

$$7 \div 0 \qquad \frac{64}{0} \qquad \frac{0}{0}$$

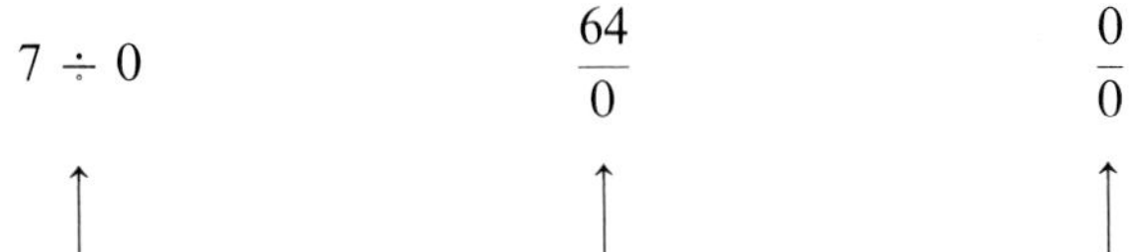

None of these operations are possible. You cannot divide by 0!

?

Do you see *why* division properties 1 and 2 are true?

You can verify these facts by rewriting a division problem as multiplication. $\frac{12}{4} = 3$ is equivalent to $12 = 4 \times 3$. Now $\frac{0}{5} = 0$ since $0 = 5 \cdot 0$. Now consider $\frac{7}{0} = ?$. Do you see that $7 = 0 \times ?$ cannot be solved? No number can be multiplied by 0 to obtain 7. Multiplication of any number by zero will result in zero (multiplication property 2).

The rules for multiplying and dividing signed numbers are somewhat different from those for addition and subtraction. The student should not confuse these new rules with the ones learned previously.

> ***Rule D***
>
> **When two numbers with the same sign are multiplied or divided, the result is positive.**
>
> ***Rule E***
>
> **When two numbers with different signs are multiplied or divided, the result is negative.**

Examples of Rule D	*Examples of Rule E*
$(5)(2) = 10$	$(-6)(3) = -18$
$\left(-\frac{3}{5}\right)\left(-\frac{2}{7}\right) = \frac{6}{35}$	$\left(\frac{2}{7}\right)(-3) = -\frac{6}{7}$
$12 \div 4 = 3$	$(-18) \div (2) = -9$
$(-25) \div (-5) = 5$	$(+100) \div (-25) = -4$

These two new rules often do not seem reasonable to students and often leave them wondering "why?" Let's examine a practical situation involving (once again) a student's checking account in a bank to illustrate why these rules seem to work.

Situation 6

You write a check for 5 weeks in a row to pay your weekly room rent of 40 dollars. You want to know the total impact on your checking balance.

Your record for situation 6

$(+5)$		(-40)	$=$	-200
The number of weekly checks you have written.	No symbol is needed here since the two parentheses together indicate multiplication.	The value of each check. Since it is a debit to your account, it is recorded as a negative number.		The total debit to your account because of the 5 checks. The answer is negative.

In this case the multiplication $(5)(-40)$ is the same as repeated addition of five (-40)'s or

$$\underbrace{(-40) + (-40) + (-40) + (-40) + (-40)}_{\text{repeated addition of five } (-40)\text{'s}} = -200$$

This example seems to show that a positive number multiplied by a negative number is negative. What if the negative number came first?

If $(5)(-40) = -200$, then $(-40)(5) = -200$, by the commutative property of multiplication. *Thus we have shown a justification that when two numbers with different signs* (one positive, one negative) *are multiplied, the result is negative.* But what if both numbers are negative? Consider the following situation.

Situation 7

Last year at college you rented a room at 40 dollars per week for 36 weeks, which included two semesters and summer school. This year you will not attend the summer session, so you will be renting the room for only 30 weeks. Thus the number of weekly rental checks will be 6 less than last year. You are making out your budget for this year. You want to know the financial impact of renting the room for 6 fewer weeks.

Your record for situation 7

(-6)		(-40)	$=$ 240
You will write 6 fewer checks this year. It is a decrease of 6, so we record it as a negative number.	No symbol is needed since the two parentheses together indicate multiplication.	The value of each check. Since it is something you pay out, it is recorded as a negative number.	The product is positive. Your financial situation will be 240 dollars better this year.

You could check that the answer is positive by calculation of the total rental expenses.

	Dollars in rent last year	$(36)(40) =$	1440
(subtract)	Dollars in rent this year	$-(30)(40) =$	-1200
	Extra dollars available this year	$=$	$+240$

This agrees with our answer above: $(-6)(-40) = +240$.

This situation allows us to see an instance where it seems logical that a negative number times a negative number should yield a positive answer. We have already known from arithmetic that a positive number times a positive number should yield a positive answer. Thus we have evidence to state that *when two numbers with the same sign* (both positive or both negative) *are multiplied, the result is positive.*

However, Rules D and E apply to division as well as multiplication. Why is that logical? The justification follows directly from the properties for multiplication. From arithmetic you know that you can use a multiplication problem to check a division problem.

We know that $(+20) \div (+4) = +5$ because $(+4)(+5) = +20$.
Similarly, $(-20) \div (-4) = +5$ because $(-4)(+5) = -20$.

In both division problems the answer is positive 5. Thus we see that *when you divide two numbers with the same sign* (both positive or both negative) *the answer is positive.*

What if the signs are different?

We know that $(-20) \div (+4) = -5$ because $(+4)(-5) = -20$.
Similarly, $(+20) \div (-4) = -5$ because $(-4)(-5) = +20$.

In these two problems the answer is negative 5. So we have reasonable evidence to see that *when you divide two numbers with different signs* (one positive and one negative) *the answer is negative.*

Rules D and E can be applied when more than two numbers are multiplied by taking two numbers at a time.

EXAMPLE A Multiply: $(-4)(-3)(-2)$.

Solution

$= (-4)(-3)(-2)$ — We first multiply the first two numbers.

$= (+12)(-2)$ — Since we are multiplying two numbers whose signs are the same, the answer is positive (Rule D).

$= -24$ — Next, we multiply two numbers whose signs are opposite, so the answer is negative (Rule E).

EXAMPLE B Multiply: $(-6)(-2)(-3)(-1)$.

Solution

$= (-6)(-2)(-3)(-1)$	We first multiply the first two numbers.
$= (+12)(-3)(-1)$	By Rule D.
$= (-36)(-1)$	By Rule E.
$= +36$ or 36	By Rule D.

See if you can see the pattern here.

When we multiply two negative numbers, we know by Rule D the answer is positive.

When we multiplied three negative numbers in Example A, the answer was negative.

When we multiplied four negative numbers in Example B, the answer was positive.

What do you think would happen if we multiplied five negative numbers? If you guessed "negative," you probably see the pattern.

The pattern can be briefly summarized as follows:

When you multiply two or more signed numbers:

1. **The result is always positive if there are an even number of negative signs. (Even numbers are integers divisible by 2, such as 2, 4, 6, 8, 10, ...)**
2. **The result is always negative if there are an odd number of negative signs. (Odd numbers are integers not divisible by 2, such as 1, 3, 5, 7, ...)**

Note that this statement applies if you are multiplying several signed numbers some of which are not negative.

EXAMPLE C Multiply: $(-3)(+2)(-1)(+6)(-3)$.

Solution

$= 3 \times 2 \times 1 \times 6 \times 3 = 108$	We multiply the numerical values without regard to sign.
The answer is negative	Since there are 3 negative signs and 3 is *odd*.

Therefore,

$$(-3)(+2)(-1)(+6)(-3) = -108$$

Now try these practice problems.

Practice Problems for Exercise 1.2

1. Multiply:

$$(-4)(+2)$$

2. Multiply:

$$(-6)\left(-\frac{3}{5}\right)$$

3. Divide:

$$\left(-\frac{12}{5}\right) \div \left(\frac{2}{3}\right)$$

4. Divide:

$$\frac{-12}{-\frac{4}{5}}$$

5. Multiply:

$$(-2)(3)(-4)(-1)(-2)$$

Now turn to the answers following Exercise 1.2 and check your solutions for these practice problems.

Exercise 1.2

Multiply or divide. Be sure to write your answer in the simplest form.

1. $3(-12)$

2. $5(-4)$

3. $(-6)(-5)$

4. $(+5)(-7)$

5. -5×12

6. $-6 \times (-3)$

7. $(-6)\left(\frac{3}{10}\right)$

8. $\left(\frac{12}{5}\right)(-10)$

9. $-8 \cdot 3$

10. $-5 \cdot 11$

11. $\left(-\frac{2}{7}\right)\left(-\frac{14}{3}\right)$

12. $\left(-\frac{3}{8}\right)\left(-\frac{5}{6}\right)$

13. $(0) \div (-3)$

14. $(12) \div (-4)$

15. $(-36) \div (-9)$

16. $(-52) \div (-2)$

17. $\frac{2}{7} \div \left(-\frac{3}{5}\right)$

18. $\left(-\frac{1}{5}\right) \div \left(\frac{2}{3}\right)$

19. $\dfrac{12}{-\frac{2}{5}}$

20. $\dfrac{-\frac{3}{7}}{-6}$

21. $\dfrac{-\frac{3}{8}}{-\frac{2}{3}}$

22. $\dfrac{-\frac{1}{7}}{\frac{3}{14}}$

23. $(-0.6) \div (0.3)$

24. $(1.2) \div (-0.03)$

25. $(-1.32)(-0.2)$

26. $(-1.8)(-0.03)$

27. $(-6)(2)(-3)(4)$

28. $(-1)(-2)(-3)(4)$

29. $(-2)(-1)(3)(-2)(2)$

30. $(-1)(-3)(-2)(-2)(3)$

31. $(-3)(2)(-4)(0)(-2)$

32. $(-3)(-2)\left(\frac{1}{3}\right)(-4)(2)$

Challenge Problems

Combine the numbers inside the parentheses first. Then perform the indicated multiplication.

33. $(5 - 7)(3 - 2)(-6 - 4)(5 + 3 - 4)$

34. $(12 - 8 - 6)(3 - 4)\left(\frac{1}{2} - \frac{1}{3}\right)$

Solutions to Practice Problems for Exercise 1.2

1. $(-4)(+2) = -8$

We are multiplying two numbers whose signs are different. The answer is negative by Rule E.

2. $(-6)\left(-\frac{3}{5}\right) = \frac{18}{5}$ or $3\frac{3}{5}$

The answer is positive by Rule D. Both signs are the same in the two numbers being multiplied.

Sometimes the students prefer to write the problem as $(-\frac{6}{1})(-\frac{3}{5})$ so that they remember to multiply the numerators and the denominators of the fractions. The denominator of -6 is 1 since $-6 = -\frac{6}{1}$.

3. $\left(-\frac{12}{5}\right) \div \left(\frac{2}{3}\right) = \left(-\frac{12}{5}\right)\left(\frac{3}{2}\right)$

since to divide two fractions we invert the second fraction and multiply by the first fraction. Now we can cancel before we multiply.

$$\left(-\frac{\overset{6}{\cancel{12}}}{5}\right)\left(\frac{3}{\underset{1}{\cancel{2}}}\right) = -\frac{18}{5} \quad \text{or} \quad -3\frac{3}{5}$$

The answer is negative by Rule E. The two numbers divided have different signs.

4. $\dfrac{-12}{-\frac{4}{5}}$ is the same as $\left(-\dfrac{12}{1}\right) \div \left(-\dfrac{4}{5}\right)$

We invert the second fraction and multiply and make use of canceling.

$$\left(-\frac{\overset{3}{\cancel{12}}}{1}\right)\left(-\frac{5}{\underset{1}{\cancel{4}}}\right) = 15$$

The answer is positive by Rule D. The two numbers being divided have the same sign.

5. $(-2)(3)(-4)(-1)(-2) = +48$ or 48

Notice that there are 4 negative signs, which is an even number. Therefore, the answer is positive.

Notice that the same result is obtained if you handle each pair of numbers using Rules D and E at each step.

$$\begin{aligned}
&(-2)(+3)(-4)(-1)(-2) \\
&\quad = (-6)(-4)(-1)(-2) && \text{By Rule E.} \\
&\quad = (+24)(-1)(-2) && \text{By Rule D.} \\
&\quad = (-24)(-2) && \text{By Rule E.} \\
&\quad = +48 && \text{By Rule D.}
\end{aligned}$$

Now do the problems in Exercise 1.2.

1.3 Use of Exponents and Variables

1.3.1 Positive Whole-Number Exponents

In mathematics, we use exponents as a way to abbreviate repeated multiplication.

Long Notation		*Exponent Notation*
$2 \cdot 2 \cdot 2 \cdot 2 \cdot 2 \cdot 2$	=	2^6
$3 \cdot 3 \cdot 3 \cdot 3$	=	3^4
$7 \cdot 7 \cdot 7 \cdot 7 \cdot 7$	=	7^5

There are two parts to exponent notation: (1) the *base* and (2) the *exponent*. The *base* tells you what number is being multiplied and the *exponent* tells you how many times this number is used as a factor.

In the first example above,

$2 \cdot 2 \cdot 2 \cdot 2 \cdot 2 \cdot 2 = 2^6$

The *base* is 2 and the *exponent* is 6

(the number being multiplied) (the number of times 2 is a factor)

One has to be careful with exponent notation if the base is negative. It is easy to make an error in determining the sign. Notice the following:

$$(-3)^2 = (-3)(-3) = +9$$

$$(-3)^3 = (-3)(-3)(-3) = -27$$

$$(-3)^4 = (-3)(-3)(-3)(-3) = +81$$

As you studied these examples you probably recalled the useful property we discussed in Section 1.2 concerning what happens when you multiply two or more signed numbers.

A *factor* is any one of the elements that is being multiplied. When we write $(-3)^4 = (-3)(-3)(-3)(-3)$ we can say the factor -3 occurs four times.

When you multiply two or more signed numbers:

1. The result is positive if there are an even number of negative signs.
2. The result is negative if there are an odd number of negative signs.

This can be generalized to the use of exponents by the following rule:

Rule F

In exponent notation if the base is negative: The expression is negative if the exponent is odd and positive if it is even.

EXAMPLE A Evaluate: $(-2)^3$.

Solution $(-2)^3 = -8$. The answer is negative since the exponent 3 is odd.

EXAMPLE B Evaluate: $(-2)^4$.

Solution $(-2)^4 = +16$. The answer is positive since the exponent 4 is even.

How do you say it?

If the value has an exponent of 2, we say "squared."

If the value has an exponent of 3, we say "cubed."

If the value has an exponent greater than 3, we say "to the (exponent)-th power."

Thus

5^2 is read "five squared."

2^3 is read "two cubed."

3^6 is read "three to the sixth power" or simply "three to the sixth."

Now try these practice problems.

Practice Problems for Exercise 1.3.1

1. Evaluate:

$$4^3$$

2. Evaluate:

$$(-2)^7$$

3. Write in exponent notation the expression that has an exponent of 4 and a base of -3. Then evaluate the expression.

Now turn to the answers following Exercise 1.3.1 and check your solutions for these practice problems.

Exercise 1.3.1

Evaluate the following expressions.

1. $2^3 =$

2. $2^4 =$

3. $3^4 =$

4. $3^3 =$

5. $7^3 =$

6. $5^4 =$

7. $(-3)^3 =$

8. $(-2)^3 =$

9. $(-2)^6 =$

10. $(-3)^4 =$

11. $(-6)^3 =$

12. $(-7)^3 =$

13. $\left(\frac{1}{4}\right)^2 =$

14. $\left(\frac{1}{2}\right)^3 =$

15. $\left(\frac{2}{5}\right)^3 =$

16. $\left(\frac{2}{3}\right)^4 =$

Write in exponent notation the expression that has the specified exponent and base and then evaluate it.

17. Base $= 4$, exponent $= 3$

18. Base $= 7$, exponent $= 2$

19. Base $= -5$, exponent $= 4$

20. Base $= -4$, exponent $= 4$

21. Base $= -7$, exponent $= 3$

22. Base $= -6$, exponent $= 3$

Challenge Problems

Evaluate.

23. $(-0.6)^5$

24. $(-0.8)^4$

Solutions to Practice Problems for Exercise 1.3.1

1. $4^3 = (4)(4)(4) = 64$

   The base is 4 and the exponent (the number of times we multiply 4) is 3.

2. $(-2)^7 = (-2)(-2)(-2)(-2)(-2)(-2)(-2) = -128$

   By Rule F, since the base (-2) is negative and the exponent is odd the result is negative.

3. If the base is -3 and the exponent is 4, we have in exponent notation

$$(-3)^4$$

   Note: You must have parentheses. -3^4 is *not* correct. We will see why later in the chapter.

   To evaluate:

$$(-3)^4 = (-3)(-3)(-3)(-3) = +81$$

   By Rule F, the sign is positive since the exponent 4 is even.

Now do the problems in Exercise 1.3.1.

1.3.2 The Idea of a Variable and Multiplication of Expressions in Exponent Notation

Many times in mathematics we do not know the value of a number. In these cases we designate the number by a letter. We call the letter a *variable*. Almost all problems in algebra use variables in one way or another. For example, if a number is cubed but we do not know the number, we could write

$$x^3$$

where x is the variable base and 3 is the exponent. Here the factor x is multiplied 3 times.

We can combine the procedure of exponent notation with our idea of a variable.

EXAMPLE A Write in exponent form:

1. $a \cdot a \cdot a$
2. $y \cdot y \cdot y \cdot y \cdot y \cdot y$
3. $(a + b)(a + b)(a + b)(a + b)$

Solution

1. a^3
2. y^6
3. $(a + b)^4$

We are now ready to multiply numbers that are in exponent form. Try to notice the pattern.

$(2^2)(2^3) = 2^5$ — Since $(2 \cdot 2)(2 \cdot 2 \cdot 2) = 2^5$. Five two's are multiplied, so the exponent here is 5.

$(3^3)(3^4) = 3^7$ — Since $(3 \cdot 3 \cdot 3)(3 \cdot 3 \cdot 3 \cdot 3) = 3^7$. Here 7 three's are multiplied, so the exponent is 7.

$(x^3)(x^5) = x^8$ — Since $(x \cdot x \cdot x)(x \cdot x \cdot x \cdot x \cdot x) = x^8$. Here the variable x is multiplied by itself. There are 8 x's being multiplied so the exponent is 8. In other words, the factor is x and it is repeated 8 times.

$(y^4)(y^2) = y^6$ Since $(y \cdot y \cdot y \cdot y)(y \cdot y) = y^6$. Here the variable y is multiplied by itself. There are 6 y's being multiplied, so the exponent is 6.

We can state the rule in words as follows:

To multiply two numbers in exponent form with the same base, *add the exponents* but keep the base unchanged.

If you try to say that rule out loud it may sound a little cumbersome. That is why we try to write the rule by using variables.

Rule G

$$x^a \cdot x^b = x^{a+b}$$

Here x represents the base (same for both numbers) while the letters a and b represent the exponents that are added.

EXAMPLE B Multiply.

1. $x^3 \cdot x^6$ **2.** $x \cdot x^5$

Solution

1. $x^3 \cdot x^6 = x^9$

2. $x \cdot x^5 = x^6$ (Note that the exponent of the first x is 1. Every variable that does not have a written exponent is understood to have an exponent of 1.)

EXAMPLE C Multiply.

1. $y^5 \cdot y^{11}$ **2.** $2^3 \cdot 2^5$ **3.** $3^5 \cdot 3^8$

4. $x^6 \cdot y^8$ **5.** $2^5 \cdot 3^8$

Solution

1. $y^5 \cdot y^{11} = y^{16}$

2. $2^3 \cdot 2^5 = 2^8$ ⟵ {Note that the *base does*

3. $3^5 \cdot 3^8 = 3^{13}$ ⟵ *not change*! Only the exponent changes.}

4. $x^6 \cdot y^8 = x^6 \cdot y^8$ (Rule G does not apply since the bases are not the same.)

5. $2^5 \cdot 3^8 = 2^5 \cdot 3^8$ (Rule G does not apply since the bases are not the same.)

We can now combine our knowledge of Rule G and our previous knowledge of this chapter to deal with expressions like

$$2x^5$$

Now $2x^5$ is a short way of saying

$$(2)(x)(x)(x)(x)(x)$$

The number 2 is called the *numerical coefficient.* When a number is multiplied by a variable the number is called a numerical coefficient.

When we multiply two such expressions we multiply the numerical coefficients and the variables with exponents separately.

EXAMPLE D Multiply: $(2x^5)(3x^6)$.

Solution

$$(2x^5)(3x^6) = 6x^{11}$$

We multiplied the numerical coefficients $2 \times 3 = 6$. And we multiplied the variables with exponents using Rule G:

$$x^5 \cdot x^6 = x^{5+6} = x^{11}$$

where we added the exponents.

EXAMPLE E Multiply: $(-3y^2)(8y^5)$.

Solution

$$(-3y^2)(8y^5) = -24y^7$$

We multiplied the numerical coefficients $(-3)(8) = -24$ and then used Rule G:

$$y^2 \cdot y^5 = y^{2+5} = y^7$$

Every variable that does not have a written numerical coefficient is understood to have a numerical coefficient of 1. A numerical coefficient of 1 is not usually written. A numerical coefficient of negative 1 is usually abbreviated by a negative sign in front of the variable.

EXAMPLE F Multiply:

1. $(5x^3)(x^6)$

2. $(2y^2)(-y^8)$

3. $(-6x)(-4x^5)$

Solution

1. $(5x^3)(x^6) = (5x^3)(1x^6) = 5x^9$ The numerical coefficient of x^6 is 1.

2. $(2y^2)(-y^8) = (2y^2)(-1y^8) = -2y^{10}$ The numerical coefficient of y^8 is -1.

3. $(-6x)(-4x^5) = 24x^6$ In the first parentheses the exponent of x is understood to be 1.

Problems of this type may involve more than one variable. The product may involve more than two factors.

EXAMPLE G Multiply:

1. $(-4x^2y^3)(-2x^5y^3)$

2. $(-3x^3)(-2x^5y)(2x^5y^2)$

Solution

1. $(-4x^2y^3)(-2x^5y^3) = 8x^7y^6$

2. $(-3x^3)(-2x^5y)(2x^5y^2) = 12x^{13}y^3$ Be careful to write down the correct sign of the numerical coefficient.

It is helpful to practice these methods carefully to be sure that you understand Rule G and that you follow carefully the rules of multiplying signed numbers (Rules D and E). Now try these practice problems.

Practice Problems for Exercise 1.3.2

1. Write in exponent form:

$$5(a)(a)(b)(b)(b)(b)$$

2. Multiply:

$$(-2x^3)(-3x^5)$$

3. Multiply:

$$(3x^5)(2xy^2)(-4y^6)$$

Now turn to the answers following Exercise 1.3.2 and check your solutions for these practice problems.

Exercise 1.3.2

Write in simplest exponent form.

1. $3 \cdot x \cdot x \cdot x \cdot y \cdot y$

2. $4 \cdot a \cdot a \cdot a \cdot a \cdot b \cdot b \cdot b$

3. $(-3)(a)(a)(b)(c)(b)(c)(c)$

4. $(-7)(x)(y)(z)(y)(x)$

State the numerical coefficient, the base(s), and the exponent(s).

5. $-8x^5y^2$

6. $6x^{11}y$

Multiply and leave your answer in exponent form.

7. $(3^8)(3^7)$

8. $(2^5)(2^8)$

9. $(5^3)(2^6)$

10. $(3^5)(8^2)$

Multiply using all the appropriate rules.

11. $-9x^2 \cdot 8x^3$

12. $-12x^2 \cdot 5x$

13. $(5x)(10x^2)$

14. $(-13x)(-2x^2)$

15. $(-2ab)(3a^2b^3)$

16. $(5xy^2)(-2x^3y)$

17. $(-3x)(2y)(5x^3y)$

18. $(8a)(2a^3b)(0)$

19. $(-20xy^3)(-4x^3y^2)$

20. $(-6xy^3)(-12x^5y^8)$

21. $(-3x^5)(20y^8)$

22. $(14a^5)(-2b^6)$

23. $(5a^3b^2)(-ab)$

24. $(-x^2y)(3x)$

25. $(-3x^2y)(0)(-5xy^3)$

26. $(3ab^3)(-2a^5b)(-2a^2b^2)$

27. $(5x^3y)(-2w^4z)$

28. $(6w^5z^6)(-4xy)$

29. $(3ab)(5a^2c)(-2b^2c^3)$

30. $(-8a^3bc)(2abc)(3b^2)$

31. $(3x)(-2y^2)(-6xy)(-2x^3)$

32. $(3y)(-5xy)(6x^2)(-2y^3)$

Challenge Problems

33. $\left(\frac{1}{2}x\right)(-3x^2y)\left(-\frac{1}{5}x\right)\left(\frac{2}{3}xy^2\right)(-12y)$

34. $(1.1ab)(1.2a^2b^2)(-1.3a)(-ab^2)$

Solutions to Practice Problems for Exercise 1.3.2

1. $5(a)(a)(b)(b)(b)(b) = 5a^2b^4$

   Not only should you know how to change the exponent notation, but it is critical that you know clearly which part is the *numerical coefficient*, which parts are *bases*, and which parts are *exponents*. These words are used extensively in algebra.

2. $(-2x^3)(-3x^5) = 6x^8$

   Two steps that you would use (in your head) are

   $(-2)(-3) = 6$ By Rule D.

   $(x^3)(x^5) = x^8$ By Rule G.

3. $(3x^5)(2xy^2)(-4y^6) = -24x^6y^8$

Here you do three steps in your head.

$(3)(2)(-4) = -24$	Note that the sign is negative.
$x^5 \cdot x = x^6$	(You can write this as $x^5 \cdot x^1 = x^6$ if you wish.)
$y^2 \cdot y^6 = y^8$	

Now do the problems in Exercise 1.3.2.

1.3.3 Division of Expressions in Exponent Notation

Frequently, we must divide expressions like the ones we have just learned to multiply. Since division by zero is undefined, in all problems in this chapter we assume that the denominator of any variable expression is not zero. We could do the division in a long fashion, but it would be time consuming.

$$\frac{x^5}{x^2} = \frac{(x)(x)(x)(\cancel{x})(\cancel{x})}{(\cancel{x})(\cancel{x})} = x^3$$

We are here using an arithmetic property of reducing fractions (see Section 0.1.1). This can be done when the values canceled are multiplied by both the numerator and the denominator of the fraction.

However, a simpler way is to subtract the exponents. The following two rules assume the restriction $x \neq 0$.

> ***Rule H***
>
> $$\frac{x^a}{x^b} = x^{a-b}$$ **if the larger exponent is in the numerator.**

EXAMPLE A Simplify.

1. $\dfrac{2^{16}}{2^{11}}$ **2.** $\dfrac{x^5}{x^3}$ **3.** $\dfrac{y^{16}}{y^7}$

Solution

1. $\dfrac{2^{16}}{2^{11}} = 2^5$ Note that the base does *not* change.

2. $\dfrac{x^5}{x^3} = x^2$

3. $\dfrac{y^{16}}{y^7} = y^9$

A similar rule can be used in those cases where the larger exponent is in the denominator.

> ***Rule I***
>
> $$\frac{x^a}{x^b} = \frac{1}{x^{b-a}}$$ **if the larger exponent is in the denominator.**

EXAMPLE B Simplify.

1. $\dfrac{b^7}{b^9}$ 2. $\dfrac{12^{17}}{12^{20}}$ 3. $\dfrac{x^{20}}{x^{24}}$

Solution

1. $\dfrac{b^7}{b^9} = \dfrac{1}{b^2}$
2. $\dfrac{12^{17}}{12^{20}} = \dfrac{1}{12^3}$ Note that the base does not change.
3. $\dfrac{x^{20}}{x^{24}} = \dfrac{1}{x^4}$

There are many times when one or more numerical coefficients not in exponent form are included and we need to divide them or reduce the fraction if they do not divide exactly. Remember to use the rules for dividing signed numbers (Rules D and E).

EXAMPLE C Simplify.

1. $\dfrac{5x^5}{25x^7}$ 2. $\dfrac{-12x^8}{4x^3}$
3. $\dfrac{2x^6}{6x^4}$ 4. $\dfrac{-16x^7}{-24x^8}$

Solution

1. $\dfrac{5x^5}{25x^7} = \dfrac{1}{5x^{7-5}} = \dfrac{1}{5x^2}$
2. $\dfrac{-12x^8}{4x^3} = -3x^{8-3} = -3x^5$
3. $\dfrac{2x^6}{6x^4} = \dfrac{x^{6-4}}{3} = \dfrac{x^2}{3}$
4. $\dfrac{-16x^7}{-24x^8} = \dfrac{2}{3x^{8-7}} = \dfrac{2}{3x}$

One has to work very carefully if two or more variables are involved.

EXAMPLE D Simplify.

1. $\dfrac{x^3y^2}{5xy^6}$ 2. $\dfrac{x^5y^3}{x^6w}$ 3. $\dfrac{-3x^2y^5}{12x^6y^8}$

Solution

1. $\dfrac{x^3y^2}{5xy^6} = \dfrac{x^2}{5y^4}$
2. $\dfrac{x^5y^3}{x^6w} = \dfrac{y^3}{xw}$
3. $\dfrac{-3x^2y^5}{12x^6y^8} = \dfrac{-1}{4x^4y^3}$

There are some cases where for a given base the exponent is the same in numerator and denominator. In these cases we can use the fact that *any nonzero number divided by itself is 1*.

EXAMPLE E Simplify.

1. $\dfrac{x^6}{x^6}$ 2. $\dfrac{3x^5}{x^5}$

Solution

1. $\dfrac{x^6}{x^6} = 1$
2. $\dfrac{3x^5}{x^5} = 3\left(\dfrac{x^5}{x^5}\right) = 3(1) = 3$

There is a way to relate this type of problem to Rules H and I. To do this we will make use of the following definition.

$$x^0 = 1 \quad \text{if } x \neq 0$$

We can write these facts as a separate rule. When $x \neq 0$:

> ***Rule J***
>
> $$\frac{x^a}{x^a} = x^0 = 1$$

We will now apply Rule J in the following example.

EXAMPLE F Simplify,

1. $\dfrac{4x^0y^2}{8^0y^5z^3}$

2. $\dfrac{5x^2y}{10x^2y^3}$

Solution

1. $\dfrac{4x^0y^2}{8^0y^5z^3} = \dfrac{4(1)y^2}{(1)y^5z^3} = \dfrac{4y^2}{y^5z^3} = \dfrac{4}{y^3z^3}$
2. $\dfrac{5x^2y}{10x^2y^3} = \dfrac{1x^0}{2y^2} = \dfrac{(1)(1)}{2y^2} = \dfrac{1}{2y^2}$

Now try these practice problems.

Practice Problems for Exercise 1.3.3

1. $\dfrac{x^7y^9}{y^{10}} =$

2. $\dfrac{13x^5y^6}{-26x^3y^8} =$

3. $\dfrac{2^3x^5y^6}{2^{12}x^8y^{12}} =$

(Leave your answer in exponent form.)

Now turn to the answers following Exercise 1.3.3 and check your solutions for these practice problems.

Exercise 1.3.3

Simplify each fraction. Leave your answer in exponent form. Assume that all variables in any denominator are nonzero.

1. $\dfrac{x^8}{x^{10}} =$

2. $\dfrac{a^{15}}{a^{13}}$

3. $\dfrac{2^{20}}{2^7} =$

4. $\dfrac{4^{15}}{4^{23}} =$

5. $\dfrac{a^{13}}{4a^5} =$

6. $\dfrac{4b^{16}}{b^{13}} =$

7. $\dfrac{x^7}{y^9} =$

8. $\dfrac{2^{20}}{5^6} =$

9. $\dfrac{-10a^5b^2}{-20a^5b^3} =$

10. $\frac{-36x^3y^6}{-18xy^8} =$

11. $\frac{12x^5y^6}{-3xy} =$

12. $\frac{-20a^3b^2}{4a^5b^2} =$

13. $\frac{12ab}{-24a^3b^2} =$

14. $\frac{a^5b^6}{a^5b^6} =$

15. $\frac{-13x^5y^6}{13x^5y^6} =$

16. $\frac{-15xy^3}{45x^6y^3} =$

17. $\frac{-27x^5y^3z}{9x^3y^6z^2} =$

18. $\frac{42m^6n^3}{4.2mn^6} =$

19. $\frac{2.3x^8y^{10}}{23x^{12}y^5} =$

20. $\frac{-51x^6y^8z^{12}}{17x^3y^8z^7}$

21. $\frac{30x^5y^4}{5x^3y^4}$

22. $\frac{27x^3y^2}{3xy^2} =$

23. $\frac{8^0x^2y^3}{16x^5y}$

24. $\frac{3^2x^3y^7}{3^0x^5y^2}$

25. $\frac{18a^6b^3c^0}{24a^5b^3}$

26. $\frac{12a^7b^8}{16a^3b^8c^0}$

27. $\frac{25x^6}{35y^8} =$

28. $\frac{39x^6y^7}{-3xy^5} =$

29. $\frac{-32x^6y^3z}{64x^5y^2} =$

30. $\frac{24y^5}{16x^3} =$

Hint: Multiply the numerator *first* in problems 31 to 34.

31. $\frac{(3x^2)(2x^3)}{3x^4} =$

32. $\frac{(4a^5)(3a^6)}{2a^3} =$

33. $\frac{(9a^2b)(2a^3b^6)}{-27a^8b^7} =$

34. $\frac{(21x^3y)(-2x^8y^4)}{-14xy} =$

Challenge Problems

35. $\dfrac{356a^3bc^3d^2e^5f^4g}{-36a^6c^2d^8e^5f^{12}g^6}$

36. $\dfrac{2^5 \cdot 3^6 \cdot 4^8x^3y^4w^6z^{12}}{2^8 \cdot 3^5 \cdot 4^6xy^{12}w^6z^{10}}$

Solutions to Practice Problems for Exercise 1.3.3

1. $\dfrac{x^7y^9}{y^{10}} = \dfrac{x^7}{y}$ By Rule I.

Notice that the rule does not apply to the base x since there is only one expression with x.

2. $\dfrac{13x^5y^6}{-26x^3y^8} = \dfrac{x^2}{-2y^2}$ or $-\dfrac{x^2}{2y^2}$

Here we use Rule H for the x variables and Rule I for the y variables.

3. $\dfrac{2^3x^5y^6}{2^{12}x^8y^{12}} = \dfrac{1}{2^9x^3y^6}$ Since all exponents are larger in the denominator, we work by Rule I.

Be sure that you do not change or divide the base 2 when doing

$$\frac{2^3}{2^{12}} = \frac{1}{2^9}$$

Now do the problems in Exercise 1.3.3.

1.3.4 Raising a Power to a Power

A problem such as $(x^4)^3$ could be done by writing

$(x^4)^3 = x^4 \cdot x^4 \cdot x^4$ By definition of raising a value to the third power.

$= x^{12}$ By adding exponents by Rule G.

However, the problem can be done more easily by using the following rule:

> ***Rule K***
>
> $$(x^a)^b = x^{ab}$$

EXAMPLE A Raise each expression to the indicated power.

1. $(x^3)^5 =$

2. $(2^7)^3 =$

Solution

1. $(x^3)^5 = x^{3 \cdot 5} = x^{15}$

2. $(2^7)^3 = 2^{7 \cdot 3} = 2^{21}$

Note that the base of 2 does not change.

There are two similar rules involving products and quotients that are very useful.

> ***Rule L***
>
> $$(xy)^a = x^ay^a$$

EXAMPLE B Raise each expression to the indicated power.

1. $(ab)^8$ **2.** $(3x)^{12}$

Solution **1.** $(ab)^8 = a^8b^8$ **2.** $(3x)^{12} = 3^{12}x^{12}$

> ***Rule M***
>
> $$\left(\frac{x}{y}\right)^a = \frac{x^a}{y^a} \quad \text{if } y \neq 0$$

EXAMPLE C Raise each expression to the indicated power.

1. $\left(\frac{x}{y}\right)^5$ **2.** $\left(\frac{7}{w}\right)^4$

Solution **1.** $\left(\frac{x}{y}\right)^5 = \frac{x^5}{y^5}$ **2.** $\left(\frac{7}{w}\right)^4 = \frac{7^4}{w^4}$

Many problems can be simplified by using Rules K, L, and M in one problem.

EXAMPLE D Raise each expression to the indicated power.

1. $\left(\frac{3x^2}{y^3}\right)^{10}$ **2.** $\left(\frac{2ab^2}{c^3}\right)^8$

Solution **1.** $\left(\frac{3x^2}{y^3}\right)^{10} = \frac{3^{10}x^{20}}{y^{30}}$ **2.** $\left(\frac{2ab^2}{c^3}\right)^8 = \frac{2^8a^8b^{16}}{c^{24}}$

You must take particular care to determine the correct sign if the numerical coefficient of the term is a negative number.

EXAMPLE E Raise each expression to the indicated power.

1. $(-3x^2yz^0)^4$ **2.** $\left(\frac{-2xw^2}{z^4}\right)^3$

Solution **1.** $(-3x^2yz^0)^4 = (-3x^2y)^4$ Since $z^0 = 1$.

$= (-3)^4x^8y^4$ Applying Rules K and L.

$= 81x^8y^4$ Since $(-3)^4 = +81$.

2. $\left(\frac{-2xw^2}{z^4}\right)^3 = \frac{(-2)^3x^3w^6}{z^{12}}$ Applying Rules K and M.

$= \frac{-8x^3w^6}{z^{12}}$ Since $(-2)^3 = -8$.

Carefully study Examples A to E and then do these practice problems.

Practice Problems for Exercise 1.3.4

Simplify each of the following using Rules K, L, and M. Assume that $x \neq 0$, $y \neq 0$, and $z \neq 0$.

1. $(3x^5)^4$

2. $(-2x^2yz^0)^5$

3. $\left(\frac{4x^8y^5}{-3z^4}\right)^3$

Now turn to the answers following Exercise 1.3.4 and check your solutions for these practice problems.

Exercise 1.3.4

Simplify each of the following using Rules K, L, and M. Assume that all variables are nonzero.

1. $(x^2)^6$

2. $(w^5)^8$

3. $(xy^2)^7$

4. $(a^3b^2c)^8$

5. $(2ab^2c^7)^4$

6. $(3^2xy^2)^4$

7. $\left(\frac{7}{w^2}\right)^8$

8. $\left(\frac{12x}{y^2}\right)^5$

9. $\left(\frac{3a}{2b}\right)^4$

10. $\left(\frac{6x}{5y^3}\right)^2$

11. $(-3a^2b^3c^0)^4$

12. $(-2a^5b^2c)^5$

13. $\frac{(2x)^4}{(2x)^5}$

14. $\frac{(4a^2b)^2}{(4ab^2)^3}$

15. $(3ab^2)^3(ab)$

16. $(-2a^2b^3)^3(ab^2)$

17. $\left(\frac{2x^2y^3}{3z^0}\right)^3$

18. $\left(\frac{4x^0y^4}{3z^3}\right)^2$

Challenge Problems

19. $[3(xy)^3]^4[2(x^2y)^4]^2$

20. $\left(\frac{-2a}{3b^2}\right)^3\left(\frac{3b^4}{-2a^3}\right)^2$

Solutions to Practice Problems for Exercise 1.3.4

1. $(3x^5)^4 = 3^4x^{20}$ — By Rules K and L.

$= 81x^{20}$ — Since $3^4 = 81$.

2. $(-2x^2yz^0)^5 = (-2x^2y)^5$ — Since $z^0 = 1$.

$= (-2)^5x^{10}y^5$ — By Rules K and L.

$= -32x^{10}y^5$ — Since $(-2)^5 = -32$.

3. $\left(\dfrac{4x^8y^5}{-3z^4}\right)^3 = \dfrac{4^3x^{24}y^{15}}{(-3)^3z^{12}}$ — By Rules K, L, and M.

$= \dfrac{64x^{24}y^{15}}{-27z^{12}}$ — Since $4^3 = 64$ and $(-3)^3 = -27$.

or $-\dfrac{64x^{24}y^{15}}{27z^{12}}$ — It is important that you see that there are three equivalent ways to write the answer. The negative sign may be placed in front of the fraction, in the numerator, or in the denominator.

Now do the problems in Exercise 1.3.4.

1.4 Addition of Expressions with Variables

We see many examples of adding and subtracting quantities that are like quantities. This is called *combining* quantities.

5 inches + 6 inches = 11 inches

20 square inches − 16 square inches = 4 square inches

However, we cannot combine things that are not the same:

16 square inches − 4 inches (cannot be done!)

Similarly, in algebra we can add or subtract like terms. However, we cannot combine terms that are not the same. A *term* is a number, a variable, or a product of numbers and variables. *Like terms* are terms that have identical variables and exponents. In other words, like terms must have exactly the same letter parts.

EXAMPLE A Simplify.

1. $4x^2 + 8x^2$

2. $5x + 3x + 2x$

Solution

1. $4x^2 + 8x^2 = 12x^2$

2. $5x + 3x + 2x = 10x$

EXAMPLE B Simplify.

1. $5a + 2b + 7a - 6b$

2. $3x^2y - 2xy^2 + 6x^2y$

3. $2a^2b + 3ab^2 - 6a^2b^2 - 8ab$

Solution

1. $5a + 2b + 7a - 6b = 12a - 4b$ — We combine the a terms and the b terms separately.
2. $3x^2y - 2xy^2 + 6x^2y = 9x^2y - 2xy^2$ — *Note:* x^2y and xy^2 are not like terms because of different powers.
3. $2a^2b + 3ab^2 - 6a^2b^2 - 8ab$ — These terms cannot be combined; there are no like terms in this expression.

The two skills in this section that a student must practice are identifying like terms and correctly adding or subtracting like terms. If a problem involves many terms, you may find it helpful to rearrange the terms so that like terms are together.

Do you really understand what a term is? A term is a number, a variable, or a product of numbers and variables. Terms are the parts of an algebraic expression separated by plus or minus signs. The sign in front of the product is considered part of the term.

EXAMPLE C

1. List the terms of this expression: $3a - 2b + 5a^2 + 6a - 8b - 12a^2$.
2. Simplify the expression.

Solution

1. There are 6 terms. They are $3a$, $-2b$, $5a^2$, $6a$, $-8b$, $-12a^2$.
2. There are three pairs of like terms.

$= \underbrace{3a + 6a}_{a \text{ terms}} \underbrace{- 2b - 8b}_{b \text{ terms}} + \underbrace{5a^2 - 12a^2}_{a^2 \text{ terms}}$ — We rearrange the terms so that like terms are together and it becomes easier to

$= 9a - 10b - 7a^2$ — collect like terms.

The order of terms in an answer in these problems is not significant. These three terms can be rearranged in a different order. (Answers such as $-10b + 9a - 7a^2$ and $-7a^2 + 9a - 10b$ are also correct.)

EXAMPLE D Simplify: $5xy - 2x^2y + 6xy^2 - xy - 3xy^2 - 7x^2y$.

Solution

$= \underbrace{5xy - xy}_{xy \text{ terms}} \underbrace{- 2x^2y - 7x^2y}_{x^2y \text{ terms}} + \underbrace{6xy^2 - 3xy^2}_{xy^2 \text{ terms}}$ — Rearrange the terms so that like terms are together.

$= 4xy - 9x^2y + 3xy^2$ — Collect like terms. (Note: If you wish to write out $5xy - 1xy$ before collecting like terms, you may. Although a coefficient of negative 1 is not usually written, some students find it helpful to place it in problems.)

There are several correct ways to express the answer. These three terms can be rearranged in a different order. (Answers such as $-9x^2y + 4xy + 3xy^2$ and $3xy^2 + 4xy - 9x^2y$ are also correct.)

Work carefully to obtain the correct answer for these practice problems.

Practice Problems for Exercise 1.4

Simplify.

1. $-x + 3a - 9x + 2a$
2. $5ab - 2ab^2 - 3a^2b + 6ab$
3. $7x^2y - 2xy^2 - 3x^2y - 4xy^2 + 5x^2y$

Now turn to the answers following Exercise 1.4 and check your solutions for these practice problems.

Exercise 1.4 Simplify each of the following.

1. $7x + 3x$
2. $2a - 8a$
3. $-12x^3 - 16x^3$
4. $-8a^2 + 4a^2$
5. $10x^4 + 8x^4 + 7x^2$
6. $3xy - 2y + 6x$
7. $5a + 2b - 7a^2$
8. $3a^3 - 6a^2 + 5a^3$
9. $2ab + 1 - 6ab - 8$
10. $2x^2 + 3x^2 - 7 - 5x^2$
11. $1.3x - 2.6y + 5.8x - 0.9y$
12. $3.1ab - 0.2b - 0.8ab + 5.3b$
13. $\frac{1}{2}x^2 - 3y - \frac{1}{3}y + \frac{1}{4}x^2$
14. $\frac{1}{5}a^2 - 2b - \frac{1}{2}a^2 - 3b$
15. $3x + 2y - 6 - 8x - 9y - 14$
16. $-a + 2b + 8 - 7a - b - 1$
17. $5x^2y - 10xy^6 + 6xy^2 - 7xy^2$
18. $5bcd - 8cd - 12bcd + cd$
19. $2ab + 5bc - 6ac - 2ab$
20. $5x^2y + 12xy^2 - 8x^2 - 12xy^2$
21. $x^3 - 3x^2 + 1 - 2x + 6x^2 - x^3$
22. $x^2 - 2x + 1 - 3x^2 - 5x + 6$
23. $2y^2 - 8y + 9 - 12y^2 - 8y + 3$
24. $5 - 2y^2 + 3y - 8y - 9y^2 - 12$

Challenge Problems

25. $\frac{1}{2}x^2 + \frac{1}{3}y^2 - \frac{1}{5}x - \frac{1}{8}y + \frac{1}{3} - \frac{1}{4}x^2 + \frac{1}{6}y^2 + \frac{1}{2}x - \frac{1}{4}y + 2$

26. $1.3a^2 + 1.4ab - 2b^2 + 3ab - 5.6a^2 + 3.2a^2b - 0.8a^2 + 1.5ab^2$

Solutions to Practice Problems for Exercise 1.4

1. $-x + 3a - 9x + 2a = 5a - 10x$ or $-10x + 5a$

We combine x terms

$$-1x - 9x = -x - 9x = -10x$$

and we combine a terms

$$3a + 2a = 5a$$

The order of terms in the final answer is not critical. $5a - 10x$ is a correct answer. $-10x + 5a$ is also a correct answer.

2. $5ab - 2ab^2 - 3a^2b + 6ab = 11ab - 2ab^2 - 3a^2b$

These three terms in the answer are not like terms, so they cannot be combined. We could only add $5ab + 6ab = 11ab$ since only the ab terms were like terms.

3. $7x^2y - 2xy^2 - 3x^2y - 4xy^2 + 5x^2y = 9x^2y - 6xy^2$

It is helpful to do this by rearranging terms so that like terms are together.

$$\underbrace{7x^2y - 3x^2y + 5x^2y}_{x^2y \text{ terms}} \underbrace{- 2xy^2 - 4xy^2}_{xy^2 \text{ terms}} = 9x^2y - 6xy^2$$

Now do the problems in Exercise 1.4.

1.5 Multiplication of Expressions with Variables

An important property of algebra is the *distributive property*. We can state it in an equation as follows:

$$a(b + c) = ab + ac$$

A numerical example shows that it does seem reasonable.

$$5(3 + 6) = 5(3) + 5(6)$$
$$5(9) = 15 + 30$$
$$45 = 45$$

We will use the distributive property extensively in algebra. The kinds of algebraic expressions that we often deal with are called *polynomials*. Polynomials are variable expressions that contain terms with nonnegative integer exponents. The following three expressions are all polynomials.

$$5x^2y + 1 \qquad 3a + 2b^2 \qquad 5a + 6b + 3c$$

There are special names for polynomials with one, two, or three terms. (You will recall from Section 1.4 that we said a *term* is a number, a variable, or a product of numbers and variables.)

A *monomial* has *one* term:

$$5a \qquad 3x^2yz^6 \qquad 10xy$$

A *binomial* has *two* terms:

$$5x^3 + 2y^2 \qquad 3x + 1 \qquad 10x + y$$

A *trinomial* has *three* terms:

$$5x^2 - 6x + 3 \qquad 2ab^3 - 6ab^2 - 8ab$$

We need to develop a method to multiply a monomial by any polynomial using the distributive law. The distributive law may be generalized to many terms: $a(b + c + d + e + \cdots) = ab + ac + ad + ae + \cdots$. In Section 1.3 we defined the word *factor*. It can be used in this case also. Two or more algebraic expressions joined by multiplication are called *factors*. Consider the following examples.

EXAMPLE A Multiply.

1. $5(a + b)$ **2.** $-1(3x + 2y)$ **3.** $6x(x + 2y)$

Solution

1. $5(a + b) = 5a + 5b$ — We multiply the factor 5 by the factor $(a + b)$

2. $-1(3x + 2y) = -1(3x) + (-1)(2y)$
$= -3x - 2y$ — We multiply the factor -1 by the factor $(3x + 2y)$

3. $6x(x + 2y) = 6x^2 + 12xy$ — We multiply the factor $6x$ by the factor $(x + 2y)$

EXAMPLE B Multiply.

1. $7a(a + 2b + 3)$ **2.** $-2ab(a^2 + a - b - 3)$

Solution

1. $7a(a + 2b + 3) = 7a^2 + 14ab + 21a$ — We multiply the factor $7a$ by the factor $(a + 2b + 3)$

2. $-2ab(a^2 + a - b - 3)$
$= -2a^3b - 2a^2b + 2ab^2 + 6ab$ — We multiply the factor $-2ab$ by the factor $(a^2 + a - b - 3)$

If the parentheses are preceded by a negative sign, we consider this to be the product of (-1) and the polynomial.

EXAMPLE C Multiply.

1. $-(a - 2b)$ **2.** $-(3x^2y - 2xy^2 + 6x)$

Solution

1. $-(a - 2b) = (-1)(a - 2b) = (-1)(a) + (-1)(-2b) = -a + 2b$

2. $-(3x^2y - 2xy^2 + 6x) = (-1)(3x^2y - 2xy^2 + 6x) = -3x^2y + 2xy^2 - 6x$

In general, we see that in all these examples we have multiplied each term of the polynomial in the parentheses by the monomial in front of the parentheses. We will see later that this distributive property also applies to more involved algebraic expressions that do not contain polynomials.

Now try these practice problems.

Practice Problems for Exercise 1.5

Multiply. Simplify your answer if possible.

1. $-2(a + 3b - 1)$

2. $-2xy^2(x - 2y)$

3. $3a^2b(2a^2 - 3ab + b^2)$

4. $ab(2a - b + 6) + a(-3ab + 4b)$ (After you multiply, collect like terms.)

Now turn to the answers following Exercise 1.5 and check your solutions for these practice problems.

Exercise 1.5

Multiply. Simplify your answer if possible.

1. $a(a - 2b)$

2. $x(2x - y)$

3. $4(-2a - 6b)$

4. $-2(3x^2 - 2)$

5. $-2x(x^2 - xy)$

6. $5a(-3a^2 - 2b^2)$

7. $-(a + 2b - c^2)$

8. $-3x(x + 2y - 1)$

9. $4(2x - 8y - 19)$

10. $-(5x^2y - 2xy^2 - xy)$

11. $-9(9x - 5y + 8)$

12. $-5(3x + 9 - 7y)$

13. $\frac{1}{3}(3x^2 + 2x - 1)$

14. $\frac{1}{4}(x^2 + 2x - 8)$

15. $\frac{x}{5}(x^2 + 10x - 4)$

16. $\frac{y}{3}(3y^2 - 4y - 6)$

17. $5x(x + 2y + z - 1)$

18. $3a(2a + b - c - 4)$

19. $-x(x^3 - 3x^2 + 2x - 1)$

20. $-ab(a^3 + 4a^2 - 2a - 1)$

21. $3x^2y(2x + 3y^2 - y)$

22. $4xy^2(-2x^2 + xy - y^2)$

23. $-0.5x(-0.3y + 1.0xy + 0.2)$

24. $-0.4a(0.3a - 0.2b + 0.02b^2)$

25. $5(a + 2b) - 6a(1 - 2b)$

26. $6(x - 2y) - 4(y - 2x)$

27. $3(x^2 - 2xy + y^2) - 6x(y - 2x)$

28. $-5y(3x - y) + 2(x^2 - xy + 2y^2)$

Challenge Problems

29. $\frac{1}{2}xy\left(y^2 - 3xy + \frac{1}{2}y^2\right) - 2\left(xy^3 + \frac{1}{2}x^2y^2 + \frac{1}{8}xy^3\right)$

30. $ab^3(1.5a - 2.7b - 1.9) + 3.2ab(6a + 2.3b - 5.1b^2)$

Solutions to Practice Problems for Exercise 1.5

1. $-2(a + 3b - 1) = -2a - 6b + 2$
   Watch your sign rules carefully.
2. $-2xy^2(x - 2y) = -2x^2y^2 + 4xy^3$
   You may find it helpful to change the original problem to $-2x^1y^2(x^1 - 2y^1)$ so that you do not make a mistake in adding exponents.
3. $3a^2b(2a^2 - 3ab + b^2) = 6a^4b - 9a^3b^2 + 3a^2b^3$
   It is best to keep track of the exponents for a and b separately when doing the mental multiplication.
4. $ab(2a - b + 6) + a(-3ab + 4b)$
   Remove parentheses first in each case by the distributive property.
   $$2a^2b - ab^2 + 6ab - 3a^2b + 4ab$$
   Now we find that the a^2b terms can be combined and the ab terms can be combined. The answer is
   $$-a^2b - ab^2 + 10ab$$

Now do the problems in Exercise 1.5.

1.6 Order of Operations and Substitution into Variable Expressions

1.6.1 Order of Arithmetic Operations

It is important to know *when* to do certain operations. Our understanding of the expression $2 - 4 \times 3$ will not be clear until we agree if we should subtract first or multiply first. The following statement of priorities will assist you.

Statement of Priority of Operations for Numbers

In involved numerical expressions written horizontally the following priority should be used in deciding which operations to perform first.

Do first ↑ **1. Combine numbers inside parentheses.**
2. Raise numbers to a power.
3. Multiply and divide numbers from left to right.
Do last ↓ **4. Add and subtract numbers from left to right.**

Let's return to the problem: $2 - 4 \times 3$. Multiplication is priority 3, so we do that first. Then we subtract since this is priority 4.

$2 - 4 \times 3$	
$= 2 - 12$	We multiply $4 \times 3 = 12$; this is priority 3.
$= -10$	We combine $2 - 12 = -10$; this is priority 4.

EXAMPLE A Evaluate: $8 \div 2 \times 3 + 4^2$.

Solution

$= 8 \div 2 \times 3 + 16$ — We first evaluate $4^2 = 16$ since the highest priority in this problem is raising to a power (priority 2).

$= 4 \times 3 + 16$

$= 12 + 16$ — Next we do multiplication and division from left to right, which is priority 3, so $8 \div 2 = 4$ and $4 \times 3 = 12$.

$= 28$ — Finally we add, which is priority 4.

Note: In priority 3 multiplication and division have equal priority. We do not do multiplication problems first. Rather, we work from left to right, doing any multiplication or division that we encounter. A similar priority status is true for addition and subtraction in priority 4.

EXAMPLE B Evaluate $(-3)^3 - 2^4$.

Solution The highest priority is to raise the expressions to the appropriate power.

$= -27 - 16$ — Be careful. In $(-3)^3$ we are cubing the number -3 to obtain -27. However we then evaluate 2^4, not $(-2)^4$! There are no parentheses around the negative sign.

$= -43$ — The last step is to do addition and subtraction from left to right.

EXAMPLE C Evaluate: $2 \times (2 - 3)^3 + 6 \div 3 + (8 - 5)^2$.

Solution

$= 2 \times (-1)^3 + 6 \div 3 + 3^2$ — Priority 1 is combining numbers inside the parentheses.

$= 2 \times (-1) + 6 \div 3 + 9$ — Priority 2 is raising to a power. Note that we need parentheses for -1 because of the negative sign, but it is not needed for 9.

$= -2 + 6 \div 3 + 9$ — Priority 3 is to multiply and divide from left to right

$= -2 + 2 + 9$

$= 9$ — Priority 4 is to add and subtract from left to right.

In problems of this type it is best to go carefully, showing your work at each step. Trying to do several steps at once greatly increases the chance that you will make a mistake. Exercise 1.6.1 should help you to review your rules of signs. First, try these practice problems.

Practice Problems for Exercise 1.6.1

Evaluate.

1. $18 - (-3)^3$

2. $12 \div (-2)(-3) - 2^4$

3. $6 - (8 - 12)^2 + 8 \div 2$

Now turn to the answers following Exercise 1.6.1 and check your solutions for these practice problems.

Exercise 1.6.1

Evaluate.

1. $4^2 + 2(4)$

2. $2 \div 2 \times 3 + 1$

3. $5 + 6 \times 2 \div 4 - 1$

4. $(2 - 8)^2 \div 6$

5. $(3 - 5)^2 \times 6 \div 4$

6. $2(3 - 5 + 6) + 5$

7. $8 - 2^3 \times 5 + 3$

8. $1 - 16 \times 2 \div 4 + 5$

9. $4 + 27 \div 3 \times 2 - 8$

10. $3 \times 5 + 7 \times 3 - 5 \times 3$

11. $5 \times 6 - (3 - 5)^2 + 8 \times 2$

12. $(-3)^2 \times 6 \div 9 + 4 \times 2$

13. $\frac{1}{2} \div \frac{2}{3} + 6 \times \frac{1}{4}$

14. $2 + \left(\frac{1}{3}\right)^2 - \left(2 - \frac{1}{2}\right)$

15. $0.8 + 0.3 \times (0.6 - 0.2)^2$

16. $0.05 + 1.4 - (0.5 - 0.7)^3$

Challenge Problems

17. $\left(\frac{1}{2}\right)^3 + \frac{1}{4} - \left(\frac{1}{6} - \frac{1}{12}\right) - \frac{2}{3} \times \left(\frac{1}{4}\right)^2$

18. $(2.4 \times 1.2)^2 - 1.6 \times 2.2 \div 4.0 - 3.6$

Solutions to Practice Problems for Exercise 1.6.1

1. $18 - (-3)^3$ — Priority 2: we raise to a power. Note that we must have parentheses in our answer at this stage.

$= 18 - (-27)$ — Priority 4 is to subtract from left to right.

$= 18 + 27$ — We rewrite the subtraction as addition to make it easier.

$= 45$

2. $12 \div (-2)(-3) - 2^4$

Note that we do *not* have $(-2)^4$. The number that is raised to the fourth power is 2, *not* -2. If it seems clearer, you may put a set of parentheses here to emphasize this.

$= 12 \div (-2)(-3) - (2^4)$

$= 12 \div (-2)(-3) - 16$ — Priority 2: we raise to a power.

$= (-6)(-3) - 16$ — Priority 3: we multiply and divide from left to right. The first operation is $12 \div (-2) = -6$. Now still in priority 3 we keep working from left to right and multiply $(-6)(-3) = +18$.

$= +18 - 16$

$= 2$ — Now with priority 4 we combine the last two values.

3. $6 - (8 - 12)^2 + 8 \div 2$

$= 6 - (-4)^2 + 8 \div 2$ — Priority 1: we combine numbers inside the parentheses. Note that the answer must be left inside parentheses at this stage, due to the other negative sign.

$= 6 - 16 + 8 \div 2$ — Priority 2: we raise to a power.

$= 6 - 16 + 4$ — Priority 3: we perform division.

$= -6$ — Priority 4: we add and subtract from left to right.

Now do the problems in Exercise 1.6.1.

1.6.2 Substitution into Variable Expressions

In algebra, the student often has to use his or her knowledge of the order of operations in the process of substitution into variable expressions.

$$\text{Evaluate } 6 + 3x \qquad \text{where } x = -4$$

Here the variable x represents some number. If we are told that x has the value -4, we can replace x by -4. Notice that when we replace the variable by a number we put a set of parentheses around the number. This will help us to identify which arithmetic operations need to be done in the next step. This yields

$6 + 3(-4)$

$= 6 + (-12)$ — We first multiply

$= -6$ — Then we add.

Often, the expressions are more involved.

EXAMPLE A Evaluate: $x^2 + x$, where $x = -4$.

Solution

$(-4)^2 + (-4)$ — We replace x with -4.

$= 16 + (-4)$ — Then we raise to a power.

$= 12$ — Finally, we add.

EXAMPLE B Evaluate: $x^2 + 3x - 5y$, where $x = -3$ and $y = 2$.

Solution

$(-3)^2 + 3(-3) - 5(2)$ — We replace x and y with -3 and 2, respectively.

$= 9 + 3(-3) - 5(2)$ — We raise to a power.

$= 9 - 9 - 10$ — We multiply.

$= -10$ — We combine the last three numbers.

A word of caution. Be careful what value you raise to a power. Consider Examples C and D. *Note*: In two of these problems we will need parentheses within parentheses. To avoid confusion we use brackets [ ] to represent the outside parentheses.

EXAMPLE C Evaluate when $x = -3$.

1. $2x^2$

2. $(2x)^2$

Solution

1. Here the value -3 is squared.

$= 2(-3)^2$

$= 2(9)$ Then we multiply.

$= 18$

2. Since $(2x)$ is squared here the highest priority is to multiply what is in the parentheses.

$= [(2)(-3)]^2$

$= (-6)^2$ Then we square.

$= 36$

EXAMPLE D Evaluate when $x = -3$.

1. $-x^4$

2. $(-x)^4$

Solution

1. In $-x^4$ when $x = -3$ the value -3 is raised to the fourth power.

$= -(-3)^4 = -(+81)$

$= -81$

2. In $(-x)^4$ when $x = -3$ we first simplify inside the parentheses.

$= [-(-3)]^4$

$= (+3)^4$ Now we raise to the fourth power.

$= 81$

Carefully study Examples C and D. You will find that taking the time to see *how* and *why* they are different is a good investment of study time.

Formulas

Frequently, the steps used in substitution of values into variable expressions are used to evaluate a formula that yields information we want. For example, the area of a triangle is found by the formula $A = \frac{1}{2}ab$, where b is the length of the base of a triangle and a is the height (or the altitude) of the triangle (see Figure 1.3).

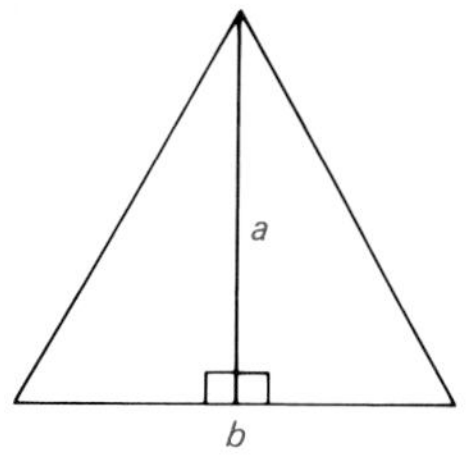

FIGURE 1.3

EXAMPLE E Find the area of a triangle with a base of 16 centimeters and a height of 12 centimeters.

Solution A triangle with a base of 16 centimeters and a height of 12 centimeters has the area

$$A = \frac{1}{2}ab = \frac{1}{2}(12)(16) = (6)(16) = 96 \text{ square centimeters}$$

The area of a circle (Figure 1.4) is given by

$$A = \pi r^2$$

We will use 3.14 as an approximation for the *irrational number* π. (An irrational number is a number that cannot be expressed as one integer divided by another integer.)

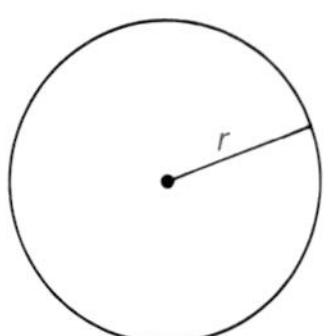

FIGURE 1.4

EXAMPLE F Find the area of the circle if the radius is 2 inches.

Solution The area

$A = \pi r^2 = (3.14)(2)^2$	We substitute the given values.
$= (3.14)(4)$	We raise to a power.
$= 12.56$ square inches	We multiply.

Now try these practice problems.

Practice Problems for Exercise 1.6.2

1. Evaluate:

$$(5x)^3 + 2x \qquad \text{when } x = -2$$

2. Evaluate:

$$2x^3 + 3xy - y^2 \qquad \text{when } x = -2, y = 3$$

3. Find the Celsius temperature when the Fahrenheit temperature is $F = 68°$ given the formula

$$C = \frac{5}{9}(F - 32)$$

Now turn to the answers following Exercise 1.6.2 and check your solutions for these practice problems.

Exercise 1.6.2

Evaluate.

1. $5x + 7$ when $x = -2$

2. $2x^2 + 3x$ when $x = -3$

3. $18 - 5x$ when $x = -3$

4. $(2x)^2 + x$ when $x = 3$

5. $2 - x^2$ when $x = -2$

6. $2 - (-x)^2$ when $x = -2$

7. $2x - 3x^2$ when $x = -4$

8. $16 - 7x + x^2$ when $x = -2$

9. $5x + (3x)^2$ when $x = -2$

10. $2x^2 + 3x + 6$ when $x = 2$

11. $x^2 + 2x + 3$
(a) when $x = 4$
(b) when $x = 0$
(c) when $x = -2$

12. $6 - 2x - 3x^2$
(a) when $x = 2$
(b) when $x = 0$
(c) when $x = -1$

13. $x^2 - 2y + 3y^2$ when $x = -3$ and $y = 4$

14. $2x^2 - 3xy + 2y$ when $x = 4$ and $y = -1$

15. $a^3 + 2abc - 3c^2$
when $a = 5$ and
$b = 0$ and
$c = -1$

16. $a^2 - 2ab + 2c^2$
when $a = +3$ and
$b = +2$ and
$c = -4$

17. Find the Fahrenheit temperature F when the Celsius temperature is 25°C.

$$F = \frac{9}{5}C + 32$$

18. Find the Fahrenheit temperature F when the Celsius temperature is -10°C.

$$F = \frac{9}{5}C + 32$$

19. Find the profit P if the initial cost $C = \$1200$, rent $R = \$300$, and sales $S = 56$ in the formula

$$P = 180S - R - C$$

20. Find the interest I if the principal $p = \$3000$, the rate $r = 0.18$, and the time $t = 2$ in the formula

$$I = prt$$

Challenge Problems

21. Find the surface area A of a cylinder using the formula $A = 2\pi r^2 + 2\pi rh$, where $\pi = 3.14$, radius $r =$ 2.0 inches, and height $h = 4.2$ inches.

22. Find the monthly retirement income R using the formula

$$R = \frac{3}{5}S\left(1 + \frac{2n}{100}\right)$$

where S (the average monthly salary of the worker) is \$1575 and n (the number of years the worker was employed) is 32.

Solutions to Practice Problems for Exercise 1.6.2

1. Evaluate: $(5x)^3 + 2x$ when $x = -2$. (Notice here that we must multiply what is inside the parentheses first *before* we cube the $5x$.)

$[(5)(-2)]^3 + 2(-2)$

$= (-10)^3 + 2(-2)$ — Now we raise to a power.

$= -1000 + 2(-2)$

$= -1000 - 4$ — Then we multiply.

$= -1004$ — Finally, we combine.

2. Evaluate: $2x^3 + 3xy - y^2$ when $x = -2$ and $y = 3$.

$2(-2)^3 + 3(-2)(3) - (3)^2$ — Substitute the values for x and y.

$2(-8) + 3(-2)(3) - 9$ — Raise to a power.

$-16 + (-18) - 9$ — Multiply.

$= -43$ — Add and subtract.

3. $C = \frac{5}{9}(F - 32)$. Find C if $F = 68°$.

$C = \frac{5}{9}(68 - 32)$	Substitute the given F value.
$C = \frac{5}{9}(36)$	Combine two numbers inside the parentheses.
$C = \frac{5}{\not 9}(\overset{4}{\not 36})$	Simplify.
$C = 20°$	The Celsius temperature value is 20° when the Fahrenheit value is 68°.

Study the solutions for these three practice problems carefully and then do the problems in Exercise 1.6.2.

1.7 Removal of Grouping Symbols

Many involved expressions in algebra use grouping symbols extensively. This often involves expressions inside other expressions. It is somewhat confusing to have more than one set of parentheses, so brackets and braces are also usually used. We deal with these problems by using this rule:

> ***Rule N***
>
> **To simplify an expression with grouping symbols inside others remove the innermost set of symbols first.**

EXAMPLE A Simplify: $3[6 - 2(x + y)]$.

Solution We want to remove the innermost parentheses first. Therefore we first use the distributive property to simplify $-2(x + y)$.

$= 3[6 - 2x - 2y]$	Now we use the distributive property again.
$= 18 - 6x - 6y$	

You recall that a negative sign in front of parentheses is equivalent to multiplying by negative 1. Removing the parentheses in this case has the effect of changing the sign of each term inside the parentheses.

$$-(x + 2y) = -1(x + 2y) = -x - 2y$$

Similarly, a positive sign in front of parentheses can be viewed as multiplication by +1.

$$+(5x - 6y) = +1(5x - 6y) = 5x - 6y$$

EXAMPLE B Simplify: $-2[3a - (b + 2c) + (d - 3e)]$.

Solution

$-2[3a - b - 2c + d - 3e]$	By Rule N, we remove the two innermost sets of parentheses. Since one is not inside the other, we remove both sets at once.
$= -6a + 2b + 4c - 2d + 6e$	Now we remove the brackets by multiplying each term by -2.

EXAMPLE C Simplify: $2[3x - (y + w)] - 3[2x + 2(3y - 2w)]$.

Solution

$2[3x - y - w] - 3[2x + 6y - 4w]$	In each set of brackets remove the inner parentheses, by Rule N.
$= 6x - 2y - 2w - 6x - 18y + 12w$	Remove each set of brackets by multiplying by the appropriate number.
$= -20y + 10w$ (or $10w - 20y$)	Collect like terms. (Note that $6x - 6x = 0x = 0$.)

Problems involving parentheses inside other grouping symbols can always be solved by the method shown above. Essentially, you just keep removing one level of grouping symbols at each step. Finally, at the end you add up the terms if possible.

These problems can be made somewhat shorter. To do this, collect like terms at each step (where this is possible).

EXAMPLE D Simplify: $-3\{7x - 2[x - (2x - 1)]\}$.

Solution

$-3\{7x - 2[x - 2x + 1]\}$	We remove the inner parentheses by multiplying each term within the parentheses by negative 1.
$= -3\{7x - 2[-x + 1]\}$	We collect like terms by combining $+x - 2x$.
$= -3\{7x + 2x - 2\}$	We remove the brackets by multiplying each term within it by -2.
$= -3\{9x - 2\}$	We add the x terms.
$= -27x + 6$	We remove the braces by multiplying each term by -3.

If a student does not wish to collect like terms at these stages it is not necessary. However, *all like terms must be combined* before the answer is complete.

Carefully study Examples A to D and then try these practice problems.

Practice Problems for Exercise 1.7

Remove grouping symbols and combine like terms.

1. $3ab - [2ab - (2 - a)]$

2. $3[4x - 2(1 - x)] - [3x + (x - 2)]$

3. $-2\{5x - 3x[2x - (x^2 - 4x)]\}$

Now turn to the answers following Exercise 1.7 and check your solutions for these practice problems.

Exercise 1.7

Remove grouping symbols and combine like terms.

1. $x - 4(x + y)$

2. $y - 3(y + z)$

3. $2(a + 3b) - 3(b - a)$

4. $4(x - y) - 2(3x + y)$

5. $x^2(x^2 - 3y^2) - 2(x^4 - x^2y^2)$

6. $ab(ab - 6b) - 8(a^2b^2 - 2ab^2)$

7. $5[3 + 2(x - 26) + 3x]$

8. $-4[-(x + 3y) - 2(y - x)]$

9. $2x + [4x^2 - 2(x - 3)]$

10. $2x[4x^2 - 2(x - 3)]$

11. $4y + [-3y^2 + 2(4 - y)]$

12. $4y[-3y^2 + 2(4 - y)]$

13. $2(x - 2y) - [3 - 2(x - y)]$

14. $5[a - b(a - b) - b^2]$

15. $x(x^2 + 2x - 3) - 2(x^3 + 6)$

16. $5x^2(x + 6) - 2[x - 2(1 + 2x^2)]$

17. $5x - \{4y - 2[2x + 3(x - y)]\}$

18. $2b^2 - 3[5b + 2b(2 - b)]$

19. $-\{4a^2 - 2[3a^2 - (b - a^2)]\}$

20. $-\{3x - x[4x + 2(x^2 - x)]\}$

Challenge Problems

21. $\frac{1}{2}\left\{3x^2 - 2x\left(x + \frac{1}{4}\right) + 4x\right\} - \frac{1}{3}\left\{5x - 2\left(x - \frac{1}{2}\right) - 3x^2\right\}$

22. $1.6\{0.2a^2 - 3(a + 2b) - 8a^2\} - 4.5\{1.2a - 0.6b - 0.2a^2\}$

Solutions to Practice Problems for Exercise 1.7

1.

$3ab - [2ab - (2 - a)]$

$= 3ab - [2ab - 2 + a]$ — We first remove the inner parentheses. Note that $+2 - a$ becomes $-2 + a$.

$= 3ab - 2ab + 2 - a$ — Next we remove the brackets.

$= ab + 2 - a$ — We collect the ab terms.

2. $3[4x - 2(1 - x)] - [3x + (x - 2)]$

$= 3[4x - 2 + 2x] - [3x + x - 2]$ — By Rule N, we can remove the parentheses in both sets of brackets at the same time.

$= 3[6x - 2] - [4x - 2]$ — If we want to, we may collect x terms *inside* the brackets. This shortens the length of the problem.

$= 18x - 6 - 4x + 2$ — Now we remove each set of brackets.

$= 14x - 4$ — The last step is to collect like terms.

3. $-2\{5x - 3x[2x - (x^2 - 4x)]\}$

$= -2\{5x - 3x[2x - x^2 + 4x]\}$ — First we remove the inner parentheses.

$= -2\{5x - 3x[6x - x^2]\}$ — Inside the brackets we may collect x terms.

$= -2\{5x - 18x^2 + 3x^3\}$ — Now we remove the brackets by multiplying each term by $-3x$. *Be careful of your exponents.*

$= -10x + 36x^2 - 6x^3$ — Finally, we remove the braces. There are no similar terms to combine.

Now do the problems in Exercise 1.7.

An Application for Chapter 1: Using Exponents

A town manager of a small southwestern town has observed that the town population is growing steadily at 8% per year. Three years ago the town had a population of 10,000 people. Two years ago the town had a population of 10,800. Last year it was 11,664 and this year it is 12,597. The manager placed the data in a table and found an interesting property. He discovered a pattern that he could use to predict the future population.

TOWN POPULATION TABLE

Time	*n*	*Population*	*Population in Exponent Form*
Available information			
3 years ago	0	10,000	10,000
2 years ago	1	10,800	$10{,}000 \times (1.08)^1$
1 year ago	2	11,664	$10{,}000 \times (1.08)^2$
Present	3	12,597	$10{,}000 \times (1.08)^3$
Predicted information			
1 year from now	4	?	$10{,}000 \times (1.08)^4$
2 years from now	5	?	$10{,}000 \times (1.08)^5$
3 years from now	6	?	$10{,}000 \times (1.08)^6$
4 years from now	7	?	$10{,}000 \times (1.08)^7$
⋮			
10 years from now	13	?	$10{,}000 \times (1.08)^{13}$
22 years from now	25	?	$10{,}000 \times (1.08)^{25}$
47 years from now	50	?	$10{,}000 \times (1.08)^{50}$
⋮			
$n - 3$ years from now	n	?	$10{,}000 \times (1.08)^n$

Assuming that the growth rate remains at 8% for any given year in the future, he could predict the population by multiplying 10,000 × 1.08 raised to a power. For larger exponents these are difficult to calculate but they can be easily found on scientific calculator with a y^x button or in a compound interest table. He found a compound interest table, a portion of which is listed below.

COMPOUND INTEREST TABLE

n	$(1.04)^n$	$(1.08)^n$	$(1.12)^n$
1	1.040000	1.080000	1.120000
2	1.081600	1.166400	1.254400
3	1.124864	1.259712	1.404928
4	1.169859	1.360489	1.573519
5	1.216653	1.469328	1.762342
6	1.265319	1.586874	1.973823
7	1.315932	1.713824	2.210681
8	1.368569	1.850930	2.475963
9	1.423312	1.999005	2.773079
10	1.480244	2.158925	3.105848
11	1.539454	2.331639	3.478550
12	1.601032	2.518170	3.895976
13	1.665074	2.719624	4.363493
14	1.731676	2.937194	4.887112
15	1.800944	3.172169	5.473566
16	1.872981	3.425943	6.130394
17	1.947900	3.700018	6.866041
18	2.025817	3.996019	7.689966
19	2.106849	4.315701	8.612762
20	2.191123	4.660957	9.646293
25	2.665836	6.848475	17.000064
30	3.243398	10.062657	29.959922
35	3.946089	14.785344	52.799620
40	4.801021	21.724521	93.050970
45	5.841176	31.920449	163.987604
50	7.106683	46.901613	289.002190
55	8.646367	68.913856	509.320606
60	10.519627	101.257064	897.596933
65	12.798735	148.779847	1,581.872491
70	15.571618	218.606406	2,787.799828
75	18.945255	321.204530	4,913.055841
80	23.049799	471.954834	8,658.483100
85	28.043605	693.456489	15,259.205681
90	34.119333	1,018.915089	26,891.934223
95	41.511386	1,497.120549	47,392.776624
100	50.504948	2,199.761256	83,522.265727

1. *What will the population be 1 year from now?*

From the town population table we see that for 1 year from now we should use $n = 4$.

$$\text{Population} = 10{,}000 \times (1.08)^4$$

The table gives $(1.08)^4 = 1.360489$, and therefore the actual population of the town is approximately

$$10{,}000 \times 1.360489 = 13{,}604.89$$

We round to the nearest unit to obtain an estimated population of 13,605.

2. *What will the population be 4 years from now?*

For 4 years from now we use $n = 7$.

$$P = 10{,}000 \times (1.08)^7 = 10{,}000 \times 1.713824 = 17{,}138.24$$

We round to obtain a population of 17,138.

The general procedure discussed here can be used to predict growth of bacteria, increase due to inflation, and a variety of other problems involving a constant rate of increase.

Supplementary Problems for Chapter 1

In problems 1 to 23, perform each operation indicated. Simplify all answers.

1. $-5 + (-2) - (-3)$

2. $6 - (-4) + (-2) + (8)$

3. $(-3)(-2) + (-2)(4)$

4. $\dfrac{-18}{-\dfrac{2}{3}}$

5. $(-3)^4$

6. $(-2)^6$

7. $(-7)(-2) \times 3 - 4 - 2^3$

8. $18 \div (-3) \times 2 - 4 \div 4$

9. $(3ab^2)(2a^3b)(-3ab^6)$

10. $5xy(-2x^3)(-4xy^5)$

11. $(-2x^2y^2)(xy^3)(-4x^6)$

12. $[3(ab)^2]^3$

13. $\dfrac{12a^3b^6}{18a^2b^{12}}$

14. $\dfrac{-25x^5y}{15x^6yz^4}$

15. $\dfrac{7x^3y^6z}{-21x^5y^6z^8}$

16. $\dfrac{-20a^8b^6}{-12a^5b^5}$

17. $3a^2b - 2bc + 6bc^2 - 8a^2b - 6bc^2 + 5bc$

18. $2xy^3 - 6x^3y - 4x^2y^2 + 3(xy^3 - 2x^2y - 3x^2y^2)$

19. $-2y^2 - 8xy + 3x^2 - 4y^2 + 3xy - 8y^2 - 2x^2 + xy$

20. $-5(x + 2y - 7) + 3x(2 - 5y)$

21. $8[a - b(3 - 4a)] - 6a[2 - a(3 - b)]$

22. $2\{x - 3(y - 2) + 4[x - 2(y + 3)]\}$

23. $-5\{2a - b[5a - b(3 + 2a)]\}$

24. Evaluate:

$$2x^2 + 3xy - 2y$$

when $x = 3$ and $y = -1$

25. Evaluate:

$$-3x^3 - 4x^2 + 2x + 6$$

when $x = -2$

26. Evaluate:

$$d = vt - \frac{1}{2}at^2$$

when $v = 24$, $t = 2$, and $a = 32$

27. Evaluate:

$$P = \frac{nRT}{V}$$

when $n = 16$, $R = -2$, $T = 4$, and $V = -20$

28. Evaluate:

$$\frac{1}{2}ab^2 - 3ab + 2b^2$$

when $a = 8$ and $b = -3$

Chapter 1 Summary

Rules for Signed Numbers

Be sure that you know how to combine signed numbers. It is easy to confuse the rules unless you practice.

Rule A: To add two numbers with the same sign, add the numbers without regard to sign and use the common sign in the answer.

Rule B: To add two numbers with opposite signs, find the difference between the larger and the smaller without regard to sign. The answer will have the sign of the larger number.

Rule C: To subtract signed numbers, add the opposite of the second number to the first.

Rule D: When two numbers with the same sign are multiplied or divided the result is positive.

Rule E: When two numbers with opposite signs are multiplied or divided the result is negative.

Rules for Exponent Notation

Rule F: In exponent notation if the base is negative: The expression is negative if the exponent is odd and positive if it is even.

Rule G: $x^a \cdot x^b = x^{a+b}$

Rule H: $\dfrac{x^a}{x^b} = x^{a-b}$ if the larger exponent is in the numerator ($x \neq 0$)

Rule I: $\dfrac{x^a}{x^b} = \dfrac{1}{x^{b-a}}$ if the larger exponent is in the denominator ($x \neq 0$)

Rule J: $\dfrac{x^a}{x^a} = x^0 = 1$ if the exponent is the same in both numerator and denominator ($x \neq 0$)

Rule K: $(x^a)^b = x^{ab}$

Rule L: $(xy)^a = x^a y^a$

Rule M: $\left(\dfrac{x}{y}\right)^a = \dfrac{x^a}{y^a}$ if $y \neq 0$

Combining Like Terms

Only like terms can be added or subtracted. Like terms are terms that have identical letters and exponents.

The Distributive Law

$a(b + c) = ab + ac$. This is the usual way to simplify multiplication of a polynomial by a monomial. There can be any number of terms in the parentheses.

Statement of Priority of Operations for Numbers

If several numbers with several indicated operations are written in a line (horizontally), remember to do the operations in the following order:

1. Combine numbers inside parentheses.
2. Raise numbers to a power.
3. Multiply and divide numbers from left to right.
4. Add and subtract numbers from left to right.

Substitution into Variable Expressions

Replace each letter by the correct numerical value and then evaluate the expression.

Removing Grouping Symbols

Remove the innermost grouping symbols first. Continue to remove all grouping symbols. At any step, like terms may be combined.

Rule N: To simplify an expression with grouping symbols inside others, remove the innermost set of symbols first.

Practice Quiz for Chapter 1

Sections 1.1-1.2

Perform the necessary operation and simplify your answer.

1. $2 - 6 + 8$

2. $7 - 6 + 5 - 2 + 3$

3. $-3 - (-8)$

4. $-2 - (+5)$

5. $7 - 6 + 5 - (-2) - 9$

6. $5 - (+9) - (-6)$

7. $-\frac{1}{3} + \frac{1}{4} - \frac{1}{2}$

8. $(-5)(+3)$

9. $\dfrac{-12}{\frac{2}{3}}$

10. $\left(-\frac{3}{4}\right) \div \left(\frac{2}{5}\right)$

11. $\left(-\frac{1}{3}\right)(6)(-4)$

12. $(-8)(-2)(-3)(-1)$

13. $1.3 - 2.7 - 0.8$

14. $(1.5)(-2.0)(-0.4)$

Sections 1.3-1.5

Follow directions for each problem and simplify your answer.

1. Write in exponent form:

$$(-5)(-5)(-5)(x)(x)(x)(x)(y)(y)$$

2. Evaluate:

$$(-3)^4$$

3. Evaluate:

$$5^3$$

4. Multiply and leave your answer in *exponent form*:

$$(2^6)(2^3)(2^8)$$

5. Multiply:

$$(-10x^2y^2)(3y^5)$$

6. Multiply:

$$(-2x^2bc)(-3b^3c^2)(2b^2c)$$

7. Divide:

$$\frac{3x^7}{6x^{10}}$$

8. Divide:

$$\frac{-15x^2y^5}{-25xy^3}$$

9. Collect like terms:

$$2xy + 6yz - 5xy - 3xz$$

10. Collect like terms:

$$x^3 + 5 - 2x^2 - 6x^3 + 4 + 5x^2 - x$$

11. Multiply:

$$-3(4x^2 - 2x + 1)$$

12. Multiply:

$$2x^2y(3 - 5xy - 7x^2y^3)$$

13. Raise to the indicated power:

$$(3x^4y)^3$$

14. Raise to the indicated power:

$$\left(\frac{a}{2b^0c^4}\right)^6$$

Sections 1.6-1.7

1. Evaluate:

$$2 - 8 \times 2 \div 4 + 7$$

2. Evaluate:

$$2 \times 3 - (5 - 7)^2 + 3$$

3. Evaluate:

$$2x^2 + x - 6$$

when $x = -3$

4. Evaluate:

$$3ab^2 - 2ab - 6a^2$$

when $a = 2$ and $b = -4$

5. Evaluate:

$$A = \frac{1}{2}h(a + b)$$

when $h = 8$, $a = 3$, and $b = 2$

6. Evaluate:

$$5 \times 2^2 - 6 + 8 \div 4 \times 5$$

7. Simplify:

$$5x(x - 2y) - 3y(3 + 2x)$$

8. Simplify:

$$3(a - b) - 2[a - 3(b + a)]$$

9. Simplify:

$$-2\{x - 2y[3 - x(1 + y)]\}$$

10. Simplify:

$$2[3x + (x - 2y)] - 4[3 - (x - 6y)]$$

Practice Test for Chapter 1

In problems 1 to 8, perform each operation and simplify each answer.

1. $-3 + (-4) + 9 + 2$
2. $\frac{1}{2} + \left(-\frac{1}{6}\right)$
3. $-0.6 - (-0.8)$
4. $(-8)(-12)$
5. $(-5)(-2)(7)(-1)$
6. $(-12) \div (-3)$
7. $(-1.8) \div (0.6)$
8. $\dfrac{-5}{\frac{3}{2}}$

9. Evaluate:

$$(-4)^3$$

10. Multiply:

$$(-2x^2y)(3x^5y^6)$$

11. Divide:

$$\frac{-35x^8y^{10}}{25x^5y^{10}}$$

12. Collect like terms:

$$2xy - 5x^2y + 3xy^2 - 8xy - 7xy^2$$

13. Multiply:

$$-2ab^2(-3a - 2b + 7ab)$$

14. Evaluate:

$$7 - (6 - 9)^2 + 5 \times 2$$

15. Evaluate:

$$x^3 - 3x^2y + 2y - 5$$

if $x = 3$, $y = -4$

16. Simplify:

$$5(3x - 2y) - (x + 6y)$$

17. Simplify:

$$3[x - (5 + 2y)] - 4[3 + (6x - 7y)]$$

18. Simplify:

$$-3\{a + b[3a - b(1 - a)]\}$$

19. Raise to the indicated power:

$$(5xy^6)^3$$

20. Raise to the indicated power:

$$\left(\frac{7a^7b^2}{3c^0}\right)^2$$

2

Solving Linear Equations and Inequalities

Diagnostic Pretest for Chapter 2

This diagnostic pretest is intended for those students who have previously studied algebra. Work as many problems as you can. Then, check your answers against those listed at the back of the book. Where you have obtained an incorrect answer or have been unable to work the problem, you will need to study the corresponding sections of the chapter.

If you have not previously studied algebra, do not take this test now. You should study the chapter and work the problems and exercises. After finishing the chapter, you may use this test as a review or as a practice test.

Follow the directions for each problem and simplify your answer.

Section 2.1

Solve for x:

1. $3x - 8 = 7x + 6$

2. $5x + 3 - 6x = 4 - 8x - 2$

Section 2.2

3. Solve for y:

$$2 - 5(y + 5) = 3(y - 2) - 1$$

4. Solve for x:

$$\frac{2}{3}(x - 2) + \frac{2}{4} = 5 - (3 + x)$$

Section 2.3

5. Solve for W:

$$P = 2(L + W)$$

6. Solve for h:

$$A = \frac{1}{2}bh + b^2$$

7. Solve for x:

$$3ax - 2y = 5ax - 13$$

Section 2.4

Solve and graph the inequality.

8. $3 - 2x < 11$

9. $\frac{1}{2}(2 + x) \leq 3(x + 1)$

Section 2.5

10. Simplify:

$$|8 - 12|$$

11. Solve for x:

$$|x + 3| = 2$$

12. Solve for x:

$$|2 - 3x| = \frac{1}{2}$$

Student Learning Objectives for Chapter 2

After successfully learning the content of each section, you will be able to:

Section 2.1

A. Solve a simple linear equation with one variable to find the value of the variable.

B. Check to verify that the solution obtained is correct in a linear equation with one variable.

Section 2.2

C. Solve an intermediate linear equation with one variable containing parentheses and/or fractions to obtain the value of the variable.

Section 2.3

D. Solve a formula or a linear equation with more than one variable for one specified variable.

Section 2.4

E. Express a relationship between two unequal values with an inequality statement.

F. Graphically represent an inequality statement with one variable.

G. Solve an intermediate linear inequality and represent the answer graphically.

Section 2.5

H. Simplify expressions involving the absolute value of numerical quantities.

I. Solve equations containing absolute values of variable quantities.

2.1 The Addition and Division Principles

2.1.1 The Addition Principle

When we use the equal sign (=) we are indicating that two expressions are equal in value. To illustrate, $3 + 5 = 8$ indicates that the sum of 3 and 5 is equal in value to 8. A mathematical statement that one value is equal to another is an equation.

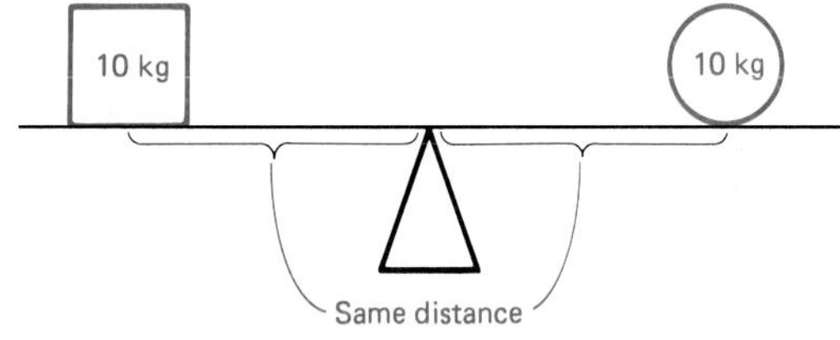

FIGURE 2.1

We observe in our everyday world that if we add the same amount to two equal values, the results are equal. To illustrate, suppose that we place a 10-kilogram empty wooden box on one side of a seesaw and a 10-kilogram stone on the other side (see Figure 2.1). If the center of the box is the same distance from the balance point as the center of the stone, we would expect the seesaw to balance. The box and the stone do not look the same, but they have the same value in weight. If we added a 2-kilogram lead weight to the center of weight of each object at the same time, the seesaw should still balance (Figure 2.2).

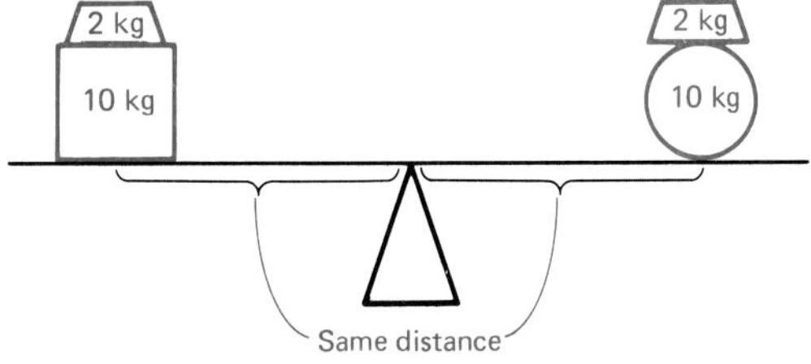

FIGURE 2.2

In mathematics, a similar property is observed. We can state it in symbols like this:

For real numbers a, b, c	**if $a = b$**	**then $a + c = b + c$**

We can state it in words thus:

> ***Rule O***
>
> **If the same number is added to both sides of an equation, the results on each side are equal in value.**

EXAMPLE A

$$\text{If } 3 = \frac{6}{2}, \text{ then } 3 + 5 = \frac{6}{2} + 5.$$

Since we added 5 to both sides, the results are still equal in value.

Now, many times an equation contains a variable for which we do not know the value. Suppose that we wanted to know the value of x in the following equation. (We are trying to find the value of x that makes the equation a true statement.)

EXAMPLE B Solve for x: $x + 16 = +20$.

Solution

$x + 16 + (-16) = +20 + (-16)$ Rule Rule O, we add -16 to both sides.

$x + 0 = +4$ Simplify.

$x = 4$ We find the value of x.

Now what we have done is to find *the solution of the equation.* The solution is a value for the variable which makes the equation true. We can easily verify this by substitution to see if each side of the equation has the same value.

$x + 16 = 20$

$4 + 16 \stackrel{?}{=} 20$ Replace x by the value 4.

$20 = 20\checkmark$ Simplify. It checks.

Since we have two numbers with the same value we believe that the mathematical equation has been correctly solved to find the solution. Note: It does not matter which side of the equation contains the variable.

EXAMPLE C Solve for x: $14 = x - 3$.

Solution

$14 + 3 = x - 3 + 3$ Add 3 to both sides.

$17 = x + 0$ Simplify.

$17 = x$ We find the value of x.

Check

$14 = x - 3$

$14 \stackrel{?}{=} 17 - 3$ Replace x by 17.

$14 = 14\checkmark$ It checks. The solution is $x = 17$.

When you are trying to solve these types of equations you notice that you must add a particular number to both sides of the equation. You always add the number that is *opposite in sign* to the number that is added to x. This is called the *additive inverse* of the number. Note carefully that before you add a number to both sides you should always simplify the equation.

EXAMPLE D Solve for x: $15 + 2 = 3 + x + 2$.

Solution

$17 = x + 5$	Simplify.
$17 + (-5) = x + 5 + (-5)$	Add the value -5 to both sides since -5 is the opposite in sign to $+5$.
$12 = x$	Simplify. The value of x is 12.

Check

$15 + 2 = 3 + x + 2$	
$15 + 2 \stackrel{?}{=} 3 + 12 + 2$	Replace x by 12 in the original equation.
$17 = 17\checkmark$	Simplify. It checks.

Take a few minutes to review Examples A to D and then try these practice problems.

Practice Problems for Exercise 2.1.1

Find the value of x that satisfies each equation.

1. $x + 0.3 = 1.2$

2. $6 + x + 3 = 9$

3. $5 - 12 = x - 3$

Now turn to the answers following Exercise 2.1.1 and check your solutions for these practice problems.

Exercise 2.1.1

Find the value of x that satisfies each equation. Check your answers for problems 1 to 12.

1. $x + 5 = 9$

2. $x + 3 = 6$

3. $11 = 13 + x$

4. $6 = x + 7$

5. $x - 16 = 5$

6. $x - 18 = 2$

7. $0 = x + 5$

8. $0 = x - 7$

9. $3 + 5 = x - 7$

10. $8 - 2 = x + 5$

11. $7 + 3 + x = 5 + 5$

12. $18 - 2 + 3 = x + 19$

13. $\frac{1}{2} + x = \frac{1}{4} + \frac{1}{3}$

14. $x - \frac{1}{5} = \frac{1}{2} + \frac{1}{10}$

15. $-2 = -7 + x$

16. $-9 + x = -4$

17. $1.3 + x + 1.8 = 0.2$

18. $3.6 + 1.2 = x + 1.3$

19. $-1.5 + x = -3.8$

20. $-0.6 + x = -1.8$

Challenge Problems

21. $\frac{1}{2} + 3x = \frac{1}{4} + 2x - 3 - \frac{1}{2}$

22. $1.6 - 5x + 3.2 = -2x + 5.6 + 4x - 8x$

Solutions to Practice Problems for Exercise 2.1.1

1.

$x + 0.3 = 1.2$	
$x + 0.3 + (-0.3) = 1.2 + (-0.3)$	Since -0.3 is opposite in sign to $+0.3$, we add -0.3 to both sides.
$x + 0 = 0.9$	Simplify.
$x = 0.9$	The value of x is 0.9.

You could subtract 0.3 from both sides and get the same result. We found in Chapter 1 that subtracting a positive 0.3 is equivalent to adding a negative 0.3. That is,

$$0.3 + (-0.3) = 0.3 - 0.3$$

2. $6 + x + 3 = 9$

$x + 9 = 9$ — Simplify.

$x + 9 + (-9) = 9 + (-9)$ — Add -9 to both sides (or subtract 9 from both sides)

$x = 0$ — We obtain the value of 0 for x.

3. $5 - 12 = x - 3$

$-7 = x - 3$ — Simplify.

$-7 + 3 = x - 3 + 3$ — Add 3 to both sides.

$-4 = x$ — The value of x is negative 4.

Note: It does not matter if you end up with x on the right side of the equal sign instead of the left.

Now do the problems in Exercise 2.1.1.

2.1.2 The Division Principle

We learned that if the same number is added to both sides of an equation, the results on each side are equal. What would happen if we divided both sides by the same number?

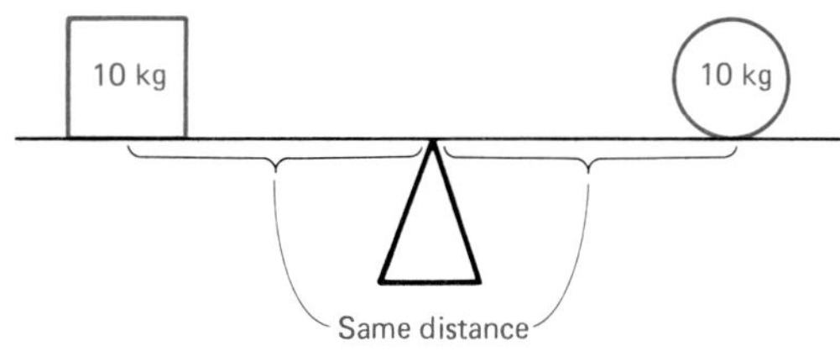

FIGURE 2.3

Let's return to our simple example of the box and the stone on a balanced seesaw (Figure 2.3). Suppose that we were to cut the two objects in half (so that the amount of weight of each was divided by 2). We then return the objects to the same places on the seesaw. We would notice that the seesaw would still balance. The "weight value" of each side remains equal (Figure 2.4).

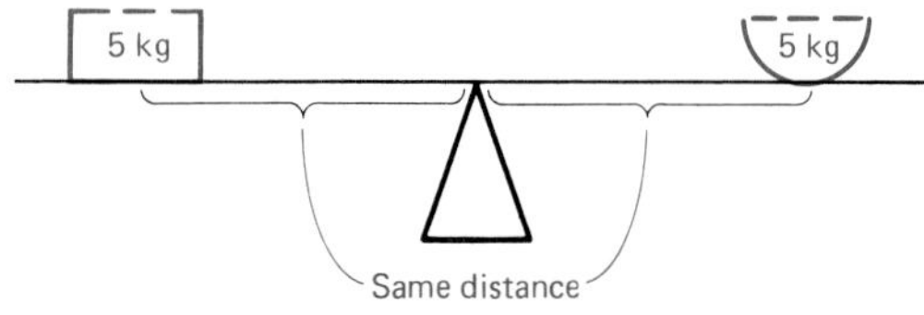

FIGURE 2.4

In symbols we can state this property thus:

> **For real numbers a, b, c where $c \neq 0$: if $a = b$ then $\frac{a}{c} = \frac{b}{c}$**

Note: We put a restriction on the number by which we are dividing. We *cannot divide* by zero. We say that expressions like $\frac{2}{0}$ are not defined. Thus we will restrict our divisor to *nonzero* numbers.

> ***Rule P***
>
> **If both sides of an equation are divided by the same nonzero number, the results on each side are equal in value.**

EXAMPLE A Solve for x: $5x = 125$.

Solution

$$\frac{5x}{5} = \frac{125}{5}$$ Divide both sides by 5.

$$x = 25$$ Simplify. The solution is 25.

Check

$$5x = 125$$

$$5(25) \stackrel{?}{=} 125$$ Replace x by 25.

$$125 = 125\checkmark$$ It checks.

EXAMPLE B Solve for x: $4x = 38$.

Solution

$$\frac{4x}{4} = \frac{38}{4}$$ Divide both sides by 4.

$$x = \frac{19}{2}$$ Simplify.

Check

$$4x = 38$$

$$2\;\cancel{4}\left(\frac{19}{\cancel{2}}\right) \stackrel{?}{=} 38$$ Replace x by $\frac{19}{2}$.

$$38 = 38\checkmark$$ It checks.

In Examples A and B we *divided by the number multiplied by x (the coefficient of x)*. This procedure is followed regardless of whether the sign of that number is positive or negative.

EXAMPLE C Solve for x: $-3x = 48$.

Solution

$$\frac{-3x}{-3} = \frac{48}{-3}$$ Divide both sides by -3.

$$x = -16$$ The solution is -16.

Check Can you check this solution?

EXAMPLE D Solve for x: $-x = -24$.

Solution

$$-1x = -24$$ Note that $-x$ is equal to $-1x$.

$$\frac{-1x}{-1} = \frac{-24}{-1}$$ Divide both sides by -1.

$$x = 24$$ The solution is 24.

Note also that the variable can be on either side of the equation.

EXAMPLE E Solve for x: $-78 = -3x$.

Solution

$$\frac{-78}{-3} = \frac{-3x}{-3}$$ Divide both sides by -3.

$26 = x$ The solution is 26.

Check

$-78 = -3x$

$-78 \stackrel{?}{=} -3(26)$ Replace x by 26.

$-78 = -78\checkmark$ It checks.

Now try these practice problems.

Practice Problems for Exercise 2.1.2

Find the value of x that satisfies each equation.

1. $-27x = 54$

2. $31.2 = 5.2x$

3. $-6x = -51$

Now turn to the answer following Exercise 2.1.2 and check your solutions for these practice problems.

Exercise 2.1.2

Find the value of the variable that satisfies each equation. Be sure to reduce your answers.

1. $7x = 28$

2. $9x = 36$

3. $-16 = 6x$

4. $-35 = 21x$

5. $1.5x = 75$

6. $2x = 0.36$

7. $-15 = -x$

8. $32 = -x$

9. $-84 = 12x$

10. $-72 = -9x$

11. $0.5x = 0.20$

12. $0.31x = 9.3$

13. $-3y = 2.4$

14. $5z = -1.8$

15. $-27 = -12z$

16. $63 = -28y$

Challenge Problems

17. $-\frac{2}{3} = -\frac{4}{7}x$

18. $5.6 = -2.7x$

Solutions to Practice Problems for Exercise 2.1.2

1. $-27x = 54$ — The coefficient of x is -27.

$\frac{-27x}{-27} = \frac{54}{-27}$ — Divide both sides by -27.

$x = -2$ — Note that the sign of the solution is -2, not $+2$.

2. $31.2 = 5.2x$ — The coefficient of x is 5.2.

$\frac{31.2}{5.2} = \frac{5.2x}{5.2}$ — Divide both sides by 5.2.

$6 = x$ — The solution is 6.

3. $-6x = -51$ — The coefficient of x is -6.

$\frac{-6x}{-6} = \frac{-51}{-6}$ — Divide both sides by -6.

$x = \frac{17}{2}$ or $8\frac{1}{2}$ — Reduce $\frac{51}{6}$ to $\frac{17}{2}$ to obtain the solution. Note that the sign of the solution is positive.

Now do the problems in Exercise 2.1.2.

2.1.3 Combining Both Principles

Many linear equations require that both Rules O and P be used in order to find the solution.

EXAMPLE A Solve for x: $5x + 3 = 18$.

Solution

$5x + 3 + (-3) = 18 + (-3)$ — Add -3 to both sides, by Rule O.

$5x = 15$ — Simplify.

$\frac{5x}{5} = \frac{15}{5}$ — Divide both sides by 5, by Rule P.

$x = 3$

Check

$5(3) + 3 \stackrel{?}{=} 18$

$15 + 3 \stackrel{?}{=} 18$

$18 = 18$ ✓ — It checks. 3 is the solution.

In some cases the variable appears on both sides of the equation. We would like to obtain all the terms containing the variable on one side. In this case it is important to add the correct value to both sides of the equation.

EXAMPLE B Solve for x:

$$9x = 6x + 15$$ x terms on both sides.

Solution

$$9x + (-6x) = 6x + (-6x) + 15$$ Add $-6x$ to both sides.

$$3x = 15$$ Collect like terms.

$$\frac{3x}{3} = \frac{15}{3}$$ Divide both sides by 3.

$$x = 5$$ The solution is 5.

Check Left to the student.

EXAMPLE C Solve for x: $-27x = -3x + 48$.

Solution

$$-27x + 3x = -3x + 3x + 48$$ Add $3x$ to both sides.

$$-24x = 48$$ Collect like terms.

$$\frac{-24x}{-24} = \frac{48}{-24}$$ Divide both sides by -24.

$$x = -2$$ The solution is -2.

Check Left to the student.

Many problems will have variable terms and constant terms on both sides of the equation. You will probably want to do separate steps to obtain all the variable terms on one side and all the constant terms on the other. What we are trying to do is to get one term with x all by itself on one side of the equation. This is called *isolating the variable.*

EXAMPLE D Solve for x: $9x + 3 = 7x - 2$.

Solution

$$9x + (-7x) + 3 = 7x + (-7x) - 2$$ Add $-7x$ to both sides of the equation.

$$2x + 3 = -2$$ Simplify.

$$2x + 3 + (-3) = -2 + (-3)$$ Add -3 to both sides.

$$2x = -5$$ Collect like terms.

$$\frac{2x}{2} = \frac{-5}{2}$$ Divide both sides by 2.

$$x = -\frac{5}{2}$$ The solution is $-\frac{5}{2}$.

Check $9x + 3 = 7x - 2$

$$9\left(-\frac{5}{2}\right) + 3 \stackrel{?}{=} 7\left(-\frac{5}{2}\right) - 2$$ Replace x by $-\frac{5}{2}$.

$$\frac{-45}{2} + 3 \stackrel{?}{=} \frac{-35}{2} - 2$$ Simplify.

$$\frac{-45}{2} + \frac{6}{2} \stackrel{?}{=} \frac{-35}{2} - \frac{4}{2}$$ Change to equivalent fractions with a common denominator.

$$\frac{-39}{2} = \frac{-39}{2} \checkmark$$ It checks. $x = -\frac{5}{2}$ is the correct solution. The statement $\frac{-32}{2} = \frac{-32}{2}$ is always true. We call this kind of statement an identity.

? What are we really doing when we "solve" an equation?

It is a good idea to examine carefully what we have been doing so far. What exactly are we accomplishing when we obtain the solution of an equation?

An *identity* is a statement that is always true. The equations $5 = 5$ and $7 = \frac{14}{2}$ are identities. A *solution of an equation* is a number which, when substituted into the given equation, yields an identity. By this definition we can see that $x = -\frac{5}{2}$ is a solution to the equation in Example D. When we "check" our answer we are actually using this definition of the solution of an equation. If it "checks" we have obtained an identity.

Sometimes you can collect like terms on one or both sides of the equation. Where this is possible it should always be done first. Note that the variable terms can be collected on the right side or the left side.

EXAMPLE E Solve for x: $5x + 26 - 6 = 9x + 12x$.

Solution

$$5x + 20 = 21x$$ Collect like terms.

$$5x + (-5x) + 20 = 21x + (-5x)$$ Add $-5x$ to both sides.

$$20 = 16x$$ Collect like terms.

$$\frac{20}{16} = \frac{16x}{16}$$ Divide both sides by 16.

$$\frac{5}{4} = x$$ Don't forget to reduce the resulting fraction.

Check Can you verify this solution?

DO I REALLY NEED ALL THESE STEPS?? No. As you become more proficient you will be able to combine or eliminate some of these steps. However, it is best to do each step completely until you are consistently obtaining the correct solution. It is much better to show every step than to take a lot of short cuts but obtain a wrong answer. This is a section of the algebra course where you need to work especially neatly and accurately.

Carefully review Examples A to E and then try these practice problems.

Practice Problems for Exercise 2.1.3

Find the value of x that satisfies each equation.

1. $3x + 2 = 5x + 2$

2. $-6x + 3 = -4x - 7$

3. Find the value of y:

$$2y + 3y = 12 - y$$

4. Find the value of z:

$$-z + 8 - z = 3z + 10 - 3$$

Now turn to the answers following Exercise 2.1.3 and check your solutions for these practice problems.

Exercise 2.1.3

Find the value of the variable that satisfies each equation. Check your answers for problems 1 to 10.

1. $7x + 7 = 31$

2. $-15 + 2x = 15$

3. $5x = 22 + 3x$

4. $30 - x = 5x$

5. $15 - 3x = -x - 5$

6. $5x + 6 = -7x - 4$

7. $x - 6 = 8 - x$

8. $2x + 5 = 4x - 5$

9. $9y - 2 = 2y - 4$

10. $5y - 7 = 3y - 9$

11. $1.2x + 4 = 3.2x - 8$

12. $9x + 1.9 = -7x - 3.1$

13. $5z + 7 - 2z = 32 - 2z$

14. $8 - 7z + 2z = 20 + 5z$

15. $-4w - 28 = -7 - w$

16. $-6w - 7 = -3 - 8w$

17. $9x - 16 + 6x = 11 + 4x - 5$

18. $7 - 9x - 12 = 3x + 5 - 8x$

19. $2x^2 - 3x - 8 = 2x^2 + 5x - 6$

20. $3x^2 + 4x - 7 = 3x^2 - 5x + 2$

21. $-3.5x + 1.3 = -2.7x + 1.5$

22. $+2.8x - 0.9 = 5.2x - 3.3$

23. $x - 0.7x + 4 = 2.6$

24. $3x + 2 - 1.7x = 0.6x + 3$

Challenge Problems

25. $6x^2 - 3x - 5 + 2x = 4x^2 - 8x + 2 - 7x + 2x^2$

26. $\frac{1}{2}x^2 - 6x + 2x^2 - 8 = \frac{5}{2}x^2 - 12x + 5 - 6x - 8$

Solutions to Practice Problems for Exercise 2.1.3

1.

$3x + 2 = 5x + 2$	We will collect x terms on the right side.
$3x + (-3x) + 2 = 5x + (-3x) + 2$	Add $-3x$ to both sides.
$2 = 2x + 2$	
$2 - 2 = 2x + 2 - 2$	Add -2 to both sides.
$0 = 2x$	
$\frac{0}{2} = \frac{2x}{2}$	Divide both sides by 2. Simplify.
$0 = x$	The solution is $x = 0$.

Alternative method: You could collect x terms on the left side by adding $-5x$ in step one. You will still obtain the same answer.

Check

$3(0) + 2 \stackrel{?}{=} 5(0) + 2$	Replace each x by 0.
$0 + 2 \stackrel{?}{=} 0 + 2$	
$2 = 2\checkmark$	It checks. $x = 0$ is correct.

2.

$-6x + 3 = -4x - 7$	We will collect x terms on the left side.
$-6x + 4x + 3 = -4x + 4x - 7$	Add $4x$ to both sides.
$-2x + 3 = -7$	
$-2x + 3 + (-3) = -7 + (-3)$	Add -3 to both sides.
$-2x = -10$	Simplify.
$\frac{-2x}{-2} = \frac{-10}{-2}$	Divide both sides by -2.
$x = 5$	The solution is $x = 5$.

Check Left to the student.

Note: There are several different ways to do this problem correctly. You could add $6x$ to both sides in the first step, for example. As long as you *consistently* get the right answer, do not be concerned if your steps look somewhat different.

3.

$2y + 3y = 12 - y$	
$5y = 12 - y$	Collect like terms.
$5y + y = 12 - y + y$	Add y to both sides.
$6y = 12$	Collect like terms.
$\frac{6y}{6} = \frac{12}{6}$	Divide both sides by 6.
$y = 2$	The solution is $y = 2$.

Check Left to the student.

4. $-z + 8 - z = 3z + 10 - 3$

$-2z + 8 = 3z + 7$	Collect like terms.
$-2z + 2z + 8 = 3z + 2z + 7$	Add $2z$ to both sides.
$8 = 5z + 7$	Collect like terms.
$8 - 7 = 5z + 7 - 7$	Subtract 7 from both sides (or add -7 to both sides).
$1 = 5z$	Collect like terms.
$\frac{1}{5} = \frac{5z}{5}$	Divide both sides by 5.
$\frac{1}{5} = z$	The solution is $\frac{1}{5}$.

Check This problem takes longer to check but you should be able to do it. You must be able to check problems with fractional answers.

$$-z + 8 - z = 3z + 10 - 3$$

$-\left(\frac{1}{5}\right) + 8 - \left(\frac{1}{5}\right) \stackrel{?}{=} 3\left(\frac{1}{5}\right) + 10 - 3$	Substitute $\frac{1}{5}$ for z.
$-\frac{2}{5} + 8 \stackrel{?}{=} \frac{3}{5} + 7$	Simplify.
$-\frac{2}{5} + \frac{40}{5} \stackrel{?}{=} \frac{3}{5} + \frac{35}{5}$	Change to equivalent fractions with a common denominator.
$\frac{38}{5} = \frac{38}{5}\checkmark$	It checks. $z = \frac{1}{5}$ is a solution.

Now do the problems in Exercise 2.1.3.

2.2 The Solution of Linear Equations with Parentheses and Fractions

2.2.1 Linear Equations with Parentheses

The type of equations that you solved in the preceding section are a simple version of another more involved type that contain parentheses. If the parentheses are first removed, the problems are similar to those previously encountered. See if you can follow each step in the following three examples.

EXAMPLE A Solve for x: $4(x + 1) - 3(x - 3) = 25$.

Solution

$4x + 4 - 3x + 9 = 25$	Remove parentheses. Be careful of the signs.
$x + 13 = 25$	Collect like terms.
$x + 13 - 13 = 25 - 13$	Subtract 13 from both sides (or add -13 to both sides).
$x = 12$	The solution is 12.

Check $4(12 + 1) - 3(12 - 3) \stackrel{?}{=} 25$ Replace x by 12.

$4(13) - 3(9) \stackrel{?}{=} 25$ Combine numbers inside parentheses.

$52 - 27 \stackrel{?}{=} 25$ Multiply.

$25 = 25\checkmark$ Simplify. It checks.

EXAMPLE B Solve for x: $5x - (3x - 2) = -4 + 12$.

Solution

$5x - 3x + 2 = -4 + 12$ Remove parentheses.

$2x + 2 = 8$ Collect like terms.

$2x + 2 - 2 = 8 - 2$ Subtract 2 from both sides (or add -2 to both sides).

$2x = 6$

$\frac{2x}{2} = \frac{6}{2}$ Divide both sides by 2.

$x = 3$ The solution is 3.

Check Left to the student.

EXAMPLE C Solve for x: $3(-x - 7) = -2(2x + 5)$.

Solution

$-3x - 21 = -4x - 10$ Remove parentheses. Watch the signs carefully.

$-3x + 4x - 21 = -4x + 4x - 10$ Add $4x$ to both sides.

$x - 21 = -10$ Simplify.

$x - 21 + 21 = -10 + 21$ Add 21 to both sides.

$x = 11$ The solution is 11.

Check Left to the student.

In problems that involve decimals, great care should be taken. In some steps you will be multiplying decimal quantities, and in other steps you will be adding them.

EXAMPLE D Solve for x: $0.3(1.2x - 3.6) = 4.2x - 16.44$.

Solution

$0.36x - 1.08 = 4.2x - 16.44$ Remove parentheses.

$0.36x - 0.36x - 1.08 = 4.2x - 0.36x - 16.44$ Subtract $0.36x$ from both sides.

$-1.08 = 3.84x - 16.44$ Collect like terms.

$-1.08 + 16.44 = 3.84x - 16.44 + 16.44$ Add 16.44 to both sides.

$15.36 = 3.84x$

$\frac{15.36}{3.84} = \frac{3.84x}{3.84}$ Divide both sides by 3.84.

$4.00 = x$ The solution is $x = 4$.

Check Left to the student.

Now try these practice problems.

Practice Problems for Exercise 2.2.1

Solve for x.

1. $5x - 2 = 2(3 - x)$

2. $4x - (x + 3) = 12 - 3(x - 2)$

3. $0.3x - 2(x + 0.1) = 0.4(x - 3) - 1.1$

Now turn to the answers following Exercise 2.2.1 and check your solutions for these practice problems.

Exercise 2.2.1

Solve for the variable. Check your answers for problems 1 to 10.

1. $4(x + 15) = 16$

2. $3(2y - 4) = 12$

3. $6(3x + 2) - 8 = -2$

4. $4(2x + 1) - 7 = 6 - 5$

5. $7x - 3(5 - x) = 10$

6. $6(3 - 4x) + 17 = 8x - 3(2 - 3x)$

7. $0.3x - 0.2(x + 1) = 0.1$

8. $3(x + 0.2) = 2(x - 0.3) + 5.2$

9. $3(x + 1) + 3 = 7(x - 2)$

10. $5x - (2x - 3) = 4(x + 9)$

11. $2(4x - x) + 6 = 2(2x + x) + 8 - x$

12. $-3(y - 3y) + 4 = -4(3y - y) + 6 + 13y$

13. $5(4 + x) = 3(3x - 1) - 9$

14. $10(x - 5) - 2(x - 9) = -28$

15. $17(y + 3) - 4(y - 10) = 13$

16. $2(5z + 4) + 19 = 4z - 3(2z + 11)$

17. $3(x + 0.2) - (2x + 0.5) = 2(x + 0.3) - 0.5$

18. $0.2(x + 3) - (x - 1.5) = 0.3(x + 2) - 2.9$

Challenge Problems

19. $2x^2 + 5(x + 1) - 4(3 - x) = x^2 + 2x - (7 - x) + x^2$

20. $3(x - 1) + 2x + 8 - 4x - 2 = 3[(2x - 1) - 2(x + 3)] + 6 + 2x + 2$

Solutions for Practice Problems for Exercise 2.2.1

1.

$5x - 2 = 2(3 - x)$	
$5x - 2 = 6 - 2x$	Remove parentheses.
$5x + 2x - 2 = 6 - 2x + 2x$	Add $2x$ to both sides.
$7x - 2 = 6$	Simplify.
$7x - 2 + 2 = 6 + 2$	Add 2 to both sides.
$7x = 8$	Simplify.
$\frac{7x}{7} = \frac{8}{7}$	Divide both sides by 7.
$x = \frac{8}{7}$	The solution is $\frac{8}{7}$.

Note: Just because the answer is a fraction, don't shy away from checking it. It is important to check fractional answers.

Check

$5\left(\frac{8}{7}\right) - 2 \stackrel{?}{=} 2\left(3 - \frac{8}{7}\right)$ Replace x by $\frac{8}{7}$.

$\frac{40}{7} - 2 \stackrel{?}{=} 2\left(\frac{21}{7} - \frac{8}{7}\right)$ Multiply 5 by $\frac{8}{7}$.

Change 3 to $\frac{21}{7}$.

$\frac{40}{7} - \frac{14}{7} \stackrel{?}{=} 2\left(\frac{13}{7}\right)$ Change 2 to $\frac{14}{7}$. Then simplify.

$\frac{26}{7} = \frac{26}{7} \checkmark$ It checks. The solution of $\frac{8}{7}$ is correct.

2. $4x - (x + 3) = 12 - 3(x - 2)$

$4x - x - 3 = 12 - 3x + 6$	Remove parentheses.
$3x - 3 = 18 - 3x$	Collect like terms.
$3x + 3x - 3 = 18 - 3x + 3x$	Add $3x$ to both sides.
$6x - 3 = 18$	Combine like terms.
$6x - 3 + 3 = 18 + 3$	Add 3 to both sides.
$6x = 21$	Combine like terms.
$\frac{6x}{6} = \frac{21}{6}$	Divide both sides by 6.
$x = \frac{7}{2}$	Reduce fraction solution to $\frac{7}{2}$.

Check Can you check this solution?

3. $0.3x - 2(x + 0.1)$
$= 0.4(x - 3) - 1.1$

$0.3x - 2x - 0.2$ $= 0.4x - 1.2 - 1.1$	Remove parentheses.
$-1.7x - 0.2 = 0.4x - 2.3$	Collect like terms.
$-1.7x - 0.4x - 0.2$ $= 0.4x - 0.4x - 2.3$	Subtract $0.4x$ from both sides.
$-2.1x - 0.2 = -2.3$	Collect like terms.
$-2.1x - 0.2 + 0.2$ $= -2.3 + 0.2$	Add 0.2 to both sides.
$-2.1x = -2.1$	Collect like terms.
$\frac{-2.1x}{-2.1} = \frac{-2.1}{-2.1}$	Divide both sides by -2.1
$x = 1$	Simplify. The solution is 1.

Check Can you check this solution?

Now do the problems in Exercise 2.2.1.

2.2.2 Linear Equations with Fractions

Linear equations with fractions can be rather difficult to solve. To avoid unnecessary work we transform the equation to an equivalent equation that does not contain fractions. To do this we multiply all terms on both sides of the equation by the lowest

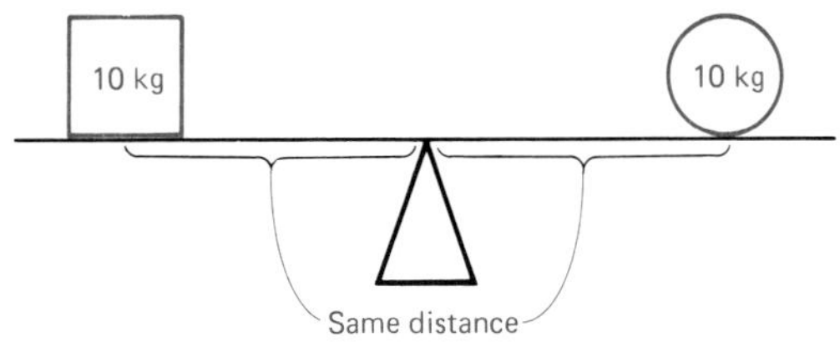

FIGURE 2.5

common denominator of all the fractions contained in the equation. Why can we multiply all the terms of an equation by a value and still obtain an equivalent equation?

Let's return to our simple example of the box and the stone on a balanced seesaw (Figure 2.5). If we triple the number of weights on each side (we are multiplying each side by 3), the seesaw should still balance. The "weight value" of each side remains equal (Figure 2.6).

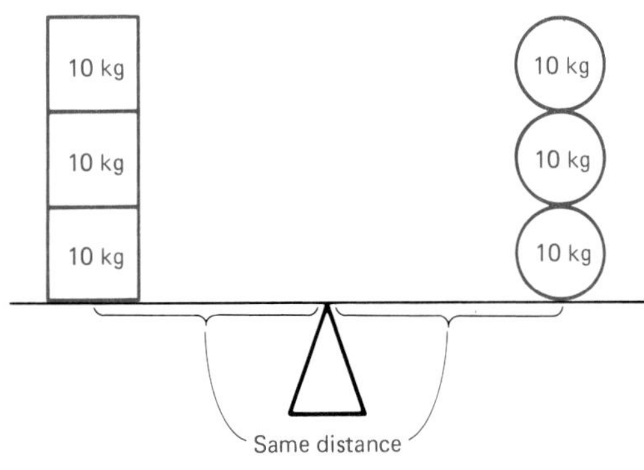

FIGURE 2.6

In symbols we can state this property thus:

For real numbers a, b, c if $a = b$ then $ca = cb$

The same principle can be applied if several terms are added on each side of the equation. To illustrate:

$$\text{If } a + b = e + f \qquad \text{then } ca + cb = ce + cf$$

For any number of terms in an equation, we can state it in words thus:

> ***Rule Q***
>
> **If both sides of an equation are multiplied by the same number, the results on each side are equal in value.**

EXAMPLE A Solve for x:

$$\frac{x}{3} + 3 = \frac{x}{5} - \frac{1}{3}$$

Solution $15\left(\frac{x}{3}\right) + 15(3) = 15\left(\frac{x}{5}\right) - 15\left(\frac{1}{3}\right)$ We observe that the lowest common denominator (LCD) is 15, so we multiply each term by 15.

$5x + 45 = 3x - 5$ Simplify.

$5x - 3x + 45 = 3x - 3x - 5$ Subtract $3x$ from both sides.

$2x + 45 = -5$

$2x + 45 - 45 = -5 - 45$ Subtract 45 from both sides.

$2x = -50$ Collect like terms.

$\frac{2x}{2} = \frac{-50}{2}$ Divide both sides by 2.

$x = -25$ The solution is -25.

Check Left to the student.

Warning

It is important to *multiply each term on both sides of the equation by the lowest common denominator*. A common student mistake is to multiply the LCD only by the fractions. Be sure that you multiply the LCD by *every term*.

EXAMPLE B Solve for x:

$$\frac{x}{7} + \frac{5}{7} = \frac{x}{4} + \frac{1}{2}$$

Solution $28\left(\frac{x}{7}\right) + 28\left(\frac{5}{7}\right) = 28\left(\frac{x}{4}\right) + 28\left(\frac{1}{2}\right)$ We observe that the LCD is 28, so we multiply each term by 28.

$4x + 20 = 7x + 14$ Simplify.

$4x - 4x + 20 = 7x - 4x + 14$ Subtract $4x$ from both sides.

$20 = 3x + 14$ Collect like terms.

$20 - 14 = 3x + 14 - 14$ Subtract 14 from both sides.

$6 = 3x$ Collect like terms.

$\frac{6}{3} = \frac{3x}{3}$ Divide both sides by 3.

$2 = x$ The solution is $x = 2$.

Check Left to the student.

If a problem contains both parentheses and fractions, it is best to remove the parentheses first. Many students find it is helpful to have a written procedure to follow in these more involved equations.

Procedure to Solve Linear Equations

1. **Remove any parentheses.**
2. **If fractions exist, multiply all terms on both sides by the lowest common denominator of all the fractions.**
3. **Collect like terms if possible.**
4. **Add or subtract terms on both sides of the equation to get all terms with the variable on one side of the equation.**
5. **Add or subtract a value on both sides of the equation to get all terms not containing the variable on the other side of the equation.**
6. **Divide both sides of the equation by the coefficient of the variable.**
7. **Simplify the solution (if possible).**
8. **Check your solution.**

Let's use each step in solving this example.

EXAMPLE C Solve for x:

$$\frac{1}{3}(x - 2) = \frac{1}{5}(x + 4) + 2$$

Solution

$\frac{x}{3} - \frac{2}{3} = \frac{x}{5} + \frac{4}{5} + 2$	Step 1.
$15\left(\frac{x}{3}\right) - 15\left(\frac{2}{3}\right) = 15\left(\frac{x}{5}\right) + 15\left(\frac{4}{5}\right) + 15(2)$	Step 2 (the LCD = 15).
$5x - 10 = 3x + 12 + 30$	Simplify results.
$5x - 10 = 3x + 42$	Step 3.
$5x - 3x - 10 = 3x - 3x + 42$	Step 4.
$2x - 10 = 42$	Simplify results.
$2x - 10 + 10 = 42 + 10$	Step 5.
$2x = 52$	Simplify results.
$\frac{2x}{2} = \frac{52}{2}$	Step 6.
$x = 26$	Step 7.

Check

$\frac{1}{3}(26 - 2) \stackrel{?}{=} \frac{1}{5}(26 + 4) + 2$	Replace x by 26.
$\frac{1}{3}(24) \stackrel{?}{=} \frac{1}{5}(30) + 2$	Combine values within parentheses.
$8 \stackrel{?}{=} 6 + 2$	Simplify.
$8 = 8\checkmark$	It checks. $x = 26$ is the solution.

It should be remembered that not every step will be needed in each problem. You can combine some steps as well, *as long as you are consistently obtaining the correct solution.* However, you are encouraged to write out every step if you tend to make careless errors.

Do you really understand fractional equations? Could you solve a fractional equation without multiplying each term by the LCD?

Let's look at Example A again. Solve for x:

$$\frac{x}{3} + 3 = \frac{x}{5} - \frac{1}{3}$$

Alternate Solution

$$\frac{x}{3} - \frac{x}{5} = -3 - \frac{1}{3}$$

Collect x terms on left, constant terms on right.

$$\frac{5x}{15} - \frac{3x}{15} = \frac{-45}{15} - \frac{5}{15}$$

Change two terms on left and two terms on right to equivalent fractions with a LCD of 15.

$$\frac{5x - 3x}{15} = \frac{-45 - 5}{15}$$

Write as one fraction on each side.

$$\frac{2x}{15} = \frac{-50}{15}$$

Simplify.

$$2x = -50$$

If two equal fractions have the same denominators their numerators are equal.

$$x = -25$$

Carefully review Examples A to C and then try these practice problems.

Practice Problems for Exercise 2.2.2

Solve for x and check your solution.

1. $\frac{5x}{4} - 1 = \frac{3x}{4} + \frac{1}{2}$

2. $\frac{5x}{6} - \frac{5}{8} = \frac{3x}{4} - \frac{1}{3}$

3. $\frac{1}{3}(x - 2) = \frac{x + 5}{4} - \frac{5}{3}$

Now turn to the answers following Exercise 2.2.2 and check your solutions for these practice problems.

Exercise 2.2.2

Solve for the variable. Check your answers for problems 1 to 8.

1. $\frac{y}{2} + \frac{y}{3} = \frac{5}{6}$

2. $\frac{x}{4} - 1 = \frac{x}{5}$

3. $20 - \frac{1}{3}x = \frac{1}{2}x$

4. $\frac{1}{5}x - \frac{1}{2} = \frac{1}{6}x$

5. $2 + \frac{y}{2} = \frac{3y}{4} - 3$

6. $\frac{x}{3} - 1 = -\frac{1}{2} - x$

7. $\frac{y - 1}{2} = 4 - \frac{y}{7}$

8. $\frac{x - 7}{6} = -\frac{1}{2}$

9. $4(3x - 2) = \frac{1}{3}(x - 1) + 4$

10. $\frac{2}{3}(x + 2) = -2(1 - 3x)$

11. $\frac{1}{3}(y + 2) = 3y - 5(y - 2)$

12. $\frac{2}{5}(y + 3) - \frac{1}{2} = \frac{1}{3}(y - 2) + \frac{1}{2}$

13. $\frac{1 + 2x}{5} + \frac{4 - x}{3} = \frac{1}{15}$

14. $\frac{x + 3}{4} = 4x - 2(x - 3)$

15. $\frac{x}{5} - \frac{2}{3}x + \frac{16}{15} = \frac{1}{3}(x - 4)$

16. $3 + \frac{1}{2}(x - 1) = 2 - \frac{1}{3}x + 3$

17. $\frac{1}{3}(x - 2) = 3x - 2(x - 1) + \frac{16}{3}$

18. $\frac{3}{4}(x - 2) + \frac{3}{5} = \frac{1}{5}(x + 1)$

Challenge Problems

19. $7x + 2[(x + 2) + 3] + 2(x + 2) = 2[2(x + 2) + 2x] + 8$

20. $\frac{7}{2} - \frac{2x - 3}{2} - (3x + 5) = \frac{3}{2} - \frac{5x}{2} - x$

Solutions to Practice Problems for Exercise 2.2.2

1. $\frac{5x}{4} - 1 = \frac{3x}{4} + \frac{1}{2}$ — We observe that the LCD = 4.

$4\left(\frac{5x}{4}\right) - 4(1) = 4\left(\frac{3x}{4}\right) + 4\left(\frac{1}{2}\right)$ — Multiply each term by 4.

$5x - 4 = 3x + 2$ — Simplify.

$5x - 3x - 4 = 3x - 3x + 2$ — Subtract $3x$ from both sides.

$2x - 4 = 2$ — Simplify.

$2x - 4 + 4 = 2 + 4$ — Add 4 to both sides.

$2x = 6$ — Simplify.

$x = 3$ — Divide both sides by 2 to obtain the solution of $x = 3$.

Check

$\frac{5(3)}{4} - 1 \stackrel{?}{=} \frac{3(3)}{4} + \frac{1}{2}$ — Replace x by 3.

$\frac{15}{4} - 1 \stackrel{?}{=} \frac{9}{4} + \frac{1}{2}$ — Simplify.

$\frac{15}{4} - \frac{4}{4} \stackrel{?}{=} \frac{9}{4} + \frac{2}{4}$ — Change the necessary terms to an equivalent fraction with a denominator of 4.

$\frac{11}{4} = \frac{11}{4}$ ✓ — It checks. The solution $x = 3$ is correct.

2. $\frac{5x}{6} - \frac{5}{8} = \frac{3x}{4} - \frac{1}{3}$ We observe that the LCD = 24.

$$\overset{4}{\cancel{24}}\left(\frac{5x}{\cancel{6}}\right) - \overset{3}{\cancel{24}}\left(\frac{5}{\cancel{8}}\right) = \overset{6}{\cancel{24}}\left(\frac{3x}{\cancel{4}}\right) - \overset{8}{\cancel{24}}\left(\frac{1}{\cancel{3}}\right)$$ Multiply each term by 24.

$20x - 15 = 18x - 8$ Simplify.

$2x - 15 = -8$ Subtract $18x$ from both sides.

$2x = 7$ Add 15 to both sides.

$x = \frac{7}{2}$ Divide both sides by 2. The solution is $x = \frac{7}{2}$.

Check Can you check this solution?

3. $\frac{1}{3}(x - 2) = \frac{x + 5}{4} - \frac{5}{3}$

$\frac{x}{3} - \frac{2}{3} = \frac{x}{4} + \frac{5}{4} - \frac{5}{3}$ We remove the parentheses and also write the right-hand side in a more convenient form.

$$\overset{4}{\cancel{12}}\left(\frac{x}{\cancel{3}}\right) - \overset{4}{\cancel{12}}\left(\frac{2}{\cancel{3}}\right) = \overset{3}{\cancel{12}}\left(\frac{x}{\cancel{4}}\right) + \overset{3}{\cancel{12}}\left(\frac{5}{\cancel{4}}\right) - \overset{4}{\cancel{12}}\left(\frac{5}{\cancel{3}}\right)$$

We multiply each term by the LCD of 12.

$4x - 8 = 3x + 15 - 20$ Simplify.

$4x - 8 = 3x - 5$ Collect like terms.

$x - 8 = -5$ Subtract $3x$ from both sides.

$x = +3$ Add 8 to both sides. The solution is $x = 3$.

Check Can you check this solution?

Now do the problems in Exercise 2.2.2.

2.3 Equations with More Than One Variable

Many linear equations contain more than one variable. Linear equations with more than one variable are often called *literal equations.* Usually, these particular types of equations are *formulas.* A formula is an equation with one or more variables that is often used to describe a real-world quantity.

To illustrate, $d = rt$ is an equation that represents a formula to find the distance d. The distance is equal to the product of r (rate) and t (time). We are often asked to isolate a particular variable in a formula. To do this we will employ similar skills to those used earlier in this chapter.

EXAMPLE A Solve for t: $d = rt$.

Solution

$$\frac{d}{r} = \frac{rt}{r}$$

Divide both sides of the equation by the coefficient of t, which is r.

$$\frac{d}{r} = t$$

You have solved for the indicated variable.

The answer may be expressed as either

$$t = \frac{d}{r} \quad \text{or} \quad \frac{d}{r} = t$$

Check Left to the student.

EXAMPLE B Solve for L: $P = 2L + 2W$.

Solution

$$P - 2W = 2L$$

We want to isolate the term that contains L. Therefore we subtract $2W$ from both sides.

$$\frac{P - 2W}{2} = \frac{2L}{2}$$

Divide both sides of the equation by 2.

$$\frac{P - 2W}{2} = L$$

You have solved for the indicated variable.

The solution to a literal equation can be checked in a similar fashion to what we have done in Sections 2.1 and 2.2. If the solution is correct we will obtain an identity.

Check

$$P = 2L + 2W$$

$$P \stackrel{?}{=} 2\left(\frac{P - 2W}{2}\right) + 2W$$

Replace L by the value obtained in our solution.

$$P \stackrel{?}{=} P - 2W + 2W$$

Remove parentheses.

$$P = P \checkmark$$

Collect like terms. We obtain an identity. Thus, $L = \dfrac{P - 2W}{2}$ is a correct solution.

If the formula has fractions, it is usually best to remove them by multiplying each term by the lowest common denominator.

EXAMPLE C Solve for b:

$$S = \frac{1}{2}bt^2$$

Solution

$$2(S) = 2\left(\frac{1}{2}bt^2\right)$$

Multiply both sides by 2.

$$2S = bt^2$$

Simplify.

$$\frac{2S}{t^2} = \frac{bt^2}{t^2}$$

Divide both sides by the coefficient of b.

$$\frac{2S}{t^2} = b$$

You have solved for the indicated variable.

Solving for a letter that represents a constant value

In many cases in an equation a variable is related to another variable in a way that involves a constant quantity. We can represent that constant quantity by a letter. If we solve the equation for that constant quantity, and we know certain corresponding values of the variables, we will be able to determine the value of that constant. This is an important technique and is used extensively in mathematics and in most of the physical sciences.

Historically, this concept was used by early civilizations to find a value for the irrational number π. Many people observed that the circumference of a circle C was related to the diameter d by the constant π in the following way (see Figure 2.7):

$$C = \pi d$$

They found an experimental value for π by substituting in measured values for the circumference and corresponding diameter of a given circle.

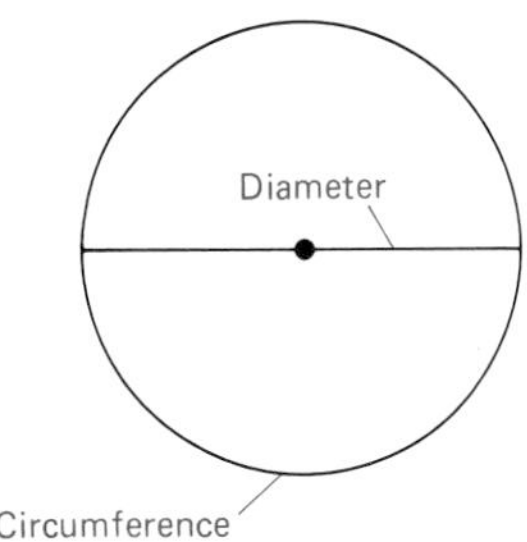

FIGURE 2.7

EXAMPLE D Solve for π: $C = \pi d$. Find an experimental value for π if C is measured to be 14.13 and d is measured to be 4.50.

Solution

$$C = \pi d$$

$$\frac{C}{d} = \frac{\pi d}{d} \qquad \text{Divide both sides by } d.$$

$$\frac{C}{d} = \pi \qquad \text{Simplify.}$$

$$\frac{14.13}{4.50} = \pi \qquad \text{Substitute measured values of } C \text{ and } d.$$

$$3.14 = \pi \qquad \text{An approximation of 3.14 for } \pi \text{ is determined.}$$

The use of $\pi = 3.14$ is accurate to two decimal places. An approximation for π accurate to 10 decimal places is $\pi = 3.1415926536$.

If the formula contains parentheses, it is usually best to eliminate them first.

EXAMPLE E A trapezoid is a four-sided figure with two parallel sides. If the parallel sides are a and b and the altitude is h (Figure 2.8), the area is given by

$$A = \frac{h}{2}(a + b)$$

Solve this equation for a.

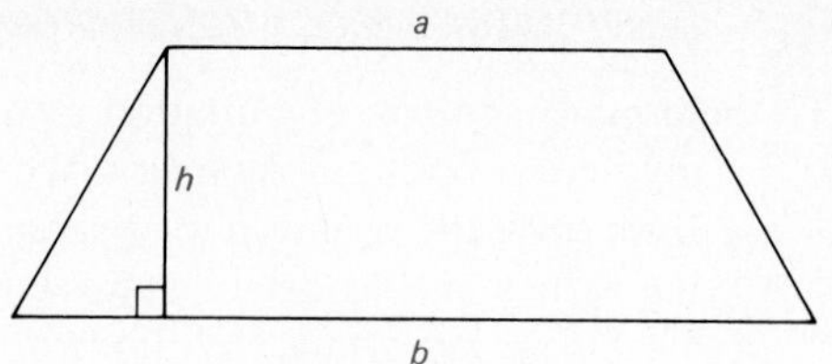

FIGURE 2.8

Solution

$$A = \frac{h}{2}(a + b)$$

$$A = \frac{ha}{2} + \frac{hb}{2}$$ Remove the parentheses.

$$2(A) = 2\left(\frac{ha}{2}\right) + 2\left(\frac{hb}{2}\right)$$ Multiply all terms by LCD of 2.

$$2A = ha + hb$$ Simplify.

$$2A - hb = ha$$ We want to isolate the term containing a. Therefore, we subtract hb from both sides.

$$\frac{2A - hb}{h} = \frac{ha}{h}$$ Divide both sides by h (the coefficient of a).

$$\frac{2A - hb}{h} = a$$ The solution is obtained.

Note: Although the solution is in simple form, it could be written in an alternative way. Since

$$\frac{2A - hb}{h} = \frac{2A}{h} - \frac{hb}{h} = \frac{2A}{h} - b$$

then we have

$$\frac{2A}{h} - b = a$$

as an alternative way of writing the answer.

There are many times in algebra when a simple linear equation with two variables is thought of as the equation for a line. It is often useful to solve for a particular variable in order to make graphing the line easier.

EXAMPLE F Solve for y: $3x - 2y = 6$.

Solution

$$-2y = 6 - 3x$$ We want to isolate the term containing y so we subtract $3x$ from both sides.

$$\frac{-2y}{-2} = \frac{6 - 3x}{-2}$$ Divide both sides by coefficient of y.

$$y = \frac{6 - 3x}{-2}$$ The solution is obtained.

Note: There are many ways to represent this answer. Since

$$y = \frac{6 - 3x}{-2} = -\frac{1}{2}(6 - 3x)$$

we could write

$$y = -3 + \frac{3}{2}x$$

by removing the parentheses. Do you see that we could also say the solution is $y = -\frac{6 - 3x}{2}$ or $y = \frac{-6 + 3x}{2}$?

Carefully review Examples A to F and then try these practice problems.

Practice Problems for Exercise 2.3

1. This is a physics equation to find energy. Solve for m:

$$E = mc^2$$

2. This is a formula for changing temperature from Celsius degrees to Fahrenheit. Solve for C:

$$F = \frac{9}{5}C + 32$$

3. Solve for y: $8 - 2y + 3x = 0$

4. Solve for x: $bx - 2a^2 = 3b(x - 3a)$

Now turn to the answers following Exercise 2.3 and check your solutions for these practice problems.

Exercise 2.3

In each formula or equation solve for the variable or constant indicated.

1. $A = \frac{1}{2}bh$ for b

2. $A = \pi r^2$ for π

3. $I = Prt$ for P

4. $S = \frac{1}{2}gt^2$ for t^2

5. $y = mx + b$ for m

6. $A = P(1 + rt)$ for t

7. $V = \frac{4}{3}\pi r^3$ for π

8. $A = \frac{\pi r^2 S}{360}$ for S

9. $I = \frac{E}{R}$ for R

10. $A = \frac{M}{fjd}$ for d

11. $L = \pi(r + e) + 2d$ for e

12. $L = a + d(n - 1)$ for n

13. $5x - 6y + 8 = 0$ for y

14. $4x + 5y - 12 = 0$ for x

15. $y = \frac{2}{3}(x - 4)$ for x

16. $x = \frac{3}{5}(2 - y)$ for y

17. $ax + 8 = 2ax - y$ for x

18. $by - 2 = x + 3by$ for y

19. $2(bt - 3) = \frac{1}{3}(4bt + x)$ for t

20. $\frac{3}{5}(x - a) = 2(x + 4)$ for x

Challenge Problems

Solve for y.

21. $\frac{2}{3}y + 5x - 18 = 2(3 - 4y - 6x) - \frac{1}{6}$

22. $2H = \frac{1}{3}(y + x) - 4xb^2 - \frac{1}{4}(3y - x)$

Solutions to Practice Problems for Exercise 2.3

1. Solve for m:

$E = mc^2$	The coefficient of m is c^2.
$\frac{E}{c^2} = \frac{mc^2}{c^2}$	Divide both sides by c^2.
$\frac{E}{c^2} = m$	Simplify.

2. Solve for C:

$F = \frac{9}{5}C + 32$	We observe that the LCD is 5.
$5F = 5\left(\frac{9}{5}C\right) + 5(32)$	Multiply each term by 5.
$5F = 9C + 160$	Simplify.
$5F - 160 = 9C$	Subtract 160 from both sides.
$\frac{5F - 160}{9} = \frac{9C}{9}$	Divide both sides by coefficient of C.
$\frac{5F - 160}{9} = C$	Simplify.

3. Solve for y:

$8 - 2y + 3x = 0$	We first want to isolate the y term.
$8 + 3x = 2y$	Add $2y$ to both sides.
$\frac{8 + 3x}{2} = \frac{2y}{2}$	Divide both sides by coefficient of y.
$\frac{8 + 3x}{2} = y$	Simplify.

Note: The problem could also be done by subtracting $(8 + 3x)$ from both sides and then dividing by -2. An equivalent answer would be obtained from

$$\frac{-8 - 3x}{-2} \quad \text{which is equal to} \quad \frac{+8 + 3x}{+2}$$

However, here it is best to use the first approach, since this allows us more easily to get the y term by itself.

4. Solve for x:

$bx - 2a^2 = 3b(x - 3a)$	
$bx - 2a^2 = 3bx - 9ab$	Remove parentheses.
$bx - bx - 2a^2 = 3bx - bx - 9ab$	We want all bx terms on one side so we subtract bx from both sides.
$-2a^2 = 2bx - 9ab$	Collect like terms.
$-2a^2 + 9ab = 2bx - 9ab + 9ab$	Add $9ab$ to both sides.
$9ab - 2a^2 = 2bx$	Simplify.
$\dfrac{9ab - 2a^2}{2b} = \dfrac{2bx}{2b}$	The coefficient of x is $2b$, so we divide both sides by $2b$.
$\dfrac{9ab - 2a^2}{2b} = x$	Simplify.

Now do the problems in Exercise 2.3.

2.4 Simple Inequalities

2.4.1 The Concept of Inequality

Many values in everyday life are not equal. We speak of one value being greater than or less than another value.

Symbolic representation of inequalities.

In mathematics we use the symbol $<$ to represent the words "is less than." We use the symbol $>$ to represent the words "is greater than."

Statement in Words	*Statement in Algebra*
5 is less than 7.	$5 < 7$
The height h is less than 27 inches.	$h < 27$
9 is greater than 4.	$9 > 4$
The weight w is greater than 2.6 kilograms.	$w > 2.6$

Note: "5 is less than 7" and "7 is greater than 5" have the same meaning. Similarly, $5 < 7$ and $7 > 5$ have the same meaning. They represent two equivalent ways of describing the same relationship between two numbers.

We can illustrate the concept of inequality graphically if we examine a number line.

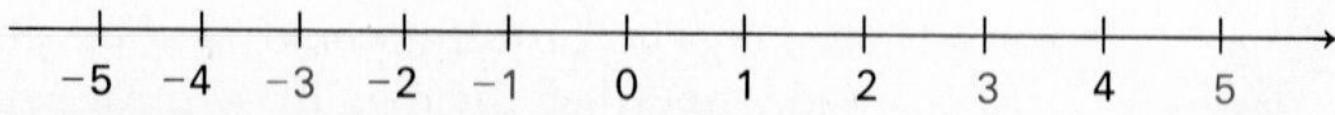

We say that *one number is greater than another* if it is to the right of the other on the number line. Thus $5 > 3$, since 5 is to the right of 3.

What about negative numbers? We can say "-1 is greater than -3" and write it in symbols $-1 > -3$ because we know that -1 lies to the right of -3 on the number line.

An interesting sidelight

It is possible to use an inequality to express relationships between irrational and rational numbers. In Section 2.3 we discussed the decimal approximation for the irrational number π. We could show graphically that $\pi > 3.14$

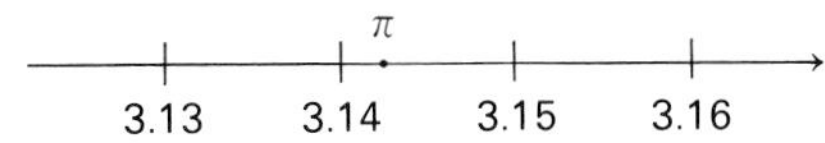

We will study irrational numbers extensively in Chapter 8. In this section we will limit our study of inequalities to inequalities between rational numbers.

EXAMPLE A Replace the blank with the symbol $<$ or $>$ in each statement.

1. $3 \_ -1$

2. $-2 \_ 1$

3. $-3 \_ -4$

4. $0 \_ 3$

5. $-3 \_ 0$

Solution

1. $3 > -1$ Since 3 is to the right of -1.

2. $-2 < 1$ Since -2 is to the left of 1 (or equivalently, we could say that 1 is to the right of -2).

3. $-3 > -4$ Since -3 is to the right of -4.

4. $0 < 3$

5. $-3 < 0$

Graphical representation of inequalities

Sometimes we express a range of values for a number by using a variable in an inequality.

The value of x is more than 3. This statement means that the value could be any number greater than 3. This is written $x > 3$. We could represent it graphically as follows:

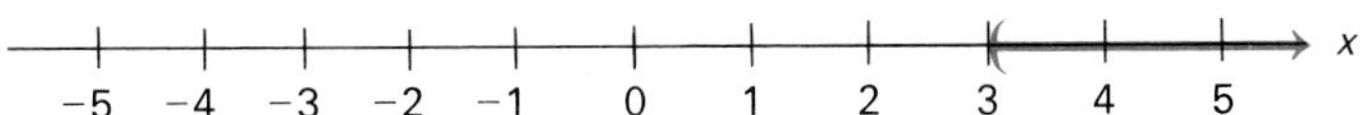

Note that the graph suggests that we do not include the point for the number 3.

Similarly, we could represent graphically $x < -2$ as follows:

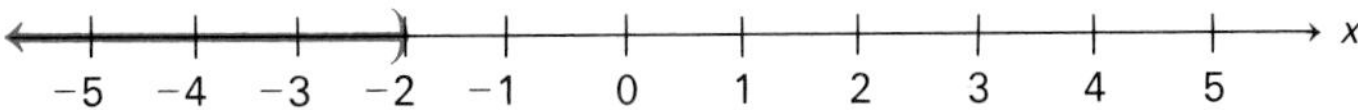

Sometimes a variable will be either greater than or equal to a certain number. In the statement "x is greater than or equal to 3" we are implying that x could have the value of 3 or any number greater than 3. We write this as $x \geq 3$. A similar symbol is used to represent "is less than or equal to." Consider these comparisons:

Statement in Words	*Statement in Algebra*
5 is less than or equal to x.	$5 \leq x$
y is less than or equal to -4.	$y \leq -4$
9 is greater than or equal to z.	$9 \geq z$
x is greater than or equal to 0.	$x \geq 0$

Graphically, we represent this by a symbol that shows that we include the end point.

$x \geq 4$ would be shown graphically as

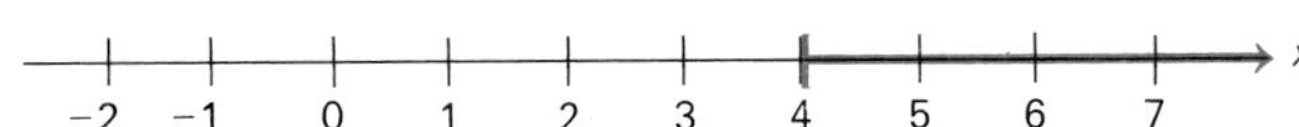

Note that the graph suggests that we *do* include the point for the number 4.

$x \leq 3$ would be shown graphically as

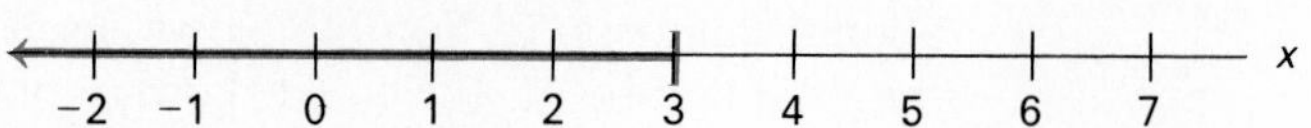

EXAMPLE B State each mathematical relationship in words and then illustrate it graphically.

1. $x < -2$ **2.** $-3 < x$

3. $x \geq -2$ **4.** $x \leq -6$

Solution **1.** We state "x is less than -2."

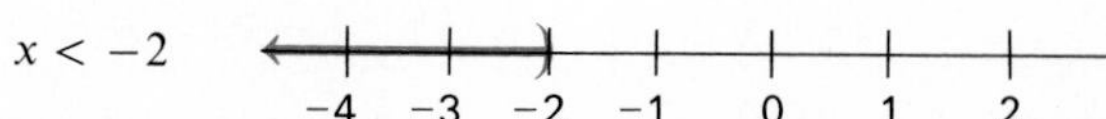

2. We can state "-3 is less than x" or, an equivalent statement, "x is greater than -3." Be sure you see that $-3 < x$ is equivalent to $x > -3$. Although both ways are correct, we *usually write the variable first* in a simple linear inequality containing a variable and a numerical value.

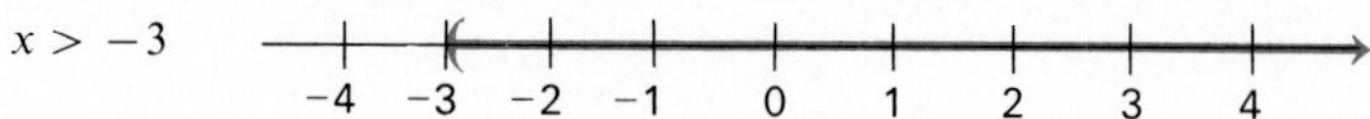

3. We state "x is greater than or equal to -2."

$x \geq -2$

4. We state "x is less than or equal to -6."

$x \leq -6$

Take a few minutes to study these examples of symbolic, graphical, and verbal representations of inequality. Now try the practice problems.

Practice Problems for Exercise 2.4.1

Fill in the blank with < or >.

1. $-6 \_\_ -4$ **2.** $+\frac{2}{5} \_\_ \frac{3}{8}$

Represent the expression graphically.

3. $x > -2$ **4.** $x \leq -\frac{1}{2}$

Now turn to the answers following Exercise 2.4.1 and check your solutions for these practice problems.

Exercise 2.4.1

Fill in the blank with < or >.

1. $9 \_\_ -3$ **2.** $-2 \_\_ 5$ **3.** $-4 \_\_ -2$

4. $-3 \_\_ -6$ **5.** $\frac{2}{3} \_\_ \frac{3}{4}$ **6.** $\frac{5}{7} \_\_ \frac{6}{8}$

7. -1.2 ___ $+2.1$

8. -3.6 ___ $+2.4$

9. $-\frac{13}{3}$ ___ -4

10. -3 ___ $-\frac{15}{4}$

11. -6.9 ___ -7.2

12. $-3\frac{1}{2}$ ___ $-3\frac{1}{4}$

Translate the graphical representation to an inequality using the variable x.

13. 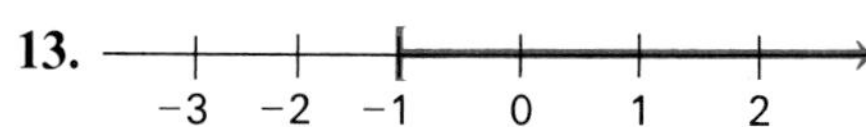

14. $-\frac{3}{2}$ -1 $-\frac{1}{2}$ 0 $\frac{1}{2}$ 1

15. 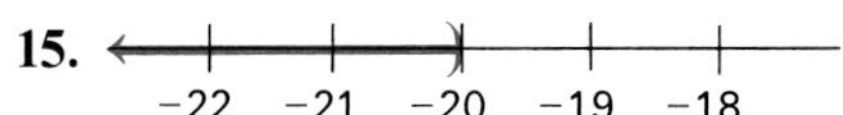

16. 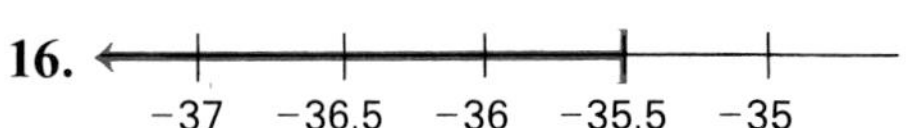

Give a graphical representation on the number line.

17. $x \geq -6$

18. $x \leq -2$

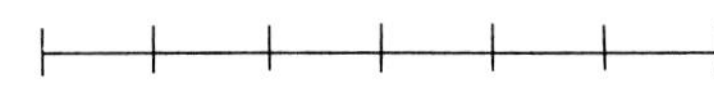

19. $x < 3$

20. $x > 5$

21. $x > \frac{3}{4}$

22. $x \geq -\frac{5}{2}$

23. $x \leq -3.6$

24. $x < -2.2$

Translate the English statement into an inequality.

25. The cost must be less than \$56. (Use the variable c for cost.)

26. The speed was greater than 580 kilometers per hour. (Use the variable V for speed.)

27. The number of hours worked in any week cannot be less than 37. (Use the variable h for hours.)

28. The number of nurses on duty on the floor can never exceed 6. (Use the variable n for the number of nurses.)

Challenge Problem

29. Suppose that the variable x must satisfy *all* of these conditions.

(a) $x \leq 4$ **(b)** $x > -3$ **(c)** $x < \frac{5}{2}$ **(d)** $x \geq -\frac{7}{2}$

Graph on a number line the region that satisfies all conditions.

Solutions for Practice Problems for Exercise 2.4.1

1. $-6 < -4$ since -6 is to the left of -4 on the number line.

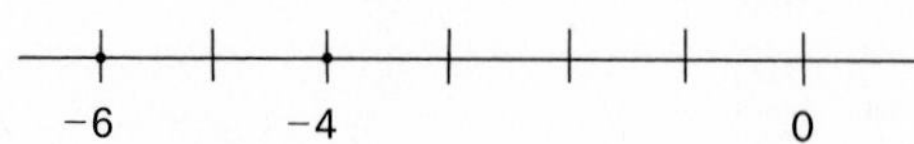

2. To compare $\frac{2}{5}$ to $\frac{3}{8}$ we need to change each fraction to an equivalent fraction with the lowest common denominator of 40.

$$\frac{2}{5} = \frac{16}{40} \quad \text{and} \quad \frac{3}{8} = \frac{15}{40}$$

Thus

$$\frac{2}{5} > \frac{3}{8} \quad \text{since} \quad \frac{16}{40} > \frac{15}{40}$$

The fraction $\frac{16}{40}$ would lie to the right of $\frac{15}{40}$ on the number line.

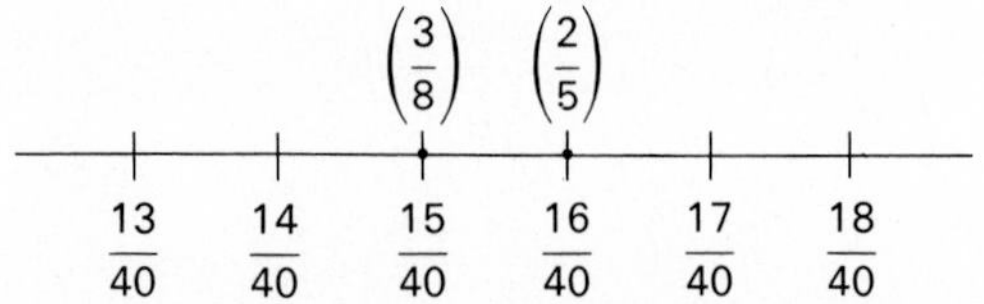

3. $x > -2$

Note we use the appropriate symbol notation to show that the point -2 is not included. If you did otherwise, your answer is not correct.

4. $x \leq -\frac{1}{2}$

In these types of problems it is helpful to label your scale to include fractional values so that you can see more clearly what values are represented by the inequality.

Now do the problems in Exercise 2.4.1.

2.4.2 Solution of Simple Inequalities

The possible values that make an inequality true are called its *solutions*. Therefore when we solve an inequality we are finding all of its solutions. Thus to solve an inequality requires that we simplify it to the point where we can clearly see the possible values for the variable. We solved equations by adding, subtracting, multiplying, and dividing a particular value on both sides of the equation. We do a similar operation with inequalities, with one important exception.

Procedure For Solving Inequalities

You may use the same procedures to solve inequalities that you did to solve equations *except* that the direction of an inequality is *reversed* if you *multiply* or *divide* both sides *by a negative number*.

We will first examine the pattern that takes place when we perform a given operation on both sides of an inequality.

EXAMPLE A

	Original Inequality				*New Inequality*
1.	$3 < 5$	$\longrightarrow$	Multiply both sides by 2.	$\longrightarrow$	$6 < 10$
2.	$-2 < -1$	$\longrightarrow$	Add -3 to both sides.	$\longrightarrow$	$-5 < -4$
3.	$0 > -4$	$\longrightarrow$	Divide both sides by 2.	$\longrightarrow$	$0 > -2$
4.	$8 > 4$	$\longrightarrow$	Subtract 6 from both sides.	$\longrightarrow$	$2 > -2$

In some problems in Example A we added or subtracted a value on each side of the inequality. In some problems we multiplied, or divided by the same positive value on both sides. The original inequality is true. After the operation we observe that the new inequality is true. Note that *we avoided multiplying or dividing by a negative number*!

Now let us examine what would happen if we did multiply or divide by a negative number.

EXAMPLE B

	Original Inequality				*New Inequality*
1.	$3 < 5$	$\longrightarrow$	Multiply by -2.	$\longrightarrow$	$-6 \ ? \ -10$

Since -6 is to the right of -10 we would expect the new inequality to be $-6 > -10$, if we wish the statement to be true.

If we follow the *procedure for solving inequalities* we would reverse the direction of the inequality from $<$ (less than) to $>$ (greater than). We would thus obtain the same result, $-6 > -10$. Thus

$$3 < 5 \quad \longrightarrow \quad \text{Multiply by } -2. \quad \longrightarrow \quad -6 > -10$$

A similar reversal takes place in the following:

	Original Inequality				*New Inequality*
2.	$-2 < -1$	$\longrightarrow$	Multiply by -3.	$\longrightarrow$	$6 > 3$
3.	$0 > -4$	$\longrightarrow$	Divide both sides by -2.	$\longrightarrow$	$0 < 2$
4.	$8 > 4$	$\longrightarrow$	Divide both sides by -4.	$\longrightarrow$	$-2 < -1$

In each problem in Example B the *direction of the inequality is reversed* when we *multiplied* or *divided* both sides *by a negative number*. Now we will use this procedure to solve the following inequalities.

EXAMPLE C Solve and graph: $3x + 7 \geq 13$.

Solution

$3x + 7 - 7 \geq 13 - 7$	Subtract 7 from both sides.
$3x \geq 6$	Collect like terms.
$\dfrac{3x}{3} \geq \dfrac{6}{3}$	Divide both sides by 3.
$x \geq 2$	Simplify. Note the direction of the inequality is not changed.

The graphical representation is:

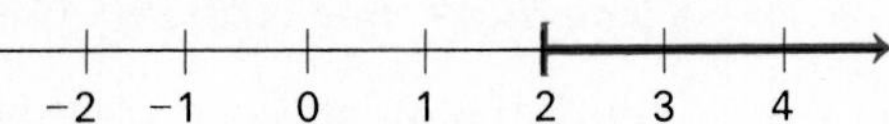

EXAMPLE D Solve and graph: $5 - 3x > 7$.

Solution

$$5 - 5 - 3x > 7 - 5$$ Subtract 5 from both sides.

$$-3x > 2$$ Collect like terms.

$$\frac{-3x}{-3} < \frac{2}{-3}$$ *Reverse inequality* since both sides are divided by negative 3.

$$x < -\frac{2}{3}$$ Note the direction of the inequality.

The graphical representation is:

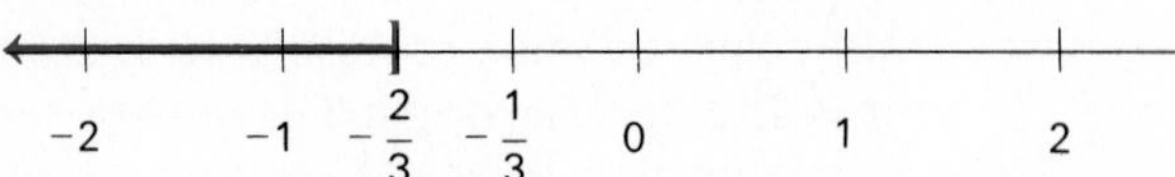

Some inequalities involve parentheses and fractions. The initial steps to solve these inequalities will be the same as those used to solve equations with parentheses and fractions.

EXAMPLE E Solve and graph:

$$\frac{3}{4} + \frac{x}{5} \leq 3x$$

Solution

$$20\left(\frac{3}{4}\right) + 20\left(\frac{x}{5}\right) \leq 20(3x)$$ Multiply all terms by LCD = 20.

$$15 + 4x \leq 60x$$ Simplify.

$$15 + 4x - 4x \leq 60x - 4x$$ Add $-4x$ to both sides.

$$15 \leq 56x$$ Simplify.

$$\frac{15}{56} \leq x$$ Divide both sides by 56. We do *not* reverse the direction of the inequality.

$$x \geq \frac{15}{56}$$ It is more convenient if we write in an equivalent form with x on the left.

Note: Either solution is correct: $\frac{15}{56} \leq x$ or $x \geq \frac{15}{56}$. However we prefer $x \geq \frac{15}{56}$.

The graphical representation is:

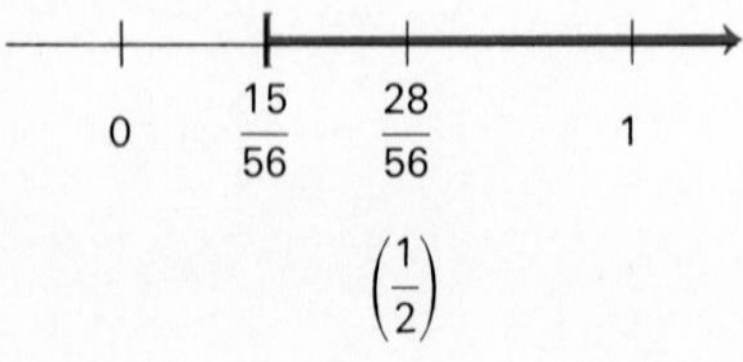

EXAMPLE F Solve and graph:

$$\frac{1}{3}(3 - 2x) \le -4(x + 1)$$

Solution

$1 - \frac{2x}{3} \le -4x - 4$ Remove parentheses.

$3(1) - 3\left(\frac{2x}{3}\right) \le 3(-4x) - 3(4)$ Multiply all terms by LCD = 3.

$3 - 2x \le -12x - 12$ Simplify.

$3 - 2x + 12x \le -12x + 12x - 12$ Add $12x$ to both sides.

$3 + 10x \le -12$ Simplify.

$3 - 3 + 10x \le -12 - 3$ Subtract 3 from both sides.

$10x \le -15$ Simplify.

$\frac{10x}{10} \le \frac{-15}{10}$ Divide both sides by 10. Since we are dividing by a positive number, the inequality is not reversed.

$x \le -\frac{3}{2}$

The graphical representation is:

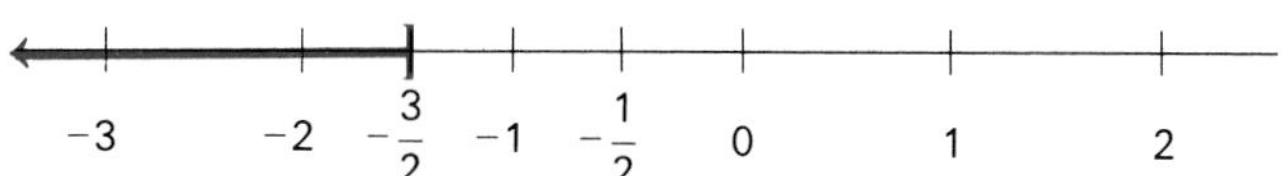

Now try these practice problems.

Practice Problems for Exercise 2.4.2

Solve each inequality and graph the result.

1. $7 + x < 4x + 28$

2. $0 \ge -6(4x - 4)$

3. $\frac{1}{2}(3 - x) \le 2x + 5$

Now turn to the answers following Exercise 2.4.2 and check your solutions for these practice problems.

Exercise 2.4.2

Solve each inequality. Graph the results for problems 1 to 8.

1. $2x - 3 < 4$

2. $3 - 3x > 12$

3. $6 - 5x \ge 3x + 14$

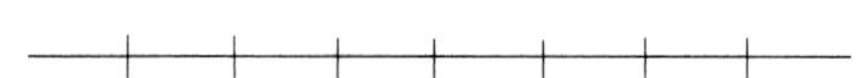

4. $7 - 8x \le -6x - 5$

5. $\frac{5x}{6} - 5 > \frac{x}{6} - 9$

6. $\frac{x}{4} - 2 < \frac{3x}{4} + 5$

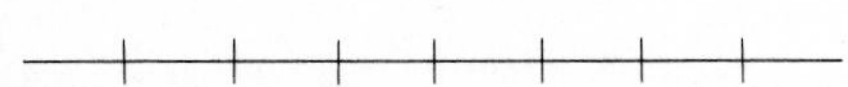

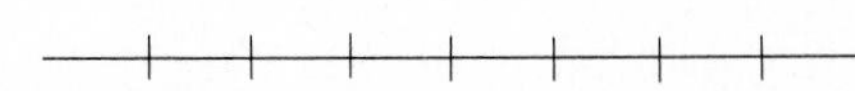

7. $3(x + 2) \leq 2x + 4$

8. $4(x + 5) \geq 3x + 7$

9. $0.3(x - 1) < 0.1x - 0.5$

10. $0.2(3 - x) + 0.1 > 0.1(x - 2)$

11. $5 - 2(3 - x) \leq 2(2x + 5) + 1$

12. $5(x + 1) + 2 \geq x - 3(2x + 1)$

13. $2x - \left(\frac{x}{2} - 6\right) \geq 1 + \frac{2x}{3}$

14. $\frac{2}{3}(x - 1) - \frac{3}{4} \leq \frac{1}{2}(x + 2)$

15. $3(0.3 + 0.1x) + 0.1 < 0.5(x + 2)$

16. $0.3 + 0.4(2 - x) \leq 6(-0.2 + 0.1x) + 0.3$

Challenge Problems

17. $\frac{1}{6} - \frac{1}{2}(3x + 2) < \frac{1}{3}\left(x - \frac{1}{2}\right)$

18. $\frac{2}{3}(2x - 5) + 3 \geq \frac{1}{4}(3x + 1) - 5$

Solutions to Practice Problems for Exercise 2.4.2

1.

$7 + x < 4x + 28$	
$7 + x - 4x < 4x - 4x + 28$	Subtract $4x$ from both sides.
$7 - 3x < 28$	Collect like terms.
$7 - 7 - 3x < 28 - 7$	Subtract 7 from both sides.
$-3x < 21$	Collect like terms.

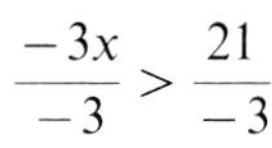

$$\frac{-3x}{-3} > \frac{21}{-3}$$ Divide both sides by -3. Note that the direction of the inequality is reversed since we divide by a negative number.

$$x > -7$$ Simplify.

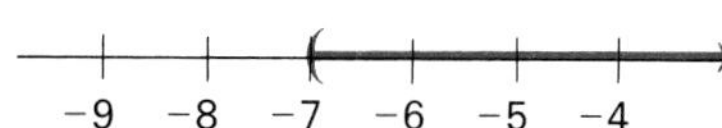

Note: There are other ways to do the problem. You could collect the x terms on the right by subtracting x from both sides in the first step. If you follow that approach you will obtain $-7 < x$ in your last step. Since $-7 < x$ and $x > -7$ are equivalent we know that either method yields a correct answer.

Remember, as a matter of practice in mathematics, when we write the solution to an inequality we usually write the x first. Thus the more common way to write the answer would be $x > -7$.

2. $$0 \geq -6(4x - 4)$$

$$0 \geq -24x + 24$$ Remove parentheses.

$$0 + 24x \geq -24x + 24x + 24$$ Add $24x$ to both sides.

$$24x \geq 24$$ Simplify.

$$\frac{24x}{24} \geq \frac{24}{24}$$ Divide both sides by 24.

$$x \geq 1$$ Simplify.

3. $$\frac{1}{2}(3 - x) \leq 2x + 5$$

$$\frac{3}{2} - \frac{x}{2} \leq 2x + 5$$ Remove parentheses.

$$2\left(\frac{3}{2}\right) - 2\left(\frac{x}{2}\right) \leq 2(2x) + 2(5)$$ Multiply all terms by LCD = 2.

$$3 - x \leq 4x + 10$$ Simplify.

$$3 - x - 4x \leq 4x - 4x + 10$$ Subtract $4x$ from both sides.

$$3 - 5x \leq 10$$ Simplify.

$$3 - 3 - 5x \leq 10 - 3$$ Subtract 3 from both sides.

$$-5x \leq 7$$ Simplify.

$$\frac{-5x}{-5} \geq \frac{7}{-5}$$ Divide both sides by -5. The direction of inequality is reversed.

$$x \geq -\frac{7}{5}$$

Now do the problems in Exercise 2.4.2.

2.5 Absolute Value

The concept of absolute value

In some situations in life we want to consider a number without regard to its sign. In mathematics we refer to this as the absolute value of a number.

Suppose that you want to examine how much the temperature changes in a given time period. Consider this situation. At 2:00 P.M. in a certain city the temperature is 30°F; at 5:00 P.M. the temperature is 10°F (see Figure 2.9).

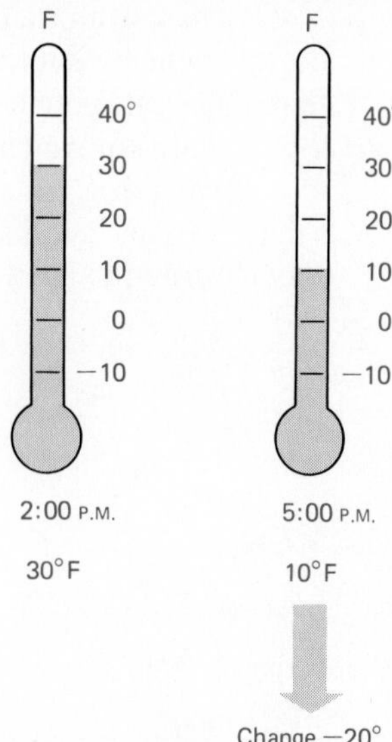

FIGURE 2.9

If someone asked, "What happened to the temperature between 2:00 P.M. and 5:00 P.M.?" we would answer that there was a drop of 20°F in the temperature. Here it is important to note if the *direction of change* is +20°F or −20°F.

However, if someone was only concerned with the *amount of change*, he would probably ask, "How much did the temperature change from 2:00 P.M. to 5:00 P.M.?" We would probably answer, "It changed 20°F." Our answer indicates a number that indicates only the *amount* of change.

Consider a number line:

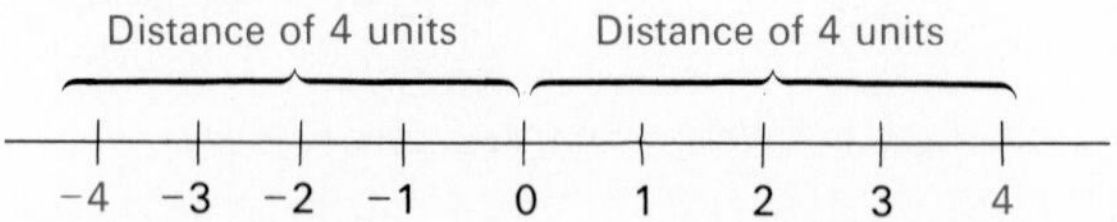

The number 4 is 4 units from the origin. The number −4 is also 4 units from the origin. This property of distance from the origin is what we call the absolute value of a number.

> ***Definition***
>
> **The absolute value of a number x is the distance between 0 and the number x on the number line. It is written as $|x|$.**

There is another way to state the definition that does not refer to the number line.

> ***Alternate Definition of Absolute Value***
>
> **The absolute value of a number x is**
>
> $$|x| = \begin{cases} x \text{ when } x \geq 0 \\ -x \text{ when } x < 0 \end{cases}$$

Example A shows the application of either of these definitions to the evaluation of numerical examples.

EXAMPLE A Evaluate. **1.** $|5|$ **2.** $|-3|$ **3.** $\left|\frac{1}{3}\right|$ **4.** $|-6.2|$ **5.** $|5-8|$ **6.** $|0|$.

Solution

1. $|5| = 5$

2. $|-3| = 3$

3. $\left|\frac{1}{3}\right| = \frac{1}{3}$

4. $|-6.2| = 6.2$

5. $|5-8| = |-3| = 3$

6. $|0| = 0$

Clearly, we see that the absolute value of any number is always 0 or positive. It is never negative.

Linear equations involving absolute value

Consider this situation. Suppose that the variable z is contained in an equation with an absolute value as follows:

$$|z| = 8$$

Do you see that there are two solutions? By the alternate definition of absolute value we know that

$$|8| = 8 \qquad \text{and also that} \qquad |-8| = 8$$

So the two solutions are $z = 8$ and $z = -8$. (If $|z| = 8$ then we have $z = 8$ or $-z = 8$.)

> ***Procedure to Solve Linear Equations that Involve an Absolute Value Expression***
>
> **1. Isolate the absolute value expression.**
>
> **2. Set the expression within the absolute value symbols equal to the value on the other side of the equal sign. Solve the resulting equation.**
>
> **3. Set the expression within the absolute value symbols equal to the *opposite* of the value on the other side of the equal sign. Solve the resulting equation.**

EXAMPLE B Find the values of x that satisfy this equation:

$$\left|\frac{1}{3}x\right| = 8$$

Solution We observe that we do not need step 1 since the absolute value expression is already by itself.

$$\frac{1}{3}x = 8 \qquad \text{Step 2}$$

$$3\left(\frac{1}{3}x\right) = 3(8) \qquad \text{Multiply by LCD} = 3.$$

$$x = 24 \qquad \text{Simplify.}$$

$$\frac{1}{3}x = -8 \qquad \text{Step 3}$$

$$3\left(\frac{1}{3}x\right) = 3(8) \qquad \text{Multiply by LCD} = 3.$$

$$x = -24 \qquad \text{Simplify.}$$

Check Left to the student.

In actual practice, most students find it is easier to write the equation for step 2 and the equation for step 3 on the same line. Usually, the same procedure can be used in the solution of each equation. This notation is used in the following two examples.

EXAMPLE C Find the values of x that satisfy this equation: $|3x - 1| = 4$. Represent the two equations that show that $(3x - 1)$ could be positive or negative. (This is really an illustration of the alternate definition of absolute value.)

Solution

$$3x - 1 = 4 \qquad 3x - 1 = -4$$

$$3x = 5 \qquad 3x = -3 \qquad \text{Add 1 to both sides of each equation.}$$

$$\frac{3x}{3} = \frac{5}{3} \qquad \frac{3x}{3} = \frac{-3}{3} \qquad \text{Divide each side by 3.}$$

$$x = \frac{5}{3} \qquad x = -1 \qquad \text{Simplify.}$$

The two values of x that satisfy the equation are

$$x = \frac{5}{3} \quad \text{or} \quad x = -1$$

Check Solutions for absolute value equations can be checked in the usual way.

For $x = \frac{5}{3}$: $\left|3\left(\frac{5}{3}\right) - 1\right| \stackrel{?}{=} 4$

$$|5 - 1| \stackrel{?}{=} 4$$

$$|4| = 4 \checkmark \qquad \text{It checks.}$$

For $x = -1$: $|3(-1) - 1| \stackrel{?}{=} 4$

$$|-4| \stackrel{?}{=} 4$$

$$4 = 4 \checkmark \qquad \text{It checks.}$$

EXAMPLE D Find the values of x that satisfy this equation: $|2x| + 1 = 9$.

Solution $|2x| + 1 - 1 = 9 - 1$ We observe that the absolute value expression is not isolated. Therefore, we subtract 1 from both sides.

$$|2x| = 8$$

Represent two equations to show that $2x$ could be positive or negative.

$2x = 8 \qquad 2x = -8$

$x = 4 \qquad x = -4$ Divide each side of each equation by 2.

The two values of x that satisfy the equation are thus

$$x = 4 \quad \text{or} \quad x = -4$$

Check Left to the student.

When you are solving more involved problems be certain to *isolate the absolute value expression first*, then follow the other necessary steps. It is a common student error to forget to get the absolute value expression by itself on one side of the equation.

EXAMPLE E Find the values of x that satisfy this equation:

$$\left|\frac{2}{3}(x + 3)\right| - 2 = 6$$

Solution $\left|\frac{2}{3}(x + 3)\right| - 2 + 2 = 6 + 2$ Add 2 to both sides of the equation to isolate the absolute value expression.

$$\left|\frac{2}{3}(x + 3)\right| = 8$$

Represent the two equations to show that $\frac{2}{3}(x + 3)$ could be positive or negative.

$\frac{2}{3}(x + 3) = 8 \qquad \frac{2}{3}(x + 3) = -8$

$\frac{2}{3}x + 2 = 8 \qquad \frac{2}{3}x + 2 = -8$ Remove parentheses.

$2x + 6 = 24 \qquad 2x + 6 = -24$ Multiply all three terms in each equation by LCD = 3.

$2x = 24 - 6 \qquad 2x = -24 - 6$ Subtract 6 from each side.

$2x = 18 \qquad 2x = -30$ Simplify.

$x = 9 \qquad x = -15$ Divide each side by 2.

The two values of x that satisfy the equation are $x = 9$ or $x = -15$.

Check For $x = 9$,

$$\left|\frac{2}{3}(9+3)\right| - 2 \stackrel{?}{=} 6$$

$$\left|\frac{2}{3}(12)\right| - 2 \stackrel{?}{=} 6$$

$$|8| - 2 \stackrel{?}{=} 6$$

$$8 - 2 \stackrel{?}{=} 6$$

$$6 = 6 \checkmark$$

For $x = -15$, the check is left to the student.

Now try these practice problems.

Practice Problems for Exercise 2.5

1. Simplify:

$$|3| - |-4|$$

Find the values of x that satisfy the equations.

2. $|4 - 2x| = 6$

3. $\left|\frac{1}{3}(x-2)\right| = 2$

4. $|5x - 3| - 4 = 8$

Now turn to the answers following Exercise 2.5 and check your solutions for these practice problems.

Exercise 2.5

Simplify.

1. $|4|$

2. $|0|$

3. $\left|-\frac{6}{7}\right|$

4. $|3 - 8|$

5. $|-3| - |5|$

6. $|6| - |-2|$

7. $\left|\frac{3}{5} - \frac{4}{7}\right|$

8. $\left|\frac{2}{3} - \frac{7}{8}\right|$

Find the values of x that satisfy the equations. Check your answers for problems 9 to 16.

9. $|2x + 1| = 9$

10. $|3x - 2| = 13$

11. $|3x| - 4 = 10$

12. $\left|\frac{1}{2}x\right| + 3 = 12$

13. $|0.3x + 0.6| = 2.4$

14. $|0.2x - 0.8| = 3.0$

15. $\left|\frac{2x}{3} + 2\right| = 8$

16. $\left|x - \frac{1}{4}\right| = \frac{5}{2}$

17. $|2 - 5x + 4| = 1.5$

18. $|-3 + 7 - 6x| = 2.6$

19. $\left|\frac{1}{2}x + 3\right| - 2 = 4$

20. $\left|3x - \frac{1}{2}\right| + 1 = 9$

21. $\left|\frac{1}{4}(2 - x) + \frac{1}{3}\right| = 2$

22. $\left|\frac{1}{5} + \frac{1}{2}(5 - x)\right| = 2$

23. $|3x - 2| + \frac{1}{4} = \frac{9}{2}$

24. $|5x + 3| - \frac{2}{3} = \frac{11}{6}$

Challenge Problems

25. $\left|\frac{1}{3}(2x - 7 - 5x)\right| - \frac{1}{4} = \frac{11}{12} + \frac{1}{2}$

26. $\left|\frac{1}{2}(2x - 3) - \frac{1}{4}(x + 2)\right| = \frac{1}{8}$

Solutions to Practice Problems for Exercise 2.5

1. $|3| - |-4|$

Since $|3| = 3$ and $|-4| = 4$ we can write this as

$$3 - 4 = -1$$

2. $|4 - 2x| = 6$

We represent the two equations that show that $(4 - 2x)$ could be positive or negative.

$4 - 2x = 6$	$4 - 2x = -6$	Two equations.
$4 - 4 - 2x = 6 - 4$	$4 - 4 - 2x = -6 - 4$	Add -4 to both sides.
$-2x = 2$	$-2x = -10$	Simplify.
$\frac{-2x}{-2} = \frac{2}{-2}$	$\frac{-2x}{-2} = \frac{-10}{-2}$	Divide by -2.
$x = -1$	$x = 5$	Simplify.

The two values of x that satisfy the equation are $x = -1$ or $x = 5$.

3. $\left|\frac{1}{3}(x - 2)\right| = 2$

We represent the two equations that show that $\frac{1}{3}(x - 2)$ could be positive or negative.

$\frac{1}{3}(x - 2) = 2$	$\frac{1}{3}(x - 2) = -2$	
$\frac{x}{3} - \frac{2}{3} = 2$	$\frac{x}{3} - \frac{2}{3} = -2$	Remove parentheses.
$\cancel{3}\left(\frac{x}{\cancel{3}}\right) - \cancel{3}\left(\frac{2}{\cancel{3}}\right) = 3(2)$	$\cancel{3}\left(\frac{x}{\cancel{3}}\right) - \cancel{3}\left(\frac{2}{\cancel{3}}\right) = 3(-2)$	Multiply each term by 3.
$x - 2 = 6$	$x - 2 = -6$	
$x - 2 + 2 = 6 + 2$	$x - 2 + 2 = -6 + 2$	Add 2 to both sides.
$x = 8$	$x = -4$	

The two values of x that satisfy the equation are $x = 8$ or $x = -4$.

4. $|5x - 3| - 4 = 8$

We must first isolate the radical by adding 4 to each side.

$$|5x - 3| - 4 + 4 = 8 + 4$$
$$|5x - 3| = 12$$

Now we represent the two equations

$5x - 3 = 12$	$5x - 3 = -12$	
$5x = 15$	$5x = -9$	Add 3 to each side.
$x = 3$	$x = -\frac{9}{5}$	Divide both sides by 5.

The two values of x that satisfy the equation are $x = 3$ or $x = -\frac{9}{5}$.

Now do the problems in Exercise 2.5.

An Application for Chapter 2

How far away is the lightning?

The most useful result of the algebraic skills you have learned in Chapter 2 is the ability to use equations to solve problems that are of interest to you. You will learn a number of ways to do this in Chapter 5.

Suppose that you see a flash of lightning and a few seconds later hear the rumble of thunder. Do you know approximately how far away the lightning is? To solve this problem we will use the relationship of motion, distance = (rate)(time), which as an equation is $d = rt$.

We are concerned with two rates of travel: the speed of light and the speed of sound. Now, light travels very rapidly, at 186,282 miles per second. That is so fast that light could travel around the world 7 times in less than a second! Since it is so very fast we can say that, for all practical purposes, we *see* the lightning the very instant the flash takes place. The speed of sound is much slower and can be measured more readily. In the normal temperature range of 50 to 90°F the speed of sound in air is approximately 1100 feet per second.

Let us assume that you see a lightning flash instantly. In 5 seconds you hear the rumble of thunder that accompanies the flash.

Question

How far away was the lightning flash?

Solution

The distance from you to the flash is

$$\text{Distance} = \text{rate} \times \text{time}$$

If the sound travels at 1100 feet per second and it took 5 seconds, we have

$d = r \cdot t$

$d = (1100)(5)$ Substituting the known rate and the known time.

$d = 5500$ feet

The distance is slightly greater than a mile. You can say that if a thunderclap is heard by a person 5 seconds after he or she sees the flash, then the original lightning was about 1 mile away. It follows that if the thunderclap is heard 10 seconds after the flash, the original lightning was about 2 miles away. For 15 seconds, it was 3 miles away; and so on.

A Challenge for You!

An observer in a firetower heard a thunderclap 1.4 seconds after seeing the flash. He heard a second thunderclap 2.5 seconds after seeing the flash. Approximately how many feet away from the observer did the lightning strike? (Find both answers.)

Supplementary Problems for Chapter 2

Solve for x.

1. $5x + 20 = 3x$

2. $7x + 3 = 4x$

3. $7(x - 4) = x + 2$

4. $2(17 - x) = 6x + 2$

5. $4x - 3(x + 2) = 4$

6. $1 - 2(6 - x) = 3x + 2$

7. $x - (0.5x + 2.6) = 17.6$

8. $-0.2(x + 1) = 0.3(x + 11)$

9. $1 = \frac{5x}{6} - \frac{2x}{3}$

10. $\frac{7x}{5} = 5 + \frac{2x}{5}$

11. $\frac{7x - 3}{2} - 4 = \frac{5x + 1}{3}$

12. $\frac{3x - 2}{2} + \frac{x}{4} = 2 + x$

13. $\frac{-3}{2}(x + 5) = 1 - x$

14. $\frac{-4}{3}(2x + 1) = -x - 2$

15. $\frac{1}{3}(x - 2) = \frac{x}{4} + 2$

16. $\frac{1}{5}(x - 3) = 2 - \frac{x}{2}$

Solve for the variable indicated.

17. Solve for y:

$$3x - y = 10$$

18. Solve for y:

$$5x + 2y + 7 = 0$$

19. Solve for r:

$$A = P(1 + rt)$$

20. Solve for h:

$$A = 4\pi r^2 + 2\pi rh$$

21. Solve for p:

$$H = \frac{1}{3}(a + 2p + 3)$$

Solve the linear inequalities and graph the result.

22. $\frac{1}{2}(2x + 3) > 10$

23. $\frac{1}{3}(x + 2) \le \frac{1}{2}(3x - 5)$

24. $4(2 - x) - (-5x + 1) \ge -8$

25. $5(1 - x) < 3(x - 1) - 2(3 - x)$

Solve.

26. $|4 - x| = 2$

27. $|2x - 3| + 4 = 7$

28. $\left|\frac{1}{2}(x - 4)\right| = 3$

29. $|4x - 3| - 1 = 8$

30. $\left|\frac{3 - 2x}{5}\right| + 2 = 4$

Chapter 2 Summary

Rules for Solving Linear Equations

To obtain the value of a variable in a linear equation, you want to isolate the variable. This is done by using the following rules:

RULE O: If the same number is added to or subtracted from both sides of an equation, the results on each side are equal in value. If $a = b$ then $a + c = b + c$.

RULE P: If both sides of an equation are divided by the same nonzero number, the results on each side are equal in value. If $a = b$ then $\frac{a}{c} = \frac{b}{c}$ if $c \neq 0$.

RULE Q: If both sides of an equation are multiplied by the same number, the results on each side are equal in value. If $a = b$ then $ca = cb$.

The order that is used in isolating the variable is listed in the following procedure:

Procedure for Solving Linear Equations

1. Remove any parentheses.
2. If fractions exist, multiply all terms on both sides by the lowest common denominator of all the fractions.
3. Collect like terms if possible.
4. Add or subtract terms on both sides of the equation to get all terms with the variable on one side of the equation.
5. Add or subtract a value on both sides of the equation to get all terms not containing the variable on the other side of the equation.
6. Divide both sides of the equation by the coefficient of the variable.
7. Simplify the solution (if possible).
8. Check the solution.

Linear Equations and Formulas with More Than One Variable

If an equation or formula has more than one variable, you can solve for a particular variable by using the *procedure for solving linear equations.* It is important to remember that the key concept is to get all terms containing the desired variable on one side of the equation and all other terms on the opposite side of the equation.

Solving Inequalities

To solve an inequality use the same steps that are outlined in the *procedure for solving linear equations* except that the direction of an inequality is *reversed* if you multiply or divide both sides of an inequality by a negative value.

Absolute Value

The absolute value of any number x, denoted $|x|$, is *always* positive or zero. To solve a linear equation containing absolute values, first isolate the absolute value expression. Then solve the two linear equations that represent the sign possibilities of the expression contained within the absolute value symbol.

Practice Quiz for Chapter 2

Sections 2.1–2.3

1. Solve for x:

$$0.2x - 0.8 = -1.4$$

2. Solve for x:

$$7x + 2 - 12x = -6 - 8x + 4$$

3. Solve for w:

$$-2 + 3w + 5(w - 2) = -2 + 14$$

4. Solve for x:

$$\frac{1}{2}(x + 2) - \frac{3}{4}(x - 4) = \frac{1}{4}$$

5. Solve for z:

$$\frac{1}{3}(z + 8) - 2z = 4 - (3 - z)$$

6. Solve for g:

$$S = \frac{1}{2}gt^2$$

7. Solve for C:

$$F = \frac{9}{5}C + 32$$

8. Solve for x:

$$2(y - 4x) = 8 - (y + 2x)$$

Sections 2.4–2.5

In problems 1 to 3, solve and graph the inequalities.

1. $-6 - 3x > 5 + 8x$

2. $5(3 + x) \le 2 - (x + 5)$

3. $\frac{2}{5}x - \left(x + \frac{7}{3}\right) \ge \frac{1}{5}(x - 1)$

4. Simplify:

$$|3| - |-6|$$

5. Solve for x:

$$|2 - (3 + 4x)| = 5$$

6. Solve for x:

$$|3x - 2| + 6 = 8$$

Practice Test for Chapter 2

Solve for the variable indicated.

1. Solve for x:

$$2(6 - x) = 3x + 2$$

2. Solve for x:

$$0.8x + 0.18 - 0.4x = 0.3(x + 0.2)$$

3. Solve for y:

$$\frac{2y}{3} + \frac{1}{5} - \frac{3y}{5} + \frac{1}{3} = 1$$

4. Solve for w:

$$\frac{2w}{3} = 4 - \frac{1}{2}(x + 6)$$

5. Solve for a:

$$A = \frac{1}{2}h(a + b)$$

6. Solve for y:

$$5ax(2 - y) = 3axy + 5$$

Solve and graph the inequalities.

7. $4(x - 1) \ge 12x$

8. $2 - 7(x + 1) - 5(x + 2) < 0$

Solve for x.

9. $\left|3 - \frac{1}{3}x\right| = 6$

10. $|3x - 2| - 6 = 12$

3 Polynomials: Multiplying and Factoring

Diagnostic Pretest for Chapter 3

This diagnostic pretest is intended for those students who have previously studied algebra. Work as many problems as you can. Then, check your answers against those listed at the back of the book. Where you have obtained an incorrect answer or have been unable to work the problem, you will need to study the corresponding sections of the chapter.

If you have not previously studied algebra, do not take this test now. You should study the chapter and work the problems and exercises. After finishing the chapter, you may use this test as a review or as a practice test.

Section 3.1

Multiply.

1. $(3x - 2y)(4x + 3y)$

2. $(x^2 - 3x + 2)(4x - 1)$

3. $(5ab - 6)^2$

4. $(8x - 11y)(8x + 11y)$

In problems 5 to 12, factor completely.

Section 3.2

5. $12x^2y^2 + 6xy^3 - 9xy^2$

6. $5a - 10b - 3ax + 6xb$

Section 3.3

7. $x^2 - 22x - 48$

8. $15x^2 - 16x + 4$

Section 3.4

9. $81x^4 - 16$

10. $49x^2 - 28xy + 4y^2$

Section 3.5

11. $6x^3 + 15x^2 - 9x$

12. $32x^2y^2 - 48xy^2 + 18y^2$

Student Learning Objectives for Chapter 3

After successfully learning the content of each section, you will be able to:

Section 3.1

A. Multiply two binomials by the FOIL method.

B. Multiply two polynomials that have three or more terms.

C. Mentally square a binomial.

D. Mentally multiply problems of the type

$$(a + b)(a - b) = a^2 - b^2$$

Section 3.2

E. Remove the common factor(s) from any polynomial.

F. Factor a four-term polynomial by the grouping method.

Section 3.3

G. Factor a general trinomial.

Section 3.4

H. Factor perfect square trinomials.

I. Factor binomials that are the difference of two squares.

Section 3.5

J. Factor polynomials in situations where two or more steps of factoring are required.

K. Identify polynomials that are prime.

3.1 Multiplication of Polynomials

In Section 1.5 we introduced the concept of a polynomial. A *polynomial* is an algebraic expression in which all the exponents of the variables are nonnegative integers and in which there is no division by a variable. We also mentioned three special kinds of polynomials:

1. A *monomial* has *one* term.
2. A *binomial* has *two* terms.
3. A *trinomial* has *three* terms.

We often refer to the degree of a term of a polynomial. The *degree of a term* of a polynomial is the sum of the exponents of its variables. The *degree of a polynomial* is the same as the degree of its highest degree term. If no variable appears in the polynomial, it is of degree zero.

EXAMPLE A

1. Examples of polynomials:

(a) $3x^2 + 5x + 6$ A polynomial of degree 2.

(b) $8a^3 - 2a^2 + 12$ A polynomial of degree 3.

(c) $3y - 1$ A polynomial of degree 1.

(d) $\frac{3}{2}x^2 + \frac{1}{5}y^2 + 6xy$ A polynomial of degree 2.

(e) 7 A polynomial of degree 0.

2. Examples of expressions that are *not* polynomials (either they have a division by a variable or an exponent that is not a positive integer):

(f) $5w^2 + 6w + \frac{3}{w}$

(g) $3x^{-2} + 6$

(h) $3x^3 + 5\sqrt{x}$

(i) $5x^{\frac{1}{2}} + 2$

3. Examples of monomials:

(j) $-5x^2$ A monomial of degree 2.

(k) $3xy^2$ A monomial of degree 3.

(l) 7 A monomial of degree 0.

4. Examples of binomials:

(m) $3c + 6d$ A binomial of degree 1.

(n) $3x^2 + \frac{1}{8}$ A binomial of degree 2.

5. Examples of trinomials:

(o) $5x^2 + 6x - 1$ A trinomial of degree 2.

(p) $3a^3 + 2a^2b - 6a^2b^2$ A trinomial of degree 4.

Our idea of a polynomial can be clarified if we use the following definition. *A polynomial in x* is an expression of the form $ax^n + bx^{n-1} + cx^{n-2} + \cdots + p$ where n is a nonnegative integer and the constants $a, b, c, \ldots, p$ are real numbers. This definition is used extensively in higher level math courses. The definition can be extended to the case of 2 or more variables.

A polynomial in x (or any one variable) is usually arranged in a specific order called descending order.

The polynomial $5x^5 - 3x^4 + 2x^3 - 6x^2 + 8x - 1$ is in *descending order* because the term with the largest exponent is first, the second term has the next largest exponent and so on. Polynomials in one variable can be rearranged so as to be in descending order by the use of the commutative property of addition and the associative property of addition.

The polynomial $-2a^2 + 5a^3 - 8 + 2a$ is not in descending order. By rearranging the terms we have $5a^3 - 2a^2 + 2a - 8$ which is now in descending order.

In this chapter we examine several cases of multiplying polynomials and factoring (which is somewhat like the reverse of multiplying) polynomials.

3.1.1 Two Types of Polynomial Multiplication

In Section 1.5 we used the distributive property in order to multiply a monomial times a polynomial. The *distributive property* can be stated that for real numbers a, b, c it is true that

$$a(b + c) = ab + ac$$

EXAMPLE B

1. Examples of multiplying a polynomial by a monomial:

(a) $3(2 - y) = 6 - 3y$ The factor 3 is multiplied by the factor $6 - 3y$.

(b) $2x(x^2 + 3x - 1) = 2x^3 + 6x^2 - 2x$ The factor $2x$ is multiplied by the factor $(x^2 + 3x - 1)$.

(c) $-2xy^2(x^2 - 2xy - 3y^2) = -2x^3y^2 + 4x^2y^3 + 6xy^4$ The factor $-2xy^2$ is multiplied by the factor $(x^2 - 2xy - 3y^2)$.

2. The distributive property also works if the multiplier is on the right.

(d) $(x + 2y + z)(3x) = 3x^2 + 6xy + 3xz$ The factor $(x + 2y + z)$ is multiplied by the factor $(3x)$.

(e) $(x^2 - 2x + 6)(-2xy) = -2x^3y + 4x^2y - 12xy$ The factor $(x^2 - 2x + 6)$ is multiplied by the factor $(-2xy)$.

3. We can use a repeated application of the distributive property to multiply two binomials as follows. Let's suppose that we want to multiply $(x + 2)(3x + 1)$ using the distributive property

$$a(b + c) = ab + ac$$

Since a represents any number, let $a = x + 2$. Then we let $b = 3x$ and $c = +1$. We now have

$a(b + c) = ab + ac$	The distributive property.
$(x + 2)(3x + 1) = (x + 2)(3x) + (x + 2)(1)$	Substitution of expressions above for a, b, and c.
$= 3x^2 + 6x + x + 2$	Using the distributive property when the multiplier is on the right.
$= 3x^2 + 7x + 2$	Collect like terms.

The distributive property shows us *how* the problem can be done and *why* it can be done. In actual practice there is a shorter method to obtain the answer. It is often referred to as the FOIL method. The letters FOIL stand for

F multiply the *F*irst terms
O multiply the *O*uter terms
I multiply the *I*nner terms
L multiply the *L*ast terms

Let's return to our original problem.

$(x + 2)(3x + 1)$	F	Multiply *first* terms and obtain $3x^2$.
$(x + 2)(3x + 1)$	O	Multiply the *outer* terms and obtain x.
$(x + 2)(3x + 1)$	I	Multiply the two *inner* terms and obtain $6x$.
$(x + 2)(3x + 1)$	L	Multiply the *last* two terms and obtain 2.

The result so far is $3x^2 + x + 6x + 2$. These four terms are the same four terms that we obtained when we multiplied using the distributive property. We can combine the like terms to obtain the final answer $3x^2 + 7x + 2$. Our result is again the same as when we used the definition of the distributive property.

Now let's study the use of the FOIL method in a few examples.

EXAMPLE C Multiply: $(2x - 1)(3x + 2)$.

Solution

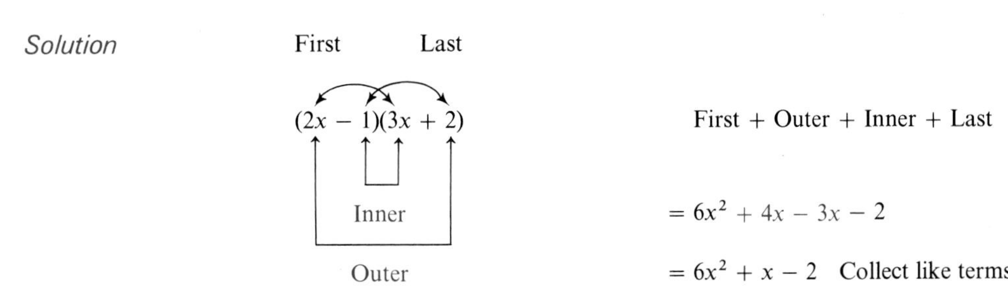

EXAMPLE D Multiply: $(3a + 2b)(4a - b)$.

Solution

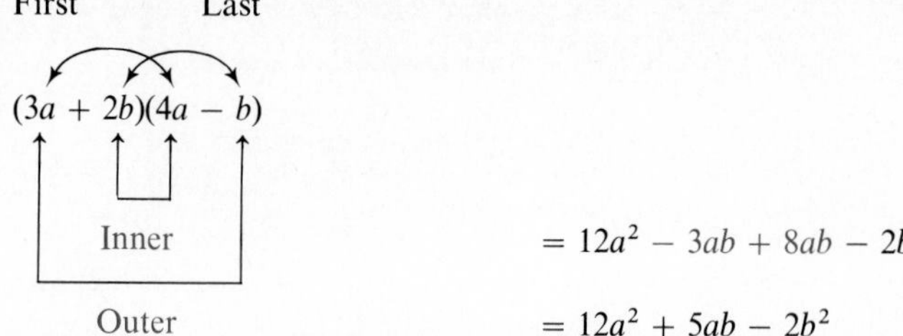

$= 12a^2 - 3ab + 8ab - 2b^2$

$= 12a^2 + 5ab - 2b^2$

After you have done several problems you may be able to combine the outer and inner products in your head.

EXAMPLE E Multiply: $(5x + 1)(x - 2)$.

Solution

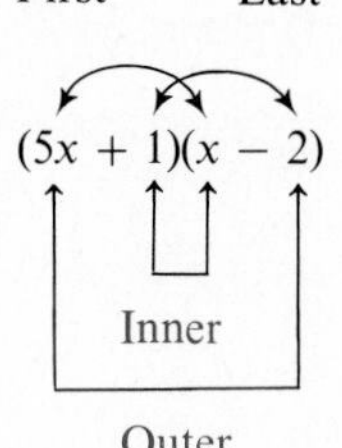

$= 5x^2 - 9x - 2$
(Were you able to mentally add $-10x + x$ and obtain $-9x$ for the middle term?)

In some problems the inner and outer products cannot be combined.

EXAMPLE F Multiply: $(3x + 2y)(5x - 3z)$.

Solution

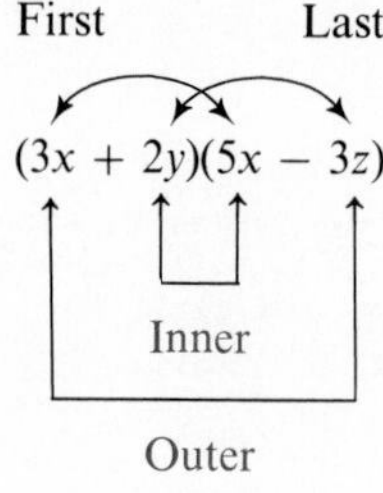

$= 15x^2 - 9xz + 10xy - 6yz$

Since there are no like terms, we cannot combine any terms.

EXAMPLE G Multiply: $(7x - 2y)^2$.

Solution $(7x - 2y)(7x - 2y)$ When we square a binomial it is the same as multiplying a binomial by itself.

$(7x - 2y)(7x - 2y)$

Inner

Outer

$= 49x^2 - 14xy - 14xy + 4y^2$

$= 49x^2 - 28xy + 4y^2$

Now try these practice problems.

Practice Problems for Exercise 3.1.1

Multiply.

1. $(5a^2 - 2ab + b^2)(-3ab)$ **2.** $(5x - 1)(x - 2)$

3. $(3a + 2b)(2a - 3b)$

4. $(8a - 5b)(3c - d)$

5. $(3x - 2y)^2$

Now turn to the answers following Exercise 3.1.1 and check your solutions for these practice problems.

Exercise 3.1.1

Multiply.

1. $2ab^2(3 - 2ab - 5b^3)$

2. $-2x^2y(x^2 - 8x + 2y^2)$

3. $(5x^3 - 2x^2 + 6x)(-3xy^2)$

4. $(3b^2 - 6b + 8ab)(5b^3)$

5. $(x + 3)(x + 4)$

6. $(x + 6)(x + 2)$

7. $(x - 8)(x + 2)$

8. $(x + 3)(x - 6)$

9. $(x - 5)(x - 4)$

10. $(x - 6)(x - 5)$

11. $(3x + 1)(-x - 4)$

12. $(5x + 1)(-x - 3)$

13. $(7x - 4)(x + 2y)$

14. $(3x - 5)(x + 3y)$

15. $(2y + 3)(5y - 2)$

16. $(3y + 2)(4y - 3)$

17. $(3y - 7)(5y - 1)$

18. $(6y - 5)(2y - 1)$

19. $(5x + 8y)(6x - y)$

20. $(3x + 8y)(5x - 2y)$

21. $(7x - 2y)(5x + 3z)$

22. $(4a - 5b)(a + 3f)$

23. $(a - 8b)(4c - 3b)$

24. $(x - 3y)(z - 8y)$

25. $(4x - 2y)(-7x - 3y)$

26. $(5x - 3y)(-8x - 9y)$

27. $(3a - 2b^2)(5a - 6b^2)$

28. $(8a - 3b^2)(2a - 5b^2)$

29. $(2x^2 - 5y^2)(2x^2 + 5y^2)$

30. $(8x^2 - 3y^2)(8x^2 - 3y^2)$

31. $(8x - 2)^2$

32. $(5x - 3)^2$

33. $(5x^2 + 2y^2)^2$

34. $(3x^2 - 7y^2)^2$

35. $(5t - 9p)(3t + 7p)$

36. $(9w - 7z)(8w + 9z)$

37. $(3ab - 5d)(2ab - 7d)$

38. $(6xy - 2z)(5xy + 4z)$

Challenge Problems

39. $\left(\frac{1}{3}x^3y^4 - \frac{1}{7}xy^6\right)\left(\frac{1}{2}x^3y^4 + \frac{1}{3}xy^6\right)$

40. $(3.2a^3b - 0.9ab^3)(5.3a^3b - 1.1ab^3)$

Solutions to Practice Problems for Exercise 3.1.1

1. $(5a^2 - 2ab + b^2)(-3ab) = -15a^3b + 6a^2b^2 - 3ab^3$

   This is an example of the distributive property with the multiplier on the right.

2. $(5x - 1)(x - 2)$

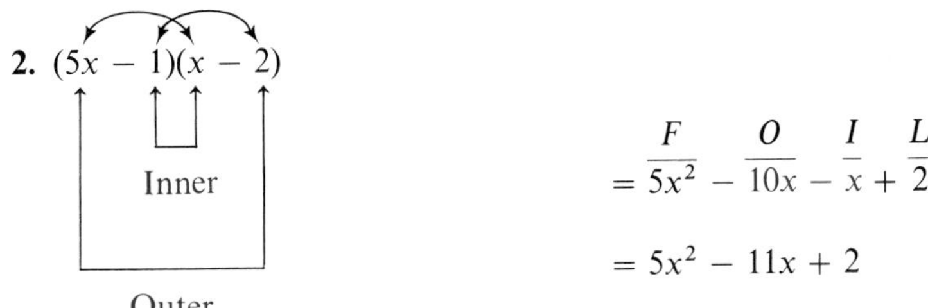

$$\overset{F}{\overline{5x^2}} - \overset{O}{\overline{10x}} - \overset{I}{\overline{x}} + \overset{L}{\overline{2}}$$

$$= 5x^2 - 11x + 2$$

3. $(3a + 2b)(2a - 3b)$

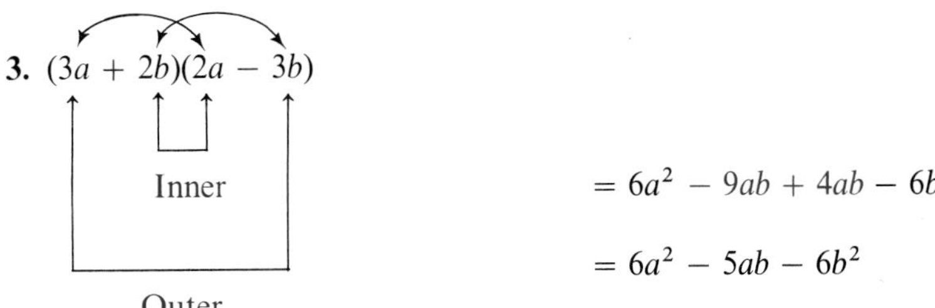

$$= 6a^2 - 9ab + 4ab - 6b^2$$

$$= 6a^2 - 5ab - 6b^2$$

4. $(8a - 5b)(3c - d)$

$$= 24ac - 8ad - 15bc + 5bd$$

Note that there are no like terms to combine.

5. $(3x - 2y)^2 = (3x - 2y)(3x - 2y)$

Inner

Outer

$$= 9x^2 - 6xy - 6xy + 4y^2$$

$$= 9x^2 - 12xy + 4y^2$$

In all these practice problems the order of the terms in your answer can be different and still be correct. However, be very careful that each of your terms has the correct sign. If just one sign is wrong, your answer is wrong.

Now do the problems in Exercise 3.1.1.

3.1.2 Special Cases of Polynomial Multiplication

The case where you multiply $(x + y)(x - y)$ is interesting and deserves special consideration. Using the FOIL method, we find

$$(x + y)(x - y) = x^2 - xy + xy - y^2 = x^2 - y^2$$

The sum of the inner and outer product is zero.

This works in all cases.

$$(5a + 2b)(5a - 2b) = 25a^2 - 10ab + 10ab - 4b^2 = 25a^2 - 4b^2$$

Again the sum of the inner and outer product is zero.

Many students find it helpful to memorize this equation.

Formula A

$$(a + b)(a - b) = a^2 - b^2$$

You may use Formula A to find the product quickly in cases where the formula applies. This happens when one set of parentheses contains the sum of two terms and the other set of parentheses contains the difference of the same two terms.

EXAMPLE A Multiply, using Formula A: $(7x + 2)(7x - 2)$.

Solution

$$(7x + 2)(7x - 2) = (7x)^2 - (2)^2 = 49x^2 - 4$$

EXAMPLE B Multiply, using Formula A: $(5x - 8y)(5x + 8y)$.

Solution

$$(5x - 8y)(5x + 8y) = (5x)^2 - (8y)^2 = 25x^2 - 64y^2$$

EXAMPLE C Multiply, using Formula A: $(2x^2 - 9)(2x^2 + 9)$.

Solution

$$(2x^2 - 9)(2x^2 + 9) = (2x^2)^2 - (9)^2 = 4x^4 - 81$$

Do you see what we are doing in Examples A, B, C? What we have actually done is to substitute into a variable expression as we did in Section 1.6 except we are substituting algebraic expressions and not just numbers.

A second case that is worth special consideration is when a binomial is squared. Consider the problem

$$\begin{aligned}(3x - 2)^2 &= (3x - 2)(3x - 2)\\ &= 9x^2 - 6x - 6x + 4\\ &= 9x^2 - 12x + 4\end{aligned}$$

If you do enough problems of this type, you will notice a pattern. The answer always contains the square of the first term added to double the product of the first and last term added to the square of the last term. This is expressed in the following formula:

Formula B

$$(a + b)^2 = a^2 + 2ab + b^2$$
$$(a - b)^2 = a^2 - 2ab + b^2$$

By using either part of Formula B, you can multiply the square of a binomial more quickly.

EXAMPLE D Multiply, using Formula B: $(2x + 3)^2$.

Solution

	Square the First Term		*Double the Product of the First and Last Terms*		*Square the Last Term*
$(a + b)^2 =$	a^2	$+$	$2ab$	$+$	b^2
$(2x + 3)^2 =$	$(2x)^2$	$+$	$2(2x)(3)$	$+$	$(3)^2$
First term is $2x$ ↑ Last term is 3 ↑ $=$	$4x^2$	$+$	$12x$	$+$	9

EXAMPLE E Multiply, using Formula B: $(5y - 2)^2$.

Solution

$$(5y - 2)^2 = (5y)^2 - (2)(5y)(2) + (2)^2$$
$$= 25y^2 - 20y + 4$$

EXAMPLE F Multiply, using Formula B: $(8x + 9y)^2$.

Solution

$$(8x + 9y)^2 = (8x)^2 + (2)(8x)(9y) + (9y)^2$$
$$= 64x^2 + 144xy + 81y^2$$

> ***Warning***
>
> **$(a + b)^2 \neq a^2 + b^2$! These are not equal! This is not a true equation! Beginning algebra students often make this error. Make sure you remember that when you square a binomial there is always a *middle term*.**
>
> $$(a + b)^2 = a^2 + 2ab + b^2$$

Multiplying polynomials with more than two terms

It is sometimes necessary to multiply polynomials such as $(3x - 2)(x^2 - 2x + 3)$. It can be quickly done using an approach similar to that used in arithmetic for multiplying whole numbers. Consider the following arithmetic problem.

128	
× 43	
384 ⟵	The product of 128 and 3.
512 ⟵	The product of 128 and 4 moved one space to the left.
5504 ⟵	The sum of the two products.

Let us follow a similar format to multiply the two polynomials.

EXAMPLE G Multiply: $(x^2 - 2x + 3)(3x - 2)$.

Solution

$x^2 - 2x + 3$	We place the polynomial with the greatest number of terms above
$3x - 2$	the other polynomial.
$-2x^2 + 4x - 6$ ⟵	The product $(x^2 - 2x + 3)(-2)$.
$3x^3 - 6x^2 + 9x$ ⟵	The product $(x^2 - 2x + 3)(3x)$ moved one space to the left so that like terms are underneath each other.
$3x^3 - 8x^2 + 13x - 6$ ⟵	The sum of the two products.

EXAMPLE H Multiply: $(3x^3 + 2x^2 + x)(x^2 - 2x - 4)$.

Solution

$$
\begin{array}{rrrrrl}
 & & 3x^3 + & 2x^2 + & x & \text{We place one polynomial over the other.}\\
 & & & x^2 - & 2x \; - \; 4 & \\
\hline
 & & -12x^3 - & 8x^2 - & 4x & \longleftarrow \text{The product } (3x^3 + 2x^2 + x)(-4).\\
 & -6x^4 - & 4x^3 - & 2x^2 & & \longleftarrow \text{The product } (3x^3 + 2x^2 + x)(-2x).\\
3x^5 + & 2x^4 + & x^3 & & & \longleftarrow \text{The product } (3x^3 + 2x^2 + x)(x^2).\\
\hline
3x^5 - & 4x^4 - & 15x^3 - & 10x^2 - & 4x & \longleftarrow \text{The sum of the three products.}
\end{array}
$$

Note that the answers for each product are placed so that similar terms are underneath each other.

Some problems may need to be done in two or more separate steps.

EXAMPLE I Multiply: $(2x - 3y)(x + 2y)(x + y)$. We first need to multiply any two binomials. Let us select the first pair.

Solution

$$(2x - 3y)(x + 2y)(x + y)$$

Do this product first

$$(2x - 3y)(x + 2y) = 2x^2 + 4xy - 3xy - 6y^2$$

$$= 2x^2 + xy - 6y^2$$

Now we replace the first two sets of parentheses with the result obtained above.

$$(2x^2 + xy - 6y^2)(x + y)$$

Result of first product

We then multiply in the same way as we did in Examples G and H.

$$
\begin{array}{rrrr}
 & 2x^2 \; + & xy \; - & 6y^2\\
 & & x \; + & y\\
\hline
 & 2x^2y + & xy^2 - & 6y^3\\
2x^3 + & x^2y - & 6xy^2 & \\
\hline
2x^3 + & 3x^2y - & 5xy^2 - & 6y^3
\end{array}
$$

Be sure to use special care in writing the exponents correctly in problems with two different variables.

Thus we have

$$(2x - 3y)(x + 2y)(x + y) = 2x^3 + 3x^2y - 5xy^2 - 6y^3$$

Note that in Example I it does not matter which two binomials are multiplied first. For example, you could first multiply $(2x - 3y)(x + y)$ to obtain $2x^2 - xy - 3y^2$ and then multiply that result by $(x + 2y)$ and still obtain the same result.

Do all the terms *always* line up so neatly?
No. Sometimes the multiplication results in several dissimilar terms. If fact, there may be no similar terms at all. As you recall from Section 1.4 we can only combine like terms.

EXAMPLE J Multiply: $(2x^2 + 3w + 2y)(x - 2y)$.

Solution

$$\begin{array}{r} 2x^2 + 3w + 2y \\ x - 2y \\ \hline -4x^2y - 6yw - 4y^2 \\ 2x^3 + 3xw + 2xy \qquad\qquad\qquad\qquad \\ \hline 2x^3 + 3xw + 2xy - 4x^2y - 6yw - 4y^2 \end{array}$$

← Note that we do not line up the product columns because there are no similar terms. We only line up like terms so they may be combined.

Now try these practice problems.

Practice Problems for Exercise 3.1.2

1. Use Formula A to multiply.

(a) $(3x + 5y)(3x - 5y)$

(b) $(7bc - 1)(7bc + 1)$

2. Use Formula B to multiply.

(a) $(5x + 4)^2$

(b) $(4a - 9b)^2$

3. Multiply the following polynomials:

$$(3x^2 - 2xy + 4y^2)(x - 2y)$$

Now turn to the answers following Exercise 3.1.2 and check your solutions for these practice problems.

Exercise 3.1.2

Use Formula A to multiply.

1. $(x + 4)(x - 4)$

2. $(y - 8)(y + 8)$

3. $(2x - 5y)(2x + 5y)$

4. $(9w - 4z)(9w + 4z)$

5. $(10p - 7q)(10p + 7q)$

6. $(11xy + 1)(11xy - 1)$

7. $(12x^2 + 7)(12x^2 - 7)$

8. $(1 - 8ab^2)(1 + 8ab^2)$

Use one of the parts of Formula B to multiply.

9. $(3y + 1)^2$

10. $(4x - 1)^2$

11. $(3 - 4y^2)^2$

12. $(8 + 3x)^2$

13. $(4a + 5b)^2$

14. $(6w + 5z)^2$

15. $(5xy - 6z)^2$

16. $(9ab - 11c)^2$

Use either Formula A or B to multiply.

17. $(7x + 3y)(7x - 3y)$

18. $(12a - 5b)(12a + 5b)$

19. $(4x^2 - 7y)^2$

20. $(3a - 10b^2)^2$

21. $(11a + 5b)(11a + 5b)$

22. $(9x - 7y^2)(9x + 7y^2)$

In problems 23 to 32, multiply out the polynomials.

23. $(x^2 - 2x + 1)(x - 2)$

24. $(x^2 + 3x - 1)(x + 3)$

25. $(3y + 2z)(3y^2 - yz - 2z^2)$

26. $(4w - 2z)(w^2 + 2wz + z^2)$

27. $(4x + 1)(x^3 - 2x^2 + x - 1)$

28. $(3x - 1)(x^3 + x^2 - 4x - 2)$

29. $(x + 2)(x - 3)(2x + 5)$

30. $(x + 6)(x - 1)(3x - 2)$

31. $(x^2 + 2x + 1)(x^2 - x - 3)$

32. $(y^2 - 3y + 1)(y^2 - 2y + 4)$

33. *Explain in words* how to obtain the answer when you multiply $(5x - 3y)^2$.

34. Find the missing binomial. $(4x + 7)(?) = 16x^2 + 56x + 49$.

35. Give an example of a problem where you multiply a binomial times a binomial and get a binomial for an answer.

Challenge Problems

Multiply.

36. $(x^2 + 3y)(x^2 - 3y)(2x^2 + 5xy - 6y^2)$

37. $(x^3 + 3x^2 - 4x + 2)(2x^3 - 4x^2 + x + 5)$

Solutions to Practice Problems for Exercise 3.1.2

1. Use *Formula A*:

$$(a + b)(a - b) = a^2 - b^2$$

(a) $(3x + 5y)(3x - 5y) = (3x)^2 - (5y)^2 = 9x^2 - 25y^2$

(b) $(7bc - 1)(7bc + 1) = (7bc)^2 - (1)^2 = 49b^2c^2 - 1$

2. Use one of the parts of *Formula B*:

$$(a + b)^2 = a^2 + 2ab + b^2$$

$$(a - b)^2 = a^2 - 2ab + b^2$$

(a) $(5x + 4)^2 = (5x)^2 + 2(5x)(4) + (4)^2 = 25x^2 + 40x + 16$

(b) $(4a - 9b)^2 = (4a)^2 - 2(4a)(9b) + (9b)^2 = 16a^2 - 72ab + 81b^2$

3.

$$\begin{array}{r} 3x^2 - 2xy + 4y^2 \\ x - 2y \\ \hline -6x^2y + 4xy^2 - 8y^3 \\ 3x^3 - 2x^2y + 4xy^2 \qquad \\ \hline 3x^3 - 8x^2y + 8xy^2 - 8y^3 \end{array}$$

Be sure that you align the terms so that xy^2 terms are underneath xy^2 terms, etc.

Thus

$$(3x^2 - 2xy + 4y^2)(x - 2y) = 3x^3 - 8x^2y + 8xy^2 - 8y^3$$

Now do the problems in Exercise 3.1.2.

3.2 Removing Common Factors

3.2.1 Removing Common Factors of One Term

In the preceding section you practiced multiplying polynomials. The two expressions that you multiplied are often referred to as factors. When two or more numbers or algebraic expressions are multiplied, each of the items that is multiplied is called a *factor*.

In the rest of this chapter we will be learning techniques to find the factors of a given polynomial. This is called factoring. *Factoring* is the process of writing a polynomial as a product of two or more polynomials. When you are asked to factor a problem you are really being asked "What things were multiplied to obtain this expression?" Factoring is simply the reverse of multiplying.

EXAMPLE A **1.** Multiply: $3(x - 2y)$.

Solution

$$3x - 6y$$

Now to reverse the process.

2. Factor: $3x - 6y$.

Solution

$$3(x - 2y)$$

What are we actually doing when we do these steps? What we have done is to determine the common factor of both terms of $3x - 6y$. Remember, our justification for doing this is the distributive property. When we remove a common factor we are actually using the distributive property.

The common factor may be a variable.

EXAMPLE B Factor: $9x + 2xy$.

Solution We observe that the factor x is common to both terms, so we obtain the result

$$x(9 + 2y)$$

The common factor may be both a number and a variable.

EXAMPLE C Factor: $5xy + 10x^2 + 25x^3$.

Solution Here we observe that 5 is a factor of each term. The variable x is a factor of each term. The result is

$$5x(y + 2x + 5x^2)$$

In each of Examples A, B, and C we can check the results of our factoring by multiplying back our result and obtaining the original problem.

Let us now make this procedure a bit more precise.

> ***To Remove a Common Factor From a Polynomial***
>
> 1. **Determine the numerical common factor by asking, "What is the largest integer that will divide into the coefficient of each term?"**
> 2. **Determine the variable common factor by asking, "What variable is common to each term and what is the largest exponent of that variable that is common?"**
> 3. **Now write the polynomial as a product of the quantities obtained in steps 1 and 2 and another polynomial.**

EXAMPLE D Factor: $12x^2 + 18y^2$.

Solution Note that the largest integer that is common to both terms is 6 (not 3 or 2).

$$6(2x^2 + 3y^2)$$

EXAMPLE E Factor: $x^2y^2 + 3xy^2 + y^3$.

Solution Although y is common to all terms we factor out y^2 since 2 is the largest exponent of y that is common to all terms. We do not factor out x since x is not common to all terms.

$$y^2(x^2 + 3x + y)$$

EXAMPLE F Factor: $8x^3y + 16x^2y^2 + 24x^3y^3$.

Solution We see that 8 is the largest integer that will divide evenly into the three numerical coefficients. We can factor an x^2 out of each term (not just x). For the other variable we factor y out of each term.

$$8x^2y(x + 2y + 3xy^2)$$

Warning

Study Example F carefully. We want to take out the largest possible common factor. Make sure that you took out $8x^2y$ and *not* $4xy$ or some other smaller quantity.

Checking the step of removing a common factor

Once you have verified that you still do not have a common factor in the parentheses, you can check any problem by multiplication. It is a good idea to check problems like Example F by multiplying your answer to see if you obtain the original problem.

$$8x^2y(x + 2y + 3xy^2) = 8x^3y + 16x^2y^2 + 24x^3y^3 \qquad \textit{It checks.}$$

EXAMPLE G Factor: $9a^3b^2 + 9a^2b^2$.

Solution We observe that both terms contain a common factor of 9. We can remove a^2 and b^2 from each term.

$$= 9a^2b^2(a + 1)$$

Warning

Don't forget to include the 1 inside the parentheses in Example G. The solution is wrong without it. You will see why if you try to check your result without the 1.

EXAMPLE H Factor: $3x(x - 4y) + 2(x - 4y)$. Be sure you understand what are terms and what are factors in this example. There are two *terms*. The expression $3x(x - 4y)$ is one term. The expression $2(x - 4y)$ is the second term.

Solution Observe that the binomial $(x - 4y)$ is a common *factor* in each term. A common factor may be a set of parentheses containing any type of polynomial. Thus

$$3x(\underline{x - 4y}) + 2(\underline{x - 4y}) = (\underline{x - 4y})(3x + 2)$$

Take a few minutes to review Examples A to H, making sure that you can check your factoring by multiplication. Then try these practice problems.

Practice Problems for Exercise 3.2.1

Remove the largest possible common factor. Check your answer by multiplication.

1. $2x^3y^2 - x^6y^3$

2. $12a^2 + 16ab^2 - 12a^2b^2$

3. $15x^4 - 20x^3 - 5x^2$

4. $3(a + b) + x(a + b)$

5. $3ax^2 + ax$

Now turn to the answers following Exercise 3.2.1 and check your solutions for these practice problems.

Exercise 3.2.1

Remove the largest possible common factor. Check your answers for problems 1 to 8 by multiplication.

1. $6x^2 - 6x$

2. $5ab - 5b$

3. $21abc - 14ab^2$

4. $18wz - 27w^2z$

5. $18abc + 12d - 6ab$

6. $9pqr - 12p^2qr - 14rx$

7. $ax^2 - 2bx^2 - cx^2 + dx$

8. $a^4y - a^3y^2 + a^2y^3 - 2a$

9. $60x^3 - 50x^2 + 25x$

10. $6x^9 - 8x^7 + 4x^5$

11. $2\pi rh + 2\pi r^2$

12. $9a^2b^2 - 36ab$

13. $14x^2y - 35xy - 63x$

14. $20z^2b - 4ab^2 + 12ab + 8a^2b^2$

Hint: In problems 15 to 22, refer to Example H.

15. $2(a + b) + 3x(a + b)$

16. $5(x - 2y) - z(x - 2y)$

17. $7a(x - 3y) - 4(x - 3y)$

18. $5b(c + 2d) - 2a(c + 2d)$

19. $3(x^2 + 1) + 2y(x^2 + 1) + w(x^2 + 1)$

20. $5a(bc - 1) + b(bc - 1) + c(bc - 1)$

21. $4y(x + 2y) + (x + 2y)$

22. $3x^2(x - 2y) - (x - 2y)$

Challenge Problems

Factor. Be sure you remove the greatest common factor.

23. $100x^6 - 25x^5 + 36x^4 - 72x^3 - 68x^2 + 120x$

24. $3x^2y(5a - 3b + 1) - 19xy(5a - 3b + 1) - y(5a - 3b + 1)$

Solutions to Practice Problems for Exercise 3.2.1

1. $2x^3y^2 - x^6y^3 = x^3y^2(2 - x^3y)$

Note that the common factor is x^3y^2. The largest exponent we can remove for the x factor is an exponent of 3. The largest exponent we can remove for the y factor is an exponent of 2. If you did not find the common factor of x^3y^2, your answer is not correct.

2. $12a^2 + 16ab^2 - 12a^2b^2 = 4a(3a + 4b^2 - 3ab^2)$

The common factor is $4a$. We cannot remove a factor of b since b is not common to all three terms.

3. $15x^4 - 20x^3 - 5x^2 = 5x^2(3x^2 - 4x - 1)$

The common factor is $5x^2$. Be sure that you remembered to include the -1 term as the third term inside the parentheses. Your answer is not correct without it.

4. $3(\underline{a + b}) + x(\underline{a + b}) = (\underline{a + b})(3 + x)$

Note: There are two terms—the $3(a + b)$ term and the $x(a + b)$ term. The $(a + b)$ is a common factor in each term.

5. $3a^2x + ax = ax(3a + \underset{\uparrow}{1})$

Note: Be sure you included a 1 here. In removing common factors, you must be alert for situations similar to this.

Now do the problems in Exercise 3.2.1.

3.2.2 Removing Common Factors of More Than One Term: Factoring By Grouping

A common factor can be just a number or a variable. However, we found in the preceding section that the common factor method applies to the common factor being an algebraic quantity in parentheses.

EXAMPLE A Factor: $x(x - 3) + 2(x - 3)$.

Solution Observe each term:

$$\underbrace{x(x - 3)}_{\text{First term}} + \underbrace{2(x - 3)}_{\text{Second term}}$$

The common factor for both the first and second terms is the quantity $(x - 3)$, so we have

$$x(\underline{x - 3}) + 2(\underline{x - 3}) = (\underline{x - 3})(x + 2)$$

However, in many cases these types of problems do not have any parentheses and we are still asked to factor the polynomial. In those cases we remove a common factor from the first two terms and a different common factor from the second two terms. If the two parentheses contain the same expressions at that step, the problem may then be completed as we did in Example A. This procedure of factoring is often called *factoring by grouping.*

EXAMPLE B Factor: $2x^2 + 3x + 6x + 9$.

Solution Remove a common factor of x from the first two terms. Remove a common factor of 3 from the second two terms.

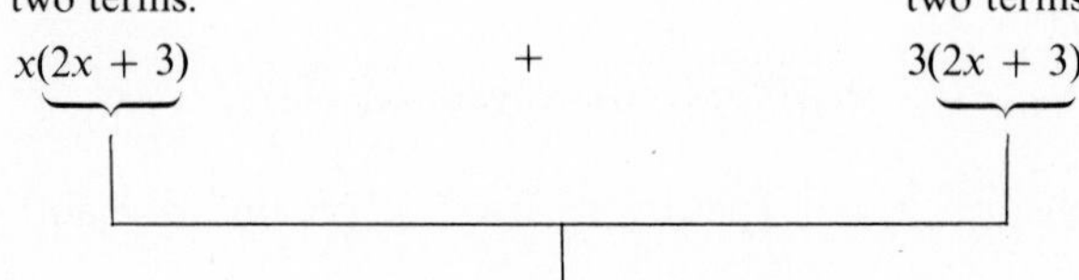

Note that the two sets of parentheses must contain the same expression at this step.

Now we finish the problem.

$$x(\underline{2x + 3}) + 3(\underline{2x + 3}) = (\underline{2x + 3})(x + 3)$$

EXAMPLE C Factor: $4x + 8y + ax + 2ay$.

Solution Remove a common factor of 4 from the first two terms.

$$4x + 8y + ax + 2ay = 4(x + 2y) + a(x + 2y)$$

Remove a common factor of a from the second two terms.

$4(\underline{x + 2y}) + a(\underline{x + 2y}) =$

$(\underline{x + 2y})(4 + a)$

Both sets of parentheses contain the same expression. The common factor of each term is the expression in parentheses, $(x + 2y)$.

In practice these problems are done in just two steps.

EXAMPLE D Factor: $cx + cy + 4x + 4y$.

Solution $c(x + y) + 4(x + y)$ — Remove a common factor of c from the first two terms. Remove a common factor of 4 from the second two terms.

$= (x + y)(c + 4)$ — Since the two parentheses are the same, we complete the problem.

In some problems the terms are not in the right order, so we have to rearrange the order of the terms first so that the first two terms have a common factor.

EXAMPLE E Factor: $bx + 4y + 4b + xy$.

Solution $bx + 4b + xy + 4y$ — Rearrange the order of the terms so that the first terms have a common factor.

$= b(x + 4) + y(x + 4)$ — Remove a common factor of b from the first two terms. Remove a common factor of y from the second two terms.

$= (x + 4)(b + y)$ — Since the two sets of parentheses contain the same expression, we can finish the problem.

Sometimes you will need to remove a negative common factor from the second two terms to obtain two sets of parentheses that contain the same expression.

EXAMPLE F Factor: $2x^2 + 5x - 4x - 10$.

Solution $x(2x + 5) - 2(2x + 5)$ — We remove a common factor of x from the first two terms and a common factor of -2 from the second two terms.

$= (2x + 5)(x - 2)$ — Since the two sets of parentheses contain the same expression, we can complete the problem.

Notice that if you removed a common factor of $+2$ in the first step, the two sets of parentheses would not contain the same expression. If the expressions inside the two sets of parentheses are not exactly the same, you cannot express the polynomial as a product of two factors!

EXAMPLE G Factor: $2ax - a - 2bx + b$.

Solution $a(2x - 1) - b(2x - 1)$ — Remove a common factor of a from the first two terms.

Remove a common factor of $-b$ from the second two terms.

$= (2x - 1)(a - b)$ — Since the two sets of parentheses contain the same expression, we can complete the problem.

Warning

Many students find that they make a factoring error in the first step of problems like Example G. Here are two questions to ask yourself to avoid a mistake at this point.

1. Do both sets of parentheses contain exactly the same expression?
2. Can I multiply these results back (or remove the parentheses) after the first step and once again obtain the original problem?

$5x^2 - 11x + 2$

$(5x - 1)(x - 2)$

$x = \frac{1}{5}$ $x = 2$

Multiplying back the results of the first step (even if you do it in your head) will usually help you to detect any error you may have made.

Now try these practice problems.

Practice Problems for Exercise 3.2.2

Factor by grouping.

1. $ax + 2a + 4bx + 8b$

2. $xz - 3yz - 5x + 15y.$

3. $3x - 10ay + 6y - 5ax$

Now turn to the answers following Exercise 3.2.2 and check your solutions for these practice problems.

Exercise 3.2.2

Factor by grouping.

1. $3x - 6 + xy - 2y$

2. $2x - 10 + xy - 5y$

3. $ax + ay - 2bx - 2by$

4. $4x + 8y - 3wx - 6wy$

5. $5a + 12bc + 10b + 6ac$

6. $2x + 15yz + 6y + 5xz$

7. $5a - 5b - 2ax + 2xb$

8. $xy - 4x - 3y + 12$

9. $y^2 - 2y - 3y + 6$

10. $x^2 + 4x - 5x - 20$

11. $3ax - y + 3ay - x$

12. $xa + 2bx - a - 2b$

13. $6ax - y + 2ay - 3x$

14. $4ab - c + 2ac - 2b$

15. $28x^2 + 8xy^2 + 21xw + 6y^2w$

16. $8xw + 10x^2 + 35xy^2 + 28y^2w$

17. $8a^2 - 6ab + 20ae - 15be$

18. $21a^2 - 6ab - 28ae + 8be$

Challenge Problems

19. $44x^3 - 63y^2w - 36xy^2 + 77x^2w$

20. $50x^7 - 40x^3y^5 - 15x^4w^2 + 12y^5w^2$

Solutions to Practice Problems for Exercise 3.2.2

Warning

There are several ways to do these problems. If you grouped your terms in a different order from these solutions, your steps may seem quite different but your answer should be the same if you have worked the problem correctly!

1. $ax + 2a + 4bx + 8b$ — There is no need to rearrange the terms.

$a(x + 2) + 4b(x + 2)$ — Make sure that you removed the common factor of $4b$ from the last two terms.

$(x + 2)(a + 4b)$ or

$(a + 4b)(x + 2)$

2. $xz - 3yz - 5x + 15y$ — The first two terms have a common factor. You do *not* need to rearrange the terms.

$z(x - 3y) - 5(x - 3y)$ Make sure that you removed -5 as the common factor from the last two terms.

$(x - 3y)(z - 5)$ or $(z - 5)(x - 3y)$ Since the two sets of parentheses contain the same expression, we can complete the factoring.

3. $3x - 10ay + 6y - 5ax$ We must rearrange the terms (since the first two terms do not have any common factor).

$3x + 6y - 5ax - 10ay$ There are several other ways to rearrange the terms. This is only one possible way!

$3(x + 2y) - 5a(x + 2y)$ Note the need to remove a common factor of $-5a$ from the last two terms.

$(x + 2y)(3 - 5a)$ or $(3 - 5a)(x + 2y)$ Since the two sets of parentheses contain the same expression, we can complete the factoring.

Now do the problems in Exercise 3.2.2.

3.3 Factoring of Trinomials

3.3.1 Factoring Trinomials of the Form $x^2 + bx + c$

Suppose that you wanted to factor $x^2 + 5x + 6$. After some trial and error you *might* obtain $(x + 2)(x + 3)$ or you might get discouraged and not get an answer. You can check this answer to this factoring problem by the FOIL method.

$$(x + 2)(x + 3) = x^2 + 3x + 2x + 6$$
$$= x^2 + 5x + 6$$

Let's see if we can determine how to obtain the answer $(x + 2)(x + 3)$. We know immediately that the answer will be of the form $(x + \_)(x + \_)$ since the first term of the polynomial is x^2. We want to know what numbers to place in the blanks. The two numbers are 2 and 3. Since $(2)(3) = 6$, their product equals the last term. Since $2 + 3 = 5$, their sum equals the coefficient of the middle term.

Let's return to our example.

The coefficient of x is the *sum* of these two numbers.

Factor: $x^2 + 5x + 6$ Solution: $(x + 2)(x + 3)$

The last term is the *product* of these two numbers.

Let's write the procedure we have observed and try a few examples.

> ***Procedure to factor trinomials of the form*** $x^2 + bx + c$
>
> **1. The answer will be of the form** $(x + \_\_)(x + \_\_)$.
>
> **2. The two numbers at the end of each set of parentheses are numbers such that:**
>
> **(a) When you multiply them you get the last term, which is** c.
>
> **(b) When you add them you get the coefficient of** x**, which is** b.

EXAMPLE A Factor: $x^2 + 7x + 12$.

Solution The answer is of the form $(x + \_)(x + \_)$. We want to find the two numbers you can multiply to get 12 but add to get 7. The numbers are 3 and 4.

$$x^2 + 7x + 12 = (x + 3)(x + 4)$$

EXAMPLE B Factor: $x^2 + 12x + 20$.

Solution We want two numbers that have a product of 20 but a sum of 12. The numbers are 10 and 2.

$$x^2 + 12x + 20 = (x + \underline{10})(x + \underline{2})$$

Note: If you cannot think of the numbers in your head, write down the possible factors whose product is 20.

Product	*Sum*	
$1 \cdot 20 = 20$	$1 + 20 = 21$	
$2 \cdot 10 = 20$	$2 + 10 = 12$	← Select this pair.
$4 \cdot 5 = 20$	$4 + 5 = 9$	

and then select the pair whose sum is 12.

EXAMPLE C Factor: $x^2 + 11x + 10$.

Solution Two numbers whose product is 10 but whose sum is 11 are the numbers 10 and 1.

$$x^2 + 11x + 10 = (x + 10)(x + 1)$$

Note: The order of the parenthetical terms is not important. You can write the answer as $(x + 1)(x + 10)$. This concept applies to all factoring problems. Remember, multiplication is commutative.

So far we have only done trinomials of the form $x^2 + bx + c$, where b and c are positive numbers. The same rule applies if b is a negative number.

EXAMPLE D Factor: $x^2 - 8x + 15$.

Solution We want two numbers that have a product of $+15$ but a sum of -8. They must be negative numbers.

(The product of two negative numbers is positive and the product here is positive. But you must add them and obtain -8. This would require two negative numbers.) Now the two numbers here are -5 and -3. When you multiply them,

$$(-5)(-3) = +15$$

When you add them,

$$-5 + (-3) = -8$$

So

$$x^2 - 8x + 15 = (x - 5)(x - 3) \quad \text{or} \quad (x - 3)(x - 5)$$

EXAMPLE E Factor: $x^2 - 9x + 14$.

Solution We want two numbers whose product is 14 but whose sum is -9. The numbers are -7 and -2. So

$$x^2 - 9x + 14 = (x - 7)(x - 2) \quad or \quad (x - 2)(x - 7)$$

> ***Stop!***
>
> **Do you understand the method we are using in Examples A to E? If not, take a few moments now to review these examples and then try practice problems 1 and 2 at the end of this section. Once you understand the solutions to these problems, do problems 1 to 8 of Exercise 3.3.1.**

If the method we are using is completely clear and you have no questions, continue reading.

All of the examples so far have had a positive last term. We can use the same rule if the last term is negative. However, we should observe that if the last term is negative, we will need one positive number and one negative number.

EXAMPLE F Factor: $x^2 + 2x - 8$.

Solution We want two numbers whose product is -8 but whose sum is $+2$. The two numbers are $+4$ and -2.

$$x^2 + 2x - 8 = (x + 4)(x - 2)$$

EXAMPLE G Factor: $x^2 - 3x - 10$.

Solution We want two numbers whose product is -10 but whose sum is -3. The two numbers are -5 and $+2$.

$$x^2 - 3x - 10 = (x - 5)(x + 2)$$

What if we made a sign error and *incorrectly* factored $x^2 - 3x - 10$ as $(x + 5)(x - 2)$? This would be wrong, of course. We could detect the error immediately since the sum of $+5$ and $-2 = 3$. Here we need instead a sum of -3! Another way to catch the error is to multiply $(x + 5)(x - 2)$ to obtain $x^2 + 3x - 10$, which does *not* coincide with the original problem.

EXAMPLE H Factor: $x^2 + 10x - 24$.

Solution The two numbers whose product is -24 but whose sum is $+10$ are the numbers $+12$ and -2.

$$x^2 + 10x - 24 = (x + 12)(x - 2)$$

Warning

It is very easy to make a sign error in these problems. Make sure that you mentally multiply your answer back to obtain the original expression. Check each sign carefully.

Check $(x + 12)(x - 2) = x^2 - 2x + 12x - 24 = x^2 + 10x - 24$ ✓ It checks.

EXAMPLE I Factor: $x^2 - 16x - 36$.

Solution We want two numbers whose product is -36 and whose sum is -16. The numbers are -18 and $+2$.

Hint: If you have difficulty finding which numbers work, try this: List all the possible factors of 36 (without regard to sign). Find the pair that has a difference of 16.

Possible Products of 36	*Difference of the Two Numbers*
$36 \cdot 1$	35
$18 \cdot 2$	16 ⟵ This is the value we want.
$12 \cdot 3$	9
$9 \cdot 4$	5
$6 \cdot 6$	0

Once we have picked the pair of numbers (18 and 2) it is easy to find the signs. For the coefficient of the middle term to be -16 we will have to add the numbers $\underline{-18}$ and $\underline{+2}$.

$$x^2 - 16x - 36 = (x - 18)(x + 2)$$

At this point you should work several problems to develop your factoring skill. This is one section where you really need to drill by doing many problems.

Feel a little confused about the signs? If you do, you may find these facts helpful.

Facts about Factoring Trinomials of the form $x^2 + bx + c$ with examples

The *two numbers* at the end of the parentheses: $x^2 + bx + c = (x \quad)(x \quad)$

1. **Will have the *same sign* if the last term is *positive*.**
   (a) **They will both be *positive* if the the *coefficient* of the *second* term is *positive*.** $x^2 + 5x + 6 = (x + 2)(x + 3)$
   (b) **They will both be *negative* if the *coefficient* of the *second* term is *negative*.** $x^2 - 5x + 6 = (x - 2)(x - 3)$
2. **Will have *opposite signs* if the last term is *negative*.**
   (a) **The *larger* of the two numbers *without regard to sign* will be given a plus sign if the coefficient of the *second term* is *positive*.** $x^2 + 6x - 7 = (x + 7)(x - 1)$
   (b) **The larger of the two numbers *without regard to sign* will be given a negative sign if the coefficient of the *second term* is *negative*.** $x^2 - 6x - 7 = (x - 7)(x + 1)$

Do not memorize these facts; rather, try to understand the pattern. Then work these practice problems.

Practice Problems for Exercise 3.3.1

Factor.

1. $x^2 + 10x + 16$

2. $x^2 - 9x + 20$

3. $x^2 + 3x - 18$

4. $x^2 - x - 30$

Now turn to the answers following Exercise 3.3.1 and check your solutions for these practice problems.

Exercise 3.3.1

Factor.

In problems 1 to 8, the last term is positive.

1. $x^2 + 9x + 8$

2. $x^2 + 8x + 15$

3. $x^2 + 8x + 12$

4. $x^2 + 21x + 20$

5. $x^2 - 4x + 3$

6. $x^2 - 6x + 8$

7. $x^2 - 13x + 22$

8. $x^2 - 15x + 14$

In problems 9 to 16, the last term is negative.

9. $x^2 + x - 12$

10. $x^2 + 2x - 8$

11. $x^2 - 13x - 14$

12. $x^2 - 6x - 16$

13. $x^2 + x - 20$

14. $x^2 + 7x - 18$

15. $x^2 - 5x - 24$

16. $x^2 - 11x - 26$

Look over your answers to problems 1 to 16 carefully. Be sure that you are clear on your sign rules. Problems 17 to 44 contain a mixture of all the types of problems in this section. Try to do them all.

17. $x^2 - 7x + 10$

18. $x^2 + 3x - 18$

19. $x^2 - x - 2$

20. $x^2 + 10x + 21$

21. $x^2 + 5x - 14$

22. $x^2 - 2x - 15$

23. $x^2 - 9x + 20$

24. $x^2 - 8x + 12$

25. $x^2 + 17x + 30$

26. $x^2 - 3x - 28$

27. $x^2 - 4x - 5$

28. $x^2 - 8x + 7$

29. $x^2 + 6x - 16$

30. $x^2 - 13x + 30$

31. $x^2 - 12x + 32$

32. $x^2 - 6x - 27$

33. $x^2 + 4x - 21$

34. $x^2 - 11x + 18$

35. $x^2 + 13x + 40$

36. $x^2 + 15x + 50$

37. $x^2 - 21x - 22$

38. $x^2 + 12x - 28$

39. $x^2 + 9x - 36$

40. $x^2 - 13x + 36$

41. $x^2 - x - 42$

42. $x^2 - x - 56$

43. $x^2 - 16x + 63$

44. $x^2 + 19x + 48$

Challenge Problems

45. $x^2 + 11x - 102$

46. $x^2 - 25x + 156$

Solutions to Practice Problems for Exercise 3.3.1

1. $x^2 + 10x + 16 = (x + 8)(x + 2)$ or $(x + 2)(x + 8)$
The two numbers whose product is 16 and whose sum is 10 are the numbers 8 and 2.

2. $x^2 - 9x + 20 = (x - 4)(x - 5)$ or $(x - 5)(x - 4)$

Both numbers must be *negative*. The two numbers whose product is 20 and whose sum is -9 are the numbers -4 and -5.

3. $x^2 + 3x - 18 = (x + 6)(x - 3)$ or $(x - 3)(x + 6)$

The two numbers will be *opposite in sign*. The two numbers whose product is -18 and whose sum is $+3$ are the numbers $+6$ and -3.

4. $x^2 - x - 30$ (Note that this is equivalent to $x^2 - 1x - 30$.)

$$x^2 - 1x - 30 = (x - 6)(x + 5) \quad \text{or} \quad (x + 5)(x - 6)$$

The two numbers whose product is -30 and whose sum is -1 are the numbers -6 and $+5$.

Now do the problems in Exercise 3.3.1.

3.3.2 Factoring Trinomials of the Form $ax^2 + bx + c$

When the coefficient of the x^2 term is not 1, the trinomial is more difficult to factor. Several possibilities must be considered.

EXAMPLE A Factor: $2x^2 + 5x + 3$.

Solution To get the coefficient of the first term to be 2, the factors would be 2 and 1. To get the last term to be 3, the factors would be 3 and 1.

Since all signs are positive we know that each set of parentheses will contain only positive signs. However, we still have two possibilities. They are:

$$(2x + 3)(x + 1)$$

$$(2x + 1)(x + 3)$$

We check them by multiplying by the FOIL method:

$$(2x + 1)(x + 3) = 2x^2 + 7x + 3 \quad \text{Wrong middle term.}$$

$$(2x + 3)(x + 1) = 2x^2 + 5x + 3 \quad \text{Right middle term.}$$

Thus the correct answer is

$$(2x + 3)(x + 1) \quad \text{or} \quad (x + 1)(2x + 3)$$

Some problems have many more possibilities. Once we know the different factors for the coefficient of the x^2 term and the different factors for the last term we can list the possibilities. We need to use the FOIL method to determine if we have the correct middle term.

EXAMPLE B Factor: $4x^2 - 13x + 3$.

Solution Let's consider the sign of the last number in each set of parentheses. Since the last term is positive, both signs are the same. Since the coefficient of the middle term is negative, both signs are negative.

The Different Factors of 4 are:	*The Factors of 3 are:*
$2 \cdot 2$	$1 \cdot 3$
$1 \cdot 4$	

Let us list the possible factoring combinations and compute the middle term by the FOIL method.

Possible Factors	*Middle Term*	*Correct?*
$(2x - 3)(2x - 1)$	$-8x$	No
$(4x - 3)(x - 1)$	$-7x$	No
$(4x - 1)(x - 3)$	$-13x$	*Yes*

The correct answer is

$$(4x - 1)(x - 3) \quad \text{or} \quad (x - 3)(4x - 1)$$

It is clear why this method is often referred to as the "trial-and-error" method. You have to keep trying possible factors until you find one that gives you the correct middle term.

EXAMPLE C Factor: $6x^2 + x - 5$.

Solution Let's consider the last number in each set of parentheses. Since the last term is negative, the two numbers will have opposite signs. Because the last number can be either positive or negative, we have more possibilities. In this case:

Factors of 6	*Factors of 5*
$3 \cdot 2$	$1 \cdot 5$
$6 \cdot 1$	

Possible Factors	*Middle Term*	*Correct?*
$(3x + 1)(2x - 5)$	$-13x$	No
$(3x - 1)(2x + 5)$	$+13x$	No
$(3x + 5)(2x - 1)$	$+7x$	No
$(3x - 5)(2x + 1)$	$-7x$	No
$(6x + 1)(x - 5)$	$-29x$	No
$(6x - 1)(x + 5)$	$+29x$	No
$(6x + 5)(x - 1)$	$-x$	No
$(6x - 5)(x + 1)$	$+x$	Yes

The correct answer is

$$(6x - 5)(x + 1) \quad \text{or} \quad (x + 1)(6x - 5)$$

As we look back on Example C we see a lot of steps. Let's see if we can shorten our work. Notice that there are four pairs of possible factors. The only difference between the two parentheses in the pair is that the last signs in each set of parentheses are reversed. This suggests that we list only half as many possibilities. Then if the coefficient of the middle term is the opposite in sign of what we want, we can just reverse signs. Let's try this shorter method.

EXAMPLE D Factor: $3x^2 - 2x - 8$.

Solution

Factors of 3	*Factors of 8*
$3 \cdot 1$	$8 \cdot 1$
	$4 \cdot 2$

Let us list only one-half of the possibilities. The last term of that first set of parentheses will always be positive.

Possible Factors	*Middle Term*	*Correct Factors?*
$(x + 8)(3x - 1)$	$+23x$	No
$(x + 1)(3x - 8)$	$-5x$	No
$(x + 4)(3x - 2)$	$+10x$	No
$(x + 2)(3x - 4)$	$+2x$	No (but it is exactly opposite in sign from what is needed)

So we just *reverse* the signs of the last term in each set of parentheses.

	Middle Term	*Correct Factor?*
$(x-2)(3x+4)$	$-2x$	Yes

The correct answer is

$$(x-2)(3x+4) \quad \text{or} \quad (3x+4)(x-2)$$

It takes a good deal of practice before you will be able to readily factor problems of this type. The more problems you do, the more proficient you will become.

Isn't there an easier way? Not that seems easier to every student! Many students find that they become quite proficient at the trial-and-error method after about 40 minutes of practice. Look over the examples and then try the Practice Problems for Exercise 3.3.2.

Still having trouble? If you attempt the trial-and-error method and it is still a slow struggle, here is a method called the grouping method that some students have found very helpful and fairly rapid. (However, you must be proficient at the problems in Exercise 3.2.2 before you try it.)

The grouping method for factoring trinomials of the form $ax^2 + bx + c$

Let's try Example A again.

EXAMPLE A (by grouping) Factor: $2x^2 + 5x + 3$.

Solution

1. Obtain a number that is the product of the *coefficient* of x^2 and the last term. This is called the *grouping number*.

   $2x^2 + 5x + 3$
   The grouping number is 6.

2. Find two numbers whose product is the grouping number and whose sum is the coefficient of the middle term.

   Product is 6.
   $2 \cdot 3 = 6$
   Sum is 5
   $2 + 3 = 5$
   The two numbers are 2 and 3.

3. Write the polynomial so that it has two terms containing x. The x term coefficients will be those two numbers obtained in step 2.

   $2x^2 + 5x + 3 = 2x^2 + \underline{2x} + \underline{3x} + 3$

4. Factor by grouping.

   $$\begin{aligned} &2x^2 + 2x + 3x + 3 \\ &= 2x(x+1) + 3(x+1) \\ &= (x+1)(2x+3) \end{aligned}$$

Let's try this on Example B.

EXAMPLE B (by grouping) Factor: $4x^2 - 13x + 3$.

Solution

1. $4x^2 - 13x + 3$
   Grouping number $= 12$

   The grouping number is 12.

2. Product $(-12)(-1) = +12$
   Sum $(-12) + (-1) = -13$

   We want two numbers whose product is 12 and whose sum is -13.

   The two numbers are -12 and -1.

3. $\begin{aligned} &4x^2 - 13x + 3 \\ &= 4x^2 - 12x - 1x + 3 \end{aligned}$

   Write polynomial as four terms with -12 and -1 as coefficients of x terms.

4. $4x^2 - 12x - 1x + 3$ — Factor by grouping.

$= 4x(x - 3) - 1(x - 3)$ — Remember to factor out a -1 from the last two terms so that both parentheses contain the same expression.

$= (x - 3)(4x - 1)$

Finally, let's redo Example D.

EXAMPLE D (by grouping) Factor: $3x^2 - 2x - 8$.

Solution

1. Grouping number $= -24$. — $(3)(-8) = -24$

2. We want two numbers whose product is -24 but whose sum is -2. They are -6 and $+4$.

3. $3x^2 - 6x + 4x - 8$ — Write as four terms using $-6x$ and $+4x$.

4. $3x(x - 2) + 4(x - 2)$ — Factor by grouping.

$= (x - 2)(3x + 4)$

It is probably best to be aware of the method that works best for you—either *trial and error* or *grouping*. However, you are encouraged to do some problems by each method in the homework. A good mathematician knows more than one way to do a problem.

Now try these practice problems, using either method.

Practice Problems for Exercise 3.3.2

Factor.

1. $2x^2 - 7x + 5$ **2.** $3x^2 + 4x - 4$ **3.** $3x^2 - x - 14$

Now turn to the answers following Exercise 3.3.2 and check your solutions for these practice problems.

Exercise 3.3.2

Factor. Check your answers for problems 1 to 10 using the FOIL method.

1. $3x^2 + 7x + 2$ **2.** $2x^2 + 7x + 3$ **3.** $2x^2 - 5x + 2$

4. $3x^2 - 8x + 4$ **5.** $2x^2 + 3x - 5$ **6.** $3x^2 + x - 2$

7. $2x^2 - 5x - 3$

8. $2x^2 - x - 6$

9. $5x^2 + 3x - 2$

10. $6x^2 + x - 2$

11. $6x^2 - 13x + 6$

12. $6x^2 - 7x + 2$

13. $2x^2 + 3x - 20$

14. $6x^2 + 11x - 10$

15. $9x^2 + 9x + 2$

16. $4x^2 + 11x + 6$

17. $6x^2 - 5x - 6$

18. $3x^2 - 13x - 10$

19. $8x^2 - 6x + 1$

20. $4x^2 - 8x + 3$

21. $4x^2 + 16x - 9$

22. $8x^2 + x - 9$

23. $9x^2 - 13x + 4$

24. $5x^2 - 11x + 2$

25. $5x^2 - 13x - 6$

26. $3x^2 - 10x - 8$

27. $12x^2 - 20x + 3$

28. $9x^2 + 5x - 4$

29. $15x^2 + 4x - 4$

30. $8x^2 - 11x + 3$

31. $10x^2 + 21x + 9$

32. $12x^2 + 11x + 2$

33. $16x^2 - 6x - 1$

34. $3x^2 - x - 14$

Challenge Problems

35. $12x^2 + 16x - 35$

36. $20x^2 - 53x + 18$

Solutions to Practice Problems for Exercise 3.3.2

Trial-and-error method

1. $2x^2 - 7x + 5$

Possible Factors	*Middle Term*	*Correct Factors?*
$(2x - 1)(x - 5)$	$-11x$	No
$(2x - 5)(x - 1)$	$-7x$	Yes

$$2x^2 - 7x + 5 = (2x - 5)(x - 1)$$

2. $3x^2 + 4x - 4$

Listing of One-Half of Possible Factors	*Middle Term*	*Correct Factors?*
$(3x + 4)(x - 1)$	$+1x$	No
$(3x + 1)(x - 4)$	$-11x$	No
$(3x + 2)(x - 2)$	$-4x$	No (but is exactly opposite in sign)

So we try

	Middle Term	*Correct Factor?*
$(3x - 2)(x + 2)$	$+4x$	Yes

$$3x^2 + 4x - 4 = (3x - 2)(x + 2)$$

3. $3x^2 - x - 14$

Listing of One-Half of Possible Factors	*Middle Term*	*Correct Factor?*
$(3x + 14)(x - 1)$	$+11x$	No
$(3x + 1)(x - 14)$	$-41x$	No
$(3x + 2)(x - 7)$	$-19x$	No
$(3x + 7)(x - 2)$	$+1x$	No (but is exactly opposite in sign)

So we try

	Middle Term	*Correct Factor?*
$(3x - 7)(x + 2)$	$-1x$	Yes

$$3x^2 - x - 14 = (3x - 7)(x + 2)$$

Note: You may take far fewer steps if you made a wise guess on these trial-and-error problems.

Grouping method

1. $2x^2 - 7x + 5$

(a) Grouping number is $+10$. $(2)(5) = 10$

(b) We want two numbers whose product is 10 and whose sum is -7. The two numbers are -2 and -5.

(c) $2x^2 - 2x - 5x + 5$ Write the polynomial as four terms with $-2x$ and $-5x$.

(d) $= 2x(x - 1) - 5(x - 1)$ Factor by grouping.
$= (x - 1)(2x - 5)$

2. $3x^2 + 4x - 4$

(a) Grouping number is -12. $(3)(-4) = -12$

(b) We want two numbers whose product is -12 and whose sum is $+4$. These two numbers are $+6$ and -2.

(c) $3x^2 + 6x - 2x - 4$ Write the polynomial as four terms with $+6x$ and $-2x$.

(d) $= 3x(x + 2) - 2(x + 2)$ Factor by grouping.
$= (x + 2)(3x - 2)$

3. $3x^2 - x - 14$

(You may write as $3x^2 - 1x - 14$.)

(a) Grouping number is -42. $(3)(-14) = -42$

(b) We want two numbers whose product is -42 and whose sum is -1. These two numbers are $+6$ and -7.

(c) $3x^2 + 6x - 7x - 14$ Write the polynomial as four terms with $+6x$ and $-7x$.

(d) $3x(x + 2) - 7(x + 2)$ Factor by grouping.

$(x + 2)(3x - 7)$

Now do the problems in Exercise 3.3.2.

3.4 Special Cases of Factoring

Certain types of problems can be identified more quickly and done more easily. You need to be very familiar with the appropriate formula in order to factor these quickly.

The difference of two squares

Recall Formula A from Section 3.1.2:

$$(a + b)(a - b) = a^2 - b^2$$

In reverse form we can use it for factoring.

> ***Difference of Two Squares Factoring Formula***
>
> $$a^2 - b^2 = (a + b)(a - b)$$

We can state it in words in this way: "The difference of two squares can be factored into the sum and difference of those values that were squared."

EXAMPLE A Factor: $9x^2 - 1$.

Solution We see that

$$9x^2 - 1 = (3x + 1)(3x - 1)$$

Here we see the values that were squared were $3x$ [since $(3x)^2 = 9x^2$] and 1 (since $1^2 = 1$).

EXAMPLE B Factor: $25x^2 - 16$.

Solution We see that

$$25x^2 - 16 = (5x + 4)(5x - 4)$$

Here we see that the values that were squared were $5x$ [since $(5x)^2 = 25x^2$] and 4 (since $4^2 = 16$).

Sometimes the problem will contain two variables.

EXAMPLE C Factor: $4x^2 - 49y^2$.

Solution We see that

$$4x^2 - 49y^2 = (2x + 7y)(2x - 7y)$$

Here we see that the values that were squared were $2x$ [since $(2x)^2 = 4x^2$] and $7y$ [since $(7y)^2 = 49y^2$].

EXAMPLE D Factor: $100x^2 - 81y^2$.

Solution We see that

$$100x^2 - 81y^2 = (10x + 9y)(10x - 9y)$$

Here we see that the values that were squared were $10x$ [since $(10x)^2 = 100x^2$] and $9y$ [since $(9y)^2 = 81y^2$].

Some problems may involve more than one step.

EXAMPLE E Factor: $81x^4 - 1$.

Solution We see that

$$81x^4 - 1 = (9x^2 + 1)(9x^2 - 1)$$

Here we see that the values that were squared were $9x^2$ [since $(9x^2)^2 = 81x^4$] and 1. However, we can factor $9x^2 - 1$ further since it is the difference of squares. (Note that $9x^2 + 1$ *cannot* be factored. A futher discussion of this case is left as an exercise.)

$$81x^4 - 1 = (9x^2 + 1)(9x^2 - 1) = (9x^2 + 1)(3x + 1)(3x - 1)$$

Perfect square trinomials

There is a formula that will help us to factor very quickly certain trinomials, called *perfect square trinomials.* Recall Formula B from Section 3.1.2:

$$(a + b)^2 = a^2 + 2ab + b^2$$

$$(a - b)^2 = a^2 - 2ab + b^2$$

In reverse form we can use these two equations for factoring.

Perfect Square Factoring Formulas

$$a^2 + 2ab + b^2 = (a + b)^2$$

$$a^2 - 2ab + b^2 = (a - b)^2$$

A perfect square trinomial is a trinomial that is the result of squaring a binomial. How can we recognize a perfect square trinomial?

1. The first and last terms are *perfect squares* (the numerical values are 1, 4, 9, 16, 25, 36, 49, ... and the variables have an even-numbered exponent such as $x^2, x^4, x^6, x^8, x^{10}, x^{12}, \ldots$.
2. The middle term is twice the product of the values whose squares are the first and last terms.

EXAMPLE F Factor: $x^2 + 6x + 9$.

Solution This *is* a perfect square trinomial.

1. The first and last terms are perfect squares because $x^2 = (x)^2$ and $9 = (3)^2$.
2. The middle term is twice the product of x and 3.

Since

$$a^2 + 2ab + b^2 = (a + b)^2$$

we have

$$x^2 + 6x + 9 = (x + 3)^2$$

EXAMPLE G Factor: $4x^2 - 20x + 25$.

Solution This *is* a perfect square trinomial.

1. The first and last terms are perfect squares because

$$4x^2 = (2x)^2 \qquad \text{and} \qquad 25 = (5)^2$$

2. The middle term is twice the product of $2x$ and 5. Note the negative sign.

Since

$$a^2 - 2ab + b^2 = (a - b)^2$$

we have

$$4x^2 - 20x + 25 = (2x - 5)^2$$

More than one variable may be involved.

EXAMPLE H Factor: $49x^2 + 42xy + 9y^2$.

Solution This *is* a perfect square trinomial.

1. The first and last terms are perfect squares because

$$(7x)^2 = 49x^2 \qquad \text{and} \qquad (3y)^2 = 9y^2$$

2. The middle term is twice the product of $7x$ and $3y$.

Since

$$a^2 + 2ab + b^2 = (a + b)^2$$

we have

$$49x^2 + 42xy + 9y^2 = (7x + 3y)^2$$

The exponents may be higher than 2. The same principles apply.

EXAMPLE I Factor: $36x^4 - 12x^2 + 1$.

Solution This *is* a perfect square trinomial.

1. The first and last terms are perfect squares. $36x^4 = (6x^2)^2$.
2. The middle term is twice the product of $6x^2$ and 1. Note the negative sign.

Since

$$a^2 - 2ab + b^2 = (a - b)^2$$

we have

$$36x^4 - 12x^2 + 1 = (6x^2 - 1)^2$$

Some problems appear to be perfect square trinomials but are not. They are factored as other trinomials in Section 3.3.

EXAMPLE J Factor: $49x^2 + 35x + 4$.

Solution This is *not* a perfect square trinomial! Although the first and last terms are perfect squares since $(7x)^2 = 49x^2$ and $(2)^2 = 4$, the middle term is not double the product of 2 and $7x$! $35x \neq 28x$! So we must factor by trial and error or by grouping to obtain

$$49x^2 + 35x + 4 = (7x + 4)(7x + 1)$$

Review Examples A to J and then try these practice problems.

Practice Problems for Exercise 3.4

Factor. Most problems can be identified as the *difference of two squares* or as a *perfect square trinomial.*

1. $1 - 64x^2$ **2.** $25x^4 - 36$ **3.** $16x^2 + 8x + 1$

4. $25x^2 - 60xy + 36y^2$ **5.** $9x^2 - 15x + 4$

Now turn to the answers following Exercise 3.4 and check your solutions for these practice problems.

Exercise 3.4

Factor.

1. $9x^2 - 16$ **2.** $36x^2 - 1$ **3.** $1 - 100s^2$

4. $144x^2 - 25$ **5.** $81x^2 - 49y^2$ **6.** $16x^4 - 1$

7. $25 - 121x^2$ **8.** $9x^2 - 49$ **9.** $81x^2 - 100y^4$

10. $25x^4 - 16y^4$ **11.** $49x^2 + 14x + 1$ **12.** $36x^2 + 12x + 1$

13. $y^2 - 6y + 9$ **14.** $y^2 - 8y + 16$ **15.** $9x^2 - 24x + 16$

16. $4x^2 + 20x + 25$ **17.** $81x^2 + 36xy + 4y^2$ **18.** $36x^2 + 60xy + 25y^2$

19. $9x^4 - 6x^2y + y^2$

20. $y^4 - 22y^2 + 121$

21. $49x^2 + 70x + 9$

22. $25x^2 - 50x + 16$

23. $x^6 - 25$

24. $x^8 - 1$

25. $x^{10} - 36y^{10}$

26. $x^4 - 49y^6$

27. $4x^8 + 12x^4 + 9$

28. $4x^6 + 4x^3 + 1$

29. When you factor Example E as $81x^4 - 1 = (9x^2 + 1)(9x^2 - 1)$, we then factor $9x^2 - 1 = (3x + 1)(3x - 1)$. Show why you *cannot* factor $9x^2 + 1$.

Challenge Problems

30. $16x^{12} - 72x^6 + 81$

31. $x^{20} - 50x^{10}y^{10} + 625y^{20}$

Solutions to Practice Problems for Exercise 3.4

1. This problem is an example of the *difference of two squares*. Since

$$a^2 - b^2 = (a + b)(a - b)$$

we have

$$1 - 64x^2 = (1 + 8x)(1 - 8x)$$

2. This problem is an example of the *difference of two squares*. Note that $(5x^2)^2 = 25x^4$. Be careful of your exponents. So since

$$a^2 - b^2 = (a + b)(a - b)$$

we have

$$25x^4 - 36 = (5x^2 + 6)(5x^2 - 6)$$

3. This is a *perfect square trinomial* using the formula

$$a^2 + 2ab + b^2 = (a + b)^2$$

We have

$$16x^2 + 8x + 1 = (4x + 1)^2$$

4. This is a *perfect square trinomial* using the formula

$$a^2 - 2ab + b^2$$

We have

$$25x^2 - 60xy + 36y^2 = (5x - 6y)^2$$

5. This is *not* a perfect square trinomial. Note the middle term of $9x^2 - 15x + 4$. If it were a perfect square trinomial, the middle term would be $-12x$ since $-(2)(3x)(2) = -12x$. We will therefore need to use the trial-and-error method or the grouping method to factor it. We will then obtain

$$9x^2 - 15x + 4 = (3x - 1)(3x - 4)$$

Now do the problems in Exercise 3.4.

3.5 Factoring Polynomials Completely

When you are asked to factor a polynomial it is expected that you will factor it completely. Many problems involve more than one step of factoring. Usually, the first step of factoring is removing the common factor. Then the problem is factored using previously learned techniques.

Not all polynomials can be factored using the methods of this chapter. So far in this text you have not encountered an instance of this. However, in this section we include a few problems that cannot be factored by the methods we have previously learned. In Chapter 8 and in subsequent math courses you will encounter polynomials that appear to be factorable by elementary methods but are not. If we cannot factor a polynomial by elementary methods, we will identify it as a *prime* polynomial.

EXAMPLE A Factor: $2x^2 - 4x - 30$.

Solution

$2(x^2 - 2x - 15)$ — First remove the common factor of 2.

$2(x - 5)(x + 3)$ — Factor the remaining trinomial.

Warning

Be sure to include the common factor of 2 in your final answer. Many times students forget to include the common factor as part of the final answer.

EXAMPLE B Factor: $25x^3 - 10x^2 + x$.

Solution

$x(25x^2 - 10x + 1)$ — First we remove the common factor of x.

$x(5x - 1)^2$ — The remaining trinomial is a perfect square trinomial and is factored using the formula $a^2 - 2ab + b^2 = (a - b)^2$.

Be sure to remove as large a common factor as possible in your first step.

EXAMPLE C Factor: $20x^2y^2 - 45y^2$.

Solution

$5y^2(4x^2 - 9)$ — First we remove the common factor of $5y^2$.

$5y^2(2x + 3)(2x - 3)$ — Factor the remaining binomial by using the difference-of-two-squares formula $a^2 - b^2 = (a + b)(a - b)$.

EXAMPLE D Factor: $3x^2 - 6xy - 24y^2$.

Solution

$3(x^2 - 2xy - 8y^2)$ — First we remove the common factor of 3.

$3(x - 4y)(x + 2y)$ — Continue factoring. Notice here the need for the two letters x and y in each set of parentheses.

EXAMPLE E Factor: $2x^3 - 5x^2y - 3xy^2$.

Solution

$x(2x^2 - 5xy - 3y^2)$ First we remove the common factor of x.

$x(2x + y)(x - 3y)$ Continue factoring by the *trial-and-error method* or the *grouping method*. Notice the need for two letters x and y in each set of parentheses.

EXAMPLE F Factor: $2ax + 4ay + 4x + 8y$.

Solution

$2[ax + 2ay + 2x + 4y]$ Remove the common factor of 2 first. This is very important.

$2[a(x + 2y) + 2(x + 2y)]$ Factor the four terms inside the bracket by the grouping method.

$2[(x + 2y)(x + 2)]$ Remove common factor of $(x + 2y)$ from each term.

$2(x + 2y)(a + 2)$ Write the answer in simple form.

How do I know which kind of factoring method to use of the methods in this chapter?

Procedure for Completely Factoring a Polynomial

1. **Remove any common factors first.**
2. **Factor further by counting the number of terms.**
   (a) **If there are two terms, look for the difference of two squares.**

$$a^2 - b^2 = (a + b)(a - b)$$

   (b) **If there are three terms, look to see if it fits the perfect square trinomial formulas.**

$$a^2 + 2ab + b^2 = (a + b)^2$$
$$a^2 - 2ab + b^2 = (a - b)^2$$

   If it does not, factor as a regular trinomial.
   (c) **If it has four terms, factor by grouping.**

If you are weak on any of these methods of factoring, review the examples in the appropriate section of the text before you go any further.

Be alert for problems that appear to be factorable but are not. Remember, if we are not able to factor a polynomial we call it a *prime* polynomial.

EXAMPLE G Factor, if possible: $x^2 + 6x + 12$.

Solution Since we are only using rational numbers we find this polynomial is prime. The rational factors of 12 are

(1)(12)

(2)(6)

(3)(4)

None of these pairs will add up to 6, the coefficient of the middle term. Thus the problem cannot be factored by the methods of this chapter.

EXAMPLE H Factor, if possible: $x^2 - 9x - 8$.

Solution Since we are only using rational numbers, we find that this polynomial is prime. The possible rational factors of -8 are

$$(-8)(1)$$

$$(-4)(2)$$

$$(-2)(4)$$

$$(-1)(8)$$

None of these pairs has a sum of -9, the coefficient of x. Thus the problem cannot be factored by the methods of this chapter.

Now try these practice problems.

Practice Problems for Exercise 3.5

Factor, if possible. Be sure to factor completely.

1. $9x^4y^2 - 9y^2$

2. $3x^2 - 36x + 108$

3. $6x^3 + 11x^2 - 10x$

4. $10x^2 - 16x - 4$

Now turn to the answers following Exercise 3.5 and check your solutions for these practice problems.

Exercise 3.5

Factor, if possible. Be sure to factor completely.

1. $3x^4 - 12$

2. $y^2 + 16y + 64$

3. $2x^2 - 10x - 132$

4. $16x^2 - 81y^2$

5. $2x^2 - 11x + 12$

6. $2xy^2 - 50x$

7. $x^2 - 3xy - 70y^2$

8. $2x^3 - 7x^2 + 4x - 14$

9. $ax + 20 - 5a - 4x$

10. $3x^2 - 9x + 15$

11. $5x^3 - 45x$

12. $18y^2 + 3y - 6$

13. $5x^3y^3 - 10x^2y^3 + 5xy^3$

14. $12x^2 - 36x + 27$

15. $27xyz^2 - 12xy$

16. $18x^3 - 12x^2 + 2x$

17. $3x^2 + 6x - 105$

18. $4x^2 - 28x - 72$

19. $5x^2 - 30x + 40$

20. $7x^2 + 3x - 2$

21. $2x^4 - 7x^2 - 4$

22. $2x^4 - 9x^2 - 5$

23. $6x^2 - 3x + 2$

24. $4x^3 + 8x^2 - 60x$

25. $8x^2 + 28x - 16$

26. $12x^2 - 30x + 12$

27. $a^2x + 2a^2y - 3ax - 6ay$

28. $2ax^2 + 2axy - 5x^2 - 5xy$

Challenge Problems

29. $-12y^3w + 2x^3y^2 - 8x^2y^3 + 3xy^2w$

30. $50x^3 - 48 - 32x + 75x^2$

Solutions to Practice Problems for Exercise 3.5

1. $9x^4y^2 - 9y^2$ — Note the common factor of $9y^2$.

$9y^2(x^4 - 1)$ — Remove the common factor of $9y^2$.

$9y^2(x^2 + 1)(x^2 - 1)$ — Factor as the difference of two squares.

$9y^2(x^2 + 1)(x + 1)(x - 1)$ — Factor $x^2 - 1$ as the difference of two squares. (Remember that you cannot factor $x^2 + 1$.)

2. $3x^2 - 36x + 108$ — Note that all coefficients are divisible by 3.

$3(x^2 - 12x + 36)$ — Remove the common factor.

$3(x - 6)^2$ — Factor as a perfect square trinomial.

3. $6x^3 + 11x^2 - 10x$ — Note the common factor of x.

$x(6x^2 + 11x - 10)$ — Remove the common factor of x.

$x(2x + 5)(3x - 2)$ — Factor the trinomial by trial and error or grouping.

4. $10x^2 - 16x - 4$ — Note that all coefficients are divisible by 2.

$2(5x^2 - 8x - 2)$ — Remove the common factor of 2.

$2(5x^2 - 8x - 2)$ — This is the final answer. $5x^2 - 8x - 2$ is prime. It cannot be factored further.

Now do the problems in Exercise 3.5.

An Application for Chapter 3: Quick Tricks in Arithmetic

Suppose that you had to multiply a lot of problems like these.

$$(103)(97) = ?$$

$$(56)(64) = ?$$

$$(398)(402) = ?$$

Without using a calculator, can you see a pattern that would help you find these products more rapidly than using long multiplication?

If you carefully look at the three problems you will perhaps observe that they can be rewritten as follows:

$$(103)(97) = (100 + 3)(100 - 3)$$

$$(56)(64) = (60 - 4)(60 + 4)$$

$$(398)(402) = (400 - 2)(400 + 2)$$

In the second form the binomials can be seen as the sum and difference of two values. We have previously developed Formula A in Section 3.1.2 for all real numbers or variables a and b. We can legitimately apply Formula A,

$$(a + b)(a - b) = a^2 - b^2$$

to this situation. Therefore the multiplication can be done thus

$$(a + b)(a - b) = a^2 - b^2$$

$$(103)(97) = (100 + 3)(100 - 3) = (100)^2 - (3)^2$$
$$= 10{,}000 - 9 = 9991$$

$$(56)(64) = (60 - 4)(60 + 4) = (60)^2 - (4)^2$$
$$= 3600 - 16 = 3584$$

$$(398)(402) = (400 - 2)(400 + 2) = (400)^2 - (2)^2$$
$$= 160{,}000 - 4 = 159{,}996$$

The application of this type of problem is quite limited since one rarely encounters products of this exact type. However, there are several "tricks" that can be used to perform arithmetic calculations that can be developed using the factoring techniques of this chapter.

A Challenge for You!

See if you can use the formula $(a + b)(a - b) = a^2 - b^2$ to quickly evaluate the following.

1. $(1005)(995) =$

2. $(76)(74) =$

3. $(504)(496) =$

4. $(19\frac{1}{2})(20\frac{1}{2}) =$

Answers:

1. 999,975

2. 5624

3. 249,984

4. $399\frac{3}{4}$

Supplementary Problems for Chapter 3

Multiply.

1. $(5a + 7b)(a - 3b)$

2. $(11x - 2y)(4x - 3y)$

3. $(x^3 + 2x^2 - x - 4)(2x - 3)$

4. $(x - 6)(2x - 3)(x + 4)$

5. $(3x^2 + 2y)(4w - 7z)$

6. $(x^2 - 4xy - 3y^2)(5x - 2y)$

7. $(x + 2y)(x - 3y)(4x + y)$

8. $(2x^2 + 3y)(x^3 + 2x^2y - xy^2 - 3y^3)$

If possible, factor completely all of the following problems. *If they cannot be factored, state "prime."*

9. $4x^2 - 9y^2$

10. $x^2 + 6x + 9$

11. $x^2 - 9x + 18$

12. $x^2 + 13x - 30$

13. $6x^2 + x - 7$

14. $10x^2 + x - 2$

15. $12x + 16$

16. $8x^2y^2 - 4xy$

17. $50x^3y^2 + 30x^2y^2 - 10x^2y^2$

18. $26a^3b - 13ab^3 + 52a^2b^4$

19. $x^3 - 16x^2 + 64x$

20. $2x^2 + 40x + 200$

21. $3x^2 - 18x + 27$

22. $25x^3 - 60x^2 + 36x$

23. $7x^2 + 3x - 10$

24. $5x^2 - 8x - 4$

25. $9x^3y - 4xy^3$

26. $3x^3a^3 - 11x^4a^2 - 20x^5a$

27. $12a^2 + 14ab - 10b^2$

28. $16a^2 - 40ab + 25b^2$

29. $7a - 7 - ab + b$

30. $3d - 4 - 3cd + 4c$

31. $2x - 1 + 2bx - b$

32. $5xb - 35x + 4by - 28y$

33. $2a^2x - 15ax + 7x$

34. $x^5 - 17x^3 + 16x$

35. $x^4 - 81y^{12}$

36. $6x^4 - x^2 - 15$

37. $28yz - 16xyz + x^2yz$

38. $12x^3 + 17x^2 + 6x$

39. $16w^2 - 2w - 5$

40. $12w^2 - 12w + 3$

41. $4y^3 + 10y^2 - 6y$

42. $10y^2 + 33y - 7$

43. $8y^{10} - 16y^8$

44. $49x^4 - 49$

45. $x^2 + 13x + 54$

46. $8x^2 - 19x - 6$

47. $8y^5 + 4y^3 - 60y$

48. $9xy^2 + 3xy - 42x$

49. $16x^4y^2 - 56x^2y + 49$

50. $128x^3y - 2xy$

Chapter 3 Summary

Multiplication of Polynomials

1. Two binomials can be multiplied by the FOIL method. You should always make sure to combine the middle terms where possible.
2. Two polynomials with two or more terms can be multiplied by a method similar to that used in arithmetic for multiplying two numbers with two or more digits. You must be sure to align like terms underneath each other so that you can combine them readily.
3. Two binomials that are the sum and difference of the same two values can be mentally multiplied using

$$\text{Formula A:} \quad (a + b)(a - b) = a^2 - b^2$$

4. To square a binomial mentally, use one of the two parts of

$$\text{Formula B:} \quad (a + b)^2 = a^2 + 2ab + b^2$$
$$(a - b)^2 = a^2 - 2ab + b^2$$

Factoring of Polynomials

1. Remove any common factors first (if the polynomial contains any common factor).
2. Factor further by counting the number of terms.
   (a) If there are two terms, look to see if it can be factored as the *difference of two squares*:

   $$a^2 - b^2 = (a + b)(a - b)$$

   (b) If there are three terms, look to see if it is a *perfect square trinomial.* If so, use one of the formulas

   $$a^2 + 2ab + b^2 = (a + b)^2$$
   $$a^2 - 2ab + b^2 = (a - b)^2$$

   If not, factor as a regular trinomial of the form

   $$x^2 + bx + c \quad \text{or} \quad ax^2 + bx + c$$

   (c) If it has four terms, factor by the *grouping method.*
3. If a polynomial is not factorable, we call it *prime.*

Practice Quiz for Chapter 3

Sections 3.1–3.2

Multiply.

1. $(a - 3b)(a + 2b)$

2. $(5x - y)(x + 4y)$

3. $(x + 2)(x - 3)(x + 4)$

4. $(x^2 + 6x - 2)(3x - 1)$

5. $(5p + 4r)^2$

6. $(7x^2 - y)(7x^2 + y)$

Factor.

7. $5a - 10b + 25ab$

8. $3x^2 - 7x^3 + 24x^2y$

9. $2x^2 + xy - 4x - 2y$

10. $3x + 6y - xy - 2y^2$

11. $a^2 + 2bc - 2ab - ac$

12. $abx - 8 - 2x + 4ab$

Sections 3.3–3.5

If possible, factor completely all of the following problems. If they cannot be factored, state "prime."

1. $25x^2 - 30x + 9$

2. $x^2 - 13x + 42$

3. $8x^2 + 24x + 10$

4. $16x^2y^2 - y^2$

5. $2x^2 - 4x - 30$

6. $9x^2y^2 + 12xy^3 + 4y^4$

7. $3x^2 + 4x + 2$

8. $2x^2 + x - 28$

9. $3x^2 - 21x + 30$

10. $acx - 2ac + bcx - 2bc$

11. $3x^4y^2 - 48y^2$

12. $4a^4 + 4a^3b - 24a^2b^2$

Practice Test for Chapter 3

Multiply.

1. $(5a - 4b)(2a + 3b)$

2. $(3x + 2)(2x + 1)(x - 3)$

3. $(7x^2 + 2y^2)^2$

4. $(9x - 2y)(9x + 2y)$

If possible, *factor completely all of the following problems. If they cannot be factored,* state "prime."

5. $x^2 + 12x - 28$

6. $9x^2 - 100y^2$

7. $10x^2 + 27x + 5$

8. $9a^2 - 30ab + 25b^2$

9. $7x - 9x^2 + 14xy$

10. $3x^2 - 4wy - 2wx + 6xy$

11. $6x^3 - 20x^2 + 16x$

12. $5a^2c - 11abc + 2b^2c$

13. $100x^4 - 16y^4$

14. $9x^2 - 15xy + 4y^2$

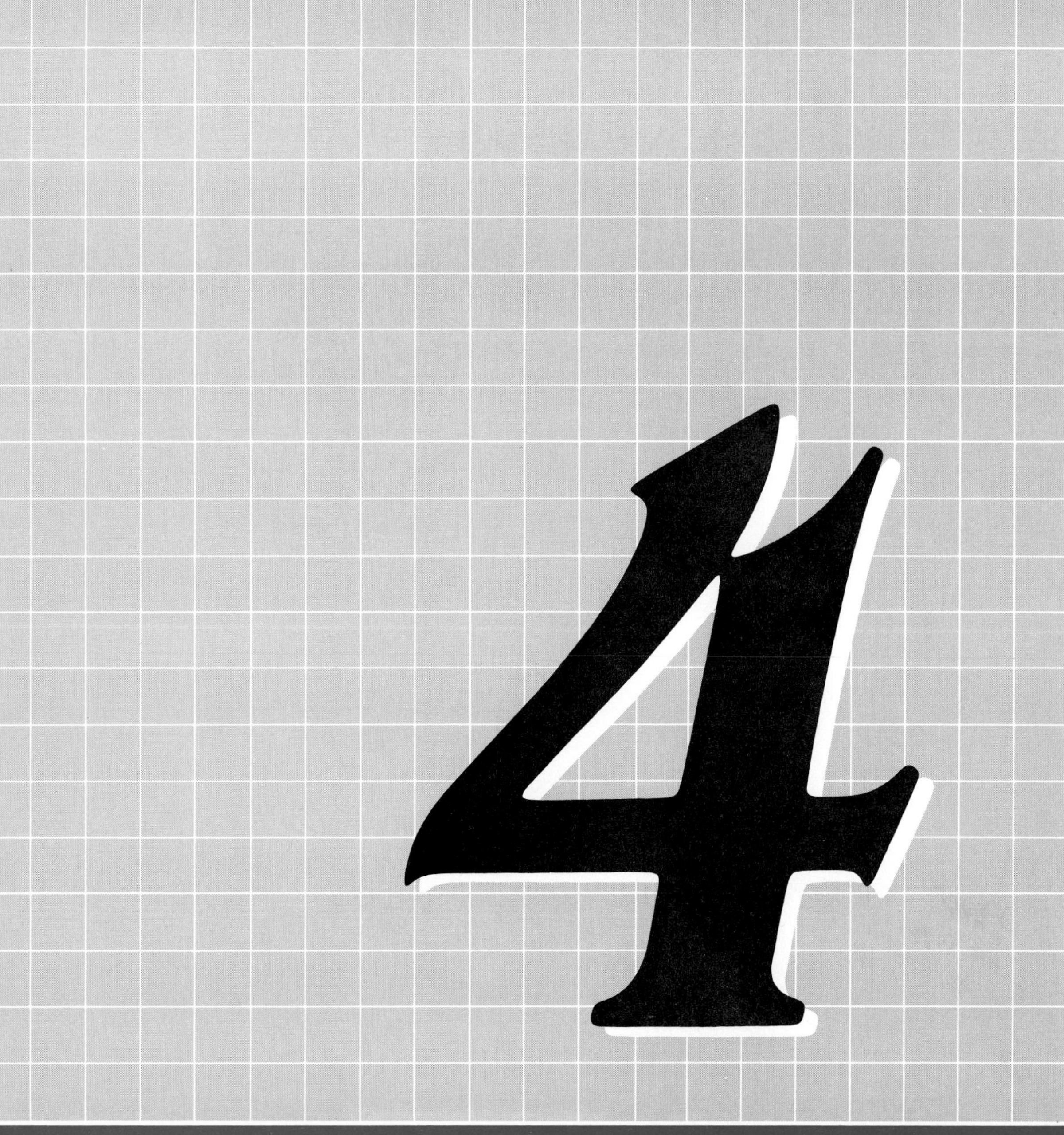

4 Fractional Algebraic Expressions and Equations

Diagnostic Pretest for Chapter 4

This diagnostic pretest is intended for those students who have previously studied algebra. Work as many problems as you can. Then, check your answers against those listed at the back of the book. Where you have obtained an incorrect answer or have been unable to work the problem, you will need to study the corresponding sections of the chapter.

If you have not previously studied algebra, do not take this test now. You should study the chapter and work the problems and exercises. After finishing the chapter, you may use this test as a review or as a practice test.

Simplify problems 1 to 10.

Section 4.1

1. $\dfrac{a^2b + 2ab^2}{2a^3 + 3a^2b - 2ab^2}$

2. $\dfrac{3x^2 - 14x + 8}{4x^2 - 13x - 12}$

Section 4.2

3. $\dfrac{4a^2 - b^2}{6a - 6b} \cdot \dfrac{3a - 3b}{6a + 3b}$

4. $\dfrac{x^2 - 6x + 9}{x^2 - x - 6} \div \dfrac{x^2 + 2x - 15}{x^2 + 2x}$

5. $\dfrac{x^2 + 5x + 6}{3x^2 + 8x + 4} \cdot \dfrac{1}{x + 3} \cdot \dfrac{6x^2 - 11x - 10}{2x^2 + x - 15}$

6. $\dfrac{xy + 3y}{x^2 - x} \div \dfrac{x + 3}{x}$

Section 4.3

7. $\dfrac{3y + 2}{2y} + \dfrac{5}{y + 2}$

8. $\dfrac{2y - 1}{2y^2 + y - 3} - \dfrac{2}{y - 1}$

Section 4.4

9. $\dfrac{\dfrac{2}{a} - \dfrac{3}{a^2}}{5 + \dfrac{1}{a}}$

10. $\dfrac{\dfrac{a}{a + 1} - \dfrac{2}{a}}{3a}$

Section 4.5

Solve for x and check your answers.

11. $\dfrac{5}{2x} = 2 - \dfrac{2x}{x + 1}$

12. $\dfrac{14}{x^2 - 3x - 10} = \dfrac{2}{x - 5} + \dfrac{3}{x + 2}$

Student Learning Objectives for Chapter 4

After successfully learning the content of each section, you will be able to:

Section 4.1

A. Reduce a fraction where the numerator and denominator are factorable polynomials.

Section 4.2

B. Multiply two or more algebraic fractions and write the answer in simplest form.

C. Divide two algebraic fractions and write the answer in simplest form.

Section 4.3

D. Add or subtract two or more algebraic fractions and write the answer in simplest form.

Section 4.4

E. Simplify a complex algebraic fraction.

Section 4.5

F. Solve a fractional linear equation where the denominator of one or more of the fractions contains a variable.

G. Check the apparent solution of a fractional equation and determine when a fractional linear equation does not have a solution.

Introduction

We usually use the word "fraction" to mean one rational number divided by another, such as $3 \div 4$ or $\frac{3}{4}$. However, we can also use this word to designate an expression such as

$$(3x + 2) \div (x + 4) \quad \text{or} \quad \frac{3x + 2}{x + 4}$$

This last fraction consists of one polynomial divided by another. This is sometimes called a *fractional algebraic expression.* There are other types of fractional algebraic expressions that are even more involved. They are called complex fractions and will be studied in Section 4.4.

There is a special restriction for all fractions. The denominator of the fraction cannot be 0. This restriction is important when working with fractional algebraic expressions. In the expression

$$\frac{3x + 2}{x + 4}$$

the denominator cannot be 0. Therefore, the value of x cannot be -4. To avoid having to mention this repeatedly throughout this chapter we will state the following:

Restriction

The denominator of an algebraic fraction cannot be zero. Any value of the variable that would make the denominator zero is not allowed.

4.1 Simplification of Fractional Algebraic Expressions

We have discovered that fractions can be simplified (or reduced) in the following way:

$$\frac{15}{25} = \frac{3 \cdot \not{5}}{5 \cdot \not{5}} = \frac{3}{5}$$

This is sometimes referred to as the *basic rule of fractions* and can be stated as follows:

Basic Rule of Fractions

For any polynomials a, b, c (where $b \neq 0$ and $c \neq 0$)

$$\frac{ac}{bc} = \frac{a}{b}$$

There is one essential property that is revealed by the basic rule of fractions. It is this: If the numerator and denominator of a given fraction are multiplied by the same quantity, then an equivalent fraction is obtained. The rule can be used two ways. You can start with $\frac{ac}{bc}$ and end with the equivalent fraction $\frac{a}{b}$. You can start with $\frac{a}{b}$ and end with the equivalent fraction $\frac{ac}{bc}$.

EXAMPLE A

1. Write an equivalent fraction for

$$\frac{3}{5}$$

with a denominator of 10:

2. Reduce:

$$\frac{21}{39}$$

Solution

1. $\frac{3}{5} = \frac{3 \cdot 2}{5 \cdot 2} = \frac{6}{10}$

Using the rule $\frac{a}{b} = \frac{ac}{bc}$, where $c = 2$, $a = 3$, $b = 5$

2. $\frac{21}{39} = \frac{7 \cdot 3}{13 \cdot 3} = \frac{7}{13}$

Using the rule $\frac{ac}{bc} = \frac{a}{b}$, where $c = 3$, $a = 7$, $b = 13$

The process of reducing the fraction shown above is often called *canceling common factors*. It can be thought of as a way of writing equivalent fractions. However, only factors of both the numerator and the denominator can be canceled. Now, to apply this rule it is usually necessary that the numerator and denominator of the fraction be completely factored. You will need to use your factoring skills from Chapter 3 to accomplish this step. When you are applying this rule you are *simplifying the fraction.*

EXAMPLE B Simplify:

$$\frac{4x + 12}{5x + 15}$$

Solution

$$= \frac{4(x + 3)}{5(x + 3)}$$

Remove a common factor of 4 from the numerator.

Remove a common factor of 5 from the denominator.

$$= \frac{4\cancel{(x + 3)}}{5\cancel{(x + 3)}}$$

Apply the basic rule of fractions.

$$= \frac{4}{5}$$

EXAMPLE C Simplify:

$$\frac{x^2 + 9x + 14}{x^2 - 4}$$

Solution

$$\frac{(x + 7)(x + 2)}{(x - 2)(x + 2)}$$ Factor the numerator. Factor the denominator.

$$= \frac{(x + 7)\cancel{(x + 2)}}{(x - 2)\cancel{(x + 2)}}$$ Apply the basic rule of fractions.

$$= \frac{x + 7}{x - 2}$$

EXAMPLE D Simplify:

$$\frac{ab + ac}{ax - 3ay}$$

Solution

$$\frac{a(b + c)}{a(x - 3y)}$$ Factor the numerator. Factor the denominator.

$$= \frac{\cancel{a}(b + c)}{\cancel{a}(x - 3y)}$$ Apply the basic rule of fractions.

$$= \frac{b + c}{x - 3y}$$

Some problems may involve more than one step of factoring. Always remember to remove any common factors first.

EXAMPLE E Simplify:

$$\frac{9x - x^3}{x^3 + x^2 - 6x}$$

Solution

$$\frac{x(9 - x^2)}{x(x^2 + x - 6)}$$ Remove a common factor for each polynomial.

$$= \frac{\cancel{x}\cancel{(3 + x)}(3 - x)}{\cancel{x}\cancel{(x + 3)}(x - 2)}$$ Factor each polynomial and apply the basic rule of fractions.

Note that $(3 + x)$ is equivalent to $(x + 3)$ since addition is commutative.

$$= \frac{3 - x}{x - 2}$$

Warning

Now you will need to watch out for a situation where *each term of one factor is opposite in sign* from each term of another factor. In such cases you should factor -1 or another negative number from one polynomial so that the expression in each set of parentheses is equivalent. Look carefully at the following two examples.

EXAMPLE F Simplify:

$$\frac{5x - 15}{6 - 2x}$$

Identify the common factors.

Solution

$$\frac{5(x - 3)}{-2(-3 + x)}$$

Remove a common factor of 5 from the numerator.

Remove a common factor of -2 from the denominator.

Note that $(x - 3)$ and $(-3 + x)$ are equivalent (since $+x - 3 = -3 + x$!).

$$= \frac{5\cancel{(x - 3)}}{-2\cancel{(-3 + x)}}$$

Apply the basic rule of fractions.

$$= \frac{5}{-2} \text{ or } -\frac{5}{2} \text{ or } \frac{-5}{2}$$

Note that the negative sign in the answer can be placed in any of three positions. The three fractions are equivalent.

EXAMPLE G Simplify:

$$\frac{2x^2 - 11x + 12}{16 - x^2}$$

Solution

$$\frac{(x - 4)(2x - 3)}{(4 - x)(4 + x)}$$

Factor numerator and denominator. Observe that $(x - 4)$ and $(4 - x)$ are opposite in sign.

$$= \frac{(x - 4)(2x - 3)}{-1(-4 + x)(4 + x)}$$

Factor -1 out of $(+4 - x)$ to obtain $-1(-4 + x)$.

$$= \frac{\cancel{(x - 4)}(2x - 3)}{-1\cancel{(-4 + x)}(4 + x)}$$

Since $(x - 4)$ and $(-4 + x)$ are equivalent, we apply the basic rule of fractions.

$$= \frac{2x - 3}{-1(4 + x)} \text{ or } -\frac{2x - 3}{4 + x}$$

Note that the answer can be expressed in a variety of ways. Each of them is correct. Do you see that each of them is equivalent?

$$\text{or } \frac{-2x + 3}{4 + x} \text{ or } \frac{3 - 2x}{4 + x}$$

Some problems will involve two or more variables. In such cases, you will need to factor carefully and make sure each set of parentheses contains the correct letters.

EXAMPLE H Simplify:

$$\frac{x^2 - 7xy + 12y^2}{2x^2 - 7xy - 4y^2}$$

Solution

$$\frac{(x - 4y)(x - 3y)}{(2x + y)(x - 4y)}$$

Factor the numerator.

Factor the denominator.

$$= \frac{\cancel{(x - 4y)}(x - 3y)}{(2x + y)\cancel{(x - 4y)}}$$ Apply the basic rule of fractions.

$$= \frac{x - 3y}{2x + y}$$

EXAMPLE I Simplify:

$$\frac{6a^2 + ab - 7b^2}{36a^2 - 49b^2}$$

Solution

$$\frac{(6a + 7b)(a - b)}{(6a + 7b)(6a - 7b)}$$ Factor the numerator. Factor the denominator.

$$= \frac{\cancel{(6a + 7b)}(a - b)}{\cancel{(6a + 7b)}(6a - 7b)}$$ Apply the basic rule of fractions.

$$= \frac{a - b}{6a - 7b}$$

Review Examples A to I and then try these practice problems.

Practice Problems for Exercise 4.1

Simplify.

1. $\dfrac{4x - 6}{2x^2 - x - 3}$ 2. $\dfrac{x^3 + 11x^2 + 30x}{3x^3 + 17x^2 - 6x}$ 3. $\dfrac{4x^2 + 3x - 10}{25 - 16x^2}$

Now turn to the answers following Exercise 4.1 and check your solutions for these practice problems.

Exercise 4.1

Simplify.

1. $\dfrac{6x - 3y}{2x - y}$ 2. $\dfrac{7a + 2b}{14a + 4b}$ 3. $\dfrac{x^2}{3x^2 - x^3}$

4. $\dfrac{xy - 2x}{3y - 6}$ 5. $\dfrac{2x - 8}{x^2 - 8x + 16}$ 6. $\dfrac{4x^2 + 4x + 1}{1 - 4x^2}$

7. $\dfrac{x^2 - 9y^2}{x + 3y}$

8. $\dfrac{xy(x + y^2)}{x^2y^2}$

9. $\dfrac{6x^2}{2x(x - 3y)}$

10. $\dfrac{x^2 + 3x - 4}{x^2 - 1}$

11. $\dfrac{x^2 + x - 2}{x^2 - x}$

12. $\dfrac{x^2 - x - 6}{x^2 - 5x + 6}$

13. $\dfrac{x^2 + 2x - 3}{4x^2 - 5x + 1}$

14. $\dfrac{4x^2 - 10x + 6}{2x^2 + x - 3}$

15. $\dfrac{3x^2 - 8x + 5}{4x^2 - 5x + 1}$

16. $\dfrac{5y^2 + 13y - 6}{5y^2 + 18y - 8}$

17. $\dfrac{2x^2 - 5x - 12}{2x^2 - x - 6}$

18. $\dfrac{2x^2 + 2x - 12}{x^2 + 3x - 4}$

19. $\dfrac{2x^3 + 11x^2 + 5x}{2x^3 - 5x^2 - 3x}$

20. $\dfrac{6x^3 + 5x^2 - 6x}{6x^3 - x^2 - 2x}$

21. $\dfrac{6 - 3x}{2x - 4}$

22. $\dfrac{5 - ay}{ax^2y - 5x^2}$

23. $\dfrac{2x^2 - 7x - 15}{25 - x^2}$

24. $\dfrac{49 - x^2}{2x^2 - 9x - 35}$

25. $\dfrac{(4x+5)^2}{8x^2+6x-5}$

26. $\dfrac{6x^2-13x-8}{(3x-8)^2}$

27. $\dfrac{2y^2-5y-12}{8+2y-y^2}$

28. $\dfrac{6+13y-5y^2}{2y^2-5y-3}$

29. $\dfrac{a^2+2ab-3b^2}{2a^2+5ab-3b^2}$

30. $\dfrac{a^2+3ab-10b^2}{3a^2-7ab+2b^2}$

31. $\dfrac{16x^2-25y^2}{4x^2+3xy-10y^2}$

32. $\dfrac{6x^2-11xy+4y^2}{9x^2-16y^2}$

Challenge Problems

33. $\dfrac{6x^4-9x^3-6x^2}{12x^3+42x^2+18x}$

34. $\dfrac{xa-yb-ya+xb}{xa-ya+2bx-2by}$

Solutions to Practice Problems for Exercise 4.1

1. $\dfrac{4x-6}{2x^2-x-3}$ — Note the common factor in the numerator.

$= \dfrac{2(2x-3)}{(2x-3)(x+1)}$ — Factor the numerator and denominator and apply the basic rule of fractions.

$= \dfrac{2}{x+1}$

2. $\dfrac{x^3+11x^2+30x}{3x^3+17x^2-6x}$ — Note a common factor of x in both numerator and denominator.

$= \dfrac{x(x^2+11x+30)}{x(3x^2+17x-6)}$ — Remove the common factor in each polynomial.

$= \dfrac{x(x+5)(x+6)}{x(3x-1)(x+6)}$ — Continue factoring and use the basic rule of fractions.

$= \dfrac{x+5}{3x-1}$

3. $\dfrac{4x^2 + 3x - 10}{25 - 16x^2}$

$= \dfrac{(4x - 5)(x + 2)}{(5 - 4x)(5 + 4x)}$

Factor both numerator and denominator. Note that each term of $(4x - 5)$ is opposite in sign to $(5 - 4x)$.

$= \dfrac{(4x - 5)(x + 2)}{-1(-5 + 4x)(5 + 4x)}$

Remove a common factor of -1 from $(5 - 4x)$.

$= \dfrac{\cancel{(4x - 5)}(x + 2)}{-1\cancel{(-5 + 4x)}(5 + 4x)}$

Since $(4x - 5)$ is equivalent to $(-5 + 4x)$, we can apply the basic rule of fractions.

$= \dfrac{x + 2}{-1(5 + 4x)}$ or $\dfrac{x + 2}{-5 - 4x}$

Note the different ways that the answer can be expressed. Examine each one to make sure that you see why they are equivalent!

or $-\dfrac{x + 2}{5 + 4x}$

Now do the problems in Exercise 4.1.

4.2 Multiplication and Division of Fractional Algebraic Expressions

> **For any two fractions with polynomials a, b, c, d where $b \neq 0$, $d \neq 0$**
>
> $$\frac{a}{b} \text{ and } \frac{c}{d}$$
>
> **the definition of multiplication of these two fractions is**
>
> $$\frac{a}{b} \cdot \frac{c}{d} = \frac{ac}{bd}$$

Simplifying or reducing fractions *prior to multiplying them* usually makes the problem easier to do. To do the problem the long way would certainly increase the chance for error! This long approach should be avoided.

EXAMPLE A Multiply using the "long way":

$$\frac{5}{7} \times \frac{49}{125}$$

Solution

$\dfrac{245}{875}$

Multiply numerators and multiply denominators.

$= \dfrac{7}{25}$

Reduce the fraction. (*Note:* It takes a bit of trial and error to discover how to reduce it.)

EXAMPLE B Multiply by simplifying *before* multiplication:

$$\frac{5}{7} \times \frac{49}{125}$$

Solution

Step 1 $\frac{5}{7} \times \frac{7 \cdot 7}{5 \cdot 5 \cdot 5} =$

It is easier to factor first. We factor the numerator and denominator of the second fraction and

Step 2 $\frac{5 \cdot 7 \cdot 7}{7 \cdot 5 \cdot 5 \cdot 5} =$

express it as one fraction (by the definition of multiplication of fractions).

Step 3 $\frac{\cancel{5} \cdot \cancel{7} \cdot 7}{\cancel{7} \cdot \cancel{5} \cdot 5 \cdot 5} = \frac{7}{25}$

Then we apply the basic rule of fractions to cancel out common factors that appear in the denominator and in the numerator.

In practice, some of these steps of Example B can be eliminated. When multiplying fractions we can directly cancel a factor of the numerator of one fraction with a factor of the denominator of any fraction.

EXAMPLE C Multiply:

$$\frac{6}{14} \times \frac{21}{10} \times \frac{25}{9}$$

Solution

$\frac{3 \cdot 2}{2 \cdot 7} \times \frac{7 \cdot 3}{5 \cdot 2} \times \frac{5 \cdot 5}{3 \cdot 3}$

Factor each numerator and denominator where possible.

$= \frac{\cancel{3} \cdot \cancel{2}}{\cancel{2} \cdot \cancel{7}} \times \frac{\cancel{7} \cdot \cancel{3}}{\cancel{5} \cdot 2} \times \frac{\cancel{5} \cdot 5}{\cancel{3} \cdot \cancel{3}}$

Now cancel a factor from the numerator of any fraction with a factor from the denominator of any fraction.

$= \frac{5}{2}$

The final answer.

A similar approach can be used with the multiplication of rational algebraic expressions. We first factor the numerator and denominator of each fraction wherever possible. Then we cancel any factor that is common to a numerator and a denominator. Finally, we multiply the remaining numerators and the remaining denominators.

EXAMPLE D Multiply:

$$\frac{x^2 - x - 12}{x^2 - 16} \cdot \frac{2x^2 + 7x - 4}{x^2 - 4x - 21}$$

Solution

$\frac{(x - 4)(x + 3)}{(x - 4)(x + 4)} \cdot \frac{(x + 4)(2x - 1)}{(x + 3)(x - 7)}$

We first factor wherever possible.

$= \frac{\cancel{(x - 4)}\cancel{(x + 3)}}{\cancel{(x - 4)}\cancel{(x + 4)}} \cdot \frac{\cancel{(x + 4)}(2x - 1)}{\cancel{(x + 3)}(x - 7)}$

Now there are three times we can cancel a factor from a numerator with the same factor in a denominator. It helps to use a different type of line each time you cancel.

$= \frac{2x - 1}{x - 7}$

The final answer.

In some cases, a given numerator can be factored more than once. You should always be sure that you remove the *common factor first* wherever possible.

EXAMPLE E Multiply:

$$\frac{x^4 - 16}{x^3 + 4x} \cdot \frac{2x^2 - 8x}{4x^2 + 2x - 12}$$

Solution

$$\frac{(x^2 + 4)(x^2 - 4)}{x(x^2 + 4)} \cdot \frac{2x(x - 4)}{2(2x^2 + x - 6)}$$

Do one step of factoring for each numerator and denominator.

$$= \frac{(x^2 + 4)(x + 2)(x - 2)}{x(x^2 + 4)} \cdot \frac{2x(x - 4)}{2(x + 2)(2x - 3)}$$

Do a second step of factoring where possible.

$$= \frac{\cancel{(x^2 + 4)}\cancel{(x + 2)}(x - 2)}{\cancel{x}\cancel{(x^2 + 4)}} \cdot \frac{\cancel{2}\cancel{x}(x - 4)}{\cancel{2}\cancel{(x + 2)}(2x - 3)}$$

Cancel a factor that appears in both numerator and denominator. (There are four such pairs to be canceled.)

$$= \frac{(x - 2)(x - 4)}{(2x - 3)} \quad \text{or} \quad \frac{x^2 - 6x + 8}{2x - 3}$$

Write the answer as one fraction. (Usually, if there is more than one set of parentheses in a numerator, the answer is left in factored form.)

For any two fractions $\frac{a}{b}$ and $\frac{c}{d}$ the operation of division can be performed by inverting the second fraction and multiplying it by the first fraction.

> **The definition for division of fractions is**
>
> $$\frac{a}{b} \div \frac{c}{d} = \frac{a}{b} \cdot \frac{d}{c}$$

This property holds whether a, b, c, d are polynomials or numerical values. (It is assumed, of course, that no denominator is zero.)

In the first step of dividing two algebraic fractions, you should invert the second fraction and write the problem as one of multiplication. Then you follow the procedure for multiplying algebraic fractions.

EXAMPLE F Divide:

$$\frac{6x + 12y}{2x - 6y} \div \frac{9x^2 - 36y^2}{4x^2 - 36y^2}$$

Solution

$$\frac{6x + 12y}{2x - 6y} \cdot \frac{4x^2 - 36y^2}{9x^2 - 36y^2}$$

Invert the second fraction and write the problem as the product of two fractions.

$$= \frac{6(x + 2y)}{2(x - 3y)} \cdot \frac{4(x^2 - 9y^2)}{9(x^2 - 4y^2)}$$

Do one step of factoring in each numerator and denominator.

$$= \frac{(3)(2)(x + 2y)}{2(x - 3y)} \cdot \frac{(2)(2)(x + 3y)(x - 3y)}{(3)(3)(x + 2y)(x - 2y)}$$

Do a second step of factoring where possible.

$$= \frac{(\cancel{3})(2)(\cancel{x+2y})}{2(\cancel{x-3y})} \cdot \frac{(2)(2)(x+3y)(\cancel{x-3y})}{(\cancel{3})(3)(\cancel{x+2y})(x-2y)}$$

Cancel a factor that appears in both numerator and denominator.

$$= \frac{(2)(2)(x+3y)}{3(x-2y)}$$

Write the result as one fraction.

$$= \frac{4(x+3y)}{3(x-2y)}$$

Simplify. Usually, answers are left in this form.

A polynomial that is not in fraction form can be written as a fraction if you write a denominator of 1.

EXAMPLE G Divide:

$$\frac{15-3x}{x+6} \div (x^2-9x+20)$$

Note that $x^2 - 9x + 20$ can be written as $\frac{x^2-9x+20}{1}$.

Solution

$$\frac{15-3x}{x+6} \cdot \frac{1}{x^2-9x+20}$$

Invert and multiply.

$$= \frac{-3(-5+x)}{x+6} \cdot \frac{1}{(x-5)(x-4)}$$

Factor where possible. Note we had to factor -3 from the first numerator so that we would have a common factor with the second denominator.

$$= \frac{-3(\cancel{-5+x})}{x+6} \cdot \frac{1}{(\cancel{x-5})(x-4)}$$

Cancel the common factor. $(-5+x)$ is equivalent to $(+x-5)$.

$$= \frac{-3}{(x+6)(x-4)}$$

The final answer.

$$\text{or} \quad -\frac{3}{(x+6)(x-4)}$$

$$\text{or} \quad \frac{3}{(x+6)(4-x)}$$

However, the answer can be written in several equivalent forms. Take a minute to study each one. Do you see that each one is equivalent to the others?

Now try these practice problems.

Practice Problems for Exercise 4.2

Multiply.

1. $\dfrac{10x}{x^2-7x+10} \cdot \dfrac{x^2+3x-10}{25x}$

2. $\dfrac{3y+3}{2y-8} \cdot \dfrac{2y^2-6y-8}{y^2-y-2} \cdot \dfrac{y^2-5y+6}{2y^2-4y-6}$

3. Divide:

$$\frac{x^2+5x+6}{x^2+8x} \div \frac{2x^2+5x+2}{2x^2+x}$$

Now turn to the answers following Exercise 4.2 and check your solutions for these practice problems.

Exercise 4.2

Perform the indicated operation.

1. $\dfrac{x+3}{x+7}\cdot\dfrac{x^2+3x-10}{x^2+x-6}$

2. $\dfrac{5x^2-5}{3x+3}\cdot\dfrac{3x+6}{5x^2-15x+10}$

3. $\dfrac{x^2+2x}{6x}\cdot\dfrac{3x^2}{x^2-4}$

4. $\dfrac{3x+12}{8x^3}\cdot\dfrac{16x^2}{9x+36}$

5. $\dfrac{x^2-x-6}{x^2-2x-8}\cdot\dfrac{x^2+7x+12}{x^2-9}$

6. $\dfrac{x^2-x-6}{x^2+x-12}\cdot\dfrac{x^2+3x-4}{x^2+2x-3}$

7. $\dfrac{x^2+7x-8}{2x^2-18}\cdot\dfrac{2x^2+20x+42}{7x^2-7x}$

8. $\dfrac{3x^2-3x-18}{x^2+2x-15}\cdot\dfrac{2x^2+6x-20}{2x^2-12x+16}$

9. $\dfrac{5x^2+6x+1}{x^2+5x+6}\div(5x+1)$

10. $(3x-2)\div\dfrac{9x^2-4}{6x+4}$

11. $\dfrac{x^2 + 2xy + y^2}{x^2 - 2xy + y^2} \div \dfrac{3x + 3y}{4x - 4y}$

12. $\dfrac{3x^2 + 8x - 3}{15x^2} \div \dfrac{3x - 1}{5x^3}$

13. $\dfrac{xy - y^2}{x^2 + 2x + 1} \div \dfrac{2x^2 + xy - 3y^2}{2x^2 + 5xy + 3y^2}$

14. $\dfrac{4a^2 - b^2}{a^2 + 4ab + 4b^2} \div \dfrac{4a - 2b}{3a + 6b}$

15. $\dfrac{x^2 + 5x - 14}{x - 5} \div \dfrac{x^2 + 12x + 35}{15 - 3x}$

16. $\dfrac{3x^2 + 13x + 4}{16 - x^2} \div \dfrac{3x^2 - 5x - 2}{3x - 12}$

17. $\dfrac{(x + 4)^2}{2x^2 - 7x - 15} \cdot \dfrac{(x - 5)^2}{x + 4}$

18. $\dfrac{3x^2 - 10x - 8}{(4x + 5)^2} \cdot \dfrac{4x + 5}{(x - 4)^2}$

19. $\dfrac{3x^2 - 7x + 2}{3x^2 + 2x - 1} \cdot \dfrac{2x^2 - 9x - 5}{x^2 + x - 6} \cdot \dfrac{4x^2 + 11x - 3}{x^2 - 11x + 30}$

20. $\dfrac{6a^2 - a - 1}{4a^2 - 1} \cdot \dfrac{10a^2 + a - 2}{3a^2 - 17a - 6} \cdot \dfrac{a^2 - 3a - 18}{5a^2 + 3a - 2}$

Challenge Problems

21. $\left(\dfrac{6xy}{3x - 12y} \cdot \dfrac{y^2 - 5xy}{4x - 4y}\right) \div \dfrac{5x - y}{x^2 - 5xy + 4y^2}$

22. $\dfrac{2y + 10}{2y + 20} \div \left(\dfrac{10y}{y^2 + 15y + 50} \cdot \dfrac{xy^2 + 10xy + 25x}{xy^2 + 10xy}\right)$

Solutions to Practice Problems for Exercise 4.2

1. $\dfrac{10x}{x^2 - 7x + 10} \cdot \dfrac{x^2 + 3x - 10}{25x}$

$\dfrac{(5)(2)(x)}{(x - 5)(x - 2)} \cdot \dfrac{(x + 5)(x - 2)}{(5)(5)(x)}$ Factor each numerator and denominator.

$\dfrac{(\cancel{5})(2)(\cancel{x})}{(x - 5)\cancel{(x - 2)}} \cdot \dfrac{(x + 5)\cancel{(x - 2)}}{(\cancel{5})(5)(\cancel{x})}$ Cancel factors common to a numerator and a denominator.

Note: We cannot cancel $(x + 5)$ with $(x - 5)$

$= \dfrac{2(x + 5)}{5(x - 5)}$ We leave the answer in factored form.

2. $\dfrac{3y + 3}{2y - 8} \cdot \dfrac{2y^2 - 6y - 8}{y^2 - y - 2} \cdot \dfrac{y^2 - 5y + 6}{2y^2 - 4y - 6}$

$= \dfrac{3(y + 1)}{2(y - 4)} \cdot \dfrac{2(y^2 - 3y - 4)}{(y - 2)(y + 1)} \cdot \dfrac{(y - 3)(y - 2)}{2(y^2 - 2y - 3)}$ Perform one step of factoring.

$= \dfrac{3\cancel{(y + 1)}}{\cancel{2}\cancel{(y - 4)}} \cdot \dfrac{\cancel{2}\cancel{(y - 4)}\cancel{(y + 1)}}{\cancel{(y - 2)}\cancel{(y + 1)}} \cdot \dfrac{\cancel{(y - 3)}\cancel{(y - 2)}}{2\cancel{(y - 3)}\cancel{(y + 1)}}$ Factor one additional step wherever possible and cancel when factor is common to a numerator and a denominator.

$= \dfrac{3}{2}$ or $1\frac{1}{2}$ The final answer.

3. $\dfrac{x^2 + 5x + 6}{x^2 + 8x} \div \dfrac{2x^2 + 5x + 2}{2x^2 + x}$

$= \dfrac{x^2 + 5x + 6}{x^2 + 8x} \cdot \dfrac{2x^2 + x}{2x^2 + 5x + 2}$ Invert and multiply.

$= \dfrac{(x + 3)\cancel{(x + 2)}}{\cancel{x}(x + 8)} \cdot \dfrac{\cancel{x}\cancel{(2x + 1)}}{\cancel{(2x + 1)}\cancel{(x + 2)}}$ Factor and cancel.

$= \dfrac{x + 3}{x + 8}$ The final answer.

Now do the problems in Exercise 4.2.

Warning

You need to be very clear on how *to find a least common denominator* and how to add and subtract fractions from arithmetic *before* you attempt this section. Please review Sections 0.1 and 0.2 if you have any weaknesses in this topic. Then you will be ready to study Section 4.3.

4.3 Addition and Subtraction of Fractional Algebraic Expressions

Fractional algebraic expressions with the same denominator can be combined in a fashion similar to that used in arithmetic. The numerators are added or subtracted and the denominator remains the same.

For any polynomials a, b, c,

$$\frac{a}{b} + \frac{c}{b} = \frac{a + c}{b} \qquad \text{where } b \neq 0$$

EXAMPLE A Add:

$$\frac{5a}{a + 2b} + \frac{6a}{a + 2b}$$

Solution

$$= \frac{5a + 6a}{a + 2b} = \frac{11a}{a + 2b}$$

For any polynomials a, b, c,

$$\frac{a}{b} - \frac{c}{b} = \frac{a - c}{b} \qquad \text{where } b \neq 0$$

EXAMPLE B Subtract:

$$\frac{3x}{(x + y)(x - 2y)} - \frac{8x}{(x + y)(x - 2y)}$$

Solution

$$\frac{3x - 8x}{(x + y)(x - 2y)}$$

$$= \frac{-5x}{(x + y)(x - 2y)}$$

When the denominators are not the same, we must find the *least common denominator* (LCD). Each fraction that does not have the LCD for a denominator must be rewritten as an equivalent fraction with the LCD as the denominator. The LCD for algebraic fractions is found in a fashion very similar to that method used in Section 0.2.

Procedure to Find the LCD

1. **Factor each denominator completely.**
2. **The LCD is the product of each *different factor*.**
3. **If a factor occurs more than once in any one denominator the LCD will contain that factor repeated the greatest number of times that it occurs in any one denominator.**

EXAMPLE C Add:

$$\frac{5}{xy} + \frac{2}{y}$$

The two factors are x and y. We observe that the LCD is xy.

Solution $\frac{5}{xy} + \frac{2}{y} \cdot \frac{x}{x}$ We multiply the second fraction by $\frac{x}{x}$.

$= \frac{5}{xy} + \frac{2x}{xy}$ Now each fraction has a common denominator of xy.

$= \frac{5 + 2x}{xy}$ We write the sum as one fraction.

EXAMPLE D Add:

$$\frac{3x}{(x + y)(x - y)} + \frac{5}{x + y}$$

The two factors are $(x + y)$ and $(x - y)$. We observe that the LCD $= (x + y)(x - y)$.

Solution $\frac{3x}{(x + y)(x - y)} + \frac{5}{x + y} \cdot \frac{x - y}{x - y}$ We multiply the second fraction by $\frac{x - y}{x - y}$.

$= \frac{3x}{(x + y)(x - y)} + \frac{5x - 5y}{(x + y)(x - y)}$ Now each fraction has a common denominator of $(x + y)(x - y)$.

$= \frac{3x + 5x - 5y}{(x + y)(x - y)}$ We write the sum of the numerators over one common denominator.

$= \frac{8x - 5y}{(x + y)(x - y)}$ Collect like terms.

It is important to remember that the LCD is the smallest algebraic expression into which each denominator can be divided. For algebraic expressions the LCD must contain *each factor* that appears in any denominator. If the factor is repeated, the LCD must contain each factor the greatest number of times that it appears in any one denominator.

EXAMPLE E Add:

$$\frac{5}{xy^2} + \frac{3}{x^2y} + \frac{2}{y^3}$$

The x factor is squared in one fraction. The y factor is cubed in one fraction. Therefore, the LCD is x^2y^3.

Solution $= \frac{5}{xy^2} \cdot \frac{xy}{xy} + \frac{3}{x^2y} \cdot \frac{y^2}{y^2} + \frac{2}{y^3} \cdot \frac{x^2}{x^2}$ We multiply each fraction by the appropriate value to obtain x^2y^3 in each denominator. *Note:* Here we *always* multiply the numerator and denominator of any one fraction by the *same value* to obtain an equivalent fraction.

$= \frac{5xy}{x^2y^3} + \frac{3y^2}{x^2y^3} + \frac{2x^2}{x^2y^3}$ Now all fractions have a common denominator.

$= \frac{5xy + 3y^2 + 2x^2}{x^2y^3}$ We write the sum as one fraction.

In many cases, the denominators are not in factored form. You must factor each denominator in order to determine the LCD. When you have collected like terms in the numerator, look to see if that final numerator can be factored. If such cases, the fraction can sometimes be simplified.

EXAMPLE F Add:

$$\frac{5}{x^2 - y^2} + \frac{3x}{x^3 + x^2y}$$

We must factor the two denominators.

Solution

$$\frac{5}{(x + y)(x - y)} + \frac{3x}{x^2(x + y)}$$

We observe that the LCD is $x^2(x + y)(x - y)$.

$$= \frac{5}{(x + y)(x - y)} \cdot \frac{x^2}{x^2} + \frac{3x}{x^2(x + y)} \cdot \frac{x - y}{x - y}$$

We multiply each fraction by the appropriate value to obtain a common denominator of $x^2(x + y)(x - y)$.

$$= \frac{5x^2}{x^2(x + y)(x - y)} + \frac{3x^2 - 3xy}{x^2(x + y)(x - y)}$$

$$= \frac{5x^2 + 3x^2 - 3xy}{x^2(x + y)(x - y)}$$

We write the sum of the numerators over one common denominator.

$$= \frac{8x^2 - 3xy}{x^2(x + y)(x - y)}$$

Collect like terms.

$$= \frac{\cancel{x}(8x - 3y)}{\cancel{x^2}(x + y)(x - y)}$$
x

Remove the common factor x in the numerator. Then cancel the common factor x in both numerator and denominator.

$$= \frac{8x - 3y}{x(x + y)(x - y)}$$

Warning

It is very easy to make a mistake in sign when subtracting two fractions. You will find it helpful to place parentheses around the numerator of the second fraction so that you will not forget to subtract the entire numerator.

EXAMPLE G Subtract:

$$\frac{3x + 4}{x - 2} - \frac{x - 3}{2x - 4}$$

We factor the second denominator.

Solution

$$\frac{3x + 4}{x - 2} - \frac{x - 3}{2(x - 2)}$$

We observe that the LCD is $2(x - 2)$.

$$= \frac{2}{2} \cdot \frac{(3x + 4)}{x - 2} - \frac{x - 3}{2(x - 2)}$$

We multiply the first fraction by $\frac{2}{2}$ so that the resulting fraction will have the common denominator.

$$= \frac{2(3x + 4) - (x - 3)}{2(x - 2)}$$

We write the indicated subtraction as one fraction. Note the parentheses around $(x - 3)$.

$$= \frac{6x + 8 - x + 3}{2(x - 2)}$$

Remove the parentheses in the numerator.

$$= \frac{5x + 11}{2(x - 2)}$$

Collect like terms.

Warning

Students often cancel before adding and subtracting the fractions. This is totally wrong. You should cancel any common factors only at the very last step of the problem, _after_ you have added or subtracted the numerators. Remember that you cannot cancel unless the value you cancel is a _factor_ of the entire numerator and the entire denominator. You cannot cancel terms. You can only cancel factors.

EXAMPLE H Subtract:

$$\frac{8x}{x^2 - 16} - \frac{4}{x - 4}$$

We factor the first denominator.

Solution

$$\frac{8x}{(x + 4)(x - 4)} - \frac{4}{x - 4}$$

We observe that the LCD is $(x + 4)(x - 4)$.

$$= \frac{8x}{(x + 4)(x - 4)} - \frac{4}{x - 4} \cdot \frac{x + 4}{x + 4}$$

We multiply the second fraction by $\frac{x + 4}{x + 4}$.

Warning

The term $(x + 4)$ is _not_ a factor of the _entire_ numerator! Be sure that you do _not_ cancel anything here. You cannot cancel terms. You can only cancel factors.

$$= \frac{8x - 4(x + 4)}{(x + 4)(x - 4)}$$

We are writing the subtraction of the numerators over one common denominator.

$$= \frac{8x - 4x - 16}{(x + 4)(x - 4)}$$

Remove parentheses.

$$= \frac{4x - 16}{(x + 4)(x - 4)}$$

Collect like terms. Notice that the numerator can be factored.

$$= \frac{4\cancel{(x - 4)}}{(x + 4)\cancel{(x - 4)}}$$

Since $(x - 4)$ is a *factor* of the numerator *and* the denominator, we may cancel.

$$= \frac{4}{x + 4}$$

The final answer.

Carefully review Examples A to H and then try these practice problems.

Practice Problems for Exercise 4.3

1. Add:

$$\frac{3}{x + 2} + \frac{5}{2x}$$

2. Add:

$$\frac{x + 6}{x - 2} + \frac{2x + 1}{x + 3}$$

3. Subtract:

$$\frac{8x}{x^2 - 36} - \frac{4}{x + 6}$$

Now turn to the answers following Exercise 4.3 and check your solutions for these practice problems.

Exercise 4.3

Perform the indicated operation. Be sure to simplify your answer.

1. $\dfrac{5}{y^2} + \dfrac{3}{y^3}$

2. $\dfrac{8}{xy} + \dfrac{2}{xy^2}$

3. $\dfrac{2}{x + 2} + \dfrac{3}{x^2 - 4}$

4. $\dfrac{5}{x^2 + 5x + 6} + \dfrac{2}{x + 3}$

5. $\dfrac{3y}{y + 2} + \dfrac{y}{y - 2}$

6. $\dfrac{2}{y - 1} + \dfrac{2}{y + 1}$

7. $\dfrac{4}{a + 3} + \dfrac{2}{3a}$

8. $\dfrac{5}{2ab} + \dfrac{1}{2a + b}$

9. $\dfrac{2}{3xy} + \dfrac{1}{6yz}$

10. $\dfrac{x - 3}{4x} + \dfrac{6}{x^2}$

11. $\dfrac{1}{x^2 + 3x + 2} + \dfrac{1}{x^2 + 4x + 3}$

12. $\dfrac{5}{x^2 - 25} + \dfrac{1}{x^2 + 10x + 25}$

13. $\dfrac{a + b}{3} - \dfrac{a - 2b}{4}$

14. $\dfrac{a + 1}{2} - \dfrac{a - 1}{3}$

15. $\dfrac{a - 1}{4a} - \dfrac{2a + 3}{a}$

16. $\dfrac{4x - 9}{3x} - \dfrac{3x - 8}{4x}$

17. $\dfrac{8}{2x - 3} - \dfrac{6}{x + 2}$

18. $\dfrac{6}{3x - 4} - \dfrac{5}{4x - 3}$

19. $\dfrac{x}{x^2 + 2x - 3} - \dfrac{x}{x^2 - 5x + 4}$

20. $\dfrac{1}{x^2 - 2x} - \dfrac{5}{x^2 - 4x + 4}$

21. $\dfrac{3y}{8y^2 + 2y - 1} - \dfrac{5y}{2y^2 - 9y - 5}$

22. $\dfrac{2x}{x^2 + 5x + 6} - \dfrac{x + 1}{x^2 + 2x - 3}$

23. $\dfrac{5}{x} + \dfrac{x}{x-5} + \dfrac{3}{5x}$

24. $\dfrac{3}{ab} + \dfrac{2}{a-b} + \dfrac{a}{b}$

25. $\dfrac{y-6}{y+6} - \dfrac{y+6}{y-6}$

26. $\dfrac{x}{x-5} - \dfrac{2x}{5-x}$

27. $\dfrac{4y}{y^2+4y+3} + \dfrac{2}{y+1}$

28. $\dfrac{y-23}{y^2-y-20} + \dfrac{2}{y-5}$

29. $\dfrac{x}{x-1} + \dfrac{x}{x+1} + \dfrac{2}{x^2-1}$

30. $\dfrac{2}{x^2-9} + \dfrac{x}{x+3} + \dfrac{2x}{x-3}$

31. $\dfrac{3x}{x^2+3x-10} + \dfrac{5}{4-2x}$

32. $\dfrac{2y}{3y^2-8y-3} + \dfrac{1}{6y-2y^2}$

33. $\dfrac{5}{3x-6} - \dfrac{2}{4x-8}$

34. $\dfrac{12}{15x-3} - \dfrac{1}{10x-2}$

35. $\dfrac{2x - 1}{x^2 + 6x + 9} - \dfrac{3}{x^2 + 4x + 3}$

36. $\dfrac{2}{x^2 + 4x + 4} - \dfrac{2x - 1}{x^2 + 5x + 6}$

37. $\dfrac{1}{x - 3y} + \dfrac{1}{x - 4y} + \dfrac{y}{x^2 - 7xy + 12y^2}$

38. $\dfrac{42x}{y^2 - 49} + \dfrac{3x}{y - 7} + \dfrac{3x}{y + 7}$

Challenge Problems

39. $\dfrac{3x}{x^2 - 9x + 14} - \dfrac{2x}{x^2 + x - 6} + \dfrac{x}{x^2 - 4x - 21}$

40. $\dfrac{5x}{x^2 + x - 6} + \dfrac{2x + 1}{x^2 + 2x - 3} - \dfrac{1 - 2x}{x^2 - 3x + 2}$

Solutions to Practice Problems for Exercise 4.3

1. $\dfrac{3}{x + 2} + \dfrac{5}{2x}$

Examine the denominators. Note that the factors are 2, x, and $(x + 2)$. Since all factors are different, the LCD = $2(x)(x + 2)$ or $2x(x + 2)$.

$= \dfrac{2x}{2x} \cdot \dfrac{3}{x + 2} + \dfrac{5}{2x} \cdot \dfrac{x + 2}{x + 2}$

We multiply to obtain two fractions with the LCD as the denominator.

$= \dfrac{6x + 5(x + 2)}{2x(x + 2)}$

We write the sum as one fraction.

$= \dfrac{6x + 5x + 10}{2x(x + 2)}$

Remove parentheses in the numerator.

$= \dfrac{11x + 10}{2x(x + 2)}$

Collect like terms.

2. $\dfrac{x + 6}{x - 2} + \dfrac{2x + 1}{x + 3}$

Note that the LCD is $(x - 2)(x + 3)$.

$= \dfrac{x + 6}{x - 2} \cdot \dfrac{x + 3}{x + 3} + \dfrac{2x + 1}{x + 3} \cdot \dfrac{x - 2}{x - 2}$

We multiply the first fraction by $\dfrac{x + 3}{x + 3}$ and the second fraction by $\dfrac{x - 2}{x - 2}$.

$= \dfrac{(x + 6)(x + 3) + (2x + 1)(x - 2)}{(x + 3)(x - 2)}$

Write the sum as one fraction. *Do not* cancel anything! There is no factor common to the entire numerator, so nothing can be canceled!

$= \dfrac{x^2 + 9x + 18 + 2x^2 - 3x - 2}{(x + 3)(x - 2)}$ — Remove the parentheses in the numerator.

$= \dfrac{3x^2 + 6x + 16}{(x + 3)(x - 2)}$ — Collector like terms in the numerator.

3. $\dfrac{8x}{x^2 - 36} - \dfrac{4}{x + 6}$ — We need to factor the first denominator.

$= \dfrac{8x}{(x + 6)(x - 6)} - \dfrac{4}{x + 6}$ — We observe that the LCD is $(x + 6)(x - 6)$.

$= \dfrac{8x}{(x + 6)(x - 6)} - \dfrac{4}{x + 6} \cdot \dfrac{x - 6}{x - 6}$ — We multiply the second fraction by $\dfrac{x - 6}{x - 6}$.

$= \dfrac{8x - 4(x - 6)}{(x + 6)(x - 6)}$ — Write the subtraction as one fraction. Do *not* cancel anything. $(x - 6)$ is not a common factor for the entire numerator!

$= \dfrac{8x - 4x + 24}{(x + 6)(x - 6)}$ — Remove the parentheses in the numerator.

$= \dfrac{4x + 24}{(x + 6)(x - 6)}$ — Collect like terms. Note that we are *not* done since the answer can be simplified.

$= \dfrac{4\cancel{(x + 6)}}{\cancel{(x + 6)}(x - 6)}$ — Factor the numerator. Note that $(x + 6)$ is a factor of the entire numerator and a factor of the entire denominator.

$= \dfrac{4}{x - 6}$ — We cancel the common factor $(x + 6)$ by the basic property of fractions.

Carefully review these practice problems and their solutions. Then do the problems in Exercise 4.3.

4.4 Complex Fractional Expressions

In Section 4.2 we discovered how to divide one fractional expression by another. However, some problems cannot be done immediately using that approach because they are more involved.

A *complex fractional expression* is one in which there is a fraction in the numerator or in the denominator, or both. For example,

$$\frac{3 + \frac{2}{x}}{\frac{x}{7} + 2} \qquad \frac{\frac{x}{y} + 1}{2} \qquad \frac{\frac{a + b}{3}}{\frac{x - 2y}{4}}$$

are complex fractional expressions.

Problems of this type can be handled in the following way.

Procedure to Simplify a Complex Fraction

1. **Add or subtract so that you have a single fraction in the numerator and in the denominator.**
2. **Divide the fraction in the numerator by the fraction in the denominator. This is done by inverting the fraction in the denominator and multiplying it by the numerator.**

EXAMPLE A Simplify:

$$\frac{\frac{1}{x}}{\frac{2}{y^2}+\frac{1}{y}}$$

Solution

$$\frac{\frac{1}{x}}{\frac{2}{y^2}+\frac{1}{y}\cdot\frac{y}{y}}$$

We want to add the two fractions in the denominator. Therefore we multiply the $\frac{1}{y}$ by $\frac{y}{y}$ so that the two fractions will have a common denominator.

$$\frac{\frac{1}{x}}{\frac{2+y}{y^2}}$$

Add the two fractions in the denominator.

$$\frac{1}{x}\cdot\frac{y^2}{2+y}$$

Invert the fraction in the denominator and multiply it by the numerator. $\frac{a}{b}\div\frac{c}{d}=\frac{a}{b}\cdot\frac{d}{c}$

$$\frac{y^2}{x(2+y)}$$

Write the answer as one fraction.

In some problems there may be two or more fractions in the numerator and the denominator.

EXAMPLE B Simplify:

$$\frac{\frac{1}{x}+\frac{1}{y}}{\frac{3}{a}-\frac{2}{b}}$$

We observe that the LCD of the fractions in the numerator is xy. The LCD of the fractions in the denominator is ab.

Solution

$$\frac{\frac{1}{x}\cdot\frac{y}{y}+\frac{1}{y}\cdot\frac{x}{x}}{\frac{3}{a}\cdot\frac{b}{b}-\frac{2}{b}\cdot\frac{a}{a}}$$

Multiply each fraction by the appropriate value to obtain common denominators.

$$\frac{\dfrac{y + x}{xy}}{\dfrac{3b - 2a}{ab}}$$

Add the two fractions in the numerator.

Subtract the two fractions in the denominator.

$$\frac{y + x}{xy} \cdot \frac{ab}{3b - 2a}$$

Invert the fraction in the denominator and multiply it by the numerator.

$$\frac{ab(y + x)}{xy(3b - 2a)}$$

Write the answer as one fraction.

In some problems, factoring may be necessary to determine the LCD and to combine fractions.

EXAMPLE C Simplify:

$$\frac{\dfrac{1}{x^2 - 1} + \dfrac{2}{x + 1}}{x}$$

We observe the need for factoring $x^2 - 1$.

Solution

$$\frac{\dfrac{1}{(x + 1)(x - 1)} + \dfrac{2}{(x + 1)}}{x}$$

The LCD for the fractions in the numerator is $(x + 1)(x - 1)$.

$$= \frac{\dfrac{1}{(x + 1)(x - 1)} + \dfrac{2}{x + 1} \cdot \dfrac{x - 1}{x - 1}}{x}$$

We multiply the second fraction in the numerator by $\dfrac{x - 1}{x - 1}$.

$$= \frac{\dfrac{1 + 2x - 2}{(x + 1)(x - 1)}}{x}$$

Add the two fractions in the numerator.

$$= \frac{2x - 1}{(x + 1)(x - 1)} \div x$$

Simplify and rewrite the division problem.

$$= \frac{2x - 1}{(x + 1)(x - 1)} \cdot \frac{1}{x}$$

Invert the second fraction and multiply.

$$= \frac{2x - 1}{x(x + 1)(x - 1)}$$

Write the answer as one fraction.

In simplifying complex fractions always be alert to see if the final fraction can be reduced or simplified.

EXAMPLE D Simplify:

$$\frac{\dfrac{3}{a + b} - \dfrac{3}{a - b}}{\dfrac{5}{a^2 - b^2}}$$

The LCD of the two fractions in the numerator is $(a + b)(a - b)$.

Solution

$$\frac{\dfrac{3}{a + b} \cdot \dfrac{a - b}{a - b} - \dfrac{3}{a - b} \cdot \dfrac{a + b}{a + b}}{\dfrac{5}{a^2 - b^2}}$$

We multiply the first fraction by $\dfrac{a - b}{a - b}$. We multiply the second fraction by $\dfrac{a + b}{a + b}$.

$$= \frac{\dfrac{3a - 3b}{(a + b)(a - b)} - \dfrac{3a + 3b}{(a + b)(a - b)}}{\dfrac{5}{a^2 - b^2}}$$

Simplify.

$$= \frac{\dfrac{-6b}{(a + b)(a - b)}}{\dfrac{5}{(a + b)(a - b)}}$$

Study carefully how we combine the two fractions in the numerator. Do you see how we obtained $-6b$?

We factor $a^2 - b^2$ as $(a + b)(a - b)$.

$$= \frac{-6b}{\cancel{(a+b)(a-b)}} \cdot \frac{\cancel{(a+b)(a-b)}}{5}$$

Invert the second fraction and multiply.

$$= \frac{-6b}{5}$$

Note that since $(a + b)(a - b)$ are *factors* in both numerator and denominator, they may be canceled.

Review Examples A to D and then try these practice problems.

Practice Problems for Exercise 4.4

Simplify.

1. $\dfrac{1 + \dfrac{2}{x}}{1 - \dfrac{4}{x^2}}$ **2.** $\dfrac{\dfrac{1}{a} + \dfrac{1}{b}}{ab}$ **3.** $\dfrac{\dfrac{3}{x} + 1}{\dfrac{3 + x}{2}}$

Now turn to the answers following Exercise 4.4 and check your solutions for these practice problems.

Exercise 4.4

Simplify.

1. $\dfrac{\dfrac{3}{x}}{\dfrac{2}{y}}$ **2.** $\dfrac{\dfrac{3}{ab}}{\dfrac{b}{a}}$ **3.** $\dfrac{\dfrac{1}{x} + \dfrac{1}{y}}{\dfrac{1}{xy}}$

4. $\dfrac{\dfrac{1}{x} + 1}{x}$ **5.** $\dfrac{\dfrac{1}{x} + \dfrac{1}{y}}{x^2 - y^2}$ **6.** $\dfrac{\dfrac{1}{x} + \dfrac{1}{y}}{x + y}$

7. $\dfrac{1 - \dfrac{9}{x^2}}{\dfrac{3}{x} + 1}$

8. $\dfrac{\dfrac{4}{x} + 1}{1 - \dfrac{16}{x^2}}$

9. $\dfrac{\dfrac{2}{x + 1} - 2}{3}$

10. $\dfrac{1 - \dfrac{3}{xy}}{x + 2y}$

11. $\dfrac{a + \dfrac{3}{a}}{\dfrac{a^2 + 2}{3}}$

12. $\dfrac{a + \dfrac{1}{a}}{\dfrac{3}{a} - a}$

13. $\dfrac{2 + \dfrac{3y}{x - y}}{2 - \dfrac{3y}{x + 2y}}$

14. $\dfrac{y - \dfrac{4}{y}}{1 + \dfrac{1}{2y + 1}}$

15. $\dfrac{\dfrac{x}{4} - \dfrac{1}{2}}{\dfrac{x}{4} + \dfrac{1}{x}}$

16. $\dfrac{\dfrac{1}{x} - \dfrac{1}{2x}}{\dfrac{1}{2} + \dfrac{1}{2x}}$

17. $\dfrac{\dfrac{1}{x^2 - 9} + \dfrac{2}{x + 3}}{\dfrac{3}{x - 3}}$

18. $\dfrac{\dfrac{5}{x + 4}}{\dfrac{1}{x - 4} - \dfrac{2}{x^2 - 16}}$

Challenge Problems

19. $\dfrac{\dfrac{y}{y + 1} - \dfrac{y - 1}{y}}{\dfrac{y + 1}{y} + \dfrac{y}{y - 1}}$

20. $\dfrac{y + 2 - \dfrac{6}{2y + 3}}{\dfrac{8y}{2y - 1} + y}$

Solutions to Practice Problems for Exercise 4.4

1. $\dfrac{1+\dfrac{2}{x}}{1-\dfrac{4}{x^2}}$

The LCD for the two expressions in the numerator is x; for those in the denominator the LCD is x^2.

$= \dfrac{1\cdot\dfrac{x}{x}+\dfrac{2}{x}}{1\cdot\dfrac{x^2}{x^2}-\dfrac{4}{x^2}}$

Multiply 1 by $\dfrac{x}{x}$ in numerator.

Multiply 1 by $\dfrac{x^2}{x^2}$ in denominator.

$= \dfrac{\dfrac{x+2}{x}}{\dfrac{x^2-4}{x^2}}$

Add two fractions in the numerator and place over the common denominator of x.

Subtract two fractions in the denominator and place over the common denominator of x^2.

$= \dfrac{x+2}{x}\cdot\dfrac{x^2}{x^2-4}$

Invert and multiply the second fraction:

$$\frac{a}{b}\div\frac{c}{d}=\frac{a}{b}\cdot\frac{d}{c}$$

$= \dfrac{\cancel{x+2}}{\cancel{x}}\cdot\dfrac{\cancel{x^2}\,x}{(\cancel{x+2})(x-2)}$

Factor x^2-4. Cancel common factors.

$= \dfrac{x}{x-2}$

The final answer.

2. $\dfrac{\dfrac{1}{a}+\dfrac{1}{b}}{ab}$

Observe that the LCD for the two fractions in the numerator is ab.

$= \dfrac{\dfrac{1}{a}\cdot\dfrac{b}{b}+\dfrac{1}{b}\cdot\dfrac{a}{a}}{ab}$

Multiply $\dfrac{1}{a}$ by $\dfrac{b}{b}$.

Multiply $\dfrac{1}{b}$ by $\dfrac{a}{a}$.

$= \dfrac{\dfrac{b}{ab}+\dfrac{a}{ab}}{ab}$

Simplify.

$= \dfrac{\dfrac{b+a}{ab}}{\dfrac{ab}{1}}$

Write the sum of the two fractions in the numerator.

You can write ab as $\dfrac{ab}{1}$ if you wish.

$= \dfrac{b+a}{ab}\cdot\dfrac{1}{ab}$

Invert and multiply the second fraction.

$= \dfrac{b+a}{a^2b^2}$

Simplify.

3. $$\frac{\frac{3}{x}+1}{\frac{3+x}{2}}$$

Observe that the LCD of the two expressions in the numerator is x.

$$=\frac{\frac{3}{x}+1\cdot\frac{x}{x}}{\frac{3+x}{2}}$$

Multiply 1 by $\frac{x}{x}$

$$=\frac{\frac{3+x}{x}}{\frac{3+x}{2}}$$

Add two expressions in numerator.

$$=\frac{\cancel{3+x}}{x}\cdot\frac{2}{\cancel{3+x}}$$

Invert and multiply the second fraction: $\frac{a}{b}\div\frac{c}{d}=\frac{a}{b}\cdot\frac{d}{c}$. Cancel the common factor.

$$=\frac{2}{x}$$

This is the final answer.

Now do the problems in Exercise 4.4.

4.5 Equations Involving Algebraic Fractions

In Section 2.2.2 we developed procedures to solve linear equations containing fractions whose denominators were numerical values. In this section we use a similar approach to solve equations containing fractions whose denominators are polynomials. It would be wise for you to review Section 2.2.2 briefly *before you begin this section.* It will be especially helpful to review Examples A and B.

To solve problems of this type, it will be necessary to find the lowest common denominator of all the fractions in the equation, and multiply that LCD by each term of the equation. The resulting linear equation can be solved using the methods employed in Chapter 2.

EXAMPLE A Solve for x:

$$\frac{5}{x}+\frac{2}{3}=2-\frac{2}{x}-\frac{1}{6}$$

We observe the LCD is $6x$.

Solution $$6x\left(\frac{5}{x}\right)+6x\left(\frac{2}{3}\right)=6x(2)-6x\left(\frac{2}{x}\right)-6x\left(\frac{1}{6}\right)$$

Multiply each term by $6x$.

$$30+4x=12x-12-x$$

Simplify. Do you see how each term is obtained?

$$30+4x=11x-12$$

Collect like terms.

$$30=7x-12$$

Subtract $4x$ from both sides.

$$42=7x$$

Add 12 to both sides.

$$6=x$$

Divide both sides by coefficient of x.

Check

$$\frac{5}{6} + \frac{2}{3} \stackrel{?}{=} 2 - \frac{2}{6} - \frac{1}{6}$$ Replace each x by 6.

$$\frac{5}{6} + \frac{4}{6} \stackrel{?}{=} \frac{12}{6} - \frac{2}{6} - \frac{1}{6}$$

$$\frac{9}{6} = \frac{9}{6} \checkmark$$ It checks.

EXAMPLE B Solve for x:

$$\frac{6}{x+3} = \frac{3}{x}$$ We observe that the LCD $= x(x + 3)$.

Solution

$$x(x+3)\left(\frac{6}{x+3}\right) = x(x+3)\left(\frac{3}{x}\right)$$ We multiply both sides by $x(x + 3)$.

$$6x = 3(x + 3)$$ Simplify. Do you see how this is done?

$$6x = 3x + 9$$ Remove parentheses.

$$3x = 9$$ Subtract $3x$ from both sides.

$$x = 3$$ Divide both sides by the coefficient of x.

Check

$$\frac{6}{3+3} \stackrel{?}{=} \frac{3}{3}$$ Replace each x by 3.

$$\frac{6}{6} = \frac{3}{3} \checkmark$$ It checks.

Special Caution

Linear equations containing fractions with variable denominators sometimes have *apparent* solutions. When an equation has been solved using correct methods but the substitution of the solution in the original equation does not make a true statement, we say that we have an *extraneous* solution. Moreover, any value of a variable that makes any denominator in the equation zero is not a solution. In such cases there is *no solution* to the equation. Students should use extra care to check solutions to this type of equation.

You can immediately identify this potential problem situation by noting in the original equation which values of the variable would result in division by zero. Your solution of any equation cannot be any of these values.

This is why we spent the extra time to define the solution of an equation in Section 2.1.3. We must obtain an identity when we substitute the solution into the equation. If we do not obtain an identity (a statement which is always true) we *do not* have a solution.

EXAMPLE C Solve for y:

$$\frac{y}{y-2} - 4 = \frac{2}{y-2}$$ We observe that the LCD $= y - 2$.

Solution $(y-2)\left(\frac{y}{y-2}\right) - (y-2)(4) = (y-2)\left(\frac{2}{y-2}\right)$ We multiply each term by $(y-2)$.

$y - 4(y-2) = 2$ Simplify. Do you see how this is done?

$y - 4y + 8 = 2$ Remove parentheses.

$-3y + 8 = 2$ Collect like terms.

$-3y = -6$ Subtract 8 from both sides.

$\frac{-3y}{-3} = \frac{-6}{-3}$ Divide both sides by negative 3.

$y = 2$ $y = 2$ is only an apparent solution.

We can see immediately that $y = 2$ is not a possible value for the original equation. The use of the value $y = 2$ in a denominator would make the expression undefined.

Check $\frac{y}{y-2} - 4 = \frac{2}{y-2}$ Suppose that you try to check the apparent solution by substituting $y = 2$.

$\frac{2}{2-2} - 4 \stackrel{?}{=} \frac{2}{2-2}$

$\frac{2}{0} - 4 = \frac{2}{0}$ This does not check since you do not obtain a real number when you divide by zero.

$\uparrow$ $\uparrow$

These expressions are not defined.

There is no such number as $2 \div 0$. We see that $y = 2$ does *not* check. *No solution*. There is no solution to this problem.

It is sometimes necessary to factor denominators before the correct LCD can be determined.

EXAMPLE D $\frac{3}{x+5} - 1 = \frac{4-x}{2x+10}$ We observe a need to factor the $2x + 10$.

Solution $\frac{3}{x+5} - 1 = \frac{4-x}{2(x+5)}$ The LCD is $2(x+5)$.

$2(x+5)\left(\frac{3}{x+5}\right) - 2(x+5)(1) = 2(x+5)\left(\frac{4-x}{2(x+5)}\right)$ Multiply each term by the LCD.

$2(3) - 2(x+5) = 4 - x$ Simplify.

$6 - 2x - 10 = 4 - x$ Remove parentheses.

$-2x - 4 = 4 - x$ Collect like terms.

$-4 = 4 + x$ Add $2x$ to both sides.

$-8 = x$ Subtract 4 from both sides.

Check $\dfrac{3}{-8+5} - 1 \stackrel{?}{=} \dfrac{4-(-8)}{2(-8)+10}$ We replace each x in the original equation by -8.

$$\frac{3}{-3} - 1 \stackrel{?}{=} \frac{4+8}{-16+10}$$

$$-1 - 1 \stackrel{?}{=} \frac{12}{-6}$$

$$-2 = -2\checkmark$$ It checks. $x = -8$ is the solution.

Carefully review the steps used in solving Examples A to D and then try these practice problems.

Practice Problems for Exercise 4.5

Solve for x and check your solution.

1. $\dfrac{5}{x} = \dfrac{3}{2-x}$

2. $\dfrac{x-1}{x^2-4} = \dfrac{2}{x+2} + \dfrac{4}{x-2}$

3. $\dfrac{5}{x} - \dfrac{2}{x+5} = \dfrac{25}{x^2+5x}$

Now turn to the answers following Exercise 4.5 and check your solutions for these practice problems.

Exercise 4.5

Solve for the variable and check your solution.

1. $\dfrac{x+1}{2x} = \dfrac{2}{3}$

2. $\dfrac{9}{2x-1} = \dfrac{3}{x}$

3. $\dfrac{1}{x} = \dfrac{1}{4-x}$

4. $\dfrac{x-2}{4x} = \dfrac{1}{6}$

5. $\dfrac{y-3}{3y+2} = \dfrac{1}{5}$

6. $\dfrac{2}{y+3} = \dfrac{5}{y}$

7. $\dfrac{2}{x} + \dfrac{x}{x+1} = 1$

8. $\dfrac{5}{2} = 3 + \dfrac{2x+7}{x+6}$

9. $\dfrac{x+1}{x} = 1 + \dfrac{x-2}{2x}$

10. $\dfrac{7x-4}{5x} = \dfrac{9}{5} - \dfrac{4}{x}$

11. $\dfrac{x+1}{x} = \dfrac{1}{2} - \dfrac{4}{3x}$

12. $\dfrac{2x-1}{3x} + \dfrac{1}{9} = \dfrac{x+2}{x} + \dfrac{1}{9x}$

13. $\dfrac{x}{x-2} - 2 = \dfrac{2}{x-2}$

14. $4 - \dfrac{8}{x+1} = \dfrac{8x}{x+1}$

15. $\dfrac{2}{x+1} - \dfrac{1}{x-1} = \dfrac{2x}{x^2-1}$

16. $\dfrac{8x}{4x^2-1} = \dfrac{3}{2x+1} + \dfrac{3}{2x-1}$

17. $\dfrac{y+1}{y^2+2y-3} = \dfrac{1}{y+3} - \dfrac{1}{y-1}$

18. $\dfrac{2y}{y+1} + \dfrac{1}{3y-2} = 2$

19. $\dfrac{3}{x+3} + \dfrac{1}{x-1} = \dfrac{8}{x^2+2x-3}$

20. $\dfrac{2}{3x+6} + \dfrac{1}{2x+4} = \dfrac{1}{6}$

21. $\dfrac{2x}{x+4} - \dfrac{8}{x-4} = \dfrac{2x^2+32}{x^2-16}$

22. $\dfrac{4x}{x+3} - \dfrac{12}{x-3} = \dfrac{4x^2+36}{x^2-9}$

Challenge Problems

23. $\dfrac{4}{x+3} - \dfrac{6x+10}{x^2+4x+3} = \dfrac{2x}{x+1}$

24. $\dfrac{10-2y^2}{y^2+3y-10} + \dfrac{4y+2}{y-2} = \dfrac{2y+2}{y+5}$

Solutions to Practice Problems for Exercise 4.5

1. $\dfrac{5}{x} = \dfrac{3}{2-x}$ — We observe that the LCD $= x(2-x)$.

$x(2-x)\left(\dfrac{5}{x}\right) = x(2-x)\left(\dfrac{3}{2-x}\right)$ — We multiply each term by the LCD.

$5(2-x) = 3x$ — Simplify.

$10 - 5x = 3x$ — Remove parentheses.

$10 = 8x$ — Add $5x$ to both sides.

$\dfrac{5}{4} = x$ — Divide both sides by 8, then reduce $\dfrac{10}{8}$ to $\dfrac{5}{4}$.

Check

$\dfrac{5}{\frac{5}{4}} \stackrel{?}{=} \dfrac{3}{2-\frac{5}{4}}$ — We replace each x in the original equation by $\dfrac{5}{4}$.

$4 \stackrel{?}{=} \dfrac{3}{\frac{8}{4}-\frac{5}{4}}$ — Use the fact that $5 \div \dfrac{5}{4} = 5 \cdot \dfrac{4}{5} = 4$.

$4 \stackrel{?}{=} \dfrac{3}{\frac{3}{4}}$ — Use the fact that $3 \div \dfrac{3}{4} = 3 \cdot \dfrac{4}{3} = 4$.

$4 = 4\checkmark$ — It checks.

2. $$\frac{x-1}{x^2-4} = \frac{2}{x+2} + \frac{4}{x-2}$$ We need to factor $x^2 - 4$.

$$\frac{x-1}{(x+2)(x-2)} = \frac{2}{x+2} + \frac{4}{x-2}$$ We observe that the LCD $= (x+2)(x-2)$.

$$(x+2)(x-2)\frac{x-1}{(x+2)(x-2)} = (x+2)(x-2)\left(\frac{2}{x+2}\right) + (x+2)(x-2)\left(\frac{4}{x-2}\right)$$

We multiply each term by the LCD.

$x - 1 = 2(x-2) + 4(x+2)$ — Simplify.

$x - 1 = 2x - 4 + 4x + 8$ — Remove parentheses.

$x - 1 = 6x + 4$ — Collect like terms.

$-1 = 5x + 4$ — Subtract x from both sides.

$-5 = 5x$ — Subtract 4 from both sides.

$-1 = x$ — Divide both sides by 5.

Check $$\frac{-1-1}{(-1)^2-4} \stackrel{?}{=} \frac{2}{-1+2} + \frac{4}{-1-2}$$ Replace each x in the original equation by -1.

$$\frac{-2}{1-4} \stackrel{?}{=} \frac{2}{1} + \frac{4}{-3}$$

$$\frac{-2}{-3} \stackrel{?}{=} 2 + \left(-\frac{4}{3}\right)$$

$$\frac{2}{3} \stackrel{?}{=} \frac{6}{3} + \left(-\frac{4}{3}\right)$$

$$\frac{2}{3} = \frac{2}{3} \checkmark$$ It checks.

3. $$\frac{5}{x} - \frac{2}{x+5} = \frac{25}{x^2+5x}$$ We will need to factor $x^2 + 5x$.

$$\frac{5}{x} - \frac{2}{x+5} = \frac{25}{x(x+5)}$$ The LCD is $x(x+5)$.

$$x(x+5)\left(\frac{5}{x}\right) - x(x+5)\left(\frac{2}{x+5}\right) = x(x+5)\frac{25}{x(x+5)}$$

We multiply each term by the LCD.

$5(x+5) - 2x = 25$ — Simplify.

$5x + 25 - 2x = 25$ — Remove parentheses.

$3x + 25 = 25$ — Collect like terms.

$3x = 0$ — Subtract 25 from both sides.

$x = 0$ — Divide both sides by 3. But can $x = 0$?

We see immediately that $x = 0$ is not a possible value since $\frac{5}{x}$ will then have a 0 denominator, which is never allowed. The fraction

$$\frac{25}{x^2 + 5x}$$

would also have a 0 denominator. There is *no solution* to this equation.

Now do the problems in Exercise 4.5.

An Application for Chapter 4: Continued Fractions

In Section 4.4 we studied the simplification of complex fractions. A related topic that has a rich history in mathematics is the topic of *continued fractions*. Consider the following continued fractions.

EXAMPLE A

$$\frac{19}{5} = 3 + \cfrac{1}{1 + \cfrac{1}{4}}$$

EXAMPLE B

$$\frac{43}{19} = 2 + \cfrac{1}{3 + \cfrac{1}{1 + \cfrac{1}{4}}}$$

EXAMPLE C

$$\frac{4}{\pi} = 1 + \cfrac{1^2}{2 + \cfrac{3^2}{2 + \cfrac{5^2}{2 + \cfrac{7^2}{2} + \cdots}}}$$

Some continued fractions like Examples A and B have a finite number of terms. Other continued fractions, like Example C, have an infinite number of terms.

In more advanced mathematics it can be shown that any rational number greater than 1 written as $\frac{a}{b}$ where a, b are positive integers can be written as a continued fraction that has a finite number of terms.

For a given continued fraction the rational number $\frac{a}{b}$ can be found by using the techniques of Section 4.4.

EXAMPLE D Evaluate:

$$\cfrac{1}{3 + \cfrac{1}{4 + \cfrac{1}{2}}}$$

Solution Since $4 + \frac{1}{2} = \frac{8}{2} + \frac{1}{2} = \frac{9}{2}$ we can write the original problem as

$$\frac{1}{3 + \frac{1}{\frac{9}{2}}}$$

But we can find

$$\frac{1}{\frac{9}{2}} = 1 \cdot \frac{2}{9} = \frac{2}{9}$$

So we can write the expression as

$$\frac{1}{3 + \frac{2}{9}}$$

Finally,

$$3 + \frac{2}{9} = \frac{27}{9} + \frac{2}{9} = \frac{29}{9}$$

so we now have

$$\frac{1}{\frac{29}{9}} = \frac{9}{29}$$

Thus

$$\frac{1}{3 + \frac{1}{4 + \frac{1}{2}}} = \frac{9}{29}$$

A Challenge for You!

Based on the discussion above evaluate the following continued fractions.

1. $$\frac{1}{2 + \frac{1}{1 + \frac{1}{4 + \frac{1}{2}}}}$$

2. $$1 + \frac{1}{1 + \frac{1}{1 + \frac{1}{5}}}$$

3. $$2 + \frac{1}{4 + \frac{1}{1 + \frac{1}{5 + \frac{1}{1 + \frac{1}{12}}}}}$$

Answers

1. $\frac{11}{31}$
2. $\frac{17}{11}$
3. $\frac{964}{437}$

Supplementary Problems for Chapter 4

Reduce, simplify, or perform the indicated operations.

1. $\frac{4x - 4y}{5y - 5x}$

2. $\frac{bx}{bx - by}$

3. $\frac{2x^2 + 5x - 3}{2x^2 - 9x + 4}$

4. $\frac{6x^2 + 7x + 2}{3x^2 + 5x + 2}$

5. $\frac{3x^2 - 13x - 10}{3x^2 + 2x} \cdot \frac{x^2 - 25x}{x^2 - 25}$

6. $\frac{x^2 + 7x + 12}{4x + 12} \cdot \frac{x^2 - 16}{x^2 + 8x + 16}$

7. $\frac{2x^2 - 18}{3y^2 + 3y} \div \frac{y^2 + 6y + 9}{y^2 + 4y + 3}$

8. $\frac{2y^2 + 3y - 2}{2y^2 + y - 1} \div \frac{2y^2 + y - 1}{2y^2 - 3y - 2}$

9. $\frac{x^3 - 36x}{12x^2 + 2x} \cdot \frac{36x + 6}{x^3 - 6x^2} \cdot \frac{x^2 + 2x}{x^2 + 11x + 30}$

10. $\frac{10x^2}{x^2 - 7x + 10} \cdot \frac{x^2 + 3x - 10}{25x^2} \cdot \frac{2x - 8}{x^2 + x - 20}$

11. $\frac{6y^2 + 13y - 5}{9y^2 + 3y} \div \frac{4y^2 + 20y + 25}{12y^2}$

12. $\frac{3xy^2 + 12y^2}{2x^2 - 11x + 5} \div \frac{2xy + 8y}{8x^2 + 2x - 3}$

13. $\frac{7}{x + 1} + \frac{4}{2x}$

14. $5 + \frac{1}{x} + \frac{1}{x + 1}$

15. $\frac{2}{x^2 - 9} + \frac{x}{x + 3}$

16. $\frac{7}{x + 2} + \frac{3}{x - 4}$

17. $\frac{x}{y} + \frac{3}{2y} + \frac{1}{y + 2}$

18. $\frac{4}{a} + \frac{2}{b} + \frac{3}{a + b}$

19. $\frac{3x + 1}{3x} - \frac{1}{x}$

20. $\frac{x + 4}{x + 2} - \frac{1}{2x}$

21. $\frac{1}{x^2 + 7x + 10} - \frac{x}{x + 5}$

22. $\frac{7}{x^2 - 9x + 20} - \frac{3x}{x - 4}$

23. $\frac{2x + 1}{x^2 + 5x + 6} + \frac{4x}{x^2 - 9}$

24. $\frac{3x}{x^2 + x - 2} - \frac{2x + 1}{x^2 + 6x + 8}$

25. $\dfrac{\frac{3}{2y} - \frac{1}{y}}{\frac{4}{y} + \frac{3}{2y}}$

26. $\dfrac{\frac{2}{x} + \frac{1}{2x}}{x + \frac{x}{2}}$

27. $\dfrac{w - \frac{4}{w}}{1 + \frac{2}{w}}$

28. $\dfrac{1 - \frac{w}{w - 1}}{1 + \frac{w}{1 - w}}$

29. $\dfrac{1 + \frac{1}{y^2 - 1}}{\frac{1}{y + 1} - \frac{1}{y - 1}}$

30. $\dfrac{\frac{1}{y} + \frac{1}{x + y}}{1 + \frac{2}{x + y}}$

31. $\dfrac{\dfrac{1}{a+b}-\dfrac{1}{a}}{b}$

32. $\dfrac{\dfrac{2}{a+b}-\dfrac{3}{b}}{\dfrac{1}{a+b}}$

33. $\left[\dfrac{1}{x+2y}-\dfrac{1}{x-y}\right]\div\dfrac{2x-4y}{x^2-3xy+2y^2}$

34. $\dfrac{x+5y}{x-6y}\div\left[\dfrac{1}{5y}-\dfrac{1}{x+5y}\right]$

Solve for the variable. If no solution is possible, so state.

35. $\dfrac{8}{a-3}=\dfrac{12}{a+3}$

36. $\dfrac{8a-1}{6a+8}=\dfrac{3}{4}$

37. $\dfrac{2x-1}{x}-\dfrac{1}{2}=-2$

38. $\dfrac{5-x}{x}-\dfrac{7}{x}=-\dfrac{3}{4}$

39. $\dfrac{5}{2}-\dfrac{2y+7}{y+6}=3$

40. $\dfrac{5}{4}-\dfrac{1}{2x}=\dfrac{1}{x}+2$

41. $\dfrac{3y-1}{3y}-\dfrac{6}{5y}=\dfrac{1}{y}-\dfrac{4}{15}$

42. $\dfrac{9}{2}-\dfrac{7y-4}{y+2}=-\dfrac{1}{4}$

43. $\dfrac{4}{x^2-1}=\dfrac{2}{x-1}+\dfrac{2}{x+1}$

44. $\dfrac{y+18}{y^2-16}=\dfrac{y}{y+4}-\dfrac{y}{y-4}$

45. $\dfrac{9y-3}{y^2+2y}-\dfrac{5}{y+2}=\dfrac{3}{y}$

46. $\dfrac{2}{3-3y}+\dfrac{2}{2y-1}=\dfrac{4}{3y-3}$

Chapter 4 Summary

Simplifying Fractions Where Numerator and Denominator Are Polynomials

1. Factor numerator and denominator, if possible.
2. Any *factor* that is common to both numerator and denominator can be canceled. This cancellation is an application of the basic rule of fractions.

The Basic Rule of Fractions

For any polynomials a, b, c (where $b \neq 0$ and $c \neq 0$),

$$\frac{ac}{bc}=\frac{a}{b}$$

Multiplying Fractional Algebraic Expressions

1. Factor all numerators and denominators, if possible.
2. Any *factor* that is common to a numerator of one fraction and the denominator of the same fraction or any fraction that is multiplied by it can be canceled.
3. Write the indicated product of the remaining factors in the numerator. Write the indicated product of the remaining factors in the denominator.

$$\frac{a}{b}\cdot\frac{c}{d}=\frac{ac}{bd}$$

Dividing Fractional Algebraic Expressions

1. Invert the second fraction and multiply it by the first fraction.

$$\frac{a}{b}\div\frac{c}{d}=\frac{a}{b}\cdot\frac{d}{c}$$

2. Apply the steps for multiplying fractional algebraic expressions.

Adding Fractional Algebraic Expressions

1. If all fractions have a common denominator, add the numerators and place the result over the common denominator.

$$\frac{a}{c} + \frac{b}{c} = \frac{a + b}{c}$$

2. If the fractions do not have a common denominator, factor the denominators (if necessary) and determine the least common denominator (LCD).
3. Multiply each fraction by the necessary value so that each fraction becomes an equivalent fraction with the LCD as the denominator.
4. Add the numerators and place the result over the common denominator.
5. Simplify, if possible.

Subtracting Fractional Algebraic Expressions

Follow the procedures of adding fractional algebraic expressions except that you subtract the second numerator from the first after each fraction has the LCD as the denominator.

$$\frac{a}{c} - \frac{b}{c} = \frac{a - b}{c}$$

Procedure to Simplify a Complex Fraction

1. Add or subtract so that you have a single fraction in the numerator and in the denominator.
2. Divide the fraction in the numerator by the fraction in the denominator. This is done by inverting the fraction in the denominator and multiplying it by the numerator.

Solving Equations Involving Algebraic Fractions

1. Determine the LCD of all denominators in the equation.
2. Multiply each term in the equation by the LCD.
3. Simplify and remove parentheses.
4. Collect any like terms.
5. Solve for the variable.
6. Check your answer. Be sure that the value you obtained does not make any fraction in the original equation have a value of 0 in the denominator. If so, there is no solution.

Practice Quiz for Chapter 4

Sections 4.1–4.3

Simplify.

1. $\dfrac{5x - 20}{x^2 - 4x}$

2. $\dfrac{2a^2 + 2a - 4}{6a^2 + 20a + 16}$

3. $\dfrac{-21a + 7}{a^2 - 2a - 48} \cdot \dfrac{a^2 - a - 42}{9a^2 - 1}$

4. $\dfrac{2x^3 - x^4}{3x^3 - x^2} \cdot \dfrac{3x^2 - 13x + 4}{x^2 - 4}$

5. $\dfrac{2x + 6}{2x - 4} \cdot \dfrac{2x^2 + 11x - 21}{x^2 + 10x + 21} \cdot \dfrac{x^2 - 8x + 12}{4x^2 - 9}$

6. $\dfrac{a^2 - ab - 6b^2}{a^2 + 2} \div (a^2 + 2ab)$

7. $\dfrac{8x^3}{2x-8} \div \dfrac{16x^4}{x^2+x-20}$

8. $\dfrac{3}{x^2-3x} + \dfrac{5x}{x-3}$

9. $\dfrac{7}{4y-8} - \dfrac{5}{10y-20}$

10. $\dfrac{3x-6}{x^2+4x+4} - \dfrac{5x}{x^2+3x+2}$

11. $\dfrac{x^2-7xy+10y^2}{3x^2-5xy-2y^2}$

12. $\dfrac{2x^2+xy-6y^2}{4x^2-12xy+9y^2}$

Sections 4.4-4.5

Simplify.

1. $\dfrac{\dfrac{x+2}{x^3}}{\dfrac{x^2-x-6}{x^2}}$

2. $\dfrac{\dfrac{x}{x}-\dfrac{1}{y}}{1-\dfrac{y}{x}}$

3. $\dfrac{4-\dfrac{1}{9y^2}}{2+\dfrac{1}{3y}}$

4. $\dfrac{\dfrac{3}{y+1}-\dfrac{3}{y-1}}{\dfrac{15}{y^2-1}}$

Solve for x. Check your answers.

5. $\dfrac{4x-7}{x-2} = 3 + \dfrac{1}{x-2}$

6. $\dfrac{1}{x} + \dfrac{2}{x+4} = \dfrac{5}{3x}$

7. $\dfrac{4}{x-2} = \dfrac{3x-2}{x^2-4} + \dfrac{4}{x+2}$

8. $\dfrac{4x+1}{x^2+3x+2} = \dfrac{x+2}{x+1} - \dfrac{x}{x+2}$

Practice Test for Chapter 4

Perform the indicated operation and simplify all answers.

1. $\dfrac{2ac+2ad}{3a^2c+3a^2d}$

2. $\dfrac{8x^2-2x^2y^2}{y^2+4y+4}$

3. $\dfrac{x^2+2x}{2x-1} \cdot \dfrac{10x^2-5x}{12x^3+24x^2}$

4. $\dfrac{a^2+ab}{b} \cdot \dfrac{a^2-2ab+b^2}{a^4-b^4} \cdot \dfrac{a^2+b^2}{3a^2b-3ab^2}$

5. $\dfrac{2a^2-3a-2}{4a^2+a-14} \div \dfrac{2a^2+5a+2}{16a^2-49}$

6. $\dfrac{9}{x-2} + \dfrac{2}{x-3} + \dfrac{1}{2x-6}$

7. $\dfrac{x-y}{xy} - \dfrac{a-y}{ay}$

8. $\dfrac{3x}{x^2-3x-18} - \dfrac{x-4}{x-6}$

9. $\dfrac{\dfrac{x}{3y}-\dfrac{1}{2}}{\dfrac{4}{3y}-\dfrac{2}{x}}$

10. $\dfrac{\dfrac{2}{x+3}+\dfrac{1}{x}}{3x+9}$

11. $\dfrac{2x^2+3xy-9y^2}{4x^2+13xy+3y^2}$

Solve for x. Check your answers.

12. $\dfrac{15}{x} + \dfrac{9x-7}{x+2} = 9$

13. $\dfrac{x-3}{x-2} = \dfrac{2x^2-15}{x^2+x-6} - \dfrac{x+1}{x+3}$

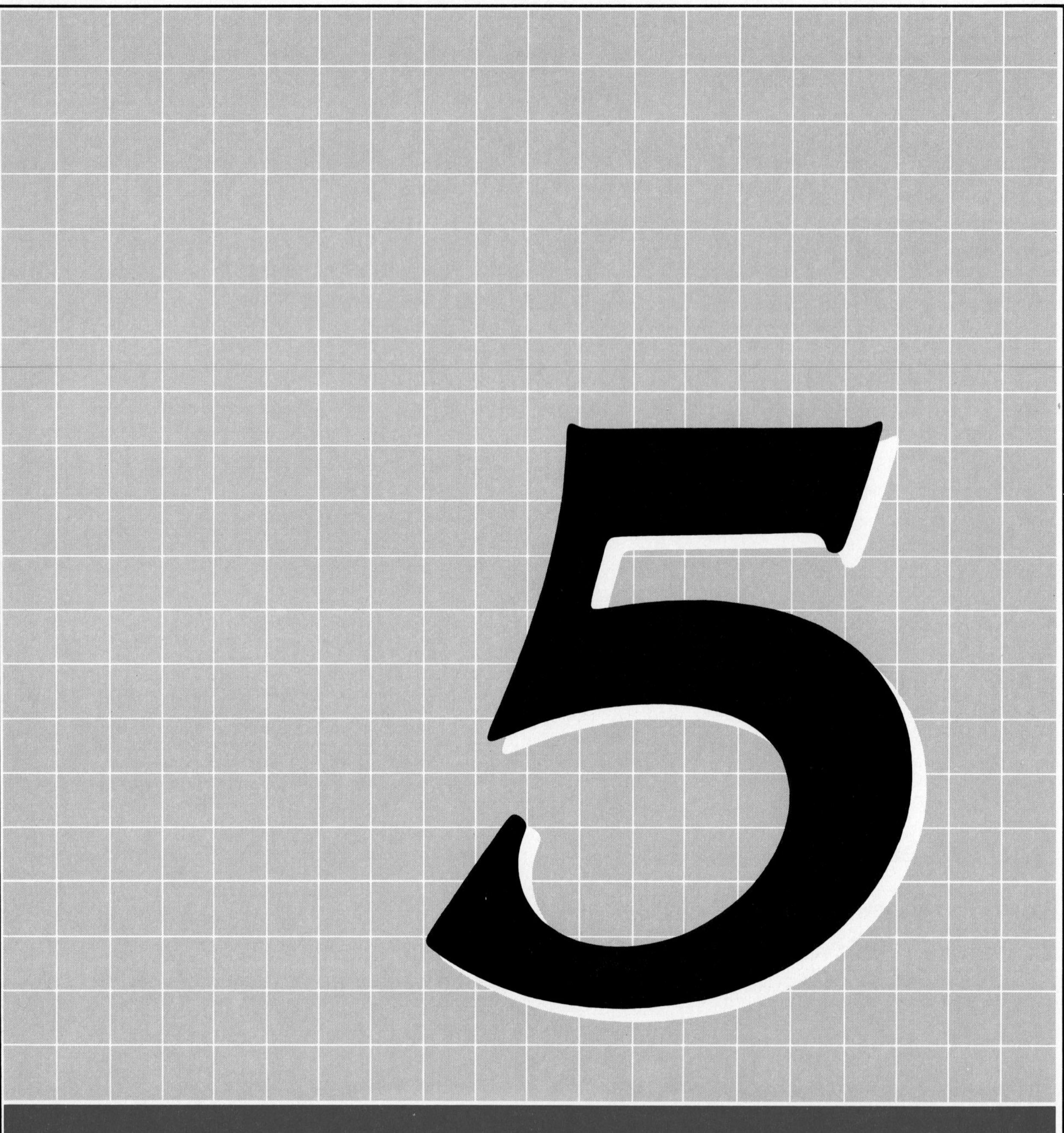

5 Word Problems

Diagnostic Pretest for Chapter 5

This diagnostic pretest is intended for those students who have previously studied algebra. Work as many problems as you can. Then, check your answers against those listed at the back of the book. Where you have obtained an incorrect answer or have been unable to work the problem, you will need to study the corresponding sections of the chapter.

If you have not previously studied algebra, do not take this test now. You should study the chapter and work the problems and exercises. After finishing the chapter, you may use this test as a review or as a practice test.

Section 5.1

In problems 1 and 2, replace each phrase by the appropriate algebraic notation. Let the number be represented by x.

1. Twice a number decreased by one-third of the same number.

2. Forty percent of a number is increased by 150.

Section 5.2

In problems 3 and 4, translate the verbal statements into an equation and find the number.

3. The product of a number and seven is increased by nine. The result is forty-four. Find the original number.

4. There are three numbers with these properties: The first number is triple a second number. A third number is one more than half of that first number. The sum of the three numbers is twenty-three. Find the three numbers.

Section 5.3

5. One side of a triangular piece of land is 3 meters less than double the second side. This one side is 3 more meters in length than the third side. The perimeter of the triangle is 41 meters. Find the length of each side of the triangle.

6. Charlene took three packages to the post office to mail. The first package was $3\frac{1}{2}$ pounds more than the second package. The third package was 2 pounds less than the first package. The total weight of the three packages was 17 pounds. How much did each package weigh?

Section 5.4

7. Two investments were made totaling \$1000. The investments gain interest once each year. The first investment yielded 12% interest after 1 year. The second investment yielded 9% interest. The total interest for both investments was \$102. How much was invested at each interest rate?

8. Enrique has \$3.35 in change in his pocket. He has four more dimes than nickels. He has three more quarters than dimes. How many coins of each type does he have?

Student Learning Objectives for Chapter 5

After successfully learning the content of each section, you will be able to:

Section 5.1

A. Replace phrases that express relationships between numbers by an algebraic expression.

Section 5.2

B. Translate the information contained in a word problem into an equivalent algebraic equation that contains the conditions specified in the word problem.

C. Solve the resulting algebraic equation to find the solution or solutions of the word problem.

D. Check the obtained solution to verify that it satisfies all the conditions of the original word problem.

Section 5.3

E. Solve a variety of word problems involving comparisons that can be solved through the use of one linear equation containing one variable.

Section 5.4

F. Solve a variety of word problems dealing with money, percent, and interest through the use of one linear equation containing one variable.

5.1 Translation of Verbal Phrases into Algebraic Expressions

In all branches of mathematics we encounter written problems that can be solved by a mathematical equation. One of the most useful applications of the algebra you have learned is the ability to solve a variety of what are commonly called "word problems." In this section we show how you can develop the important skill of being able to analyze and solve many different types of verbal problems.

First we need to learn how to express common English expressions using the symbols of algebra. Frequently, we use English phrases to describe the operation of addition.

English Phrases Describing Addition	*Notation in Algebra*
Three *more than* a number The *sum of* a number and three A number *increased by* three Three is *added to* a number Three *greater than* a number A number *plus* three	$x + 3$

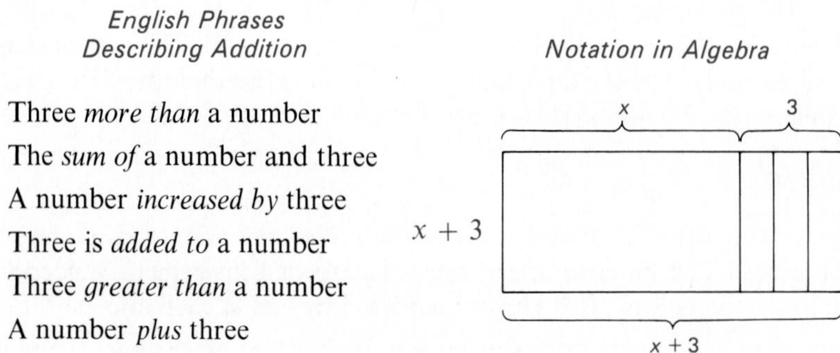

If we represent the unknown number by the variable x, all of the phrases above can be expressed in algebra by

$$x + 3$$

In a similar way we can use algebra to express English phrases that describe the operations of subtraction, multiplication, and division.

English Phrases Describing Subtraction	*Notation in Algebra*
A number *decreased by* four Four *less than* a number The *difference between* a number and four Four *smaller than* a number A number *diminished by* four	$x - 4$:

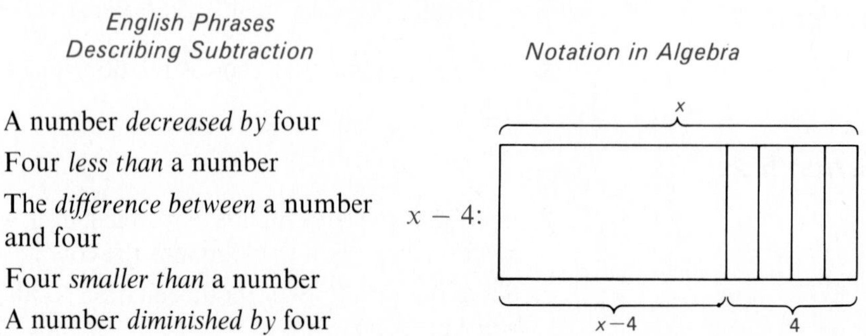

English Phrases Describing Multiplication	*Notation in Algebra*
Double a number *Twice* a number The *product* of two and a number Two *times* a number Two *of* a number	$2x$: 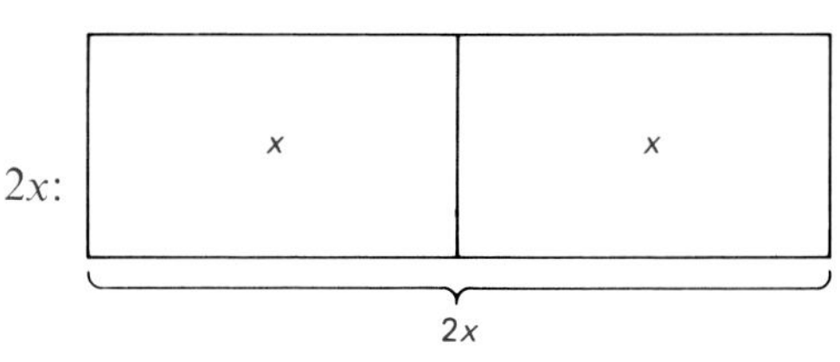

English Phrases Describing Division	*Notation in Algebra*
A number *divided by* five The *quotient* of a number and five One-*fifth* of a number	$\frac{x}{5}$: 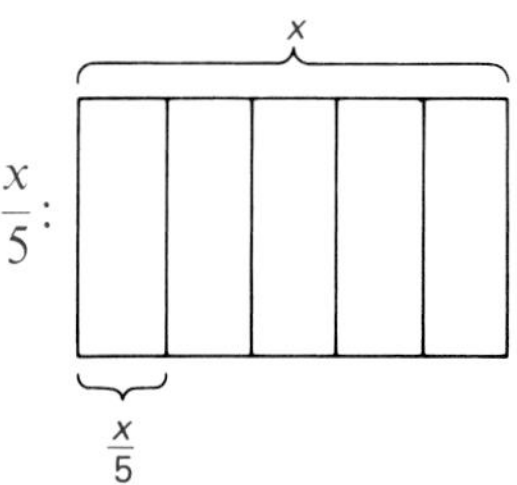

It is important to review these phrases and to see how they can be represented in algebraic notation. Often other words are used in English instead of the word "number".

EXAMPLE A

English Phrase	*Notation in Algebra*
1. A quantity is increased by five	$x + 5$
2. The product of eight and the value	$8x$
3. The difference between the measured item and seven	$x - 7$
4. One-third of the weight	$\frac{x}{3}$ or $\frac{1}{3}x$

Finally, we see that more than one operation can be described in an English phrase or group of phrases.

EXAMPLE B

English Phrase	*Notation in Algebra*
1. Seven more than double a number	$2x + 7$
	Note that these are *not* the same.
2. The value of the number is increased by seven and then doubled	$2(x + 7)$
3. A number decreased by five is divided by eight	$\frac{x - 5}{8}$
	These are *not* the same.
4. The quotient of a number and eight is decreased by five	$\frac{x}{8} - 5$
5. A number increased by five is subtracted from three times the same number	$3x - (x + 5)$ or $3x - x - 5$ You must use parentheses here. Do you see that $3x - x + 5$ is wrong?

6. Four times the sum of a number and six $4(x + 6)$

or $4x + 24$

Again we need parentheses. Do you see that $4x + 6$ is wrong?

Carefully review these examples and you will not find the exercises difficult. Now try these practice problems.

Practice Problems for Exercise 5.1

Replace each phrase by the appropriate algebraic notation. Let x represent the unknown value.

1. Four less than three times a number.

2. Two-thirds of the sum of a number and five.

3. The quotient of a value and seven is diminished by five.

Now turn to the answers following Exercise 5.1 and check your solutions for these practice problems.

Exercise 5.1

Replace each phrase by the appropriate algebraic notation. Let x represent the unknown value.

1. The product of twelve and a number.

2. One-eighth of a quantity.

3. The difference between a number and five.

4. Fifteen greater than a value.

5. Two-sevenths of a value is decreased by three.

6. Double a quantity increased by four.

7. The difference of two times a value and fifteen.

8. Twice a number added to one-third of the same number.

9. Four times a number decreased by one-half the same number.

10. The quotient of a number and six is subtracted from the same number.

11. Five more than double the length.

12. Twelve percent of the number is increased by 50.

13. A number is added to triple the same number and further increased by one-half of the same number.

14. The difference of a number and twelve is then doubled.

15. Eighteen percent of half of the weight.

16. Five-ninths the result of subtracting thirty-two from the Fahrenheit temperature (see Figure 5.1.).

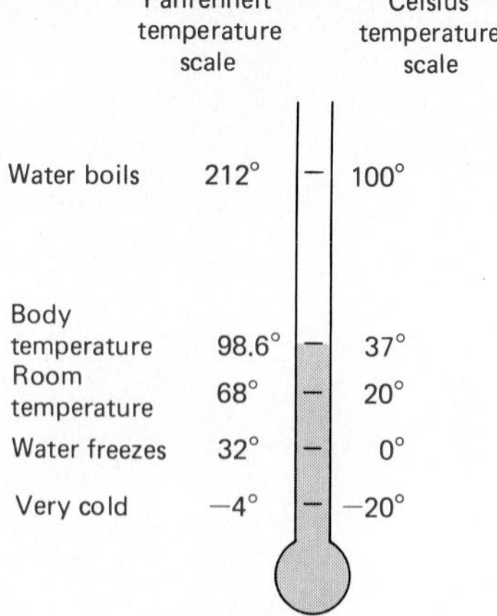

FIGURE 5.1

17. The sum of nine times the Celsius temperature and one hundred sixty is divided by five (see Figure 5.1).

18. Five kilograms less than three-fourths of his weight.

19. The cost was increased by an extra $100 and by a surcharge of twenty percent of the original cost.

20. Triple the profit diminished by $400 was further diminished by fifty percent of the original profit.

Solutions to Practice Problems for Exercise 5.1

1. Four less than three times a number.

$3x - 4$ — Note that we multiply the variable x by 3 before we subtract.

2. Two-thirds of the sum of a number and five.

$$\frac{2}{3}(x + 5) \quad \text{or} \quad \frac{2}{3}x + \frac{10}{3}$$

It would not be correct to say $\frac{2}{3}x + 5$. We must take two-thirds of the *sum*. Thus it is necessary to put the sum $x + 5$ inside parentheses.

3. The quotient of a value and seven is diminished by five.

$\frac{x}{7} - 5$ — The order is important!

Be sure that you write $\frac{x}{7}$ and not $\frac{7}{x}$

Also note that it would *not* be correct to say $\frac{x - 5}{7}$.

We must find the quotient first; then we subtract five.

5.2 Translation of Verbal Problems into Equations

The skills you have just developed will enable you to solve a variety of problems. You will need to find an algebraic equation that corresponds to the problem. Often, an English word appears in a sentence that can be replaced by the equality symbol in algebra.

English Words Describing Equality	*Notation in Algebra*
are, is, will be, equals, is equal to, yields, gives, makes, has a value of	$=$

By studying the statement or statements in the problem and by looking for these English words you can usually write an equation.

EXAMPLE A

English Statement	*Equation*
1. Three times a number increased by one is seventeen.	let $x =$ the number $3x + 1 = 17$
2. Twice a number is the same as the original number increased by five.	let $x =$ the number $2x = x + 5$
3. The difference of five times a number and three times a number is eighteen.	let $x =$ the number $5x - 3x = 18$
4. When four is subtracted from one-half of a number the result is six.	let $x =$ the number $\frac{x}{2} - 4 = 6$

EXAMPLE B Two-thirds of a number is eighty-four. What is the number?

Solution Let $x =$ the number. (See Figure 5.2). Although a sketch is not always needed, it is sometimes helpful to draw some visual representation of the situation. Often a sketch will help you to understand the relationship described in the verbal problem.

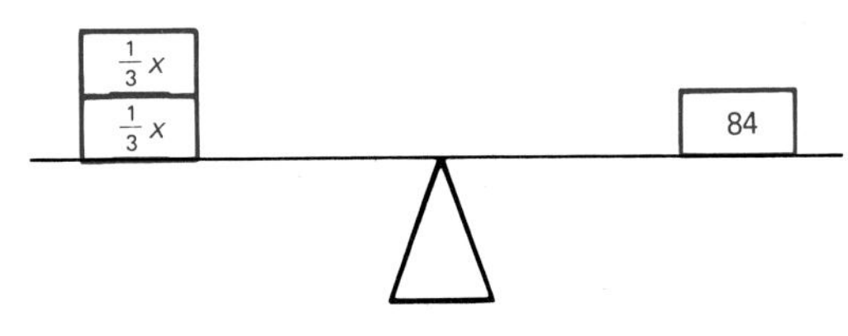

FIGURE 5.2

$$\frac{2}{3}x = 84 \qquad \text{Two-thirds of } x \text{ is eighty-four.}$$

$$3\left(\frac{2}{3}x\right) = 3(84) \qquad \text{Multiply both sides of the equation by 3.}$$

$$2x = 252 \qquad \text{Simplify.}$$

$$\frac{2x}{2} = \frac{252}{2} \qquad \text{Divide by the coefficient of } x.$$

$$x = 126$$

The original number is 126.

EXAMPLE C Five more than six times a number is three hundred and five. Find the number.

Solution Let x = the number.

$$6x + 5 = 305$$

$$6x = 300 \qquad \text{Subtract 5 from both sides.}$$

$$\frac{6x}{6} = \frac{300}{6} \qquad \text{Divide by the coefficient of } x.$$

$$x = 50 \qquad \text{Simplify.}$$

The original number is 50.

EXAMPLE D Six times a number decreased by twenty is four times the number. What is the number?

Solution Let x = the number.

$$6x - 20 = 4x$$

$$6x - 4x = 20 \qquad \text{Collect } x \text{ terms on one side, numerical values on the other side.}$$

$$2x = 20 \qquad \text{Simplify.}$$

$$x = 10 \qquad \text{The original number is 10.}$$

If the problem refers to two numbers, we must write an algebraic expression for each before writing the equation. This should be done as two separate steps.

EXAMPLE E The larger of two numbers is three more than twice the smaller. The sum of the numbers is thirty-nine. Find each number.

Solution Let x = the smaller number. Then $2x + 3$ = the larger number. Then we write

$$(x) + (2x + 3) = 39 \qquad \text{The sum of the numbers is 39.}$$

$$3x + 3 = 39 \qquad \text{Collect like terms.}$$

$$3x = 36 \qquad \text{Subtract 3 from each side.}$$

$$x = 12 \qquad \text{Divide both sides by 3.}$$

The smaller number is 12. Since the larger number is $2x + 3$, we substitute the known value of 12.

$$\text{Larger number} = 2(12) + 3$$

$$= 24 + 3$$

$$= 27$$

The larger number is 27.

? How do you know what number to represent by x?
How do you get started solving these problems?

Many students ask those questions. There is no magic answer that works for every problem. However, there is one basic principle that helps analyze most problems.

> **Usually it is best to let x represent that number that is used for a basis of comparison.**

Let's examine the following example.

EXAMPLE F We are studying the relationship between two numbers. The first number is four less than triple the second number. The sum of the two numbers is forty-eight. Find each number.

Solution The first number is being compared to the second number. The basis of comparison is "*the second number*." Since "the first number" is being compared to "the second number" it is logical to let x represent "the second number." *So let x = the second number.*

Now the first number is 4 less than triple the second number. So let $3x - 4$ = "the first number." The sum of these two numbers is 48.

(the first number) + (the second number) = 48	Express the sum as an equation with words.
$(3x - 4) + (x) = 48$	Express the sum as an equation with variables.
$4x - 4 = 48$	Collect like terms.
$4x = 52$	Add 4 to each side.
$x = 13$	Divide both sides by the coefficient of x.

So 13 is "the second number."

Since the first number is $3x - 4$,

$3x - 4 = 3(13) - 4$	we replace x by 13.
$= 39 - 4$	
$= 35$	The first number is 35.

Check Does this make sense? Does this satisfy the conditions of the problem?

Is the first number 4 less than triple the second number?

$$35 \stackrel{?}{=} 3(13) - 4$$

$$35 \stackrel{?}{=} 39 - 4$$

$$35 = 35 \checkmark$$

Is the sum of the numbers 48?

$$35 + 13 \stackrel{?}{=} 48$$

$$48 = 48 \checkmark$$

So we are confident that the first number is 35 and the second number is 13.

It will be worthwhile to review Examples A to F before trying the practice problems.

Practice Problems for Exercise 5.2

Solve these word problems.

1. The sum of a number, double the number, and five times the number is seventy-two. Find the original number.
2. When one-third of a number is increased by twelve the result is eight. Find the number.
3. If you double a number and add twenty you obtain one-third of the original number. Find the original number.
4. Consider two numbers. The second number is twelve less than triple the first number. The sum of the two numbers is twenty-four. Find each number.

Now turn to the answers following Exercise 5.2 and check your solutions for these practice problems.

Exercise 5.2

Solve these word problems.

1. When one-fourth of a number is decreased by five the result is ten. Find the original number.

2. One-half of a number is multiplied by seven to obtain a result of ninety-one. Find the original number.

3. The product of a number and twelve is increased by nine. The result is one hundred seventeen. Find the original number.

4. The product of a number and seven is added to triple the original number. The result is one hundred thirty. Find the original number.

5. Four times a number decreased by thirty is seven times the original number. What is that number?

6. Thirty is added to five times a number. The result is double the original number. What is that number?

7. One number is two more than triple a smaller number. The sum of the two numbers is thirty-four. Find each number.

8. One number is six less than four times a smaller number. The sum of the two numbers is forty-nine. Find each number.

9. Consider three numbers. The first number is one-half of the second number. A third number is triple that second number. The sum of the three numbers is sixty-three. Find each number.

10. Consider three numbers. The first number is four times the second number. A third number is two-thirds that second number. The sum of the three numbers is thirty-four. Find each number.

11. One-fourth of a number, one-eighth of a number, and one-fifth of a number are added. The result is forty-six. Find the original number.

12. One-eighth of a number, one-third of a number, and one-sixth of a number are added. The result is sixty. Find the original number.

Solutions to Practice Problems for Exercise 5.2

1. The sum of a number, double the number, and five times the number is seventy-two. Find the original number. Let $x =$ the original number.

$(x) + (2x) + (5x) = 72$	Write the equation for the sum.
$8x = 72$	Collect like terms.
$x = 9$	Divide both sides by 8.

The original number is 9.

2. When one-third of a number is increased by twelve the result is eight. Find the number.

$\frac{x}{3} + 12 = 8$	Note we take one-third of x, *not* one-third of $x + 12$.
$3\left(\frac{x}{3}\right) + 3(12) = 3(8)$	Multiply all terms by 3.

$$x + 36 = 24 \qquad \text{Simplify.}$$

$$x = 24 - 36 \qquad \text{Subtract 36 from both sides.}$$

$$x = -12$$

The original number is -12.

3. If you double a number and add twenty you obtain one-third of the original number. Find the original number. Let $x =$ the original number.

$$2x + 20 = \frac{x}{3} \qquad \left(\text{One-third of the number may be written } \frac{x}{3} \text{ or } \frac{1}{3}x.\right)$$

$$3(2x) + 3(20) = 3\left(\frac{x}{3}\right) \qquad \text{Multiply all terms by 3.}$$

$$6x + 60 = x \qquad \text{Simplify.}$$

$$60 = x - 6x \qquad \text{Subtract } 6x \text{ from both sides.}$$

$$60 = -5x \qquad \text{Simplify.}$$

$$\frac{60}{-5} = \frac{-5x}{-5} \qquad \text{Divide both sides by } -5.$$

$$-12 = x \qquad \text{Simplify.}$$

The original number is -12.

4. Consider two numbers. The second number is twelve less than triple the first number. The sum of the two numbers is twenty-four. Find each number. Let $x =$ the first number; then $3x - 12 =$ the second number.

$$(x) + (3x - 12) = 24 \qquad \text{We write an equation for the sum.}$$

$$4x - 12 = 24 \qquad \text{Collect like terms.}$$

$$4x = 36 \qquad \text{Add 12 to both sides.}$$

$$x = 9 \qquad \text{Divide both sides by 4.}$$

The first number is 9.

The second number is obtained by substituting $x = 9$ into the expression $3x - 12$.

$$\begin{aligned} 3x - 12 &= 3(9) - 12 \\ &= 27 - 12 \\ &= 15 \end{aligned}$$

The second number is 15.

Now do the problems in Exercise 5.2.

5.3 Word Problems Involving Comparisons or Formulas

A student doing the problems in the preceding section is likely to wonder, "What good is all this stuff? When will I ever need to do problems like that?" Frankly, the preceding section is primarily drill and practice with basic algebraic skills that are necessary to solve applied word problems.

The following problems relate more directly to real-life situations. They will help you to begin to see the power of algebra. Algebra provides the tools to solve a variety of problems for people in almost every occupation of life.

EXAMPLE A In a hospital, a storage vial is filled with 1700 milligrams of medication which will be used for two patient dosages. The first patient is given 530 milligrams more than the second. How much will each patient receive?

Solution The amount of medication to be given to the first patient is being *compared to* the amount of medication to be given to the second patient. The basis of comparison is the amount given to the second patient, so we let

$$x = \text{the dosage in milligrams to be given the second patient}$$

Since the first patient will be given 530 milligrams more, let $x + 530$ = dosage in milligrams to be given the first patient (see Figure 5.3).

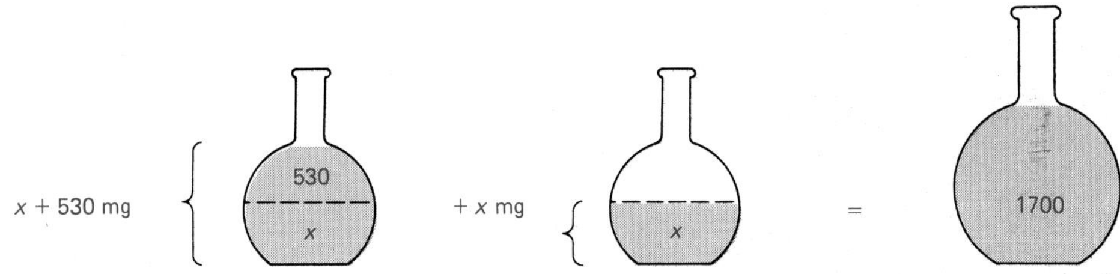

FIGURE 5.3

$(x + 530) + (x) = 1700$	The dosage of each patient is added to obtain the total.
$2x + 530 = 1700$	Collect like terms.
$2x = 1170$	Subtract 530 from each side.
$x = 585$	Divide both sides by 2.

We know the second patient receives 585 milligrams.

We then must find the dosage the first patient receives by evaluating $(x + 530)$ when $x = 585$.

So

$$x + 530 = 585 + 530 = 1115$$

The first patient receives 1115 milligrams.

Check Do these answers seem reasonable? We can verify these answers by substituting the values into equations that relate to the word problem. Do the two dosages total 1700 milligrams?

$$\text{Yes:}\quad 585 + 1115 = 1700\checkmark$$

Is the first patient given a dosage 530 milligrams more than the second?

$$\text{Yes:}\quad 1115 - 585 = 530\checkmark$$

Checking Word Problems

It is always best to follow the steps of this example to verify that the solutions you obtain are correct. You should ask yourself two questions:

1. **Are these answers reasonable (do they seem to represent values that I would estimate for the answers)?**
2. **Do these values satisfy the statements of the word problem?**

If you can say yes to both these questions, you have probably done the problem correctly.

If the word problem contains three unknown quantities, first determine what is the basis of comparison for the two quantities mentioned first in the problem. Describe each of them in terms of x. Then seek to describe the third quantity in terms of one of the others.

EXAMPLE B The city airport has filed a report showing the number of plane departures that took off from the airport during each month last year. The number of departures in March was 50 more than the number of departures in January. In July the number of departures was 150 less than triple the number of departures in January. In those three months, the airport had 2250 departures. How many departures were recorded for each month?

Solution Look at the statement, "The number of departures in March was 50 more than the number of departures in January." The first comparison is between the number of departures in March and the number of departures in January. The *basis of comparison* is the number of departures in January. So we let x = the number of departures in Janurary and $x + 50$ = the number of departures in March.

Now we examine the next sentence, "In July the number of departures was 150 less than triple the number of departures in January." Again the basis of comparison is the number of departures in January. So we write $3x - 150$ = the number of departures in July.

Now we can write the equation:

Number of departures in January		number of departures in March		number of departures in July		3 months total departures
(x)	$+$	$(x + 50)$	$+$	$(3x - 150)$	$=$	2250

$5x - 100 = 2250$ Collect like terms.

$5x = 2350$ Add 100 to each side.

$x = 470$ Divide both sides by 5.

Now if $x = 470$, then

$$x + 50 = 470 + 50 = 520$$

and

$$3x - 150 = 3(470) - 150 = 1410 - 150 = 1260$$

Thus

The number of departures in January was 470.

The number of departures in March was 520.

The number of departures in July was 1260.

Check Do these answers seem reasonable? Yes. Do these answers verify all the statements in the word problem?

1. Is the number of departures in March 50 more than those in January?

$$520 \stackrel{?}{=} 50 + 470$$

$$520 = 520 \checkmark$$

2. Is the number of departures in July 150 less than triple those in January?

$$1260 \stackrel{?}{=} 3(470) - 150$$

$$1260 \stackrel{?}{=} 1410 - 150$$

$$1260 = 1260 \checkmark$$

3. Is the total number of departures in the 3 months equal to 2250?

$$470 + 520 + 1260 \stackrel{?}{=} 2250$$

$$2250 = 2250 \checkmark$$

Yes, all conditions are satisfied. The three answers are correct.

Sometimes the solution to a problem requires that you know a formula or a certain property. In such cases it is best to write down the equation as your first step. Then you will be better able to see how to apply the formula in the given situation.

EXAMPLE C A small plot of land is in the shape of a rectangle. The length is 7 meters longer than twice the width. The perimeter of the rectangle is 86 meters. Find the width and length.

Solution Here we use the fact that the perimeter of a rectangle is twice the length added to twice the width.

$$P = 2(\text{length}) + 2(\text{width})$$

$$\boldsymbol{P = 2L + 2W}$$

Now we can proceed to solve the problem. The length is compared to the width. Since the width is the basis of comparison, we represent the width by x. Let $x =$ the width of the rectangle measured in meters, and let $2x + 7 =$ the length of the rectangle measured in meters.

$$P = 2L + 2W$$

$L = 2x + 7$

$W = x$ $W = x$

$L = 2x + 7$

$86 = 2(2x + 7) + 2(x)$	We substitute a numerical value for the perimeter. We use variable expressions for length and width. Note carefully how parentheses are used here.
$86 = 4x + 14 + 2x$	Remove parentheses.
$86 = 6x + 14$	Collect like terms.
$72 = 6x$	Subtract 14 from both sides.
$12 = x$	Divide both sides by 6.

Therefore, the width is 12 meters.

Next, to find the length we evaluate $2x + 7$ when $x = 12$.

$$2x + 7 = 2(12) + 7 = 31$$

Thus we see that the length is 31 meters.

Check Do these values satisfy the statements in the problem?

Is the length 7 meters longer than twice the width?

$$2(12) + 7 \stackrel{?}{=} 31$$

$$24 + 7 \stackrel{?}{=} 31$$

$$31 = 31\checkmark$$

Is the perimeter of a rectangle of width 12 meters and length 31 meters going to total 86 meters?

$$2(12) + 2(31) \stackrel{?}{=} 86$$

$$24 + 62 \stackrel{?}{=} 86$$

$$86 = 86\checkmark \qquad \text{That condition is also satisfied.}$$

Since both statements in the original problem are satisfied, we are confident that the width is 12 meters and the length is 31 meters.

EXAMPLE D Two people travel by car a distance of 330 miles on an interstate highway. To maximize fuel economy Fred travels at exactly 50 miles per hour. Sam travels at exactly 55 miles per hour. How much time did the trip take each person?

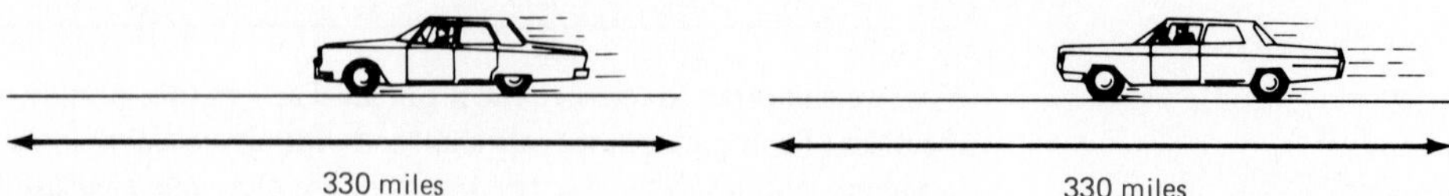

Solution Here we use the property

$$\text{distance} = (\text{rate}) \times (\text{time}) \qquad \text{or} \qquad d = rt$$

In the case of Fred's trip the distance d is 330 miles and the rate is 50 miles per hour. Since we are dealing with two different times it may cause confusion if we use t each time. We will use *subscripts* to denote the different values of t. (The use of subscripts becomes more common in higher level mathematics and science courses.) Let t_f = the amount of time in hours it took Fred to drive the trip.

$330 = 50t_f$ — Substituting the value for d and r into $d = rt$.

$6.6 = t_f$ — Divide both sides by 50.

It took Fred 6.6 hours or 6 hours and 36 minutes.

$$\left[\begin{aligned} 0.6 \text{ hour} &= \left(\frac{6}{10}\right)(1 \text{ hour}) \\ &= \left(\frac{6}{10}\right)(60 \text{ minutes}) = 36 \text{ minutes} \end{aligned}\right]$$

⟵ (Make sure you can change 0.6 hour to 36 minutes.)

In the case of Sam's trip, let t_s = amount of time in hours it took Same to drive the trip.

$d = rt$

$330 = 55t_s$ — Substitute rate = 55 mph, same distance = 330 miles.

$\frac{330}{55} = t_s$ — Divide by 55.

$6 = t_s$

It took Sam 6 hours.

Check See if (rate) × (time) = distance traveled for each person.

Fred's trip: $6.6 \times 50 \stackrel{?}{=} 330$

$330 = 330\checkmark$

Sam's trip: $6 \times 55 \stackrel{?}{=} 330$

$330 = 330\checkmark$

Thus we have verified the two answers.

EXAMPLE E A teacher told Melinda that she had a course average of 78 based on her 6 math tests. When she got home Melinda found 5 of her tests. She had scores of 87, 63, 79, 71, and 96 on the 5 tests. She could not find her sixth test. What score did she obtain on that test?

Solution We obtain an *average* by adding all the test scores and dividing by the number of tests. In this case since we do not know one score we can represent it by x. Let x = the score on the sixth text. We now write the equation to obtain an average.

$$\frac{87 + 63 + 79 + 71 + 96 + x}{6} = 78$$

$\frac{396 + x}{6} = 78$ — Add the numbers in the numerator.

$\not{6}\left(\frac{396 + x}{\not{6}}\right) = 6(78)$ — Multiply both sides of the equation by 6 to remove the fraction.

$$396 + x = 468 \quad \text{Simplify.}$$

$$x = 72 \quad \text{Subtract 396 from both sides to find } x.$$

Check To verify that this is correct, we see if the average of the 6 tests is 78.

$$\frac{87 + 63 + 79 + 71 + 96 + 72}{6} \stackrel{?}{=} 78$$

$$\frac{468}{6} \stackrel{?}{=} 78$$

$$78 = 78 \checkmark$$

So the problem checks. We know that the score on the sixth test was 72.

Now try these practice problems.

Practice Problems for Exercise 5.3

Solve these word problems and check your solutions.

1. A farmer purchased 720 meters of wire fencing. He wishes to fence in a rectangular plot of land that will hold several head of cattle prior to milking. He would like the length of the rectangle to be 40 meters longer than triple the width of the rectangle. What should be the dimensions of this rectangular holding area for the cattle?
2. Three families compared the cost of heating their homes for the year. The first family found that their annual heating bill was \$360 more than that of the second family. The third family found that their heating bill was \$200 less than double the heating bill of the second family. The total annual heating bill for the three families was \$3960. What was the annual heating bill for each family?
3. Sarah left the city to visit her aunt and uncle, who live in a rural area north of the city. She traveled the 220-mile trip in 4 hours. When she went home she took a slightly longer route which measured 225 miles on the car odometer. The return trip took 4.5 hours.
   (a) What was her average speed on the trip leaving the city?
   (b) What was her average speed on the return trip?
   (c) On which trip did she travel faster, and by how much?

Now turn to the answers following Exercise 5.3 and check your solutions for these practice problems.

Exercise 5.3

Solve these word problems and check your solutions. Use an algebraic equation to find the solution in each case. Do not guess or try random methods to find the solution.

1. The length of a rectangular piece of tin is 3 centimeters less than double the width. The perimeter of the rectangle is 42 centimeters. Find the dimensions in centimeters of the width and length of the piece of tin.

2. A rectangular piece of land is being used by the college for an experimental student cooperative garden this semester. The garden will have a perimeter of 1760 feet. The length of the garden is 200 feet less than twice the width of the garden. Find the dimensions of the garden in feet.

3. A small region in front of a building is triangular in shape. The perimeter of the triangle is 37 meters. The second side is one-half of the first side in length. The second side is 5 meters less than the third side. Find the length in meters of each side of the triangular region.

4. An electronics firm needs to use a copper plate that is triangular in shape. The perimeter of the copper plate measures 78 centimeters. The longest side is 10 centimeters longer than the length of the second longest side. The shortest side is 4 centimeters shorter than the second longest side. Find the length in centimeters of each side.

5. A state park in Colorado has a perimeter of 92 miles. The park is in the shape of a rectangle. The length of the park is 30 miles less than triple the width of the park. What are the dimensions of the park?

6. The Topsfield Fair set records this year for total attendance at the fairgrounds. On Monday there were 350 more people attending than on Tuesday. The crowds on Monday were 3100 people fewer than the crowds on Wednesday. The total attendance for Monday, Tuesday, and Wednesday was 64,550. How many people attended each of those three days?

7. Alice bought a new car that cost $1368 more than one purchased by Michael. Michael's car cost $852 more than the new car that Bettina bought. The three cars cost $27,588. Can you determine how much each car cost?

8. Three overweight people went to a weight-reducing class for one whole year. During that time the total weight lost by all three people was 156 pounds. Rita lost two-thirds the number of pounds that Margaret lost. Tony lost two more pounds than Rita. How many pounds did each lose?

9. An air sample was taken in an industrial city. It was found to contain 12.8 ppm (parts per million) of three harmful pollutants. The amount of the first pollutant was three times as great as that of the second pollutant. The amount of the third pollutant was four times as great as that of the first pollutant. How many parts per million of each pollutant were found?

10. The sum of the electric current in three transistors is 0.027 amperes (a measure of the amount of electric current). By performing a comparison test, a technician found that the electric current in the first transistor is half of that in the second. The technician also found that the amount of current in the third transistor is one and one-half times the amount of current in the second transistor. How much current is in each transistor?

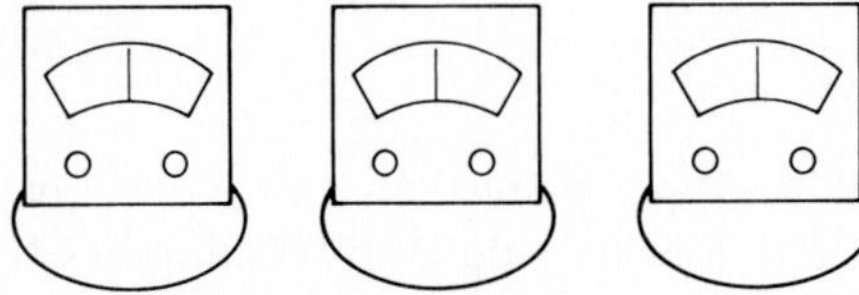

11. The sum of the electric current in three transistors is 0.1445 amperes. A technician found that the electric current in the first transistor is one-fourth the amount of current in the second transistor. However, the amount of current in the third transistor is twelve times the amount in the first transistor. How much current is in each of the three transistors?

12. Martha traveled on a highway a distance of 154 miles at exactly 55 miles per hour. An ambulance traveled the same distance moving at exactly 70 miles per hour. How much longer did it take Martha than the ambulance driver to make the trip?

13. Alfredo traveled from his house to the airport traveling at exactly 50 miles per hour. On the way home going over the same distance he traveled at exactly 47.5 miles per hour. If the distance from the airport to his house is 38 miles, how much longer was the amount of time for his return trip than for his first trip out to the airport?

14. William drove to a camp in Maine. He took a back road and found that the trip took him 4 hours to cover 88 miles. He was anxious to make better time coming home and took the Maine Turnpike for part of the trip even though that distance was 114 miles when you travel that way. Using the turnpike the total trip took 3.8 hours. How much faster (in miles per hour) did William average on the trip home than he did on the trip to the camp?

15. The Ramirez family has five licensed drivers and three cars. The smallest family car is a subcompact and gets excellent fuel mileage of 38 mpg in city driving. The second car is a compact and it gets 21 mpg in city driving. The third car is an ancient heavy station wagon. Dad Ramirez does not like to admit the mileage rating of the "old beast." Rita calculated that if all three cars were driven the same number of miles in city driving each year, the *average* mpg rating of the three Ramirez cars would be $22\frac{2}{3}$ mpg in city driving. What is the mpg rating of the old station wagon?

16. Four friends rented a house this semester while going to college. All four own cars and carry the same type of insurance coverage. Gregg pays \$350 per year to insure his car. Rich pays double that amount each year. Hector pays \$420 per year for his car insurance. Constanos purchased a new Corvette and won't tell anyone the huge amount he pays each year to insure the car. The four roommates have the same insurance agent. The agent said that if you took the average amount each of the four boys paid for car insurance it would be \$855. What is the yearly amount Constanos pays to insure his Corvette?

Challenge Problem

17. In warmer climates approximate temperature prediction can be made by counting the number of chirps of a cricket during a minute. The Fahrenheit temperature decreased by forty is equivalent to one-fourth of the number of cricket chirps.
   (a) Write an equation for this relationship.
   (b) Approximately how many chirps should be recorded if the temperature is 90°F?
   (c) If a person recorded 148 cricket chirps in a minute, what would be the Fahrenheit temperature according to this formula?

Solutions to Practice Problems for Exercise 5.3

1. A farmer purchased 720 meters of wire fencing. He wishes to fence in a rectangular plot of land that will hold several head of cattle prior to milking. He would like the length of the rectangle to be 40 meters longer than triple the width of the rectangle. What should be the dimensions of this rectangular holding area for the cattle? The perimeter of the rectangle is 720 meters.

$$P = 2L + 2W$$

L

W W

L

The basis of comparison is the width, so we represent the width by x. Let $x =$ the width of the rectangle in meters. Then $3x + 40 =$ the length of the rectangle in meters.

$P = 2L + 2W$	The perimeter equation for a rectangle.
$720 = 2(x) + 2(3x + 40)$	Substitute expressions for length and width and the value of the perimeter.
$720 = 2x + 6x + 80$	Remove parentheses.
$720 = 8x + 80$	Collect x terms.
$640 = 8x$	Subtract 80 from both sides.
$80 = x$	Divide both sides by 8.

The width is 80 meters. Thus if $x = 80$, then

$$3x + 40 = 280$$

Thus the length is 280 meters.

Check Does the perimeter equal 720?

$$2(80) + 2(280) \stackrel{?}{=} 720$$

$$160 + 560 \stackrel{?}{=} 720$$

$$720 = 720\checkmark$$

Does the length equal 40 meters longer than triple the width?

$$280 \stackrel{?}{=} 3(80) + 40$$

$$280 \stackrel{?}{=} 240 + 40$$

$$280 = 280\checkmark$$

2. Three families compared the cost of heating their homes for the year. The first family found that their annual heating bill was \$360 more than that of the second family. The third family found that their heating bill was \$200 less than double the heating bill of the second family. The total annual heating bill for the three families was \$3960. What was the annual heating bill for each family? The basis of comparison is the amount of money used for the annual heating cost of the second family. Let

x = the cost in dollars of the heating cost of the second family

Then

$x + 360$ = the cost in dollars of the heating cost of the first family

$2x - 200$ = the cost in dollars of the heating cost of the third family

Now since the total annual cost for all three families is \$3960, we write the equation

$$(x + 360) + (x) + (2x - 200) = \$3960$$

$4x + 160 = 3960$ Remove parentheses and collect like terms.

$4x = 3800$ Subtract 160 from both sides.

$x = 950$ Divide both sides by 4.

The heating cost of the second family is \$950. Then

$$x + 360 = 950 + 360 = 1310$$

The heating cost of the first family is \$1310. Finally,

$$2x - 200 = 2(950) - 200 = 1900 - 200 = 1700$$

The heating cost of the third family is \$1700.

Check Does the total equal \$3960?

$$\$1310 + \$950 + \$1700 \stackrel{?}{=} \$3960$$

$$\$3960 = \$3960\checkmark$$

Does the amount of the first family come to \$360 more than that of the second family?

$$\$950 + \$360 \stackrel{?}{=} \$1310$$

$$\$1310 = \$1310 \checkmark$$

Does the amount of the third family come to \$200 less than double the amount of the second family?

$$\$1700 \stackrel{?}{=} 2(\$950) - \$200$$

$$\$1700 \stackrel{?}{=} \$1900 - \$200$$

$$\$1700 = \$1700 \checkmark$$

3. Sarah left the city to vist her aunt and uncle, who live in a rural area north of the city. She traveled the 220-mile trip in 4 hours. When she went home she took a slightly longer route, which measured 225 miles on the car odometer. The return trip took 4.5 hours.

(a) What was her average speed on the trip leaving the city?

(b) What was her average speed on the return trip?

(c) On which trip did she travel faster and by how much?

(a) Let us examine the 220-mile trip from the city, which took 4 hours.

$R \cdot T = D$	The basic formula.
$(R_{\text{out}})(4) = 220$	Substitute known values.
$R_{\text{out}} = 55$	Divide both sides by 4.

The average rate of speed going out of the city was 55 mph.

(b) The trip into the city was 225 miles and took 4.5 hours.

$R \cdot T = D$	The basic formula.
$(R_{\text{in}})(4.5) = 225$	Substitute known values.
$R_{\text{in}} = 50$	Divide both sides by 4.5.

The average rate of speed coming into the city was 50 mph.

(c) She traveled 5 mph faster on the trip out.

Check Can you verify these answers?

Now do the problems in Exercise 5.3.

5.4 Word Problems Involving the Value of Money and Percents

Certain applied problems are a bit more involved than those of Section 5.3. Some of these problems are encountered in business applications. Many problems deal with the value of certain monetary amounts. In this section we examine a number of these types of problems.

Many applications require an understanding of the use of percentage and decimals. Please review Sections 0.4 and 0.5 if you are weak in these skills.

EXAMPLE A A business executive rented a car. The Car Rental Agency charged \$39 per day and \$0.28 per mile. The executive rented the car for two days and the total rental cost was computed to be \$176.00. How many miles did the executive drive the rented car?

Solution Our basic concern is to determine the number of miles he drove. So we represent that by the variable x. Let x = number of miles driven in the rented car. Now we analyze the total cost in terms of a fixed charge of \$39 for each day added to a variable cost based on \$0.28 for each mile (x) that was driven.

$$\text{Per day cost} + \text{mileage cost} = \text{total cost}$$

For 2 days at \$0.28 per mile and a bill of \$176.00, we have

$$(39)(2) + (0.28)(x) = 176$$

$$78 + 0.28x = 176 \qquad \text{Simplify the equation.}$$

$$0.28x = 98 \qquad \text{Subtract 78 from both sides.}$$

$$\frac{0.28x}{0.28} = \frac{98}{0.28} \qquad \text{Divide both sides by 0.28.}$$

$$x = 350 \qquad \text{Simplify.}$$

The executive drove 350 miles.

Check If you drive the car 350 miles in 2 days, would it cost \$176?

$$(\text{Cost of \$39 per day for 2 days}) + (\text{cost of \$0.28 per mile for 350 miles}) \stackrel{?}{=} \text{total cost of \$176}$$

$$(\$39)(2) + (350)(\$0.28) \stackrel{?}{=} \$176.00$$

$$\$78.00 + \$98.00 \stackrel{?}{=} \$176.00$$

$$\$176.00 = \$176.00 \checkmark$$

The topic of interest and high interest rates is frequently discussed in the news media and by people in general (see Figure 5.4). A student of mathematics should be able to solve problems involving interest. There are two basic types of interest.

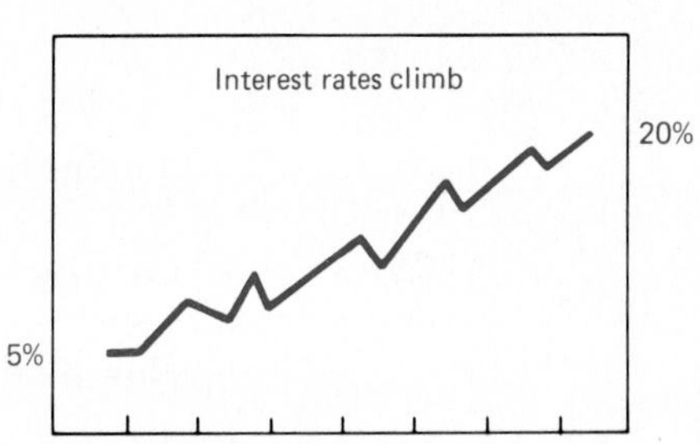

FIGURE 5.4

Simple interest

Simple interest is a charge for borrowing money or an income from investing money. It is computed by multiplying the amount of money borrowed or invested (which is called the *principal*) times the rate of interest times the period of time it is borrowed or invested (usually measured in years unless otherwise stated).

$$\text{Interest} = \text{principal} \times \text{rate} \times \text{time}$$

$$I = prt$$

> ***All exercises discussed in this chapter will involve simple interest.***

Compound interest

You often hear of banks offering a certain interest rate compounded quarterly, monthly, weekly, or daily. In *compound interest* the amount of interest is added to the amount of the original principal at the end of each time period, so that future interest is based on the sum of both principal and previous interest. Most financial institutions use compound interest in their transactions.

Problems involving compound interest may be solved by:

1. Repeated calculations using the simple interest formula
2. Using a compound interest table (see page 94)
3. Using exponential functions, a topic that is usually covered in a higher-level college algebra course

We will *not* use compound interest in the exercises of this chapter.

Let us now apply the concept of *simple interest* to the following examples.

EXAMPLE B Find the interest on \$3000 borrowed at an interest rate of 18% for 1 year.

Solution

$I = prt$	The simple interest formula.
$I = (3000)(0.18)(1)$	Substitute the value of the principal = 3000; the rate = 18% = 0.18; the time = 1 year.
$I = 540$	

Thus the interest charge for borrowing \$3000 for 1 year at a simple interest rate of 18% is \$540.

EXAMPLE C Write an expression for the interest earned if the principal of x dollars is invested at 12% for 1 year.

Solution

$I = prt$	The simple interest formula.
$I = (x)(0.12)(1)$	Substitute known values and represent principal by x.
$I = 0.12x$	Simplify.

Now we apply this concept to a more involved problem.

EXAMPLE D A woman invested an amount of money in two places for 1 year. She invested some at 8% interest and the rest at 12% interest. Her total amount of investment was \$1250. At the end of the year she had earned \$128.00 in interest. How much money did she invest in each place?

Solution There are two amounts of money. The total value of the two amounts is \$1250. We do not know either amount. If we represent one amount by x, do you see that the other amount could be represented by $1250 - x$? (This is because the total of the two amounts must be \$1250! Look at Figure 5.5.)

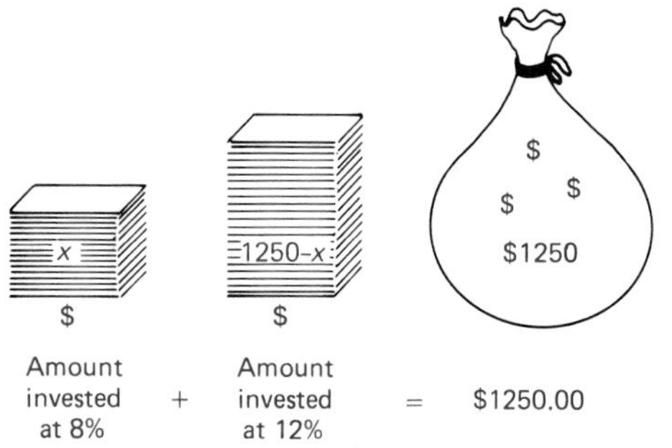

FIGURE 5.5

Therefore,

let $x =$ the amount of money invested at 8% interest

$1250 - x =$ the amount of money invested at 12% interest

How much interest is earned at 8%?

$I = prt$	The simple interest formula
$I = (x)(0.08)(1)$	Substitute 0.08 for rate, 1 for time of 1 year, and represent principal by x.
$I = 0.08x$	Simplify.

How much interest is earned at 12%?

$I = prt$	The simple interest formula.
$I = (1250 - x)(0.12)(1)$	Substitute 0.12 for rate, 1 for time of 1 year, and represent principal by $1250 - x$.
$I = 0.12(1250 - x)$	Simplify.

Now we look at the total interest:

$$\begin{matrix} \text{Interest earned at } 8\% & + & \text{interest earned at } 12\% & = & \text{total interest earned during the year.} \\ 0.08x & + & 0.12(1250 - x) & = & \$128.00 \end{matrix}$$

In writing this equation, we substituted the two expressions for interest that we developed above.

$0.08x + 150 - 0.12x = 128$	Remove parentheses.
$-0.04x + 150 = 128$	Collect like terms.
$-0.4x = -22$	Subtract 150 from both sides.
$\frac{-0.04x}{-0.04} = \frac{-22}{-0.04}$	Divide both sides by -0.04.
$x = 550$	The amount invested at 8% interest.
$1250 - x = 1250 - 550 = 700$	The amount invested at 12% interest.

So \$550 was invested at the 8% interest rate and \$700 was invested at the 12% interest rate.

Check Do the two amounts equal \$1250?

$$\$550 + \$700 \stackrel{?}{=} \$1250$$

$$\$1250 = \$1250 \checkmark$$

Would these amounts earn \$128 interest in 1 year invested at the specified rates?

$$0.08(\$550) + 0.12(\$700) \stackrel{?}{=} \$128$$

$$\$44 + \$84 \stackrel{?}{=} \$128$$

$$\$128 = \$128 \checkmark$$

Solving verbal problems involving the value of money

Suppose that we know enough information to write an equation about the value of some coins, but we do not know how many coins we have. Consider this example.

EXAMPLE E When Bob got out of math class he had to make a long-distance call. He had exactly enough dimes and quarters to make a phone call that would cost \$2.55. He had one less quarter than he had dimes. How many coins of each type did he have?

Solution The number of quarters is compared to the number of dimes. Since the point of comparison is the number of dimes, we represent the number of dimes by x. Let x = the number of dimes. Therefore, $x - 1$ = the number of quarters. How much are these coins worth in value?

Each dime is worth \$0.10.

If we had five dimes they would be worth $(5)(0.10) = 0.50$.

If we had x dimes they would be worth $(x)(0.10) = 0.10x$.

Each quarter is worth \$0.25.

If we had 8 quarters they would be worth $(8)(0.25) = 2.00$.

If we had $(x - 1)$ quarters they would be worth $0.25(x - 1)$.

Now we can write an equation for the total value.

$$(\text{Value of dimes}) + (\text{value of quarters}) = \$2.55$$

$$0.10x + 0.25(x - 1) = \$2.55$$

$0.10x + 0.25x - 0.25 = 2.55$ Remove parentheses.

$0.35x - 0.25 = 2.55$ Collect like terms.

$0.35x = 2.80$ Add 0.25 to both sides.

$\frac{0.35x}{0.35} = \frac{2.80}{0.35}$ Divide both sides by 0.35.

$x = 8$ Simplify.

$x - 1 = 7$

Thus Bob had 8 dimes and 7 quarters.

Check Does Bob have 1 less quarter than he has dimes?

$$8 - 7 = 1 \checkmark$$

Are 8 dimes and 7 quarters worth \$2.55?

$$8(\$0.10) + 7(\$0.25) \stackrel{?}{=} \$2.55$$

$$\$0.80 + \$1.75 \stackrel{?}{=} \$2.55$$

$$\$2.55 = \$2.55 \checkmark$$

EXAMPLE F Michele returned from the store with \$2.80 in change. She had twice as many quarters as nickels. She had 2 more dimes than nickels. How many nickels, dimes, and quarters did she have?

Solution Let

$$x = \text{the number of nickels}$$

$$2x = \text{the number of quarters}$$

$$x + 2 = \text{the number of dimes}$$

Now we must find the *value* of the coins.

Since 1 nickel = \$0.05, we let $0.05x$ = *value* of the nickels.
Since 1 dime = \$0.10 we, let $0.10(x + 2)$ = *value* of the dimes.
Since 1 quarter = \$0.25, we let $0.25(2x)$ = *value* of the quarters.

We can now write an equation for total value.

$$(\text{Value of nickels}) + (\text{value of dimes}) + (\text{value of quarters}) = \$2.80$$

$$0.05x \quad + \quad 0.10(x + 2) \quad + \quad 0.25(2x) \quad = \$2.80$$

$$0.05x + 0.10x + 0.20 + 0.50x = 2.80 \qquad \text{Remove parentheses.}$$

$$0.65x + 0.20 = 2.80 \qquad \text{Collect like terms.}$$

$$0.65x = 2.60 \qquad \text{Subtract 0.20 from both sides.}$$

$$\frac{0.65x}{0.65} = \frac{2.60}{0.65} \qquad \text{Divide both sides by 0.65.}$$

$$x = 4 \qquad \text{Simplify.}$$

$$2x = 8 \qquad \text{Michele had 4 nickels, 8 quarters,}$$

$$2 + x = 6 \qquad \text{and 6 dimes.}$$

Check Does Michele have twice as many quarters as nickels?

$$(4)(2) \stackrel{?}{=} 8 \qquad 8 = 8\checkmark$$

Does she have 2 more dimes than nickels?

$$4 + 2 \stackrel{?}{=} 6 \qquad 6 = 6\checkmark$$

Do 4 nickels, 8 quarters, and 6 dimes have a value of \$2.80?

$$4(\$0.05) + 8(\$0.25) + 6(\$0.10) \stackrel{?}{=} \$2.80$$

$$\$0.20 + \$2.00 + \$0.60 \stackrel{?}{=} \$2.80$$

$$\$2.80 = \$2.80\checkmark$$

Practice Problems for Exercise 5.4

Solve these word problems.

1. Alfredo rented a truck to move to Florida. He determined that the cheapest rental rates are from a local company that will charge him \$22 per day and \$0.18 per mile for a truck of the size that he will need. He has not yet completed an estimate of the mileage of the trip, but he knows that he will need the truck for 3 days. He has allowed $363 in his moving budget for the truck. How far can he travel if the truck rental cost is exactly $363?

2. An elderly lady invested her savings of $8000 in two accounts that calculate interest only once per year. She placed one amount in a special notice account that yields 15% annual interest. The remainder she placed in a tax-free All-Savers account that yields 12% annual interest. At the end of the year she earned $1155 in interest from the two accounts together. How much did she invest in each account?

3. A young boy told his friend that he had twice as many nickels as dimes in his pocket. He also said he had 4 more quarters than dimes. He said he had $2.35 in change in his pocket. Can you determine how many nickels, dimes, and quarters he had?

Now turn to the answers following Exercise 5.4 and check your solutions for these practice problems.

Exercise 5.4

Solve these word problems.

1. Barbara rented a car for $18 per day. She was charged a mileage rate of $0.08 per mile. She rented the car for 4 days. Her total bill to the rental agency was $100.80. How many miles did she drive?

2. Richard rented a truck for $38 per day. He used the truck for 2 days. The mileage charge was $0.20 per mile. He paid a total of $107.60 to the truck rental company. How many miles did he drive the truck?

3. In the summer Nancy has a sales job where she is paid a salary plus a commission. She earns $10 per day plus $8 for every dictionary she sells. Last week she worked 5 days and earned $186. How many dictionaries did she sell?

4. Marybelle is contemplating a job as a waitress. She would be paid $2 per hour plus tips. The other waitresses have told her that an average tip at that restaurant is $3 per table served. If she works 20 hours per week, how many tables would she have to service in order to make $151 during the week?

5. Little Melinda has nickels and quarters in her bank. She has 4 less nickels than quarters. She has \$3.70 in the bank. How many coins of each type does she have?

6. Fred's younger brother had several coins when he returned from his paper route. He had in dimes and quarters a total of \$5.35. He said he had 6 more quarters than he had dimes. How many of each coin did he have?

7. A newspaper boy has \$3.75 in change. He has 3 more quarters than dimes but twice as many nickels as quarters. How many coins of each type does he have?

8. Alice collected a number of nickels, dimes, and quarters from her room. She found 3 more dimes than nickels but twice as many quarters as dimes. The value of her coins was \$5.05. How many coins of each type did she have?

9. Terry has 2 less quarters than nickels and 1 more half-dollar than twice the number of nickels. She has \$9.10. How many of each coin does she have?

10. Mike collected coins from a vending machine. He found 3 times as many nickels as dimes. He had 8 more quarters than dimes. He collected \$4.50 from one machine. How many coins of each type did he collect?

11. Linda has a job as a loan counselor in a bank. She advised a customer to invest part of her money in a 12% money market fund and the rest in a 14% investment fund. The customer had \$4000 to invest. If she earned \$508 in interest after 1 year, how much did she invest in each fund?

12. A retired couple earned \$656 last year from investments made in tax-free bonds. They invested some of the money at 8% and the rest at 10%. The total amount they invested was \$7200. If the couple invested the \$7200 for exactly 1 year, how much did they invest at each interest rate?

13. A man invested some money at 14% interest. At the end of the year the total amount of his original principal and the interest was $5700. How much did he originally invest?

14. The cost of living last year went up 13%. Fred was fortunate to get a 13% increase in salary. He now makes $15,820 per year. How much did he make last year?

Challenge Problems

The following problems are more unusual and significantly more challenging. Be sure that you understand the solutions to problems 1 to 14 before you attempt them.

15. The sum of the measure of the three interior angles of a triangle is always 180°. In a certain triangle the measure of one angle is double the measure of a second angle but is 5 degrees less than the measure of the third angle. Can you find the measure of each angle?
16. A pumping station pumps 2000 gallons per hour into a reservoir tank for a town's drinking water. The tank is initially empty. The station pumps for 3 hours. Then a leak in the reservoir tank is created by a large crack. Some water flows out of the reservoir tank at a constant rate. The pumping station continues pumping for 6 more hours while the leak is undetected. At the end of 9 hours the reservoir has 17,640 gallons. During the last 6 hours how many gallons per hour were leaking from the reservoir?
17. The Superior Car Rental Agency will rent a compact car for $35.00 per day and an additional charge of $0.24 per mile. The Peerless Car Rental Agency only charges $0.16 per mile but charges $41.00 per day. If a salesperson wanted to rent a car for 3 days, how many miles would that person have to drive to make the Peerless Car Rental Agency car a better bargain?
18. The length of a spring increases 1.32 centimeters for each kilogram it supports. The spring is 12.00 centimeters long when it is supporting 3.5 kilograms. Find the original length of the spring.

Super-Challenge Problems

19. Wong Lin does a lot of city driving. He is considering buying a new compact car. The model that costs $6400 has a gasoline engine and is EPA rated at 28 mpg in the city. The model that costs $8200 has a diesel engine and is EPA rated at 36 mpg in the city. Wong Lin plans to own this new car for 4 years. How many miles would he have to drive during the 4 years to make the purchase of the diesel car worthwhile? (Assume that gasoline is $1.40 per gallon and the diesel fuel is $1.35 per gallon.)
20. An airline travel service for the Grand Canyon has fixed costs of $150,000 per year. For flying one passenger in an open airplane the operating costs to the airline travel service are $12 per trip. For each sightseeing trip the airline travel service charges the passenger $32. For each 10 trips the airline has to pay for a $150 preventive maintenance servicing of the plane engine. How many trips were made last year if the airline travel service made a profit of $50,000?

Solutions to Practice Problems for Exercise 5.4

1. Alfredo rented a truck to move his family to Florida. He determined that the cheapest rental rates are from a local company that will charge him $22 per day and $0.18 per mile for a truck of the size that he will need. He has not yet completed an estimate of the mileage of the trip, but he knows he will need the truck for 3 days. He has allowed $363 in his moving budget for the truck. How far can he travel if the truck rental cost is exactly $363?

   The major unknown quantity is the number of miles driven, so let $x =$ the number of miles driven in the rental truck.

$$\text{(Per day cost)} + \text{(mileage cost)} = \text{total cost}$$

Three days at \$22 per day and a mileage charge of \$0.18 per mile will amount to a total of \$363. We have this equation:

$$(3)(22) + (0.18)(x) = 363 \qquad \text{Substitution of known values.}$$

$$66 + 0.18x = 363 \qquad \text{Simplify.}$$

$$0.18x = 297 \qquad \text{Subtract 66 from each side.}$$

$$\frac{0.18x}{0.18} = \frac{297}{0.18} \qquad \text{Divide both sides by 0.18.}$$

$$x = 1650 \qquad \text{Simplify.}$$

The truck could be driven 1650 miles.

Check Is the cost exactly \$363 to drive the truck 1650 miles in the 3 days?

(\$22 per day for 3 days) + (\$0.18 per mile for 1650 miles) $\stackrel{?}{=}$ total cost of \$363

$$(22)(3) + (0.18)(1650) \stackrel{?}{=} 363$$

$$66 + 297 \stackrel{?}{=} 363$$

$$363 = 363\checkmark$$

2. An elderly lady invested her savings of \$8000 in two accounts that calculate interest only once per year. She placed one amount in a special notice account that yields 15% annual interest. The remainder she placed in a tax-free All-Savers account that yields 12% annual interest. At the end of the year she earned \$1155 in interest from the two accounts together. How much did she invest in each account?

The two investment amounts total \$8000. Let $x =$ the amount invested at 15%. Then $8000 - x =$ the remaining amount invested at 12%.

$$\begin{pmatrix}\text{Interest from}\\ 15\%\text{ investment}\end{pmatrix} + \begin{pmatrix}\text{interest from}\\ 12\%\text{ investment}\end{pmatrix} = \begin{matrix}\text{total amount of}\\ \text{interest income}\\ \text{earned}\end{matrix}$$

$$(0.15)(x) \quad + \quad 0.12(8000 - x) \quad = \quad \$1155 \qquad \text{Substitution.}$$

$$0.15x + 960 - 0.12x = 1155 \qquad \text{Remove parentheses.}$$

$$0.03x + 960 = 1155 \qquad \text{Collect like terms.}$$

$$0.03x = 195 \qquad \text{Subtract 960 from both sides.}$$

$$\frac{0.03x}{0.03} = \frac{195}{0.03} \qquad \text{Divide both sides by 0.03.}$$

$$x = 6500$$

Therefore,

$$8000 - x = 8000 - 6500 = 1500$$

The lady invested \$6500 at 15% and 1500 at 12%.

Check Do the two amounts total \$8000?

$$1500 + 6500 = \$8000\checkmark$$

Do the two amounts invested at the specified rates yield \$1155 in interest?

$$(0.15)(6500) + 0.12(1500) \stackrel{?}{=} 1155$$

$$975 + 180 \stackrel{?}{=} 1155$$

$$1155 = 1155\checkmark$$

3. A young boy told his friend that he had twice as many nickels as dimes in his pocket. He also said he had 4 more quarters than dimes. He said he had \$2.35 in change in his pocket. Can you determine how many nickels, dimes, and quarters he had?

Since the number of dimes appears to be the point of comparison for the number of coins we begin by representing the number of dimes as x. So let

$$x = \text{the number of dimes}$$

$$2x = \text{the number of nickels}$$

$$x + 4 = \text{the number of quarters}$$

Now we must find *the value of the coins.*

Since 1 dime = \$0.10, we let $0.10x$ = value of the dimes.

Since 1 nickel = \$0.05, we let $0.05(2x)$ = value of the nickels.

Since 1 quarter = \$0.25, we let $0.25(x + 4)$ = value of the quarters.

Now we look at the total value of \$2.35.

(Value of dimes) + (value of nickels) + (value of quarters) = total value

$$0.10x + 0.05(2x) + 0.25(x + 4) = 2.35$$

$$0.10x + 0.10x + 0.25x + 1.00 = 2.35 \qquad \text{Remove parentheses.}$$

$$0.45x + 1.00 = 2.35 \qquad \text{Collect like terms.}$$

$$0.45x = 1.35 \qquad \text{Subtract 1.00 from both sides.}$$

$$\frac{0.45x}{0.45} = \frac{1.35}{0.45} \qquad \text{Divide both sides by 0.45.}$$

$$x = 3 \qquad \text{Simplify.}$$

$$2x = 6$$

$$x + 4 = 7$$

So the boy had 3 dimes, 6 nickels, and 7 quarters.

Check Do 3 dimes, 6 nickels, and 7 quarters yield \$2.35?

$$3(\$0.10) + 6(\$0.05) + 7(\$0.25) \stackrel{?}{=} \$2.35$$

$$\$0.30 + \$0.30 + \$1.75 \stackrel{?}{=} \$2.35$$

$$\$2.35 = \$2.35\checkmark$$

Does he have twice as many nickels as dimes?

$$2(3) = 6\checkmark$$

Does he have 4 more quarters than dimes?

$$7 - 3 = 4\checkmark$$

Now do the problems in Exercise 5.4.

An Application for Chapter 5: Response Time to a Distress Call

Today computer programs have been written to assist the Coast Guard and other rescue personnel in making decisions and predictions concerning rescue operations. Many of those decisions can be made using mathematical techniques that you have learned in this chapter. However, in practice the information is needed very quickly, so usually rescue personnel do not sit down and leisurely solve mathematical word problems. If they do not have access to a computer, they sometimes use special tables and charts to determine the required values needed in the rescue operation.

A challenge for you!

See if you can solve the following problem.

During a winter storm, a fishing boat develops a leak and the ship's pumps fail. The sinking boat has made emergency radio contact with the Coast Guard. The captain reports that he is 100 miles from the coastal harbor station and that his ship is leaking. The boat is able to head directly toward the harbor, proceeding slowly at 5 miles per hour. The captain warned that he can stay afloat only for 1 more hour. A rescue helicopter leaves the Coast Guard harbor station 5 minutes ($\frac{1}{12}$ hour) later. The helicopter is flying at 120 miles per hour. Will the helicopter reach the boat before it sinks?

Although the answer follows, see if you can solve this problem on your own.

Answer

Try to visualize the distance and the directions of travel of both the boat and the helicopter (Figure 5.6). We use the formula $d = rt$ for both. Let

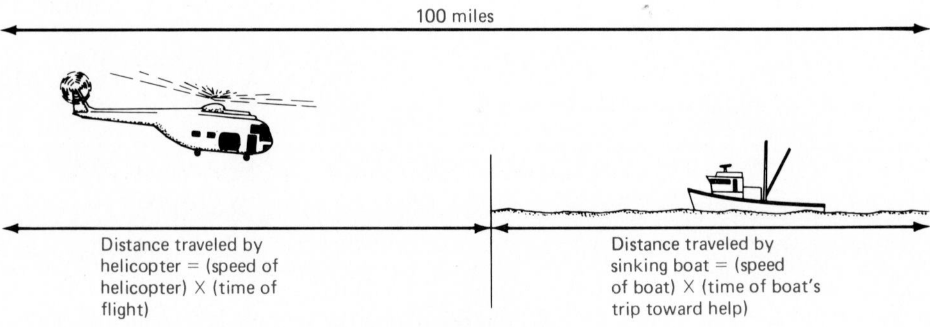

FIGURE 5.6

t = the time in hours of the boat traveling (starting with the time of the distress call)

$t - \dfrac{1}{12}$ = the time in hours that the helicopter flies to reach the boat (note that since the helicopter left 5 minutes or $\frac{1}{12}$ of an hour after the distress call the time is $t - \frac{1}{12}$ hours)

$$\text{Distance by helicopter} + \text{distance by boat} = 100 \text{ miles}$$

$$(r_{\text{helicopter}})(t_{\text{helicopter}}) + (r_{\text{boat}})(t_{\text{boat}}) = 100$$

$$120\left(t - \frac{1}{12}\right) + (5)(t) = 100$$ We substitute rate and time of both boat and helicopter.

$$120t - 10 + 5t = 100$$ Simplify.

$$125t - 10 = 100 \qquad \text{Collect like terms.}$$

$$125t = 110 \qquad \text{Add 10 to both sides.}$$

$$t = \frac{110}{125} = \frac{22}{25} \qquad \text{Divide both sides by 125.}$$

The time the boat travels before the helicopter arrives is $\frac{22}{25}$ hour or about 52.8 minutes. If the captain's opinion that his boat will stay afloat for 1 hour is correct, the helicopter will reach the boat with approximately 7.2 minutes to spare before it sinks!

Supplementary Problems for Chapter 5

Solve these word problems.

1. A man invests \$5000 in a savings bank. He places part of it in a checking account which earns 4.5% and the rest in a regular savings account which earns 6%. His total annual income from this investment is \$270. How much was invested in each account?

2. Mary has \$3.75 in nickels, dimes, and quarters. She has 3 more quarters than dimes. She has twice as many nickels as quarters. How many of each coin does she have?

3. The electric bill at Jane's house this month was \$71.50. The charge is based on a flat rate of \$25 per month plus a charge of \$0.15 per kilowatt-hour of electricity used. How many kilowatt-hours of electricity were used?

4. Susan borrowed \$333 from her uncle to help purchase a car. She promised to pay him back \$45 in June and then \$18 per month until the loan is paid back. Starting with the June payment, how many months will it take to pay off the loan?

5. The perimeter of a triangle is 40 meters. The length of the second side is 4 meters longer than twice the first side. The length of the third side is 8 meters less than the length of the second side. Find the length of each side of the triangle.

6. The length of a rectangle is $2\frac{1}{2}$ times its width. The rectangle has a perimeter of 84 feet. Find the length and width of the rectangle.

7. If the length of one side of a square is increased by 3 inches and the length of the adjacent side is decreased by 2 inches, a rectangle is formed. The rectangle has an area equal to the area of the original square. Find the length of one side in the original square.

8. Fred timed a popular TV show that was scheduled for 30 minutes. The entertainment portion of the show lasted 4 minutes longer than four times the amount of time that was devoted to commercials. How many minutes long was the actual entertainment portion of the show?

9. Faye conducted an experiment. To verify the weight of 4 identical steel balls she placed them on a balance. She placed 3 of the balls on one side of the balance. On the other side she placed an 8-ounce weight and one of the balls. She discovered that these two sides exactly balanced on the balance beam. How much does each steel ball weigh?

10. A student took five math tests. She obtained scores of 85, 78, 65, 92, and 70. She needs to complete one more test. If she wants to get an average of 80 on all six tests, what score does she need on her last test?

11. In an election for town manager 7179 votes were cast for two people running for the office. If only six people who voted for the winner had changed their minds and voted for the loser, then the person who lost would have actually won by one vote. How many votes were given to each candidate?

12. Two salesmen work for different companies. Richard's company pays him only a 10% commission on all the items he sells. Gregg's company pays him \$75 per week plus a 6% commission on all the items he sells. Last week the two salesmen earned exactly the same amount of money. Richard made weekly gross sales of \$125 more in merchandise than Gregg did. What was the amount of weekly gross sales for each person?

13. The college baseball team has won seven of their first nine games. They will play an additional 16 games. The coach desires that they win at least 84% of their games in order to be eligible for the regional playoffs. How many of the upcoming 16 games will they need to win?

14. Twenty-five percent of a number is added to 30% of 4 more than that number. The sum is 21. What is the original number?

15. A piece of rope 50 yards long is cut into two pieces. One piece is three-fifths as long as the other. Find the length of each piece.

16. The measure of the second angle of a triangle is three times the measure of the first angle. The measure of the third angle is 12 degrees less than twice the measure of the first. Find the measure of each of the three angles.

Chapter 5 Summary

Solving Word Problems

1. Read the problem to get an overview. Draw a sketch of quantities or objects described in the problem if such a sketch will help you visualize the situation.
2. Carefully reread the problem to determine what is required and what quantity you will represent by a variable. Write down an equation that indicates exactly what the variable is representing.
3. Determine the relationship between this variable and other quantities in the problem and express this as an algebraic expression.
4. Write an equation that best describes the given information with the variable you have selected.
5. Solve the resulting equation and determine the values asked for in the problem.

Checking Word Problems

1. Ask if the answers obtained are reasonable. Do the values you obtain seem to be close to what you would have estimated them to be?
2. Check to see that your values obtained actually satisfy all given requirements in the problem. This will usually involve at least two separate sets of calculations.

Practice Quiz for Chapter 5

Sections 5.1–5.2

Replace each phrase by the appropriate algebraic notation. Let the number be represented by x.

1. The quotient of a number and eight is increased by five.
2. Two-thirds of the difference between a number and six.

Translate each word problem into an equation and find the information requested.

3. Five times a number is increased by three. The result is the same as if the original number were multiplied by four and then decreased by seven. Find the original number.
4. One number is five less than four times a smaller number. The sum of the numbers is twenty-four. Find the two numbers.

Sections 5.3–5.4

Solve these word problems.

1. The length of a rectangle is four more than twice the width. The perimeter of the rectangle is 44 meters. Find the length and width of the rectangle.

2. Three boys earned \$856 doing part-time jobs during the year. The first boy earned \$110 more than the second boy. However, the first boy earned \$36 less than the third boy. How much did each boy earn?
3. Alicia rented a car from an agency that charged \$19 per day and \$0.24 per mile. She rented the car for three days and was charged \$129. How many miles did she drive the car?
4. Fred has \$3.95 in change in his pocket. He has two-thirds as many nickels as he has dimes. He has two less dimes than he has quarters. How many of each type of coin does he have?

Practice Test for Chapter 5

Solve these word problems.

1. The sum of one-half of a number, one-ninth of a number, and one-twelfth of a number is twenty-five. Find the original number.
2. Triple a number is increased by six. The result is the same as when the original number was diminished by three and then doubled. Find the original number.
3. Ramon invested \$4000 in money market funds. Part was invested at 14% interest, the rest at 11% interest. At the end of each year the fund company pays interest. After one year he earned \$482 in interest. How much was invested at each interest rate?
4. Three harmful pollutants were measured by a consumer group in the city. The total sample contained 15 parts per million of three harmful pollutants. The amount of the first pollutant was double the second. The amount of the third pollutant was 75% of the second. How many parts per million of each pollutant were found?
5. Mary has \$3.50 in change. She has twice as many nickels as quarters. She has one more quarter than she has dimes. How many of each coin does she have?
6. A triangular region has a perimeter of 66 meters. The first side is two-thirds of the second side. The third side is 4 meters shorter than the first side. What are the lengths of the three sides of the triangular region?

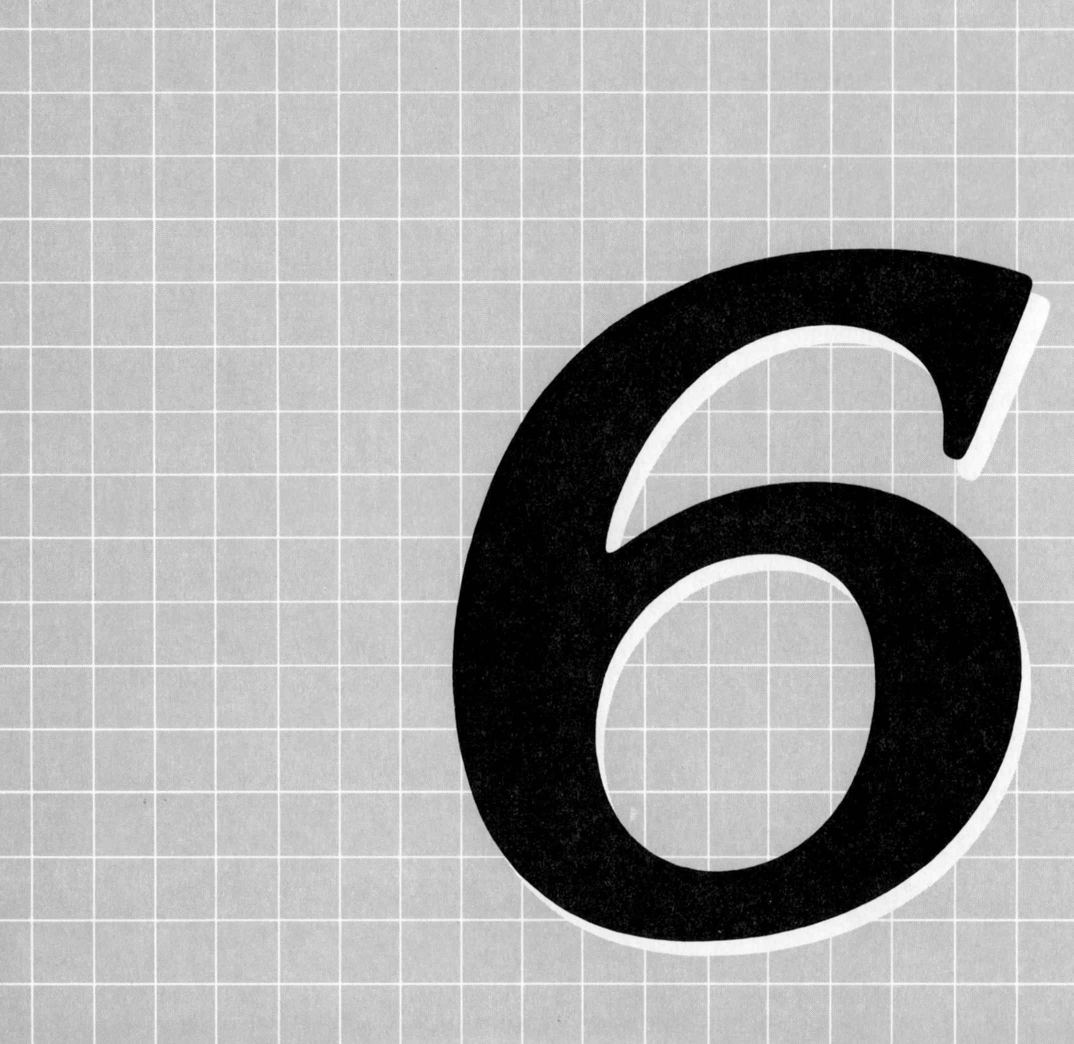

6

Graphing and Geometry

Diagnostic Pretest for Chapter 6

This diagnostic pretest is intended for those students who have previously studied algebra. Work as many problems as you can. Then, check your answers against those listed at the back of the book. Where you have obtained an incorrect answer or have been unable to work the problem, you will need to study the corresponding sections of the chapter.

If you have not previously studied algebra, do not take this test now. You should study the chapter and work the problems and exercises. After finishing the chapter, you may use this test as a review or as a practice test.

Follow the directions and solve each problem.

Section 6.1

1. Graph the points $A = (-3, 1)$, $B = (4, -2)$, and $C = (-7, -6)$.

2. Connect the points $A = (-1, -2)$, $B = (3, -2)$, and $C = (-1, 2)$ by straight lines. What shape is obtained?

Section 6.2

Graph the straight lines defined by the following equations.

3. $5x - 2y = -20$

4. $y = \frac{2}{3}x + 4$

Section 6.3

5. What is the slope of the line passing through $(4, -2)$ and $(-3, -4)$?

6. Find the slope of the line given by the equation $4x - 3y - 7 = 0$.

7. Write the equation of the line with y intercept -5 that passes through the point $(2, 3)$.

Section 6.4

8. Graph the region described by $y \geq 3x - 2$.

9. Graph the region described by $4x + 2y < -12$.

Section 6.5

10. Find the area of a triangle whose base is 4 meters and whose altitude is 17 meters.

11. A parallelogram has a perimeter of 100 inches and two sides of length 12.5 inches. What is the length of each of the other sides?

12. Write an expression for the area of a trapezoid whose altitude is 5 centimeters and whose longer base is 14 centimeters, but whose smaller base is not known.

Student Learning Objectives for Chapter 6

After successfully learning the content of each section, you will be able to:

Section 6.1

A. Plot and label points in a rectangular coordinate system.

B. Use the terms x-coordinate, y-coordinate, x-axis, y-axis, abscissa, ordinate, coordinates, and origin.

Section 6.2

C. Graph a straight line when given the equation.

D. Determine algebraically the y-intercept and x-intercept from the equation of a straight line.

Section 6.3

E. Find the slope of a line that passes through two points.

F. Find the slope and y-intercept of a line given the equation of the line.

G. Write the equation of a line when given the y-intercept and a point on the line.

H. Write the equation of a line when given two points on the line.

Section 6.4

I. Graph a linear inequality in two variables.

Section 6.5

J. Find the area, perimeter, circumference, or volume of common geometric objects.

K. Apply the knowledge of common geometrical formulas to simple applied geometrical problems.

6.1 The Rectangular Coordinate System

Many things in everyday life are clearer if we can see a picture of the object of interest. Similarly, we often find a drawing is helpful in many areas of mathematics. For algebraic relationships we will need to be able to draw a *graph*. In order to draw a graph we need a frame of reference. We will use the *rectangular coordinate system* as the basis for graphs in elementary algebra.

We discussed in Chapter 1 how any real number could be represented on a number line. The arrow indicates the positive direction on the number line.

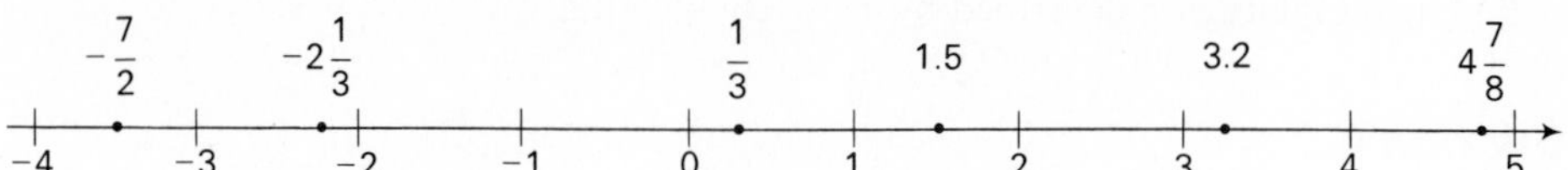

To form a rectangular coordinate system (Figure 6.1) we draw a second number line vertically. We construct it so that the 0 point on each number line is exactly at the same place. We refer to this location as the *origin*. The horizontal axis is often called the x-axis. The vertical axis is often called the y-axis. The arrows indicate the positive direction for each axis.

Using this reference system we can represent a point in the plane of this rectangular coordinate system by a pair of real numbers. We will use an *ordered pair* of numbers. The first number of the pair represents the distance measured on the horizontal or x-axis. The second number of the pair represents the distance measured on the y-axis or a line parallel to the y-axis.

EXAMPLE A What ordered pair of numbers would represent the point A in Figure 6.2?

Solution We see that the point A is represented by the ordered pair (5, 4). We first counted 5 units to the right on the x-axis. Thus we obtain 5 as the first number of the ordered pair. Then we counted 4 units upward on a line parallel to the y-axis. Thus we obtain 4 as the second number of the ordered pair.

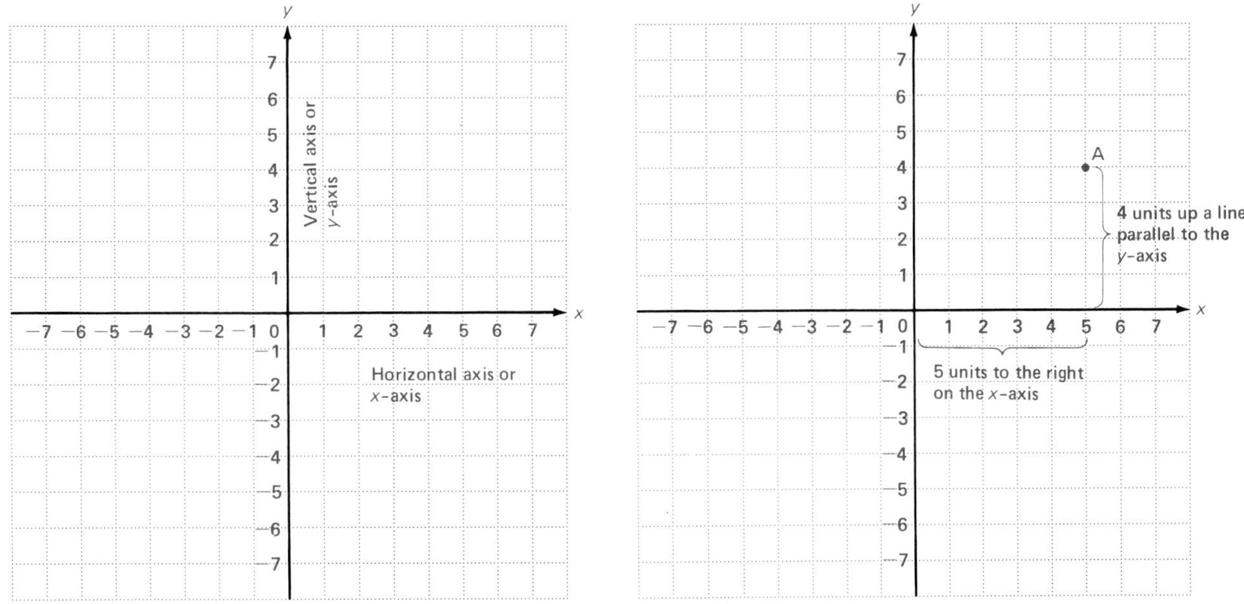

FIGURE 6.1

FIGURE 6.2

EXAMPLE B Graph the point (2, 5) on a rectangular coordinate system.

Solution As shown in Figure 6.3, we count 2 units to the right for the x value of the point. Then we count 5 units upward for the y value of the point. We represent a dot at B as the location of the ordered pair (2, 5).

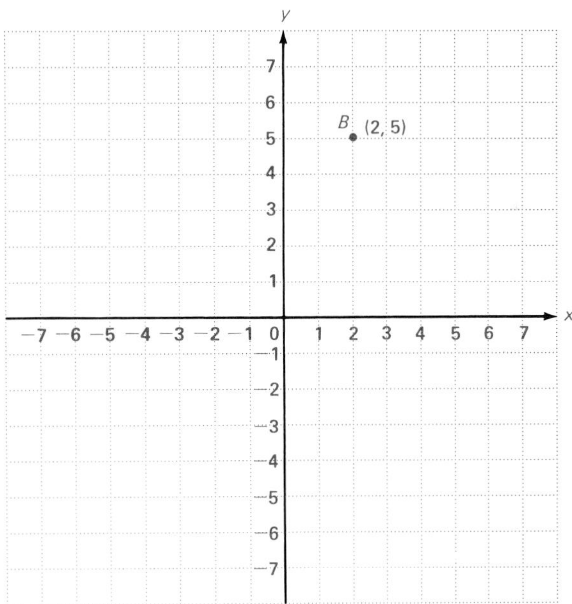

FIGURE 6.3

The ordered pair of numbers that represent a point are often referred to as the *coordinates of a point*. The first value is called the x-coordinate. The second value is called the y-coordinate. If the x-coordinate is positive, we count the proper number of squares to the right. If it is negative, we count to the left. If the y-coordinate is positive, we count the proper number of squares upward. If the y-coordinate is negative, we count downward.

EXAMPLE C Graph the point C whose coordinates are $(-5, 3)$ on a rectangular coordinate system.

Solution As we see in Figure 6.4, the x-coordinate is negative so we first count the proper number of units to the left. The y-coordinate is positive, so we then count the proper number of units upward.

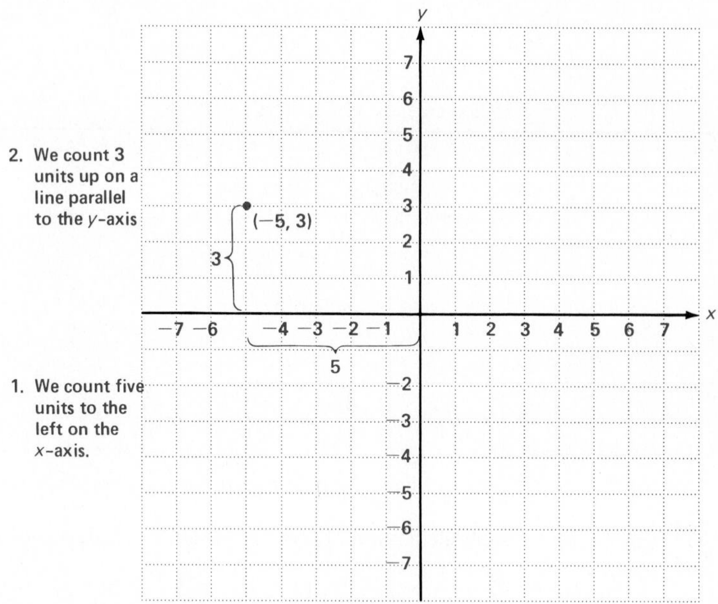

FIGURE 6.4

EXAMPLE D Graph the point whose coordinates are $(2, -6)$ on a rectangular coordinate system.

Solution This point is shown plotted in Figure 6.5. Note that when plotting points to represent ordered pairs of numbers great care should be taken to count *first* along the x-axis and then to count *second* along the y-axis or a line parallel to the y-axis.

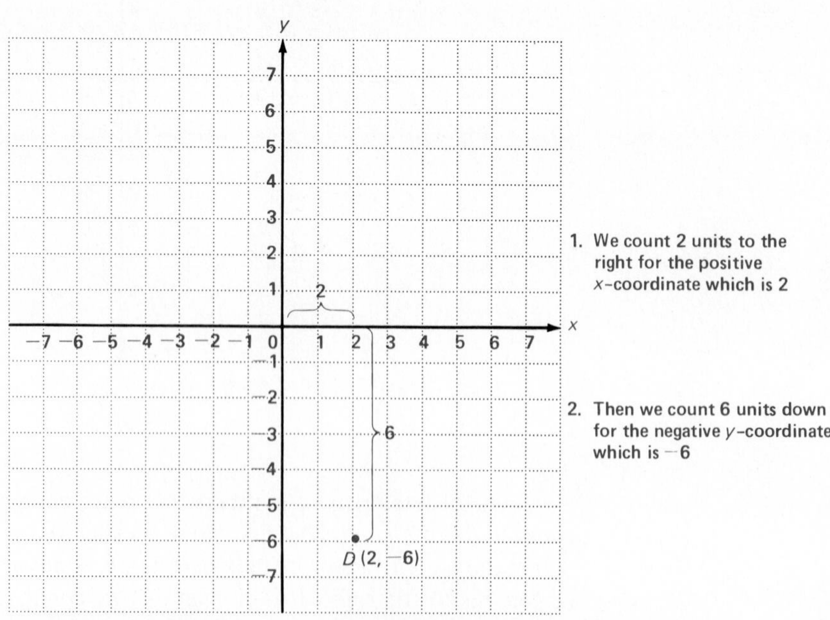

FIGURE 6.5

EXAMPLE E Plot the following points whose coordinates are given:

$F = (0, 5)$ $H = (-6, 4)$ $K = (2, -3)$

$G = \left(3, \frac{3}{2}\right)$ $I = (-3, -4)$ $M = (6.5, -7.2)$

$J = (-4, 0)$

Solution These points are plotted in Figure 6.6.

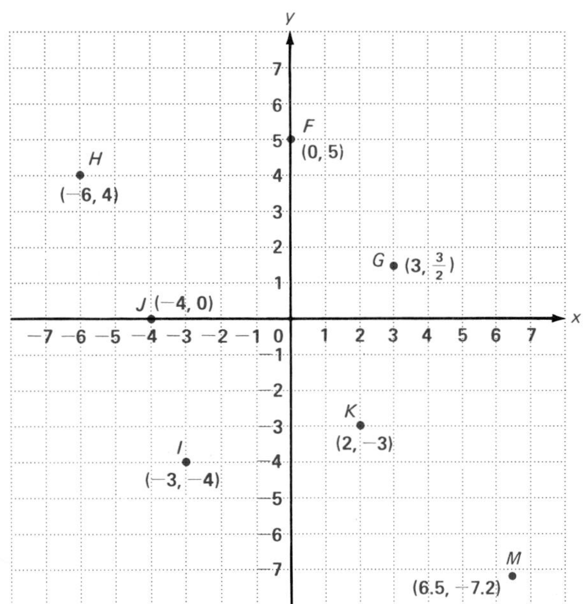

FIGURE 6.6

Note: When you are plotting decimal values like (6.5, −7.2) the location of the plotted point is approximate.

There are two names that are frequently used in mathematics to represent each of the coordinates of a point. The x-coordinate is often called the *abscissa.* The y-coordinate is often called the *ordinate.*

Take a few minutes to review this discussion and Examples A to E. Then try these practice problems.

Practice Problems for Exercise 6.1

1. Plot the following points whose coordinates are given. Label each point with both the letter and the ordered pair. Place your answer on the coordinate system provided (Figure 6.7).

$A = (3, 7)$

$B = (0, -6)$

$C = (-5, -1)$

$D = \left(-\frac{7}{2}, 2\right)$

$E = (5.5, 4.3)$

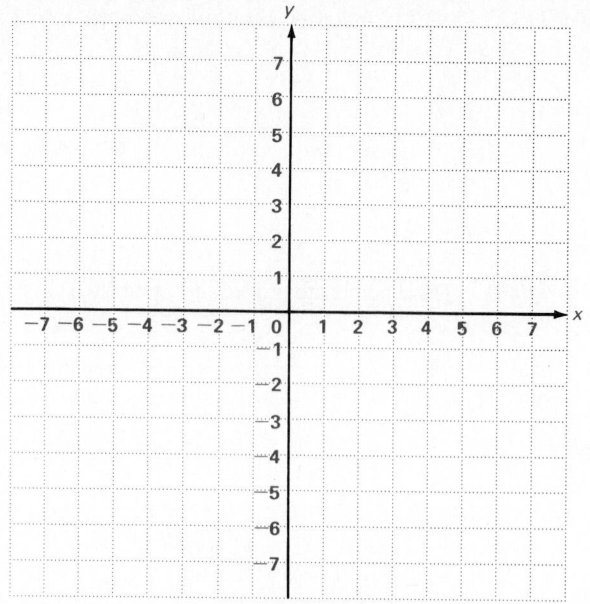

FIGURE 6.7

2. Write the proper ordered pair for each point on the coordinate system shown in Figure 6.8.

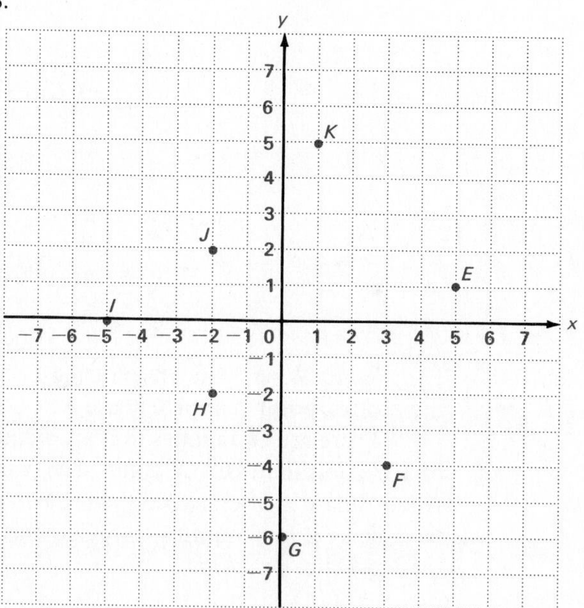

FIGURE 6.8

Now turn to the answers following Exercise 6.1 and check your solutions for these practice problems.

Exercise 6.1

1. Draw a coordinate system and plot the following points whose coordinates are given:

$$A = (-2, -7) \qquad C = (0, 8)$$

$$B = (3, -4) \qquad D = (4, -8)$$

2. Draw a coordinate system and plot the following points whose coordinates are given:

$$E = (-6, 2) \qquad G = (-6, -1)$$

$$F = (3\tfrac{1}{2}, 5) \qquad H = (-5, 0)$$

Consider points I, J, K, and L plotted in Figure 6.9.

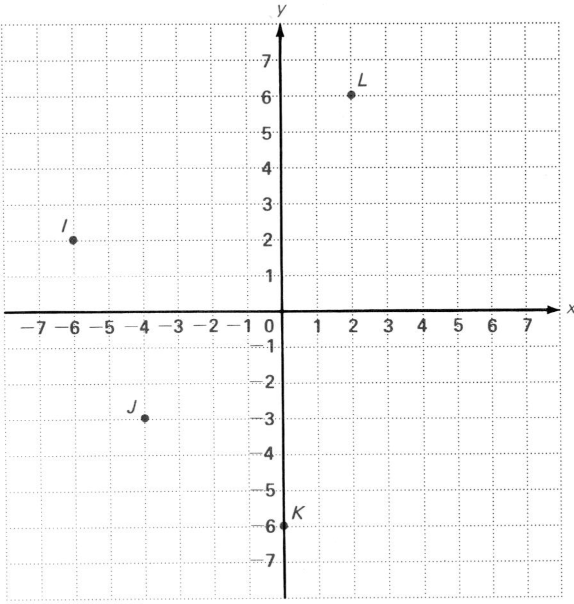

FIGURE 6.9

3. What is the abscissa of point I? What is the ordinate of point J?

4. What is the abscissa of point K? What is the ordinate of point L?

For questions 5 and 6 on the next page, consider points M, N, P, and Q plotted in Figure 6.10.

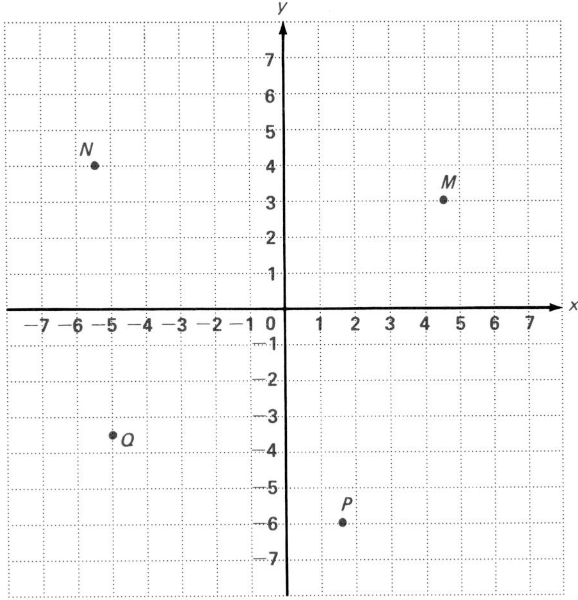

FIGURE 6.10

5. Refer to points M and N in Figure 6.10. What is the y-coordinate of M? What is the x-coordinate of N?

6. Refer to points P and Q in Figure 6.10. What is the y-coordinate of P? What is the x-coordinate of Q?

7. Give the coordinates for points R, S, X, and Y plotted in Figure 6.11.

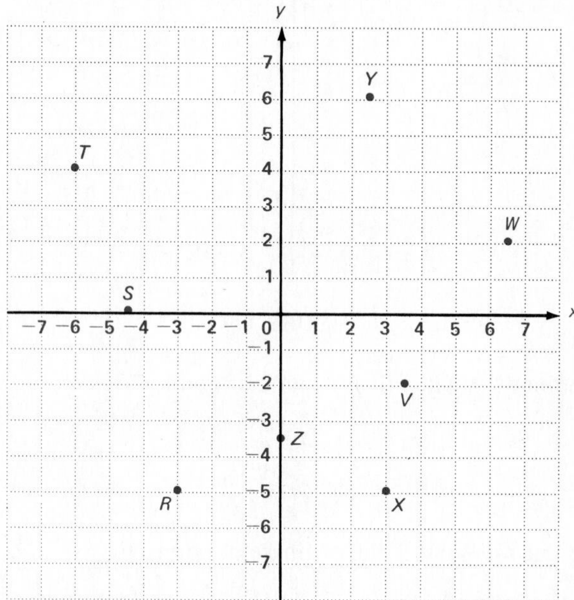

FIGURE 6.11

8. Give the coordinates for points T, V, W, and Z plotted in Figure 6.11.

9. What is the x-coordinate of the origin?

10. What is the y-coordinate of the origin?

11. In Figure 6.12 ten points are plotted. List all the ordered pairs needed to represent these points.

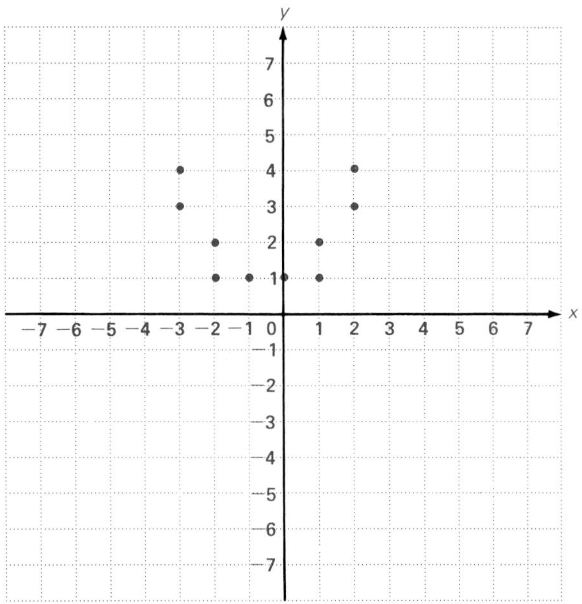

FIGURE 6.12

12. In Figure 6.13 ten points are plotted. List all the ordered pairs needed to represent these points.

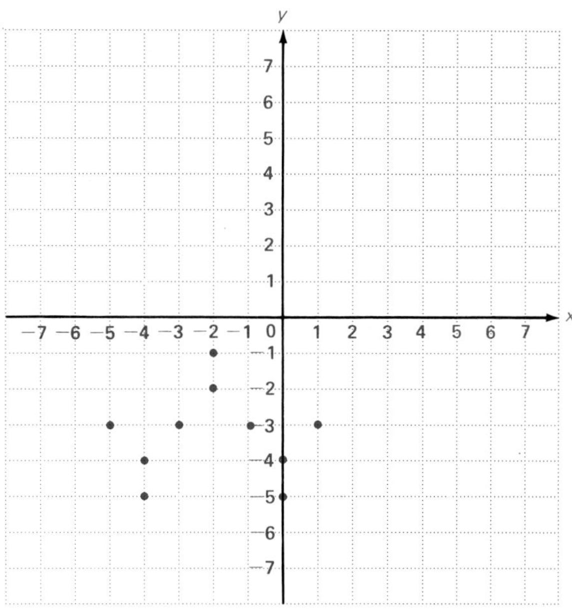

FIGURE 6.13

Challenge Problems

For each problem select a convenient scale for each axis. Then locate all the listed ordered pairs.

13. The following points are described in terms of a constant a. The letter a is understood to represent a real number. Plot the following set of points:

$$(0, a), \quad (3a, 0), \quad (1.5a, -2a), \quad \left(\frac{a}{3}, \frac{a}{6}\right)$$

14. The following points are described in terms of a constant b. The letter b is understood to represent a real number. Plot the following set of points:

$$(-2b, -3b), \quad (0.4b, 0.8b), \quad \left(0, \frac{3}{2}b\right), \quad \left(5b, -\frac{b}{5}\right)$$

Solutions to Practice Problems for Exercise 6.1

1. $A = (3, 7)$
$B = (0, -6)$
$C = (-5, -1)$
$D = \left(-\frac{7}{2}, 2\right)$
$E = (5.5, 4.3)$
These points are shown plotted in Figure 6.14.

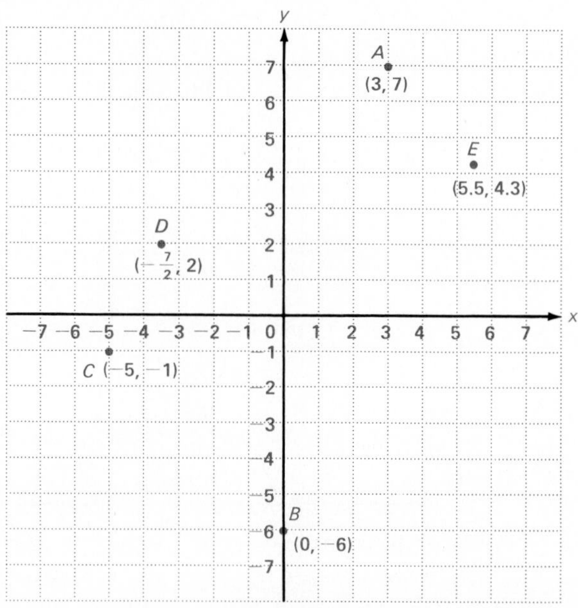

FIGURE 6.14

Note on Point D: When plotting fraction values such as $-\frac{7}{2}$ for the x-coordinate for point D you may want to write $-\frac{7}{2}$ as $-3\frac{1}{2}$ to speed your work.
Note on Point E: The location of point E is an approximation.

2. The ordered pairs for each point are:

$E = (5, 1)$
$F = (3, -4)$
$G = (0, -6)$
$H = (-2, -2)$
$I = (-5, 0)$
$J = (-2, 2)$
$K = (1, 5)$

Be very careful that you put the x-coordinate first and the y-coordinate second. Be careful that each sign is correct.

Now do the problems in Exercise 6.1.

6.2 Graphing Linear Equations

All equations of the form $Ax + By = C$, where A, B, C are constant values, are called *linear equations*. The graphs of all such linear equations are straight lines. We need to know how to graph straight lines if we are given an equation.

An infinite number of ordered pairs will satisfy a linear equation with two variables. Consider the equation $2x + y = 4$. We can find an ordered pair by choosing a value for x. We can then solve the equation for y.

EXAMPLE A Find three ordered pairs that satisfy $2x + y = 4$. Then graph the resulting straight line.

Solution

$$2x + y = 4$$

Let $x = 0$.

$$2(0) + y = 4$$

$$y = 4 \quad \text{Since } 2(0) = 0.$$

Thus if $x = 0$, then $y = 4$. This gives us the ordered pair (0, 4).

If $x = 1$, we have

$$2(1) + y = 4$$

$$2 + y = 4$$

$$y = 2$$

A second ordered pair is (1, 2).

If $x = 3$, we have

$$2(3) + y = 4$$

$$6 + y = 4$$

$$y = -2$$

A third ordered pair is (3, −2).

It is often helpful to record a table of values that summarizes the results obtained.

Table of Values For
$2x + y = 4$

x	y
0	4
1	2
3	−2

If we plot these ordered pairs and connect the three points together by a straight line we will have a *graph* of the line $2x + y = 4$, as shown in Figure 6.15. Since the straight line continues on in both directions it is common practice to extend the line beyond the points plotted and place arrows at the ends of the line.

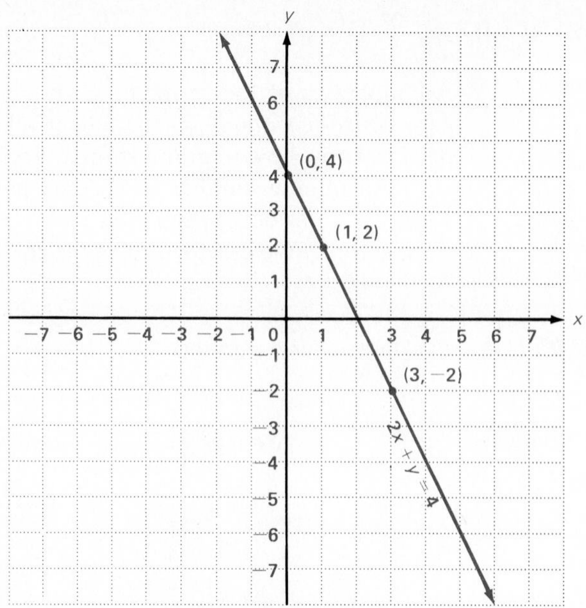

FIGURE 6.15

Why Did You Pick These Values for x?

The values that we picked for x were arbitrary. Usually, you will want to pick small numerical values that are easy to work with. Although two points are enough to draw a straight line we will always plot three points as a check to be more confident that we have drawn the line carefully. If when graphing a linear equation the three points we plot do not give us a straight line, we know that we have made an error.

Can You Start by Picking y Values First?

When graphing linear equations, we can arbitrarily pick values for y and solve for the corresponding values of x to obtain some ordered pairs.

EXAMPLE B Graph: $3x - 4y = 12$.

Solution We first find three ordered pairs by arbitrarily picking three values for y and in each case solving for x.

$$3x - 4y = 12$$

Let $y = 0$ $\quad 3x - 4(0) = 12$ Replace y by 0.

$3x = 12$ Since $4(0) = 0$.

$x = 4$ Divide both sides by 3.

We thus obtain (4, 0).

Let $y = 1$ $\quad 3x - 4(1) = 12$ Replace y by 1.

$3x - 4 = 12$

$3x = 16$ Add 4 to both sides.

$x = \frac{16}{3}$ Divide both sides by 3.

or $5\frac{1}{3}$

We thus obtain $\left(5\frac{1}{3}, 1\right)$.

$$\text{Let } y = 3 \qquad 3x - 4(3) = 12$$

$$3x - 12 = 12$$

$$3x = 24 \qquad \text{Add 12 to both sides.}$$

$$x = 8 \qquad \text{Divide both sides by 8.}$$

We thus obtain (8, 3).

Our table of values for $3x - 4y = 12$ is:

x	y
4	0
$5\frac{1}{3}$	1
8	3

We plot the three points and sketch the line shown in Figure 6.16.

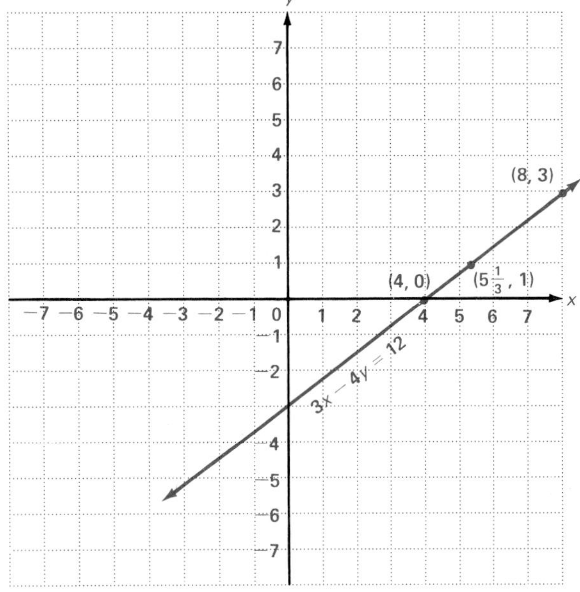

FIGURE 6.16

We observe that the coordinates $(5\frac{1}{3}, 1)$ were more difficult to plot. Probably we cannot plot fractional values like this as easily as integer values.

What values should we pick for x and y? Which points should we use for plotting? For many straight lines it is easiest to pick the two *intercepts. The intercepts* are the two points where the line crosses the two axes. A few lines have only one intercept. We will discuss these separately.

The number a is the *x-intercept* of a line if the line crosses the x-axis at $(a, 0)$. The number b is the *y-intercept* of a line if the line crosses the y-axis at $(0, b)$.

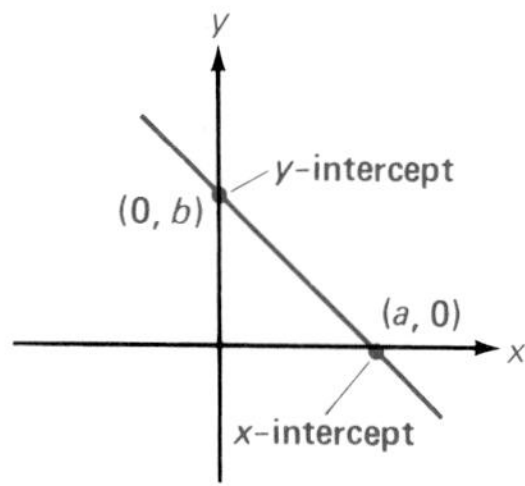

From the graph of the line $3x - 4y = 12$ shown in Figure 6.16 we see the x-intercept is 4 and the y-intercept is -3.

> **To graph an equation by using intercepts, we:**
>
> **1. Find the x-intercept by letting $y = 0$ and solving for x.**
>
> **2. Find the y-intercept by letting $x = 0$ and solving for y.**
>
> **3. Find one additional ordered pair so that we have three points with which to plot the line.**

EXAMPLE C Graph by the intercept method: $5y - 3x = 15$.

Solution

Let $x = 0$ $\quad 5y - 3(0) = 15$

$5y = 15$ — Divide both sides by 5.

$y = 3$ — The y-intercept is 3.

The ordered pair is (0, 3).

Let $y = 0$ $\quad 5(0) - 3x = 15$

$-3x = 15$ — Divide both sides by -3.

$x = -5$ — The x-intercept is -5.

The ordered pair is $(-5, 0)$.

We pick another pair to have a third point.

Let $y = 1$ $\quad 5(1) - 3x = 15$

$5 - 3x = 15$ — Simplify.

$-3x = 10$ — Subtract 5 from both sides.

$x = -\frac{10}{3}$ — Divide both sides by -3.

or $-3\frac{1}{3}$

The ordered pair is $(-3\frac{1}{3}, 1)$.

Our table of values is:

x	y
0	3
-5	0
$-3\frac{1}{3}$	1

We plot the three points and connect them to form a straight line as shown in Figure 6.17.

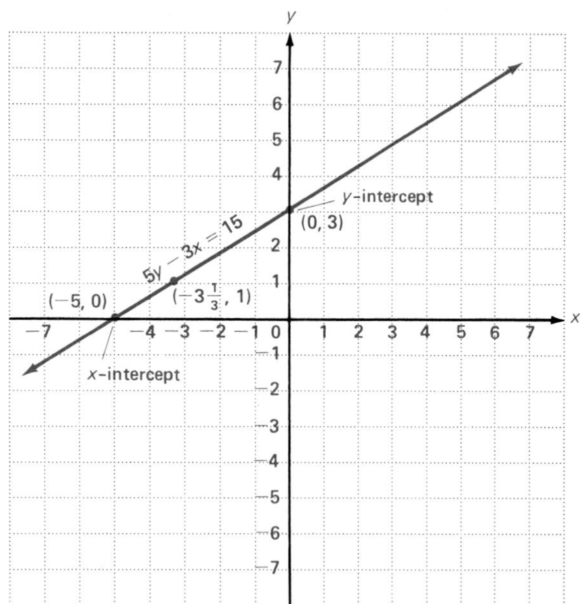

FIGURE 6.17

? Can you draw all straight lines by the intercept method?

Not really. Some straight lines have only one intercept. They may go through the origin or they may be parallel to one of the axes. If it goes through the origin, the straight line will have an equation of the form $Ax + By = 0$ (where $A \neq 0$ or $B \neq 0$ or both). In such cases you should plot two additional points besides the origin.

EXAMPLE D Graph: $5x - 4y = 0$.

Solution We see at once that (0, 0) satisfies the equation so that we know the line passes through the origin. Now we find two additional points.

Let $x = 4$ $\quad 5(4) - 4y = 0$

$20 = 4y$ $\quad$ Add $4y$ to both sides.

$5 = y$ $\quad$ Divide both sides by 4.

Thus (4, 5) is a point on the line.

Let $x = 2$ $\quad 5(2) - 4y = 0$

$10 = 4y$

$\frac{5}{2} = y$

Thus $(2, \frac{5}{2})$ is a point on the line.

We now construct a table of values, and then plot the three points and draw a graph of the line as shown in Figure 6.18.

x	y
0	0
4	5
2	$\frac{5}{2}$

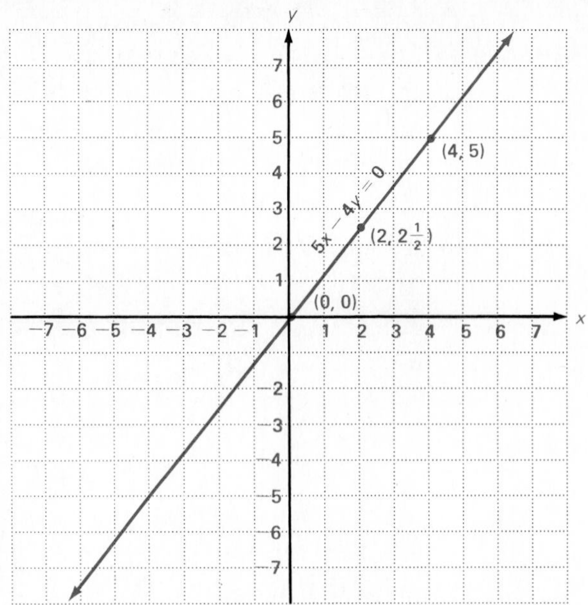

FIGURE 6.18

> **If an equation is a straight line that is parallel to the x-axis, it will be of the form $y = b$, where b is some real number.**

EXAMPLE E Graph: $y = -3$.

Solution You could write the equation as $0x + y = -3$. Then it is clear that for any value of x you substitute you will always obtain $y = -3$. Thus as shown in Figure 6.19, $(4, -3)$, $(0, -3)$, and $(-3, -3)$ are all ordered pairs that satisfy the equation $y = -3$. Since the y-coordinate of every point on this line is -3, it is easy to see that the line will be 3 units below the x-axis.

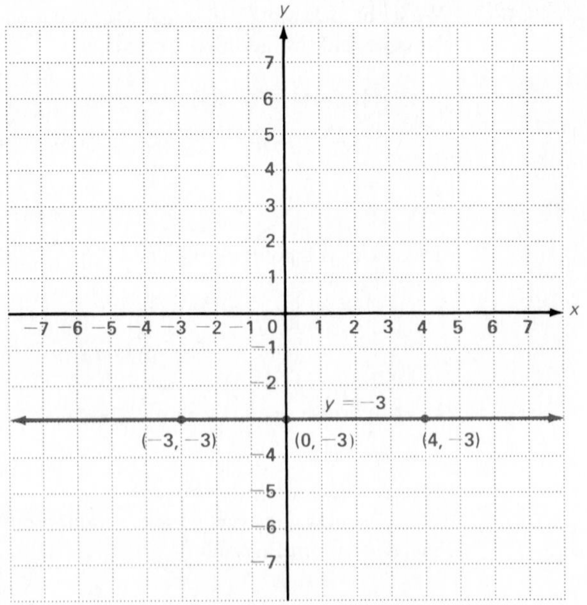

FIGURE 6.19

In actual practice you will not need to plot points to draw a graph of lines of this types.

> **If an equation is a straight line that is parallel to the y-axis, it will be of the form $x = a$, where a is some real number.**

EXAMPLE F Graph: $x = 5$.

Solution This can be done immediately by drawing a vertical line 5 units to the right of the origin (Figure 6.20). This line contains all points whose abscissa is 5. The x-coordinate of every point on this line is 5.

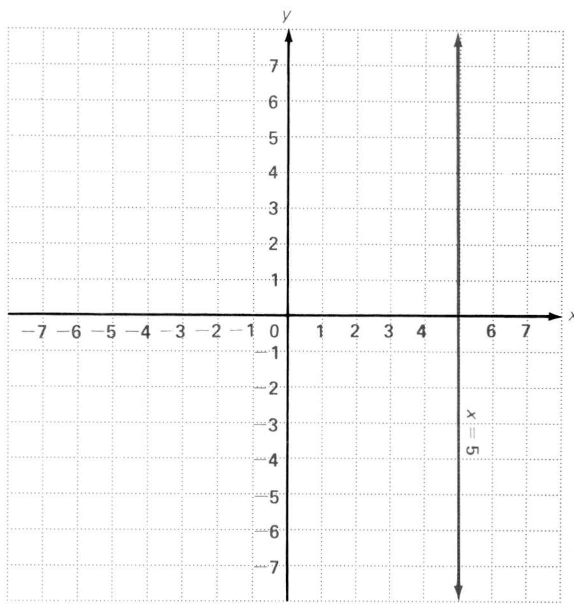

FIGURE 6.20

Carefully review Examples A to F and then try these practice problems.

Practice Problems for Exercise 6.2

For each equation plot three points to obtain a graph.

1. $2y - x = 5$ **2.** $2x - 3y = 4$ **3.** $7x = -2y$

Now turn to the answers following Exercise 6.2 and check your solutions for these practice problems.

Exercise 6.2

For each equation plot three points to obtain a graph.

1. $x + 2y = 5$

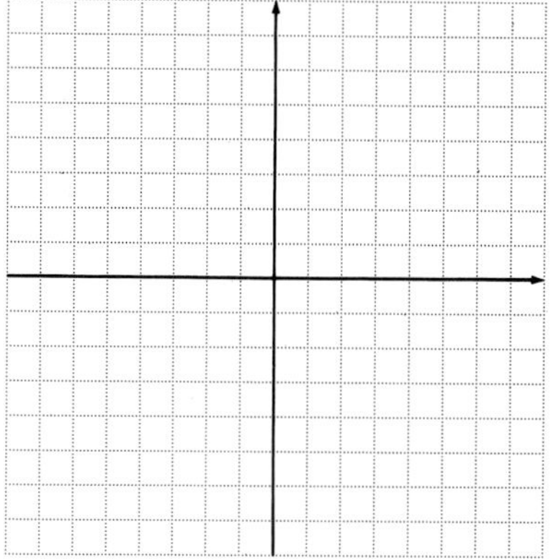

2. $4x + y = 5$

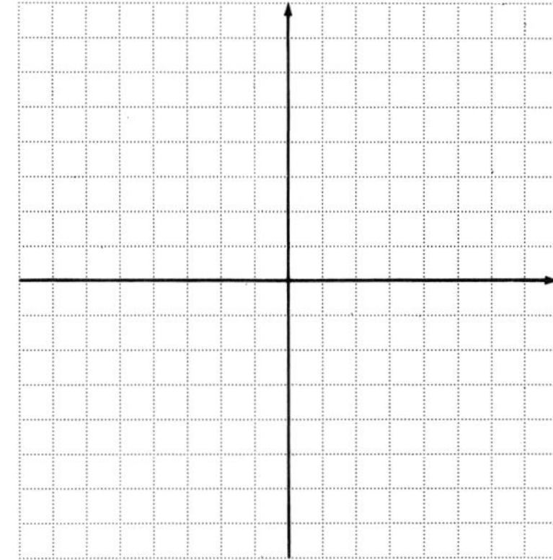

3. $y = 3x + 2$

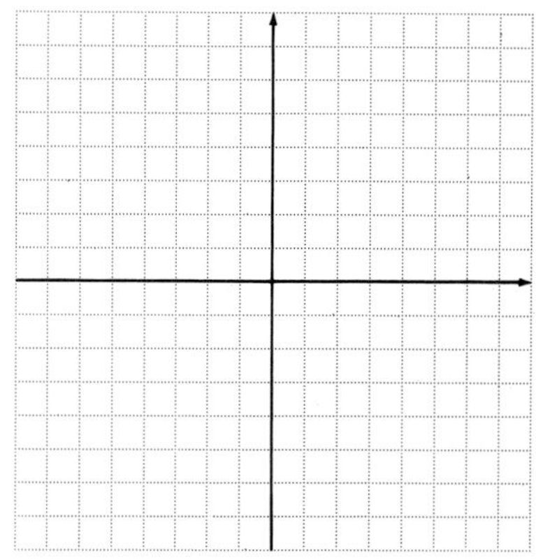

4. $y = 2x + 1$

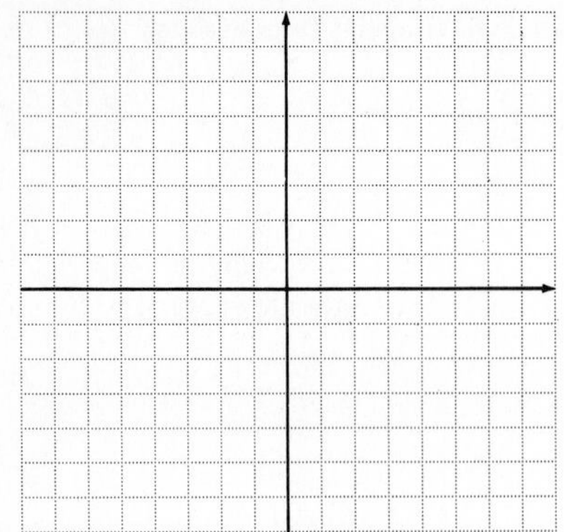

5. $y = 5 - 3x$

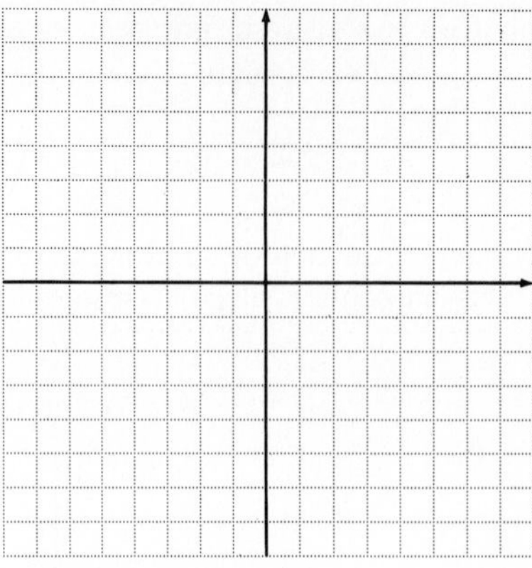

6. $y = 4 - 2x$

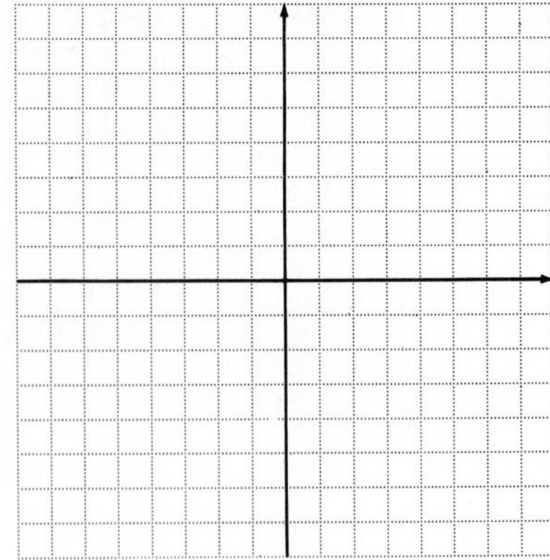

7. $3x = 5y - 9$

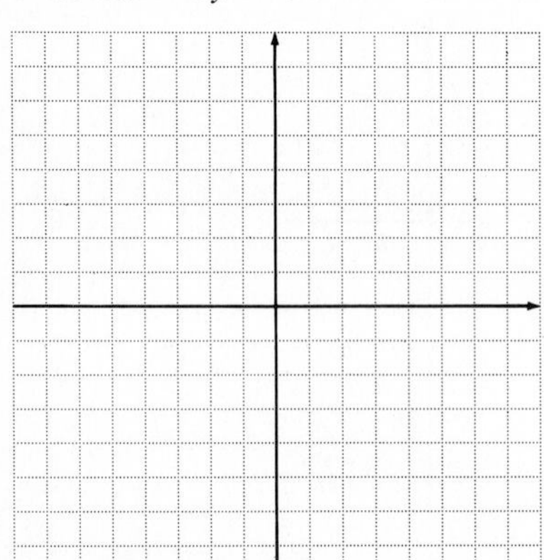

8. $2y = 4 - 3x$

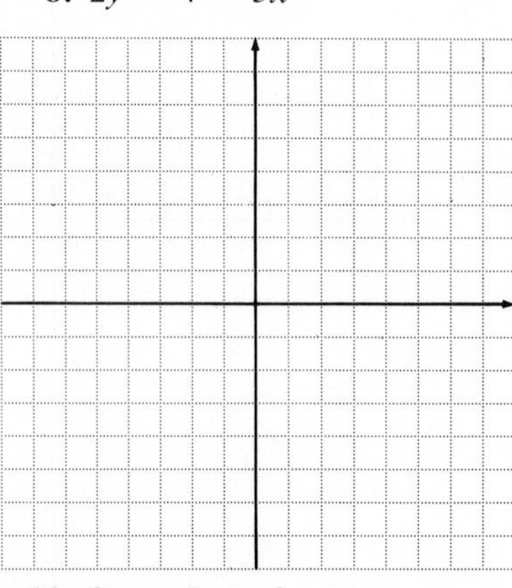

9. $x - 4 = 2y$

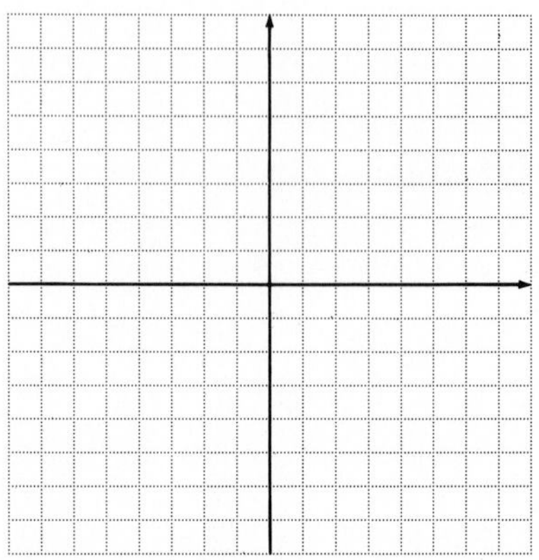

10. $2y - 4 = -3x$

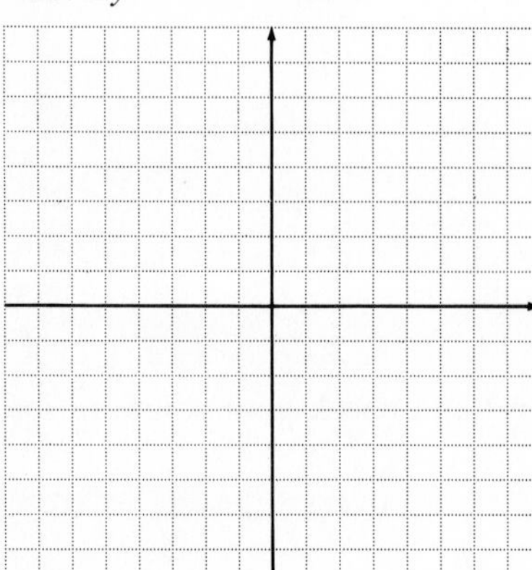

11. $3x - 5y = 0$

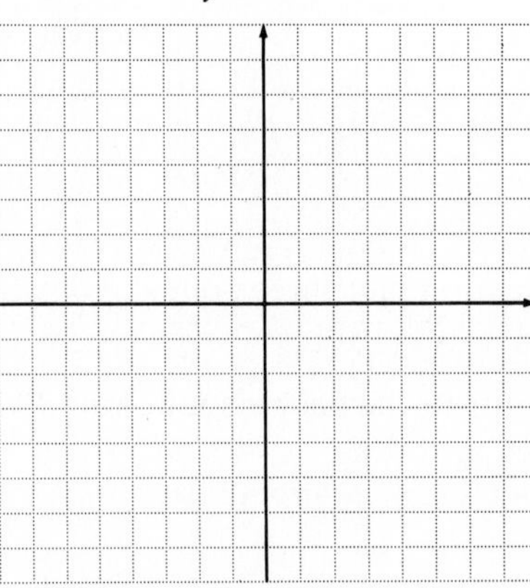

12. $2y - 4x = 0$

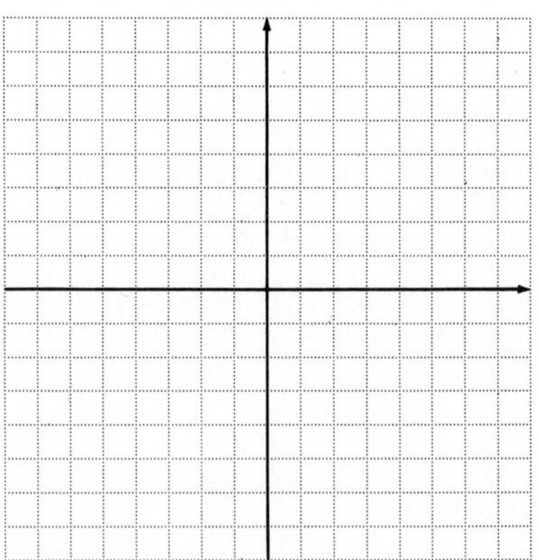

13. $x + 5 = 0$

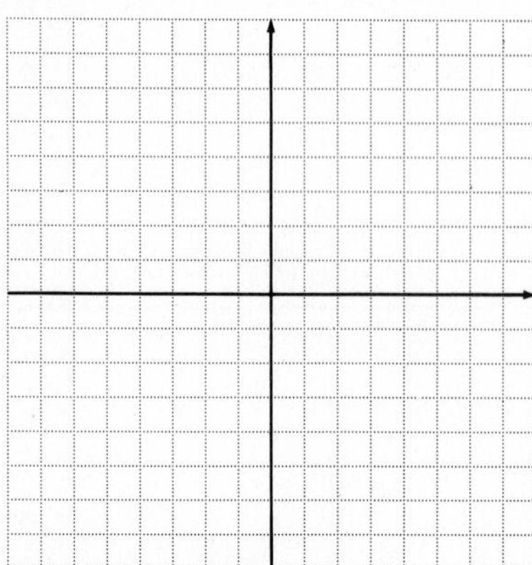

14. $y - 3 = 0$

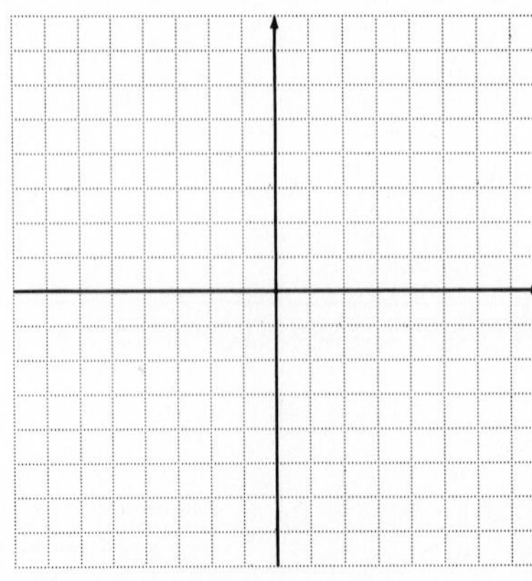

15. $y = \frac{1}{2}x - 2$

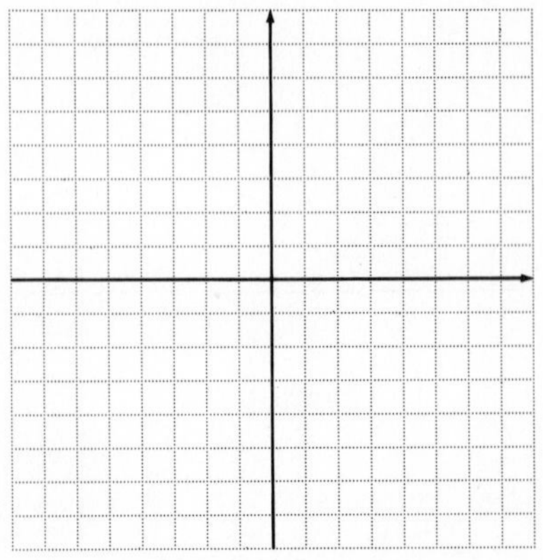

16. $y = \frac{2}{3}x + 1$

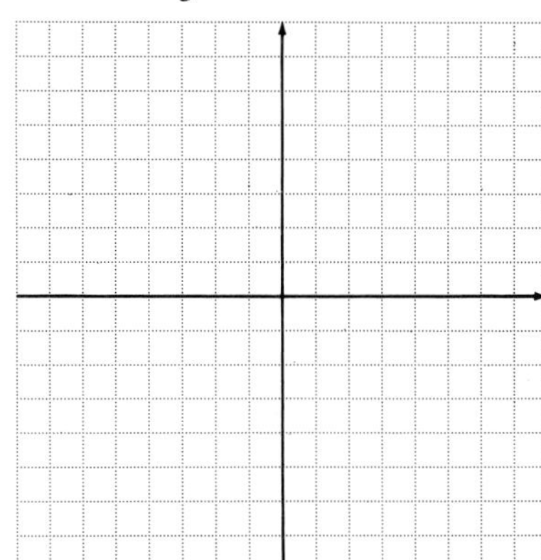

17. $x - y = 0$

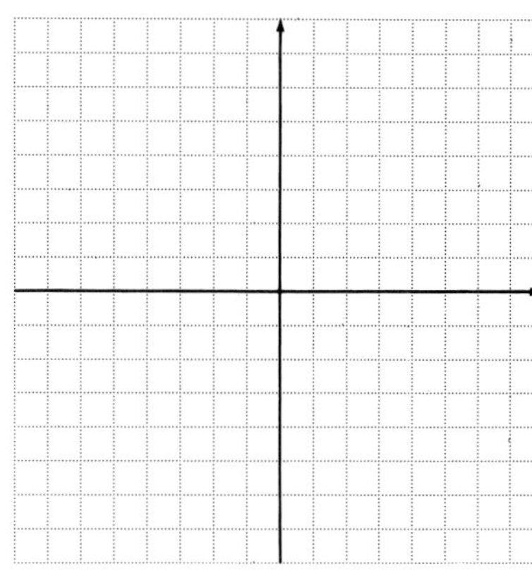

18. $x + 5 = -2$

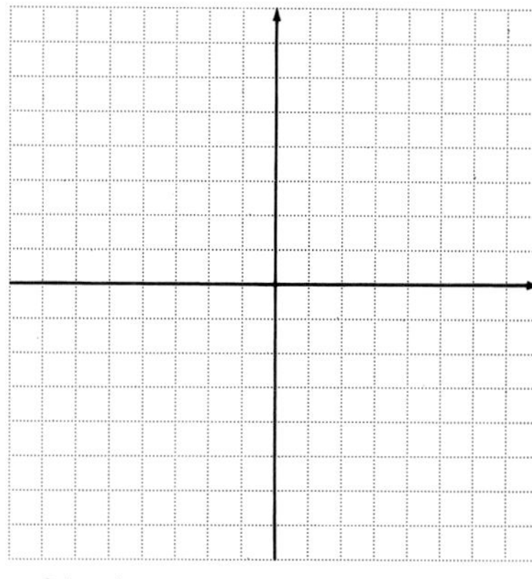

19. $21 - 7x = 3y$

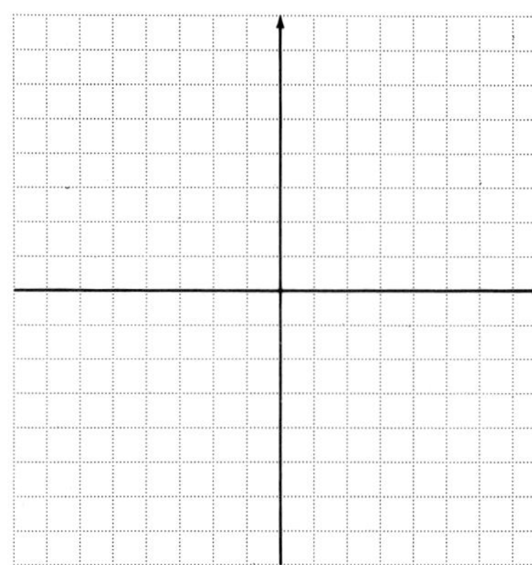

20. $3x + 2y = 7 + 2x$

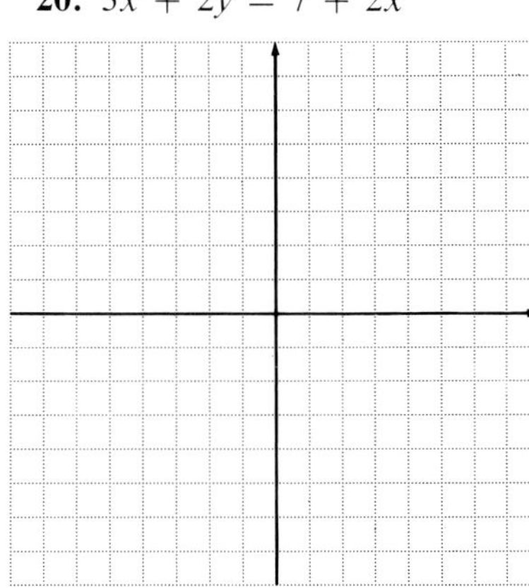

21. $4x + y = 2x + 3$

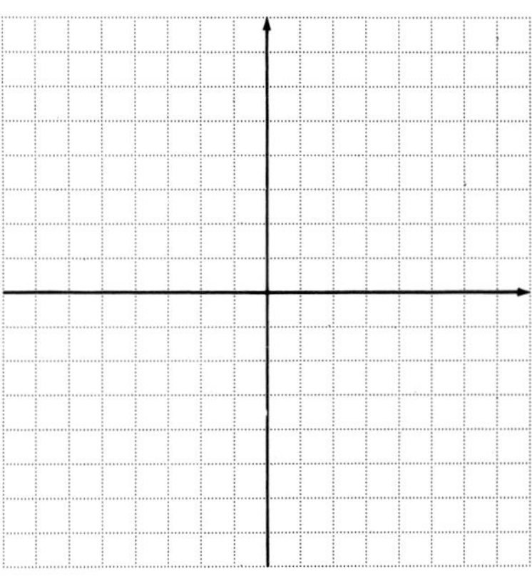

22. $5x + 2y = 3y$

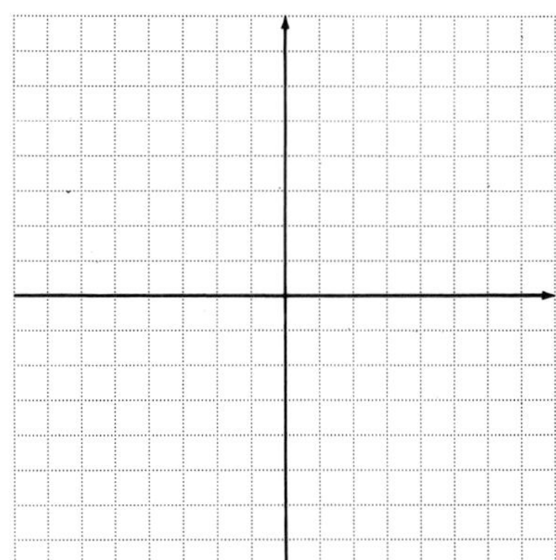

Challenge Problems

Plot three points for each equation and obtain a graph. (Be sure to pick a convenient scale for each axis.)

23. $250y = 135x + 28$

24. $64x + 48y - 200 = 0$

25. $3.4x - 1.9y = 6.0$

Solutions to Practice Problems for Exercise 6.2

1. Graph: $2y - x = 5$.

Find the two intercepts.

Let $x = 0$

$$2y - (0) = 5$$
$$2y = 5$$
$$y = \frac{5}{2}$$

Let $y = 0$ $\quad 2(0) - x = 5$

$$-x = 5$$

$$x = -5$$

Arbitrarily pick another point: say, let $x = 1$

$$2y - (1) = 5$$

$$2y - 1 = 5$$

$$2y = 6$$

$$y = 3$$

x	y
0	$\frac{5}{2}$
-5	0
1	3

Graph the line using these three points (Figure 6.21).

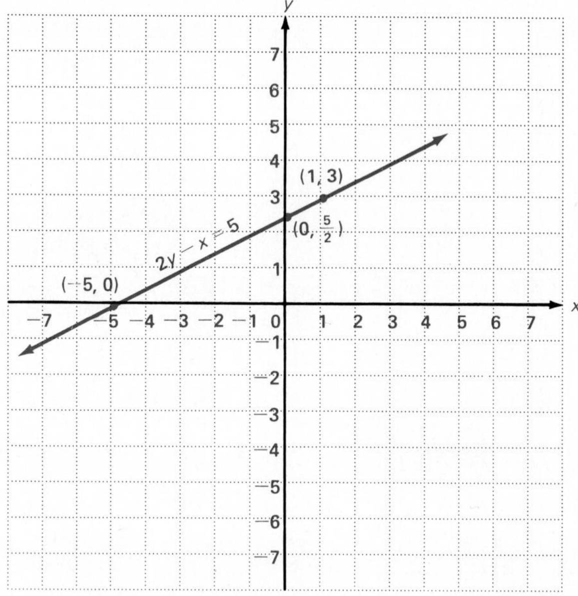

FIGURE 6.21

2. Graph: $2x - 3y = 4$.

Find the two intercepts.

Let $x = 0$ $\quad 2(0) - 3y = 4$

$$-3y = 4$$

$$y = -\frac{4}{3}$$

Let $y = 0$ $\quad 2x - 3(0) = 4$

$$2x = 4$$

$$x = 2$$

Pick an arbitrary point: say, $x = -7$.

$$2(-7) - 3y = 4$$
$$-14 - 3y = 4$$
$$-3y = 4 + 14$$
$$-3y = 18$$
$$y = -6$$

x	y
0	$-\frac{4}{3}$
2	0
−7	−6

Plot the three points and plot the line (Figure 6.22).

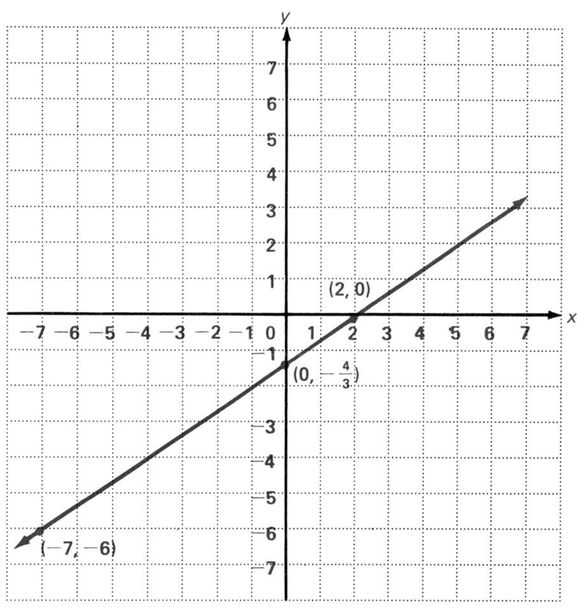

FIGURE 6.22

3. Graph: $7x = -2y$.

Since $7x + 2y = 0$ is of the form $Ax + By = 0$, we know that the line goes through the origin. Another approach is to let $x = 0$ and solve for y. Since we then obtain $y = 0$, we know that the line goes through origin.

Pick another two points.

Record these points in a table of values:

Let $x = 2$

$$7(2) = -2y$$
$$14 = -2y$$
$$-7 = y$$

Let $x = -2$

$$7(-2) = -2y$$
$$-14 = -2y$$
$$7 = y$$

x	y
0	0
2	−7
−2	7

You can use any values for x, but it is to your advantage to pick values that avoid fractions. Fractional values tend to be difficult to plot accurately. The graph is shown in Figure 6.23.

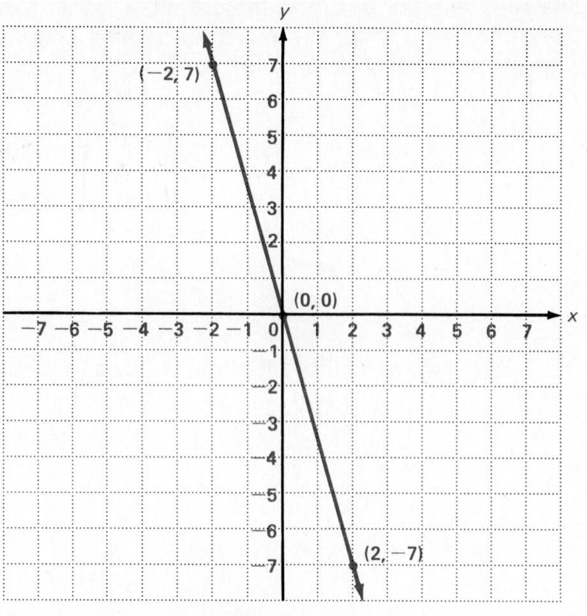

FIGURE 6.23

Now do the problems in Exercise 6.2.

6.3 Properties of Linear Equations

We often use the word "slope" to describe such things as the amount of incline of a hill. The term is used in mathematics to describe the incline of a straight line.

The slope of a straight line is defined by the change in y divided by the change in x.

$$\text{Slope} = \frac{\text{change in } y}{\text{change in } x}$$

Consider the line drawn through points A and B in Figure 6.24. If we measure the change from point A to point B in the x-direction and the y-direction, we would

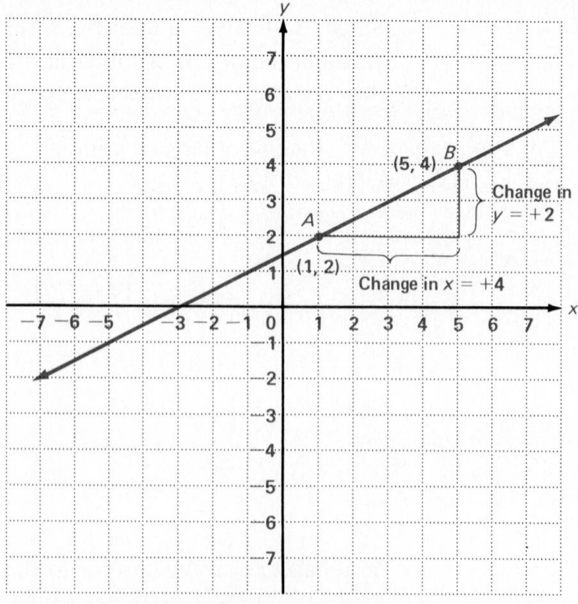

FIGURE 6.24

have an idea of the steepness (or the slope) of the line. From point A to point B the change in y values was from 2 to 4, a *change of* 2. From point A to point B the change in x values was from 1 to 5, a *change of* 4. Thus

$$\text{Slope} = \frac{\text{change in } y}{\text{change in } x} = \frac{2}{4} = \frac{1}{2}$$

We now state a more formal (and more frequently used) definition.

***The slope of any straight line* that contains the points with coordinates (x_1, y_1) and (x_2, y_2) has a slope defined by**

$$\textbf{Slope} = m = \frac{y_2 - y_1}{x_2 - x_1} \qquad \textbf{where } x_2 \neq x_1$$

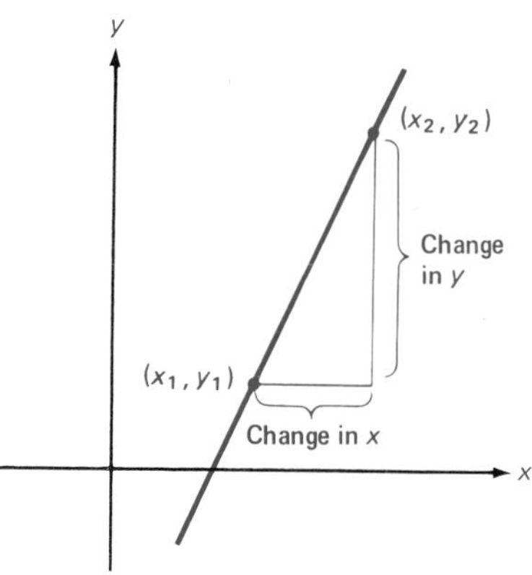

The use of subscripted terms such as x_1, x_2, etc., is just a way of indicating that the first x value is x_1 and the second x value is x_2. Thus (x_1, y_1) are the coordinates of the first point and (x_2, y_2) are the coordinates of the second point.

EXAMPLE A Find the slope of the *line n* that passes through (2, 0) and (4, 2).

Solution Let

(2, 0) be the first point (x_1, y_1)

(4, 2) be the second point (x_2, y_2)

Then by the formula,

$$m = \frac{y_2 - y_1}{x_2 - x_1} = \frac{2 - 0}{4 - 2} = \frac{2}{2} = 1$$

The sketch of line n is shown in Figure 6.25.

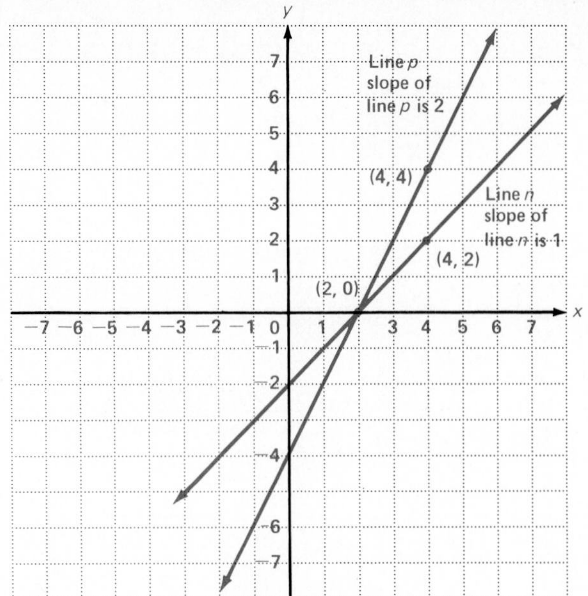

FIGURE 6.25

> **The slope of a line is not changed if the points (x_1, y_1) and (x_2, y_2) are interchanged.**

Alternative Solution Let

(4, 2) be the first point (x_1, y_1)

(2, 0) be the second point (x_2, y_2)

Then by the formula,

$$m = \frac{y_2 - y_1}{x_2 - x_1} = \frac{0 - 2}{2 - 4} = \frac{-2}{-2} = 1$$

The result is the same. Either way we obtain a slope of 1. Thus if you are given two points *it does not matter which one you call (x_1, y_1) and which one you call (x_2, y_2).*

EXAMPLE B Find the slope of the *line p* that passes through (2, 0) and (4, 4).

Solution Let

(2, 0) be the first point (x_1, y_1) $\quad m = \frac{4 - 0}{4 - 2} = \frac{4}{2} = 2$

(4, 4) be the second point (x_2, y_2) $\quad$ The slope of line p is 2.

The slope of line p can be compared to that of line n in Figure 6.25. As we expect, the line with the larger slope has a steeper incline. The slope of line p is twice that of line n. The rate of change of y of line p is twice as great as that of line n as x increases from 2 to 4.

Note:

1. If y values *increase* as x *increases*, the slope of the line is positive (Figure 6.26a).
2. If the y values *decrease* as x *increases*, the slope of the line is negative (Figure 6.26b).

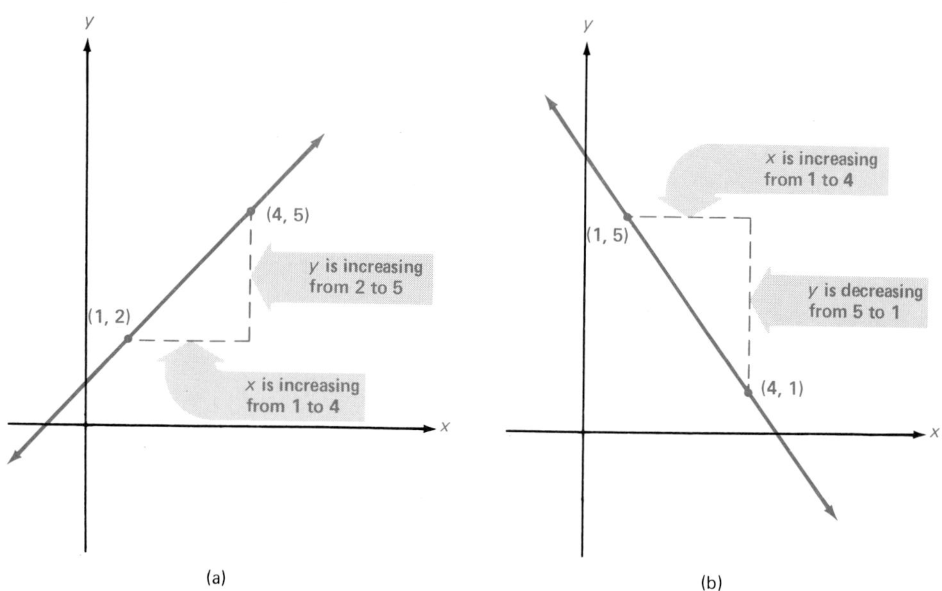

FIGURE 6.26

In the following example, the slope of the line is negative.

EXAMPLE C Find the slope of the line through $(-3, 2)$ and $(2, -4)$.

Solution Let

$$(x_1, y_1) \text{ be } (-3, 2)$$

$$(x_2, y_2) \text{ be } (2, -4)$$

$$\text{Slope} = m = \frac{y_2 - y_1}{x_2 - x_1} = \frac{-4 - (2)}{2 - (-3)} = \frac{-4 - 2}{2 + 3} = \frac{-6}{5} = -\frac{6}{5}$$

The slope of this line is *negative*. As the x value increased from -3 to 2, the y value decreased from 2 to -4.

The line is graphed in Figure 6.27.

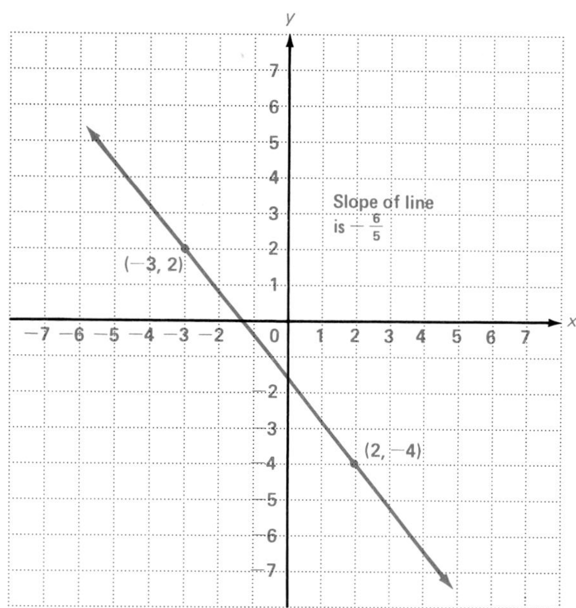

FIGURE 6.27

It is good to visualize the idea of the different possible values for the slope of a line.

1. *Lines with positive slopes go upward as x increases* (Figure 6.28a). The larger the value of m, the more steeply the line is inclined upward.
2. *Lines with negative slopes go downward as x increases* (Figure 6.28b). The smaller the value of m, the more steeply the line is inclined downward.

The slope of a horizontal line is zero.

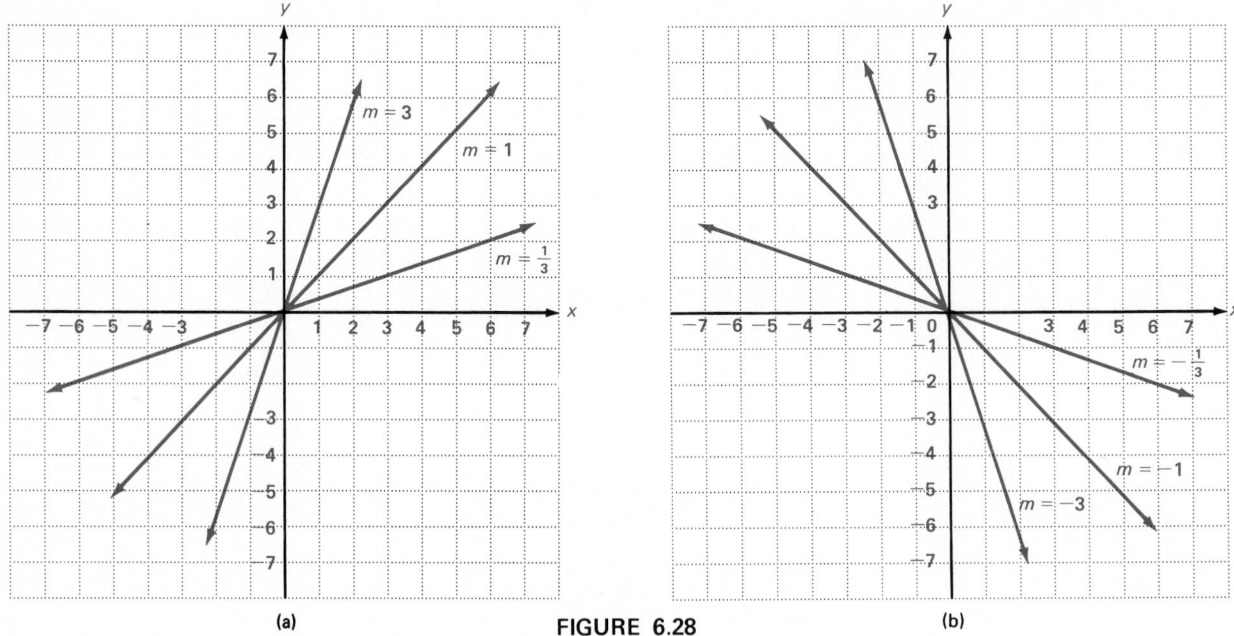

FIGURE 6.28

EXAMPLE D Find the slope of the line through (0, 2) and (5, 2).

Solution

$$\text{Slope} = m = \frac{y_2 - y_1}{x_2 - x_1} = \frac{2-2}{5-0} = \frac{0}{5} = 0$$

The line is shown in Figure 6.29.

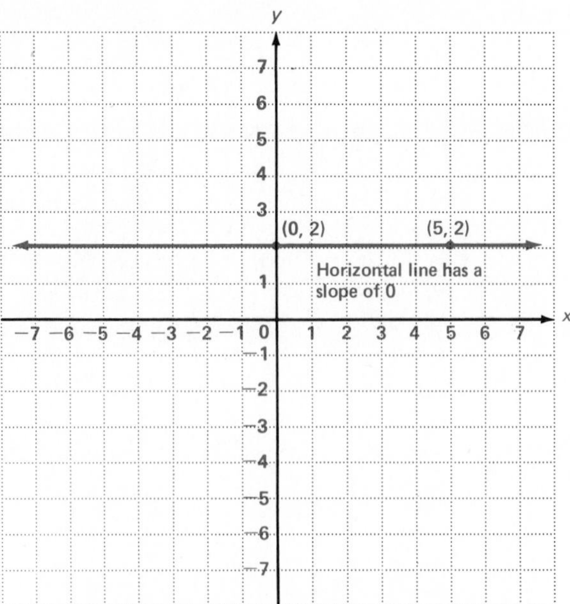

FIGURE 6.29

A vertical line is said to have no slope. In other words, *the slope of a vertical line does not exist.*

EXAMPLE E Find the slope of a line through $(-4, 0)$ and $(-4, -4)$.

Solution There is a restriction on the definition of slope. The formula does not hold if $x_1 = x_2$. When we try the formula we see that the

$$\text{Slope } m = \frac{y_2 - y_1}{x_2 - x_1} = \frac{-4 - 0}{-4 - (-4)} = \frac{-4}{-4 + 4} = \frac{-4}{0} = \text{not defined}$$

($-4 \div 0$ cannot be done). Therefore, it seems logical to state that the slope of a vertical line *does not exist*. The line passing through $(-4, 0)$ and $(-4, -4)$ has *no slope*.

Slope-intercept form of the equation of a line

There is a very useful form of the equation of a straight line which is called the slope-intercept form. The form reveals immediately what the slope of a line is and where it intercepts the y-axis. Since every straight line except a vertical line crosses the y-axis somewhere, this form can be used for all nonvertical straight lines.

Suppose, as shown in Figure 6.30, that a straight line with slope m crossed the y-axis at a point $(0, b)$. The y-coordinate b would be the y-intercept. Now consider any point on the line and label that point with coordinates (x, y).

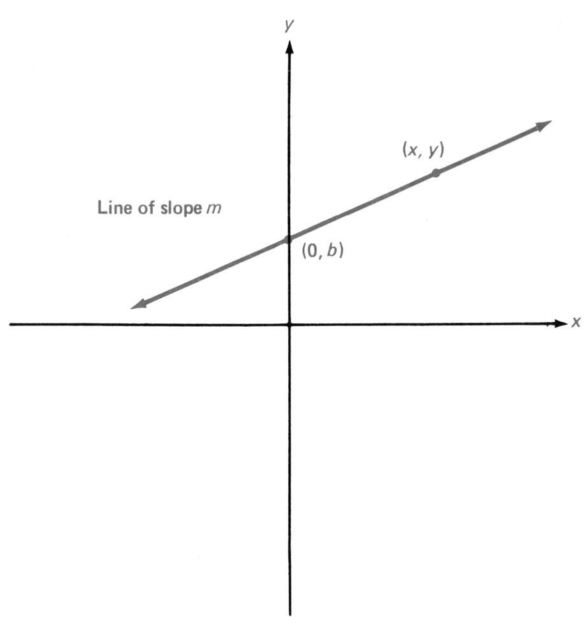

FIGURE 6.30

$\dfrac{y_2 - y_1}{x_2 - x_1} = m$	We use the definition of slope.
$\dfrac{y - b}{x - 0} = m$	and substitute in the two points $(0, b)$ for (x_1, y_1) and (x, y) for (x_2, y_2).
$\dfrac{y - b}{x} = m$	Simplify.
$y - b = mx$	Do you see that this is a fractional equation similar to those covered in Section 4.5? Multiply both sides by x to remove the fraction.
$y = mx + b$	Now we merely add b to both sides.

The slope-intercept equation of a line that has slope m and the y-intercept at $(0, b)$ is given by $y = mx + b$.

This form allows us immediately to write the equation of a straight line if we know the slope and the y-intercept.

EXAMPLE F Find the equation of a line with slope $\frac{2}{5}$ and y-intercept -3.

1. Express the answer in the slope-intercept form.
2. Express the answer in the form $Ax + By = C$.

Solution **1.** $y = mx + b$

Here $m = \frac{2}{5}$ and $b = -3$.

$$y = \frac{2}{5}x + (-3)$$

$$y = \frac{2}{5}x - 3$$

2.

$5y = 5\left(\frac{2x}{5}\right) - 5(3)$	Multiply each term by 5.
$5y = 2x - 15$	Simplify.
$-2x + 5y = -15$	Subtract $2x$ from each side.
$-2x + 5y = -15$	This is in the form $Ax + By = C$ where $A = -2$, $B = 5$, $C = -15$.

By using algebraic operations we can write any nonvertical line in the slope-intercept form and use this form immediately to identify the slope and the y-intercept.

EXAMPLE G What is the slope and the y-intercept of the line $5x + 3y = 2$?

Solution We want to get the equation in the form $y = mx + b$.

$5x + 3y = 2$	We need to isolate the y term.
$3y = -5x + 2$	Subtract $5x$ from both sides.
$y = \frac{-5x + 2}{3}$	Divide both sides by 3.
$y = -\frac{5}{3}x + \frac{2}{3}$	Write the right-hand side as two fractions using the property $\frac{a + b}{c} = \frac{a}{c} + \frac{b}{c}$.

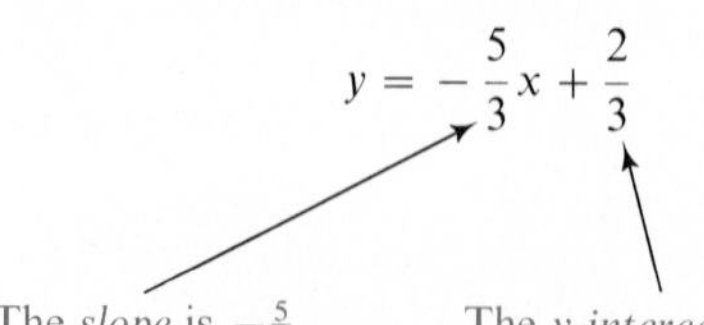

It is possible to write the equation of a line using this form if you know one point on the line and the slope.

EXAMPLE H Find the equation of a line with a slope of 4 that passes through the point (3, 1).

Solution Since the point (3, 1) is one of the points (x, y) that satisfies the equation and since we know that $m = 4$, we can substitute the values x, y, and m into the equation.

$$y = mx + b$$

$$1 = (4)(3) + b \qquad \text{Replace } y \text{ by 1, } m \text{ by 4, and } x \text{ by 3.}$$

$$1 = 12 + b \qquad \text{Simplify.}$$

$$-11 = b \qquad \text{Subtract 12 from both sides.}$$

The y-intercept is -11.

$$y = (4)x + (-11) \qquad \text{Replace } m \text{ by 4 and } b \text{ by } -11 \text{ in } y = mx + b.$$

$$y = 4x - 11 \qquad \text{This is the equation of the line.}$$

It is possible to find the equation of a line if it is graphed accurately by using the slope-intercept equation of a line. This is usually done only to get an approximation since most graphs drawn are slightly in error. However, it is a useful skill in checking answers.

EXAMPLE I What is the equation of the line graphed in Figure 6.31?

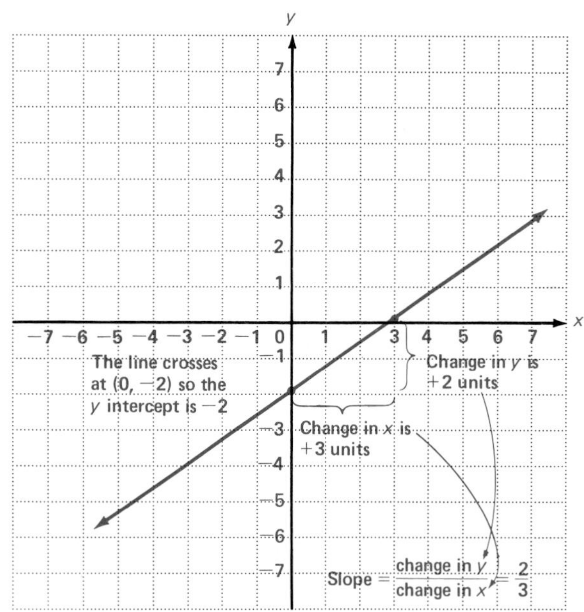

FIGURE 6.31

Solution The equation of the pictured line is

$$y = mx + b \quad \text{where } m = \frac{2}{3} \quad \text{and} \quad b = -2$$

so

$$y = \frac{2}{3}x + (-2) \quad \text{or} \quad y = \frac{2}{3}x - 2$$

EXAMPLE J Find the equation of a line that passes through (2, 5) and (6, 3).

Solution We first find the slope of the line. Then we proceed as in Example H.

$$m = \frac{y_2 - y_1}{x_2 - x_1}$$ Equation of the slope.

$$m = \frac{3 - 5}{6 - 2}$$ Substitute $(x_1, y_1) = (2, 5)$ and $(x_2, y_2) = (6, 3)$ into the formula.

$$m = \frac{-2}{4} = -\frac{1}{2}$$ Simplify.

We now know the slope is $-\frac{1}{2}$. We substitute this into the equation $y = mx + b$ using (2, 5) as one of the points (x, y) that satisfies the equation.

$$y = mx + b$$

$$5 = \left(-\frac{1}{2}\right)(2) + b$$ Substitute $x = 2$, $y = 5$, $m = -\frac{1}{2}$

$$5 = -1 + b$$ Solve for the y intercept, which is 6.

$$6 = b$$

Since we have $m = -\frac{1}{2}$ and $b = 6$ we can write the equation

$$y = mx + b$$

$$y = -\frac{1}{2}x + 6$$

Review the concepts discussed in Examples A to J and then try these practice problems.

Practice Problems for Exercise 6.3

1. Find the slope of the line that passes through (2, 0) and (−1, 4).
2. Find the slope of the line that passes through (−5, −2) and (3, −6).
3. What is the slope and the y-intercept of the line $4x - 2y = -5$?
4. Find the equation of the line that passes through (−2, 4) and has a slope of $\frac{1}{2}$.

Now turn to the answers following Exercise 6.3 and check your solutions for these practice problems.

Exercise 6.3

In problems 1 to 12, find the slope of a straight line that passes through the given pair of points.

1. (3, 7) and (9, 6)

2. (5, 2) and (8, 5)

3. $(-2, 1)$ and $(3, 4)$

4. $(5, 6)$ and $(-3, 1)$

5. $(-7, -6)$ and $(-3, 2)$

6. $(2, -4)$ and $(-6, -3)$

7. $(3, -4)$ and $(-6, -4)$

8. $(7, -2)$ and $(-12, -2)$

9. $\left(\frac{1}{3}, -2\right)$ and $\left(\frac{2}{3}, -5\right)$

10. $(4, -3)$ and $\left(\frac{5}{2}, 5\right)$

11. $(-3, 4)$ and $\left(\frac{3}{2}, -1\right)$

12. $\left(-3, \frac{7}{2}\right)$ and $\left(-6, -\frac{3}{2}\right)$

13. Can you find the slope of the line passing through $(5, -12)$ and $(5, -6)$? Why?

14. Write the equation of a line with slope $-\frac{3}{4}$ and a y-intercept of 6.

15. Write the equation of a line with slope -7 and a y-intercept of $\frac{2}{5}$.

Find the slope and the y-intercept of the following lines.

16. $3x + y = -2$

17. $-4x + y = 5$

18. $5x + 2y = 3$

19. $7x + 3y = 4$

20. $4x - 3y = 12$

21. $18x - 3y = 6$

22. $7x - 2y + 3 = 0$

23. $4x - 5y - 2 = 0$

24. $3x + \frac{1}{2}y = -6$

25. $2x + \frac{1}{3}y = 4$

Find the equation of the line that passes through the given point and has the given slope.

26. $(-3, 2)$, $m = 6$

27. $(4, -1)$, $m = 5$

28. $(4, 3)$, $m = -2$

29. $(5, 7)$, $m = -4$

30. $(-3, -2)$, $m = \frac{2}{3}$

31. $(-4, -2)$, $m = \frac{2}{5}$

32. $\left(\frac{1}{2}, 2\right)$, $m = -3$

33. $\left(3, \frac{1}{3}\right)$, $m = -2$

34. $(7, -4)$, $m = 0$

35. $(-3, 2)$, no slope

You will need to estimate the slope and the y-intercept from the graph of lines A, B, C, D in Figures 6.32 and 6.33 in order to answer problems 36 to 39.

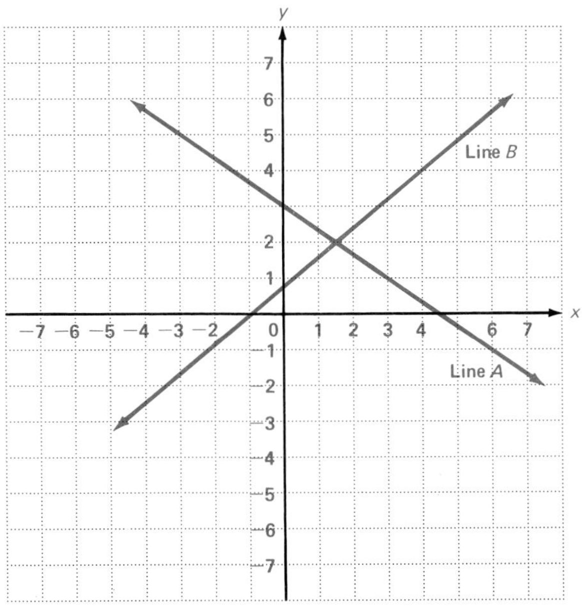

FIGURE 6.32

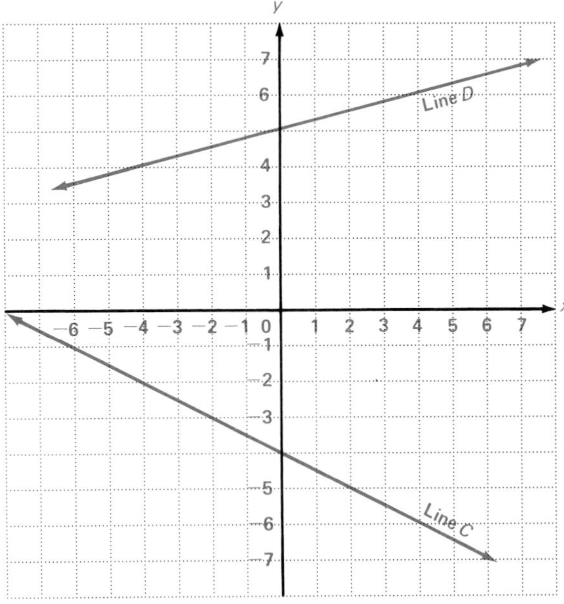

FIGURE 6.33

36. Write an equation for line B in Figure 6.32.

37. Write an equation for line D in Figure 6.33.

38. Write an equation for line A in Figure 6.32.

39. Write an equation for line C in Figure 6.33.

Write the equation of a line passing through the given two points. (Hint: Use the method of Example J)

40. $(1, 5)$ and $(4, 11)$

41. $(-1, -5)$ and $(2, 7)$

42. $(1, -2)$ and $(3, -8)$

43. $(2, -7)$ and $(-1, 8)$

Challenge Problems

First find the slope m. Then follow the method you used to work problems 40 to 43.

44. Find the equation of the line passing through:

$$\left(-\frac{2}{3}, \frac{3}{4}\right) \text{ and } \left(\frac{1}{5}, -\frac{3}{5}\right)$$

45. Find the equation of the line passing through:

$$\left(-\frac{1}{2}, -\frac{3}{2}\right) \text{ and } \left(\frac{2}{7}, \frac{1}{5}\right)$$

Solutions to Practice Problems for Exercise 6.3

1. $m = \dfrac{y_2 - y_1}{x_2 - x_1}$ — Let $(2, 0)$ be (x_1, y_1); let $(-1, 4)$ be (x_2, y_2).

$m = \dfrac{4 - 0}{-1 - 2}$ — Substitute the values into the equation.

$m = \dfrac{4}{-3} = -\dfrac{4}{3}$ — The slope $= -\dfrac{4}{3}$.

Notice that it does not matter which point you let be (x_1, y_1). If you selected them in the opposite order, you obtain

$$m = \frac{0 - 4}{2 - (-1)} = \frac{-4}{2 + 1} = \frac{-4}{3} = -\frac{4}{3}$$

2. $m = \dfrac{y_2 - y_1}{x_2 - x_1}$ — Let $(-5, -2)$ be (x_1, y_1); let $(3, -6)$ be (x_2, y_2).

$m = \dfrac{-6 - (-2)}{3 - (-5)}$ — Substitute the values into the equation.

$m = \dfrac{-6 + 2}{3 + 5}$ — Simplify.

$m = \dfrac{-4}{8} = -\dfrac{1}{2}$ — Reduce. The slope $= -\frac{1}{2}$.

3. We want to write the equation $4x - 2y = -5$ in the form $y = mx + b$.

$-2y = -4x - 5$ — Subtract $4x$ from both sides.

$y = \dfrac{-4x - 5}{-2}$ — Divide both sides by negative 2.

$y = \dfrac{-4x}{-2} + \dfrac{-5}{-2}$ — Write the fraction as sum of two fractions.

$y = 2x + \dfrac{5}{2}$ — Simplify.

The *slope* is 2. The *y-intercept* is $\dfrac{5}{2}$.

4. The line passes through $(-2, 4)$ and has a slope of $\frac{1}{2}$. Thus we know that $m = \frac{1}{2}$ and that one pair of coordinates (x, y) has the value $(-2, 4)$.

$y = mx + b$

$4 = \dfrac{1}{2}(-2) + b$ — Replace y by 4, m by $\dfrac{1}{2}$, x by -2.

$4 = -1 + b$ — Simplify.

$5 = b$ — Add $+1$ to both sides.

The y-intercept is 5 and we were given a slope of $\frac{1}{2}$, so the equation $y = mx + b$ becomes $y = \frac{1}{2}x + 5$.

Now do the problems in Exercise 6.3.

6.4 Graphing Linear Inequalities

In Section 2.4 we discussed linear inequalities in one variable and the graphical representation of these linear inequalities. You may want to return to that section to refresh your memory on those concepts.

We saw in Section 2.4 that when you write an expression such as $x < -2$ you divide the number line into two sections.

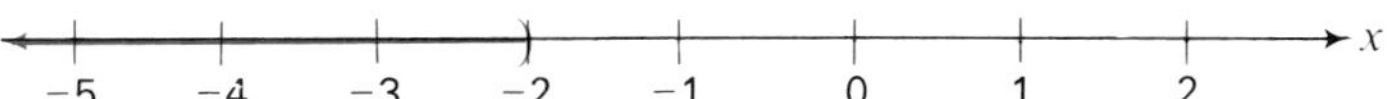

One section is all the values of x to the left of -2 on the number line. This region is indicated by the darker arrow to the left of -2. The other section is -2 and all values of x to the right of -2 on the number line. This section is not shaded.

Linear equations with two dimensions are graphed on a rectangular coordinate system called a plane. A straight line cuts across the plane. A linear inequality in two variables divides the plane into two half-planes.

Consider the inequality $y \geq x$. The region described by this inequality is shown in Figure 6.34. We can speculate that this would be the region above the line $y = x$. If we think of some ordered pairs that are above that line, for example (0, 6), (−2, 1), and (3, 5), we notice in each case that the y-coordinate is greater than the x-coordinate. Thus all the points in the half-plane above the line $y = x$ satisfy $y \geq x$. In addition, all the points on the line are values where $y = x$ such as [(0, 0) or (2, 2)], so these points are also included.

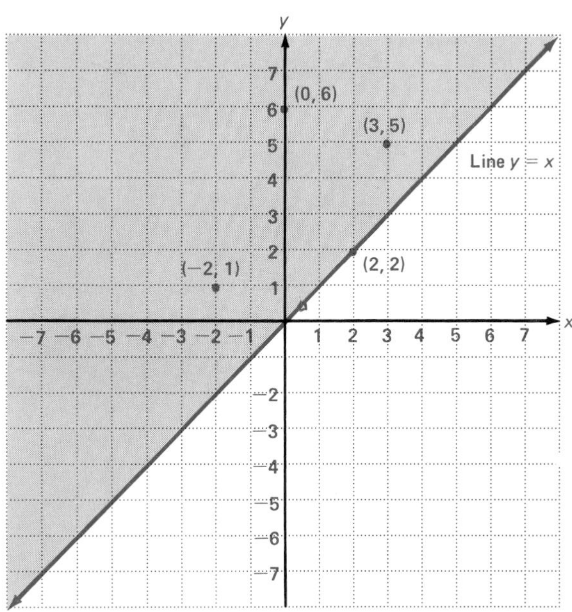

FIGURE 6.34 Graph of $y \geq x$.

The *solution of an inequality with two variables* is the set of all possible or... pairs that when substituted into the inequality will yield an identity. Obvious... cannot list them all, but we indicate where they are by shading in a region on ... coordinate plane.

Conclusion: The region $y \geq x$ is described by all points either on the line $y = x$ or above that line. It is shown by the line and the shaded region above it in Figure 6.34.

EXAMPLE A Graph: $y > x$.

Solution The shaded region in Figure 6.35 is *above* the line $y = x$. The dashed line indicates that points on the line are not included.

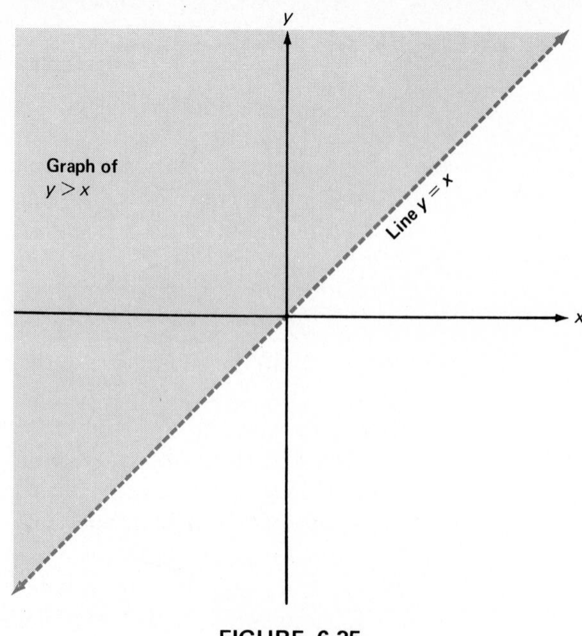

FIGURE 6.35

> ***Procedure for Graphing Linear Inequalities***
>
> **1. Replace the inequality symbol by the equality symbol and draw the line represented by that equation. This line will be a boundary line.**
>
> **2. The line itself will be a solid line and is part of the answer if the inequality uses the $\geq$ or the $\leq$ symbols. A dashed line is used if the inequality involves the $<$ or $>$ symbols.**
>
> **3. The solution to the inequality is represented by a shaded region. To determine if you should shade above the line or below the line, you should pick a point that does not lie on the line. Substitute the x and y coordinates of that point into the inequality. If you obtain an inequality that is true, shade in the half-plane that includes that point. If you obtain a false inequality, shade in the opposite half-plane. Unless the line goes through the origin, the point (0, 0) is usually a good test point.**

EXAMPLE B Graph: $5x + 3y \geq 15$.

Solution

1. Since the greater-than-or-equal-to ($\geq$) symbol is used, the line is included. We use a solid line to graph $5x + 3y = 15$, which will form the boundary line. The line may be written in the form $y = -\frac{5}{3}x + 5$ so that we see the y-intercept is 5 and the slope is $-\frac{5}{3}$. This will aid us in graphing the line.
2. Since the line does not pass through (0, 0), we use that point as a test point.
3. We now examine the inequality at the test point.

$5x + 3y \geq 15$	Original inequality.
$5(0) + 3(0) \geq 15$	Replace x by 0 and y by 0.
$0 + 0 \geq 15$	Simplify.
$0 \geq 15$	A false inequality is obtained. ($0 < 15$!)

4. We therefore shade on the other side of the line from the point (0, 0).

The desired region is shown in Figure 6.36. It is the region above the line and the line itself.

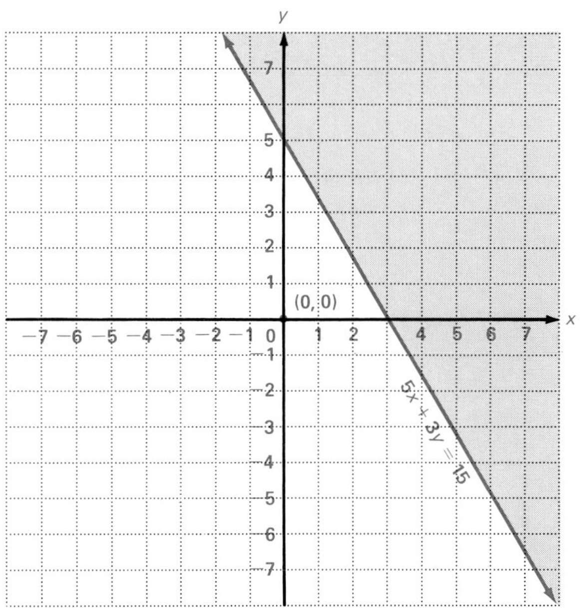

FIGURE 6.36

EXAMPLE C Graph: $y - 2x < 8$.

Solution

1. Since the less-than (<) symbol is used, the line is *not* included. We graph the line $y - 2x = 8$, using a dashed line to show that points on the line are *not* included.
2. Since the line does not pass through (0, 0), we use that point as a test point.
3. We now examine the inequality at the test point.

$y - 2x < 8$	
$0 - 2(0) < 8$	Replace x by 0 and y by 0.
$0 - 0 < 8$	Simplify.
$0 < 8$	A true inequality is obtained.

4. We therefore shade on the same side of the line as the point (0, 0). The desired region is shown in Figure 6.37. It is the half-plane below the line but not including the line.

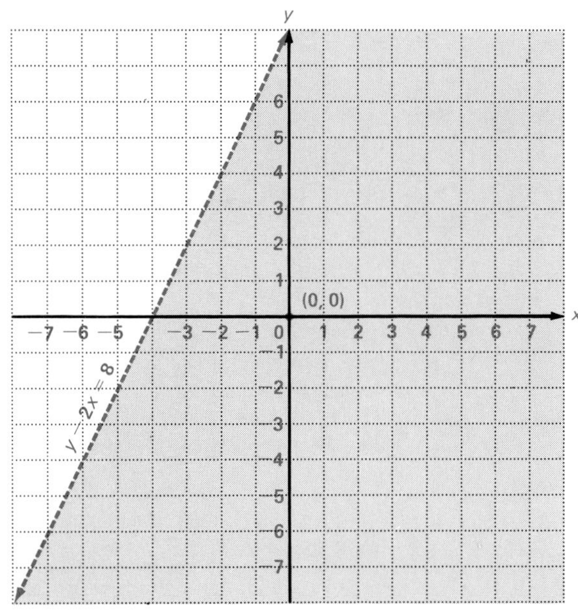

FIGURE 6.37

If the line passes through the origin or very close to it, use another test point that is clearly above or below the boundary line.

EXAMPLE D Graph: $-3x \leq 2y$.

Solution

1. Since $\leq$ is used, we know the boundary line is included. We graph $-3x = 2y$ using a solid line.

2. By inspection of the equation $-3x = 2y$ we see that the line passes through (0, 0), therefore we must pick a different point that is clearly not on the line. The point (2, 2) is a convenient one.

3. $-3x \leq 2y$ We use the original inequality and test it with the point (2, 2).

$-3(2) \leq 2(2)$ Replace x by 2 and y by 2.

$-6 \leq 4$ A true inequality is obtained.

4. We therefore shade in the region above the line since that is the half-plane that contains (2, 2). The region is shown in Figure 6.38. The solution is the region above the line and the line itself.

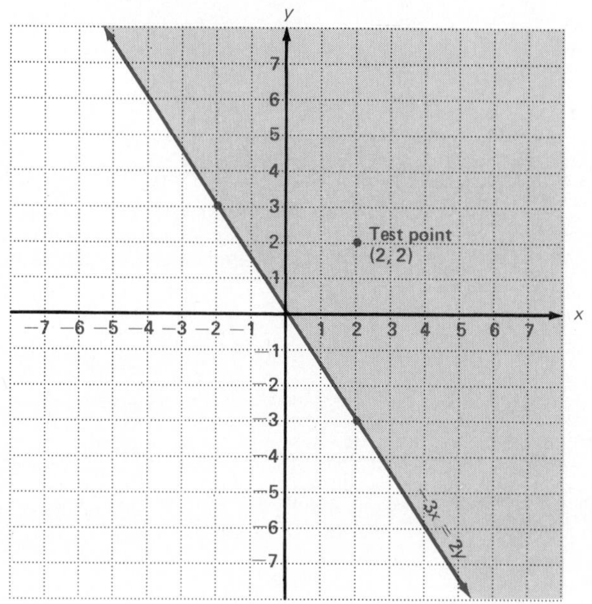

FIGURE 6.38

EXAMPLE E Graph: $4y - 11x < 1$.

Solution

1. Since $<$ is used, we know that the boundary line is not included. We graph the line $4y - 11x = 1$, using a dashed line to show that points on the line are *not* included.

2. The line does not pass through (0, 0). However, it is so close to the line that it is best to pick a point clearly not on the line. A convenient point is $(-1, 2)$, which is certainly above the line.

3. $4y - 11x < 1$ We use the original inequality to test the point $(-1, 2)$.

$4(2) - 11(-1) < 1$ We replace x by -1 and y by 2.

$8 + 11 < 1$ Simplify.

$19 < 1$ We obtain a false inequality; certainly, 19 is not less than 1.

4. We thus shade in the region below the line, which is the half-plane on the opposite side of the line from the test point $(-1, 2)$. The region is shown in Figure 6.39.

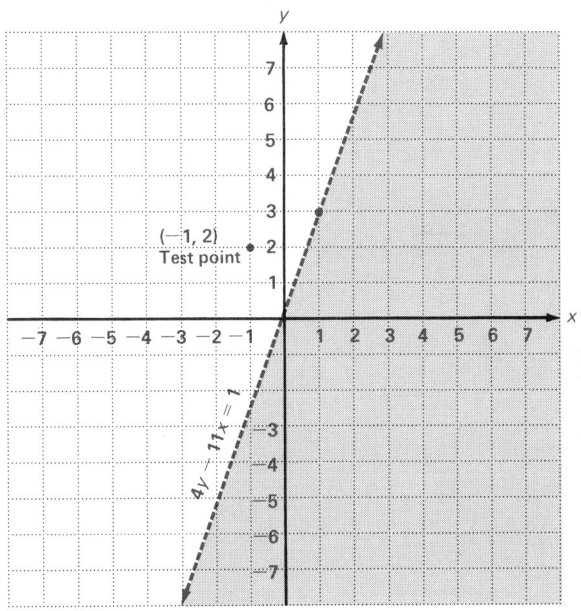

FIGURE 6.39

Carefully review Examples A to E and then try these practice problems.

Practice Problems for Exercise 6.4

Graph the region described by the inequality.

1. $5x - 2y > 0$
2. $x \leq -3$
3. $3x + 4y \leq -16$

Now turn to the answers following Exercise 6.4 and check your solutions for these practice problems.

Exercise 6.4

Graph the region described by the inequality.

1. $y > 2x - 2$

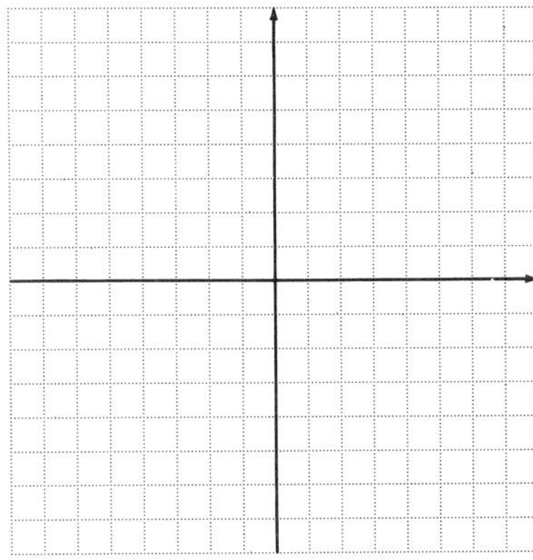

2. $y > 3x - 1$

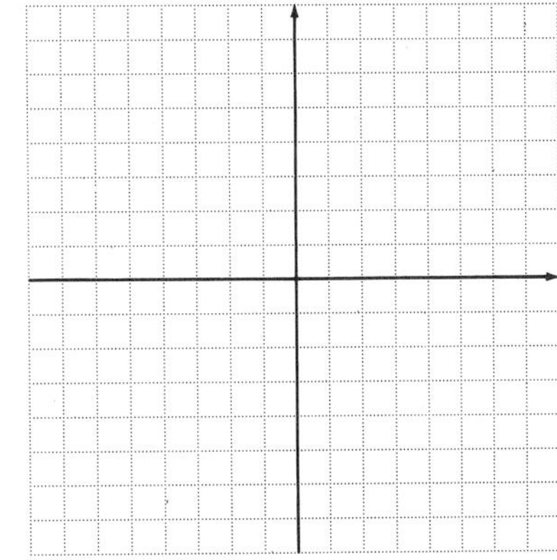

3. $2x - 3y < 6$

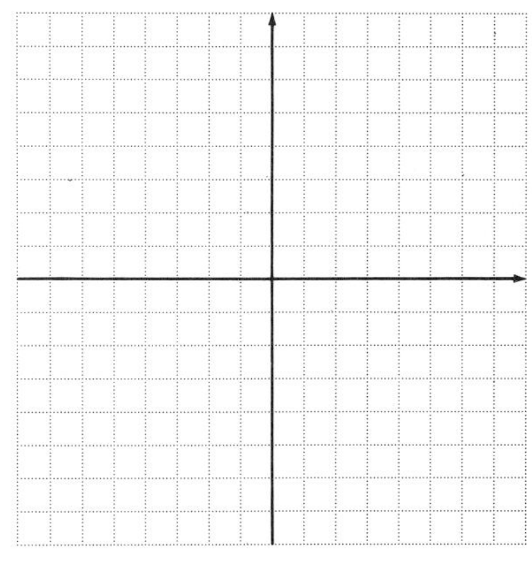

4. $3x + 2y < -6$

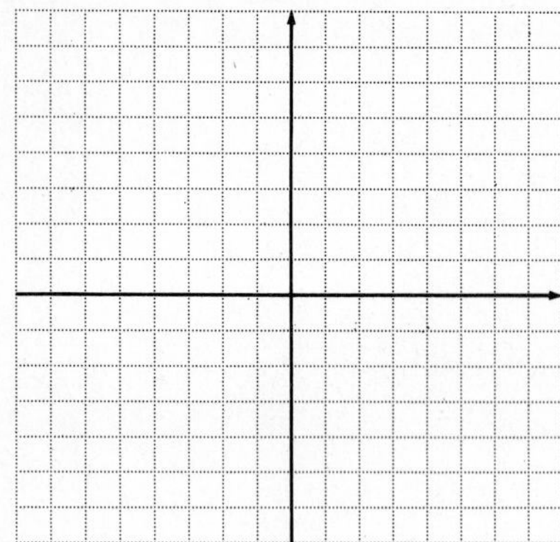

5. $x - y \geq 2$

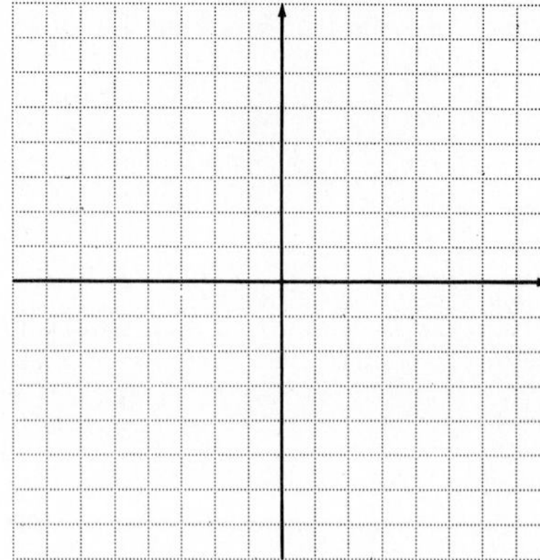

6. $x - 6 \geq 3$

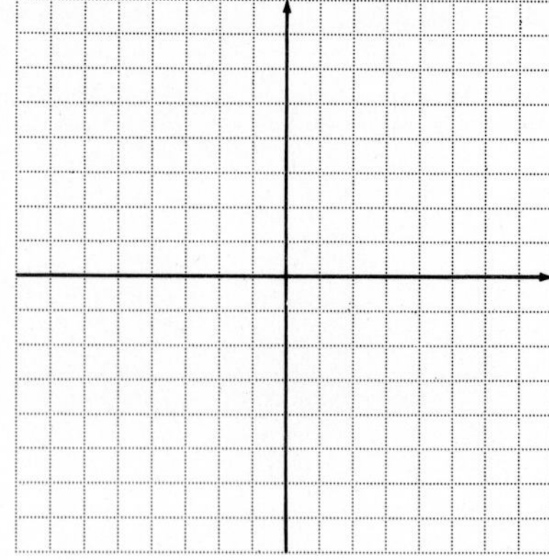

7. $5x - y - 5 \leq 0$

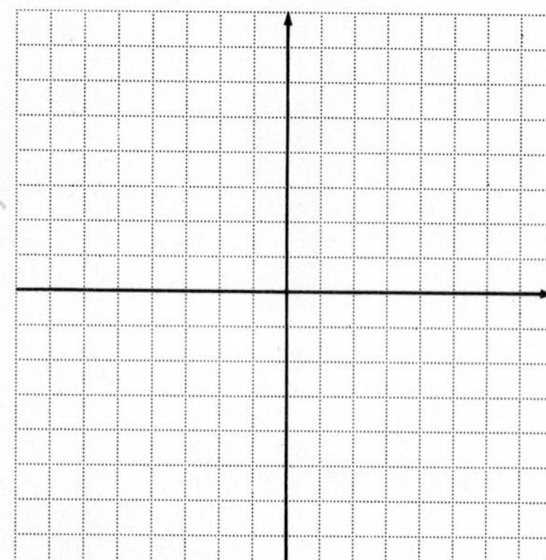

8. $4x + y - 4 \geq 0$

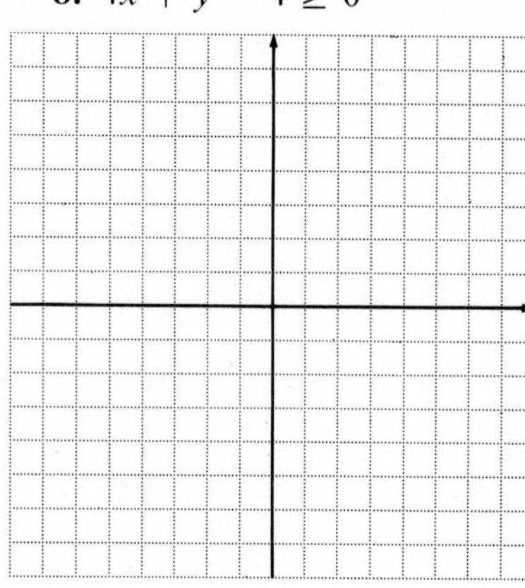

9. $y \geq -2$

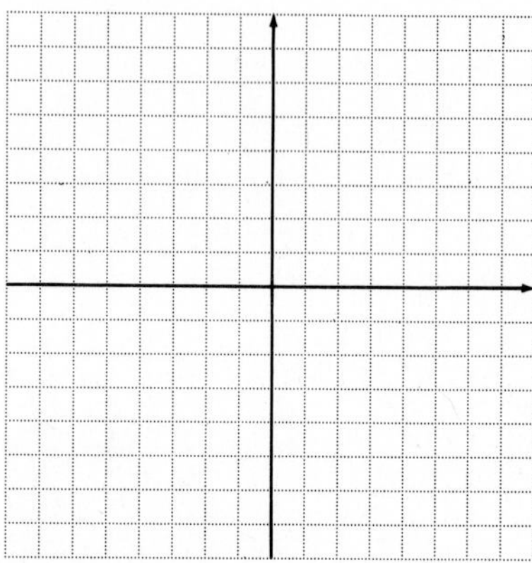

10. $x \leq 0$

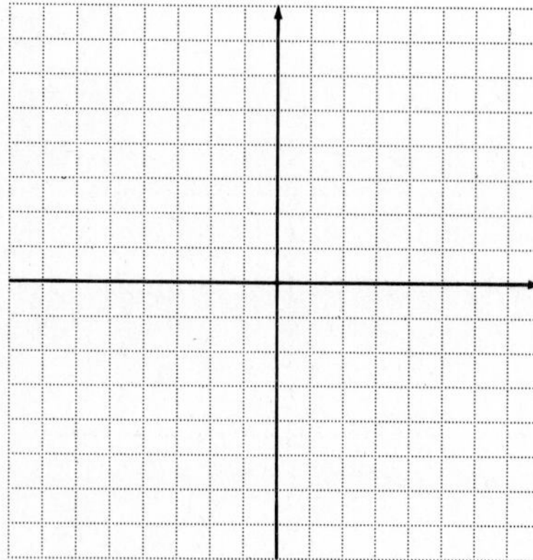

11. $2x > 3 - y$

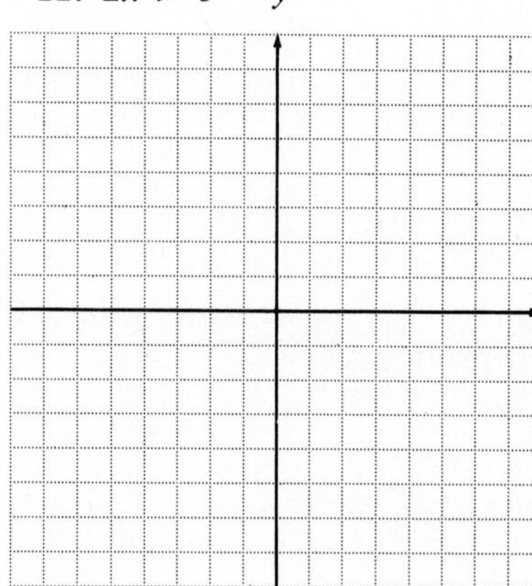

12. $3x > 1 + y$

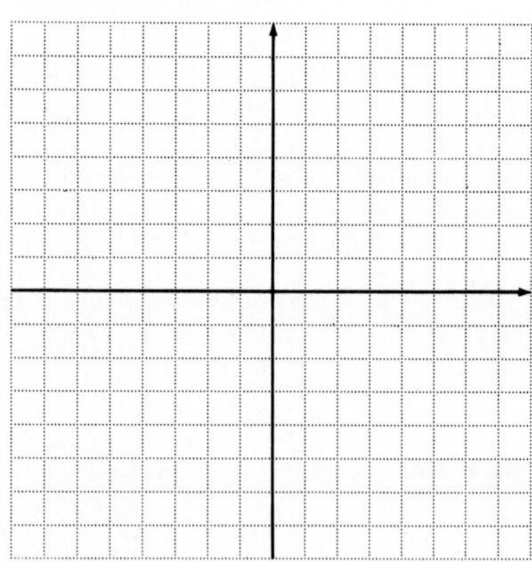

13. $5y + 2 < 7x + 2$

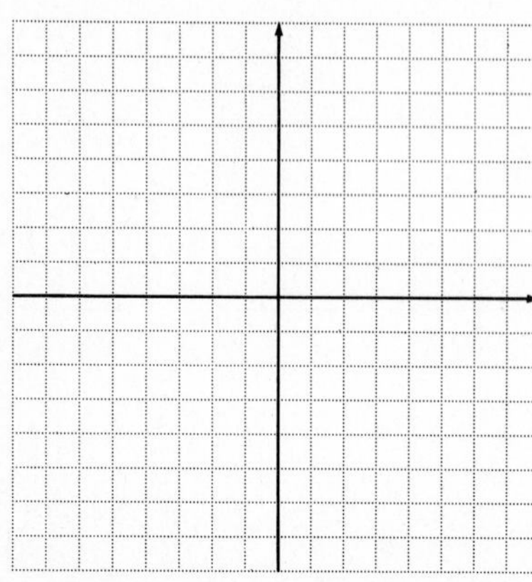

14. $6x - 3 < 4y - 3$

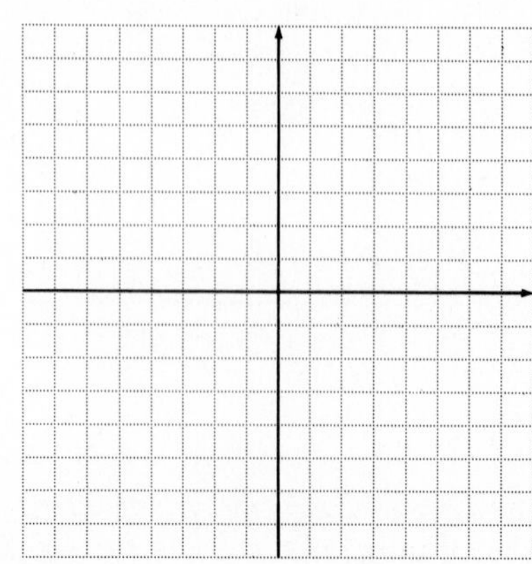

15. $y + \frac{1}{2}x \geq -2$

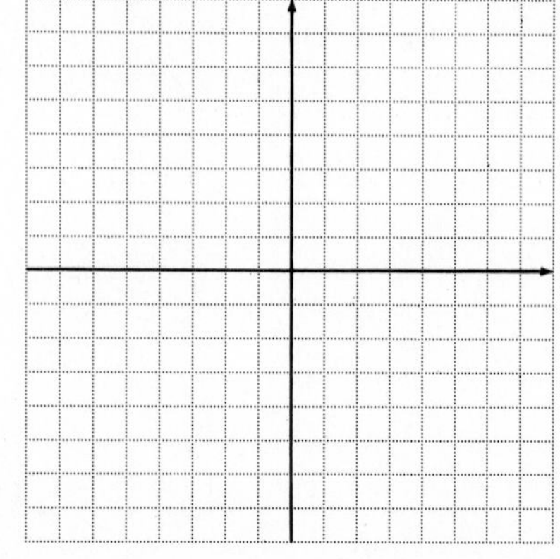

16. $y + \frac{1}{3}x \geq 4$

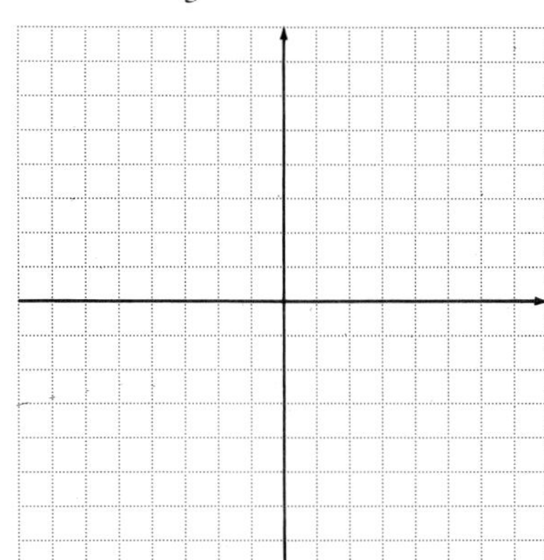

17. $2x - 3y \leq 1$

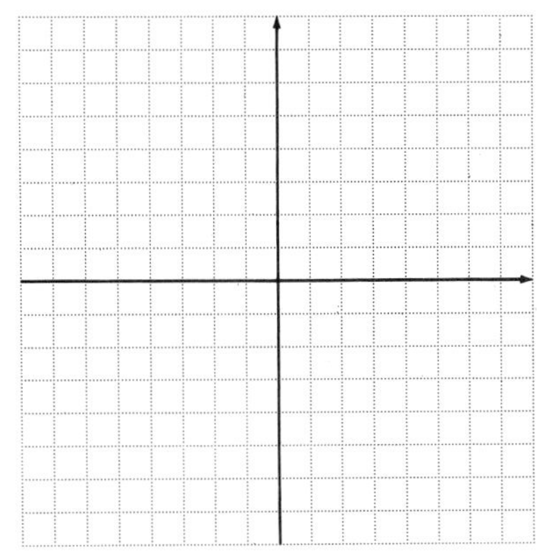

18. $3x - 4y \geq 10$

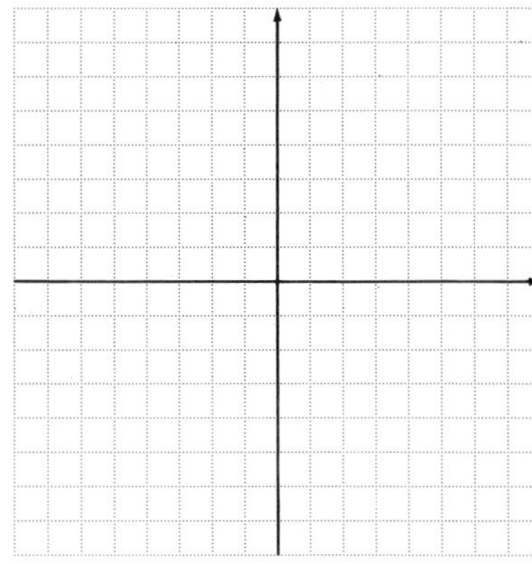

19. $y \geq \frac{1}{3}x$

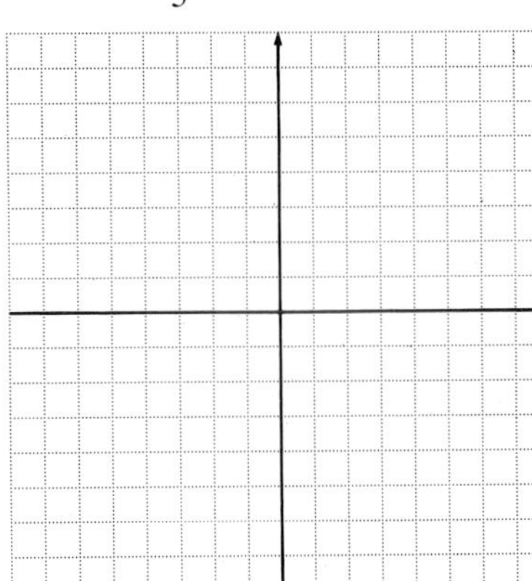

20. $y \leq \frac{2}{3}x$

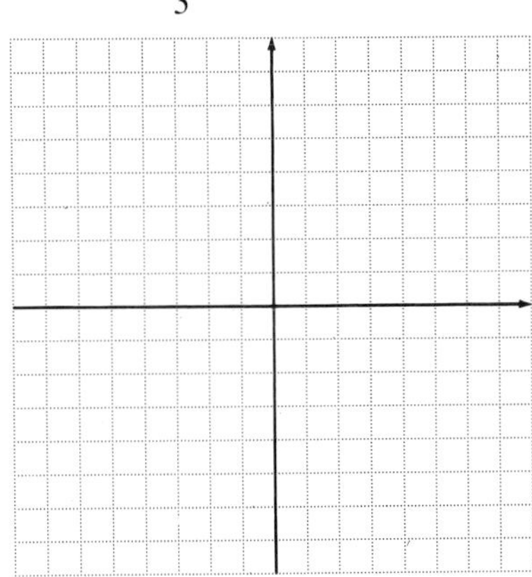

21. $y < -\frac{3}{4}x + 2$

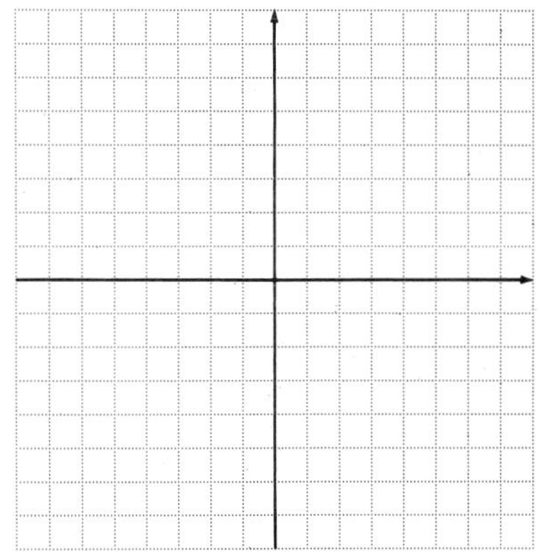

22. $y > -\frac{2}{3}x - 1$

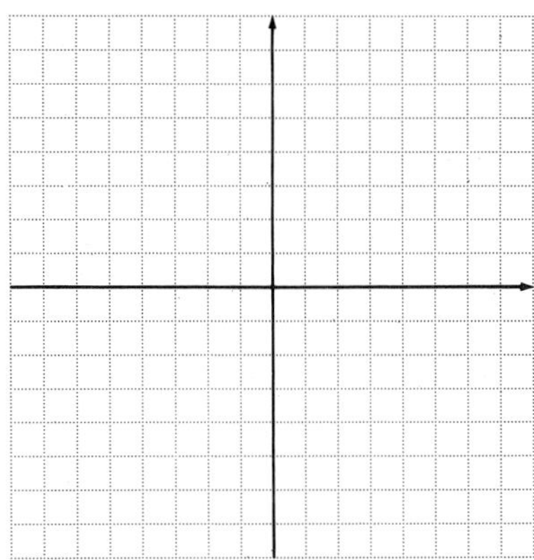

23. $y \geq 3 - \frac{3}{5}x$

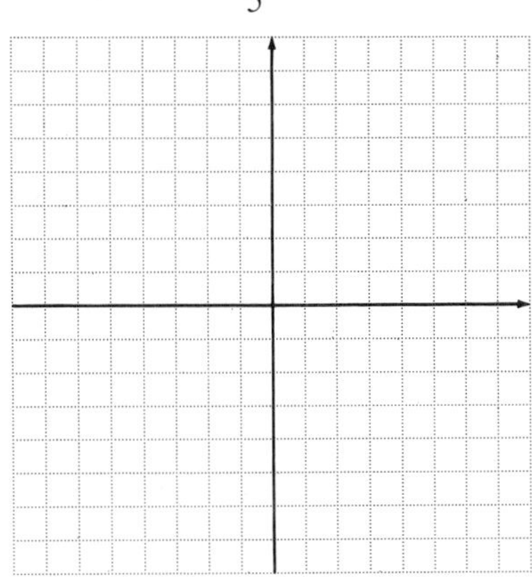

24. $y \leq 4 - \frac{1}{5}x$

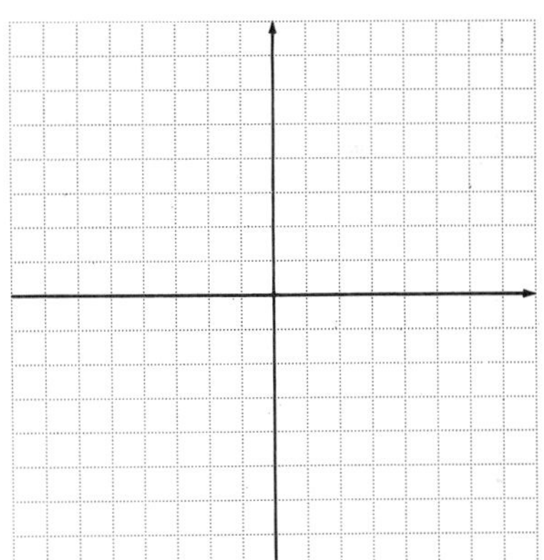

Challenge Problems

Simplify and then graph the inequality.

25. $17x + 3y + 2 \geq 5(x - 3y)$

26. $\frac{1}{2}(x + 3y) - 4 < 3x - 4y$

Solutions to Practice Problems for Exercise 6.4

1. Graph: $5x - 2y > 0$.

(1) Graph $5x - 2y = 0$ using a dashed line (Figure 6.40). The line is *not* included in the desired region.

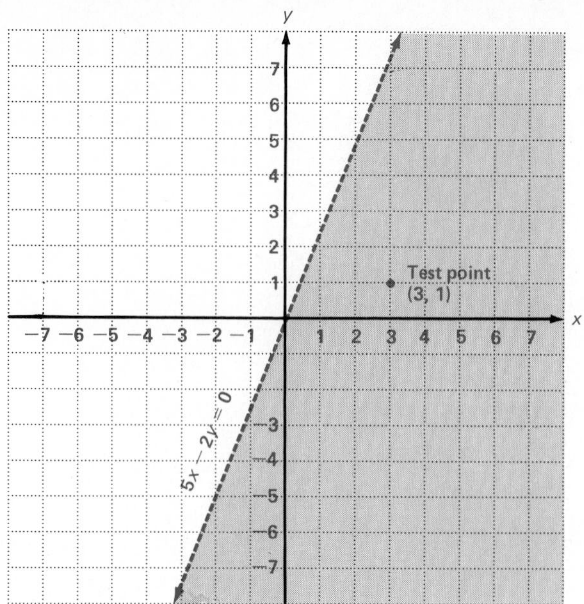

FIGURE 6.40

(2) Since $5x - 2y = 0$ passes through the origin we use another test point. (3, 1) is a convenient point.

(3) $5x - 2y > 0$

$5(3) - 2(1) > 0$ — Substitute $x = 3$, $y = 1$ into the inequality.

$15 - 2 > 0$

$13 > 0$ — A true inequality is obtained.

(4) We shade in the region on the same side of the line as (3, 1). We shade the region below the line.

2. Graph: $x \leq -3$.

A solid line is used (Figure 6.41). The desired region is the line $x = -3$ and all the region to the left of the line. A test point is not actually needed in the case of simple horizontal and vertical lines. Usually the region can be determined by inspection.

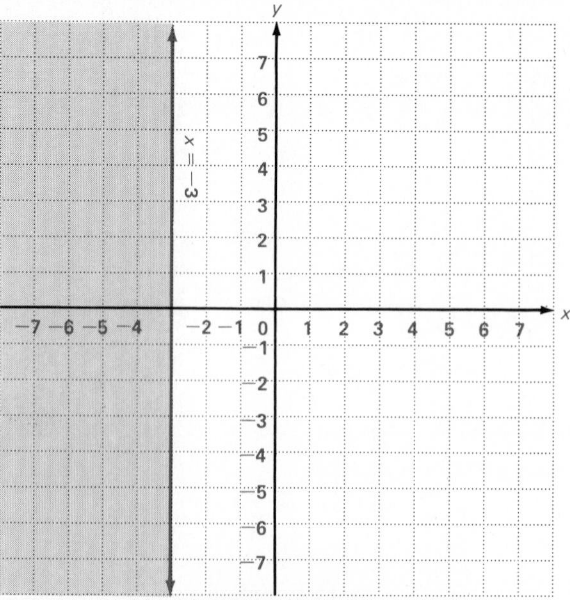

FIGURE 6.41

3. Graph: $3x + 4y \leq -16$.

(1) Graph $3x + 4y = -16$ using a solid line (Figure 6.42). The line *is* included in the final answer. You will find that the graph is easier if you plot a few points like $(-4, -1)$, $(0, -4)$, and $(4, -7)$.

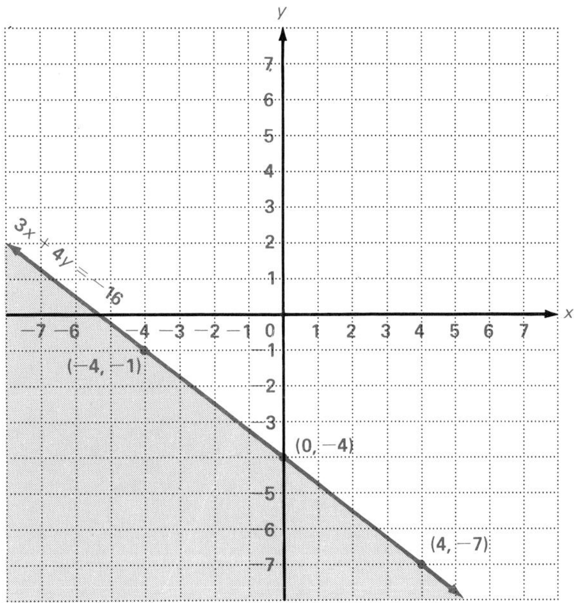

FIGURE 6.42

(2) We use (0, 0) as a test point since the line does not go through the origin.

(3) $3(0) + 4(0) \leq -16$ — Replace x by 0 and y by 0.

$0 \leq -16$ — A false inequality is obtained. $(0 > -16!)$

(4) We shade the region below the line. The final answer is the line $3x + 4y = -16$ and the half-plane below that line.

When you are sure that you understand these practice examples, sketch the graphs in Exercise 6.4.

6.5 Geometric Formulas

In mathematics it is often helpful to draw a graph of an equation to examine some properties of the equation. If we need to evaluate a particular quantity of a two-dimensional object such as a square or a three-dimensional object such as a sphere, it is helpful to draw a sketch of that object.

You have no doubt previously encountered some formulas for evaluating areas, perimeters, and volumes of certain geometrical objects. These formulas are reviewed for you here. They are very useful in a variety of mathematics courses as well as in many situations in everyday life.

Common cases where perimeter and area are required

1. A *square* is a rectangle with four equal sides.

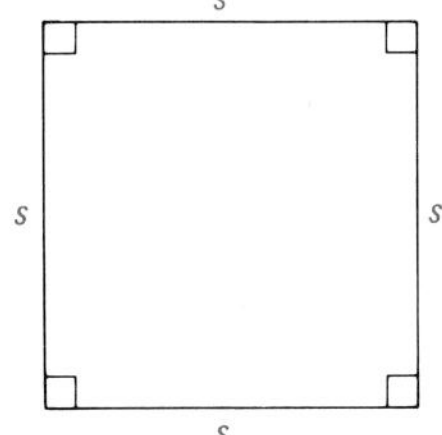

Area $= s^2$

Perimeter $= 4s$

2. A *rectangle* is a four-sided figure with opposite sides parallel and all interior angles measuring 90°.

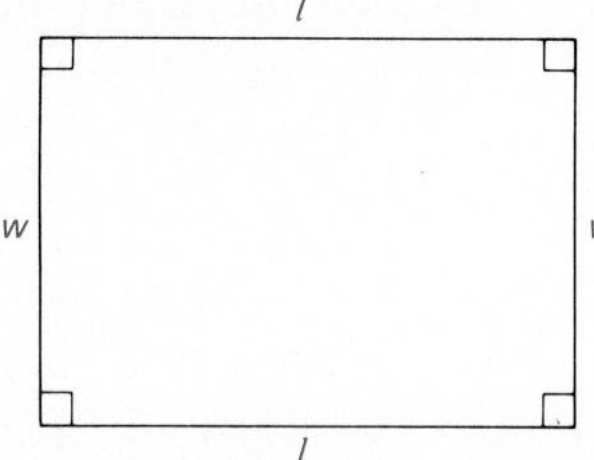

Area $= lw$

Perimeter $= 2l + 2w$

Angles that measure 90° are called *right angles*. They are often labeled with small squares in a sketch to indicate they are right angles.

3. A *triangle* is a closed plane figure with three sides.

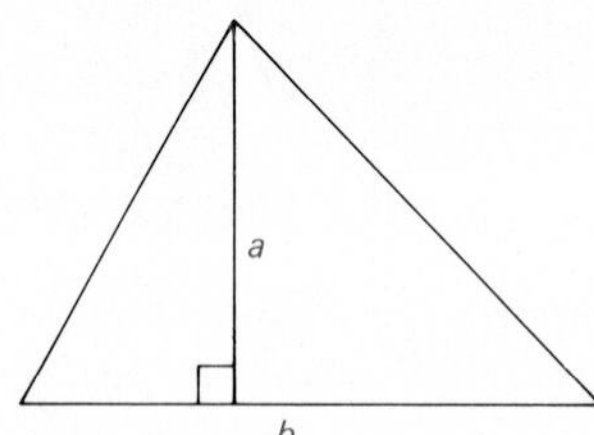

$$\text{Area} = \frac{1}{2}ab \quad \text{or} \quad \frac{ab}{2}$$

where $a =$ the altitude

$b =$ the base

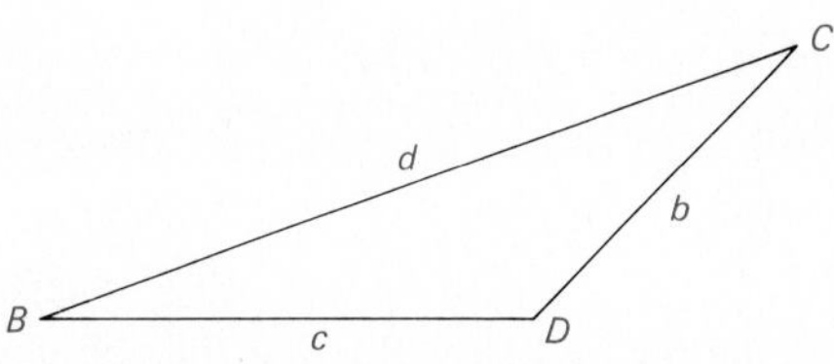

Perimeter $=$ sum of the three sides

$= b + c + d$

Sum of interior angles $= 180°$

Measure of $\angle B$ + measure of $\angle C$ + measure of $\angle D = 180°$

4. Special triangles include:

(a) *Equilateral Triangle*: a triangle with three sides equal in length and three angles which measure 60°.

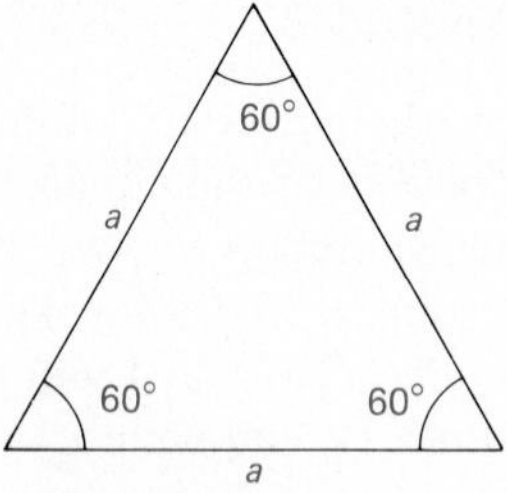

(b) *Isosceles Triangle*: a triangle with two equal sides and the two angles opposite the equal sides also equal.

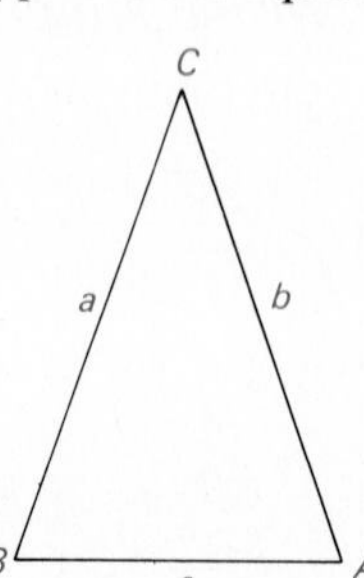

Side a = side b

Measure of $\angle A$ = measure of $\angle B$

5. A *parallelogram* is a four-sided figure with opposite sides parallel. In a parallelogram opposite sides are equal and opposite angles are equal.

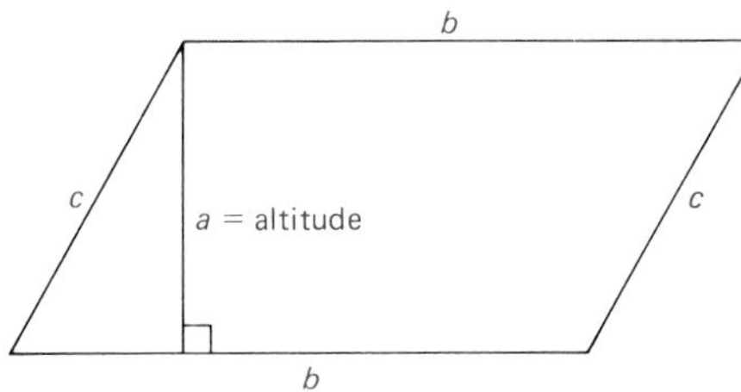

For a parallelogram with sides b and c:

$$\text{Perimeter} = 2b + 2c$$

For a parallelogram with base b and altitude a:

$$\text{Area} = ab$$

6. A *rhombus* is a parallelogram with four equal sides.

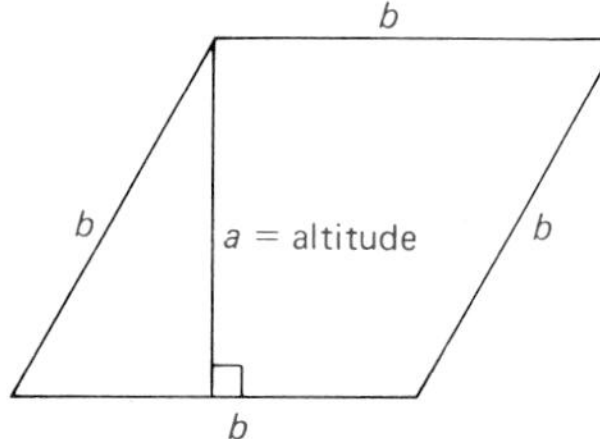

$$\text{Perimeter} = 4b$$

$$\text{Area} = ab$$

7. A *trapezoid* is a four-sided figure with two sides parallel. The parallel sides are called the bases of the trapezoid.

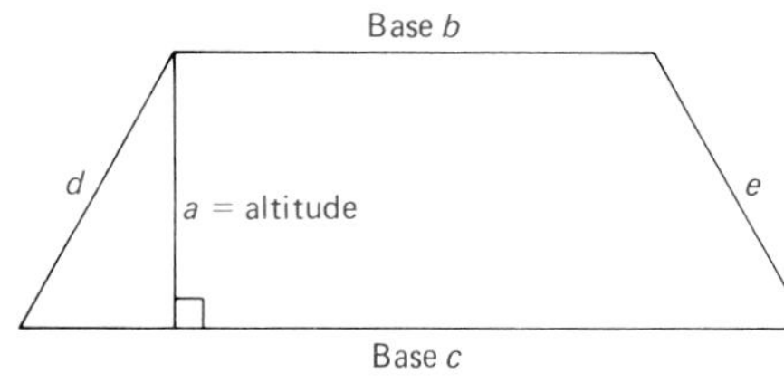

$$\text{Perimeter} = b + c + d + e$$

$$= \text{sum of 4 sides}$$

$$\text{Area} = \frac{1}{2}a(b + c)$$

$$= \frac{1}{2}\text{(altitude)(sum of the bases)}$$

8. A *circle* is a plane curve consisting of all points at an equal distance from a given point called the center.

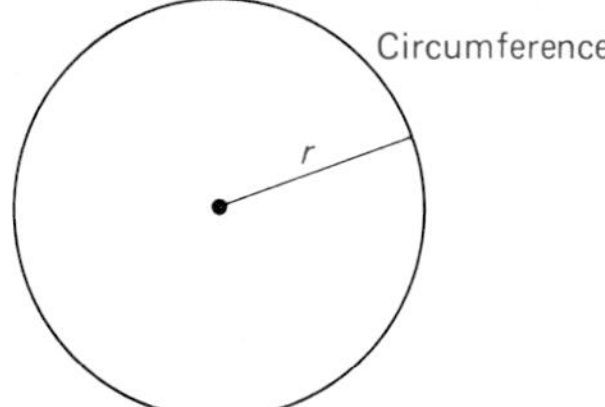

r = radius

Diameter $d = 2r$

Circumference $= \pi d$ or $2\pi r$

Area $= \pi r^2$

π is an irrational number which is approximately 3.141592654. We usually use 3.14 as a sufficiently accurate approximation.

Common cases where volume is required

1. *Rectangular Solid*

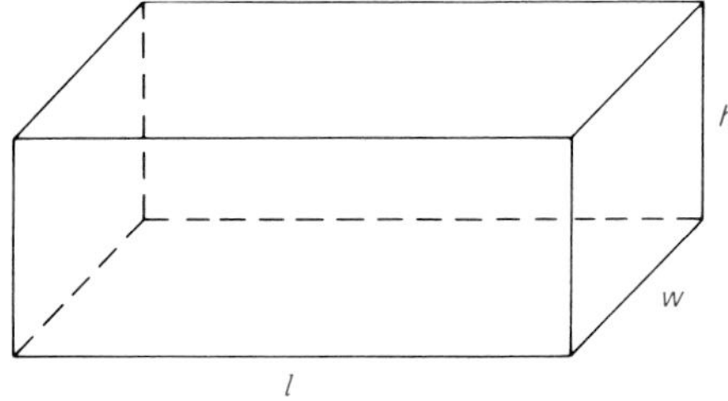

Volume $V = lwh$

Total surface area $A = 2lw + 2wh + 2lh$

2. *Sphere*

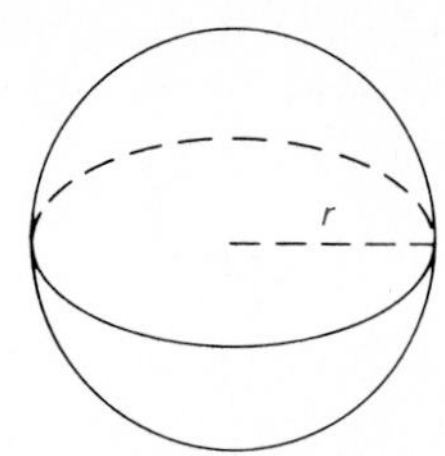

Volume $V = \frac{4}{3}\pi r^3$ (π is approximately 3.14)

Total surface area $A = 4\pi r^2$

3. *Right Circular Cylinder*

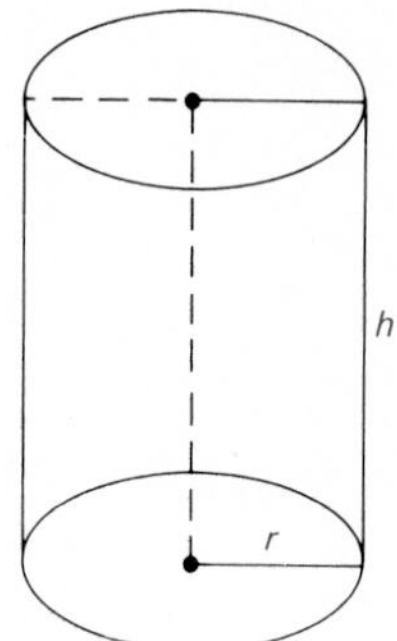

Volume $V = \pi r^2 h$

Total surface area $A = 2\pi rh + 2\pi r^2$

Use of the Formulas

These formulas can be used directly to find a given area, perimeter, circumference, or volume if all other values are given.

EXAMPLE A Find the area of a triangle whose base is 5 centimeters and whose altitude is 12 centimeters (Figure 6.43).

Solution

$$A = \frac{1}{2}ab$$

$$A = \frac{1}{2}(12)(5)$$

$$A = 30 \textit{ square centimeters}$$

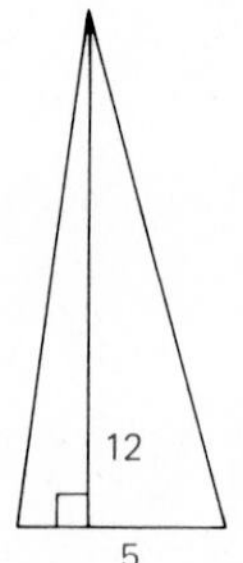

FIGURE 6.43

Use of Formulas

These formulas can be used to write a specific formula for finding a value when only some information is given.

EXAMPLE B Write an expression for the area of a rectangular field whose width is 8 meters (Figure 6.44).

Solution

$$A = lw$$

$$A = (l)(8)$$

$$A = 8l$$

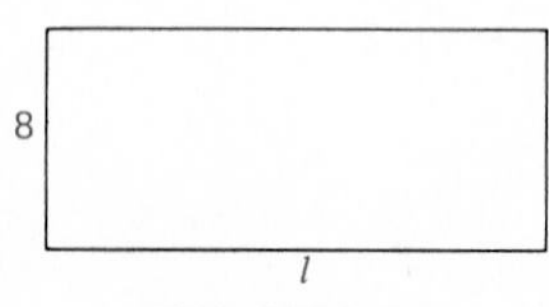

FIGURE 6.44

> ### *Use of Formulas*
>
> **These formulas can be used to find one part of the formula if all other parts of the physical object are known.**

EXAMPLE C Find the altitude of a trapezoid whose area is 80 square inches and whose bases are 3 inches and 5 inches, respectively (Figure 6.45).

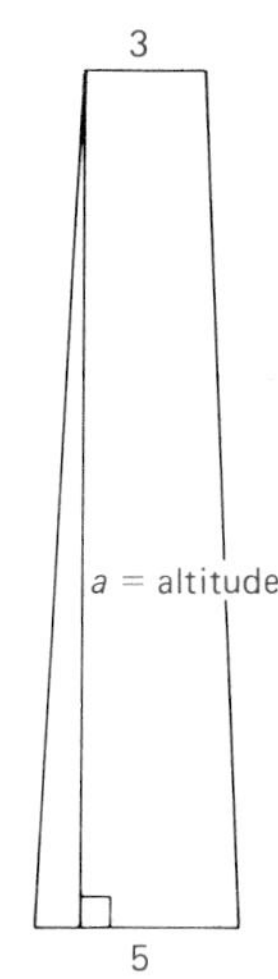

FIGURE 6.45

Solution

$A = \frac{1}{2}a(b + c)$	We write the formula for the area of a trapezoid.
$80 = \frac{1}{2}a(3 + 5)$	Substitute known parts.
$160 = a(3 + 5)$	Multiply both sides by 2 to clear fractions.
$160 = a(8) = 8a$	Simplify
$\frac{160}{8} = \frac{8a}{8}$	Divide both sides by 8.
$20 = a$	The altitude is 20 inches.

Alternate Solution First solve for the desired letter a in the equation.

$A = \frac{1}{2}a(b + c)$	
$2A = a(b + c)$	Multiply both sides by 2.
$\frac{2A}{b + c} = a$	Divide both sides by $(b + c)$.
$\frac{2(80)}{3 + 5} = a$	Substitute known values.
$\frac{160}{8} = 20 = a$	Evaluate the quantity.

Now try these practice problems.

Practice Problems for Exercise 6.5

1. Find the volume of a sphere (Figure 6.46) with radius 4 centimeters. (Use $3.14 = \pi$.)

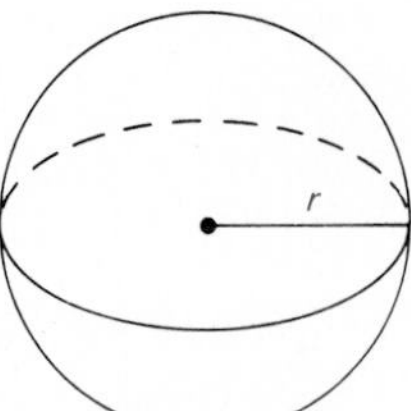

FIGURE 6.46

2. Write an expression for the perimeter of a parallelogram (Figure 6.47) whose one known side is 4 feet long.

FIGURE 6.47

3. Find the radius of a circle whose circumference is approximately 28.26 meters. (Use $\pi = 3.14$.)
4. Figure 6.48 shows a quarter-circle (of radius 1.5 yards) connected to two rectangles with dimensions labeled on the sketch. You need to lay a strip of carpet in your house according to this sketch. How many square yards of carpeting will be needed on the floor? (Use 3.14 for π.)

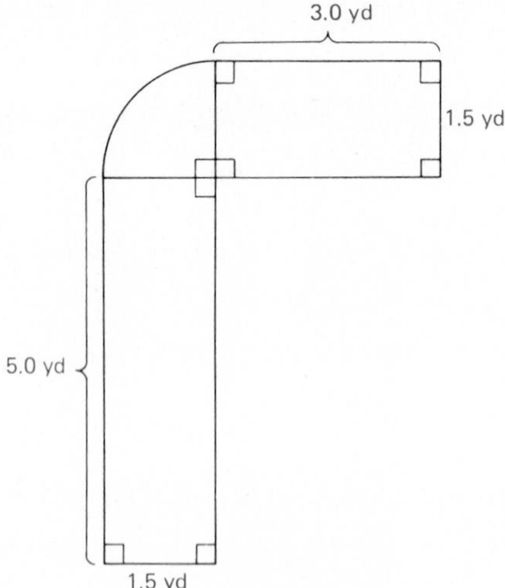

FIGURE 6.48

Now turn to the answers following Exercise 6.5 and check your solutions for these practice problems.

Exercise 6.5

1. Find the perimeter of a rectangle whose width is 12 feet and whose length is 18.5 feet.

2. Find the perimeter of a triangle whose sides are 12 feet, 27 feet, and 31 feet, respectively.

3. Find the perimeter of a trapezoid whose sides are 7 meters, 5 meters, 15 meters, and 5 meters, respectively.

4. Find the circumference of a circle whose diameter is 20 meters. (Use $\pi = 3.14$ as an approximate value.)

5. Find the area of a triangle whose altitude is 12 inches and whose base is 7 inches.

6. Find the area of a parallelogram whose altitude is 14 inches and whose base is 5 inches.

7. Write an expression for the area of a rhombus whose base is 10 centimeters long.

8. Write an expression for the volume of a rectangular solid whose height is 12 centimeters and whose width is 5 centimeters.

9. What is the width of a square whose perimeter is 22 kilometers?

10. What is the length of a rectangle if the width is 4.5 kilometers and the perimeter is 20 kilometers?

11. Two angles of a triangle measure 50° and 65°, respectively. What is the measure of the third angle?

12. An equilateral triangle has a side that measures 2.6 meters. What is the perimeter of the triangle?

13. An isosceles triangle has one side that measures 4 feet and another that measures 7.5 feet. The third side was not measured. Can you find one unique perimeter for this type of triangle?

14. Bob wanted to know the circumference of a circular flower bed. The radius of the circle is 2 feet (Figure 6.49). He made a rough guess by using 3 as an approximation for π. This caused him to obtain a value that was significantly shorter for the circumference than what he should have obtained. The next day he decided to use 3.14 to obtain a more accurate approximation. How much shorter was the value for the circumference obtained on the first day than that obtained the next day?

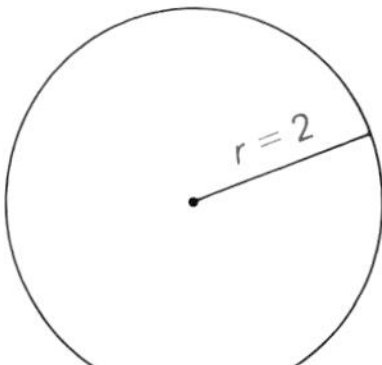

FIGURE 6.49

15. Find the area of the object shown in Figure 6.50. The diameter of each half-circle is 4.00 feet. (The four-sided object drawn with dashed lines is a square.)

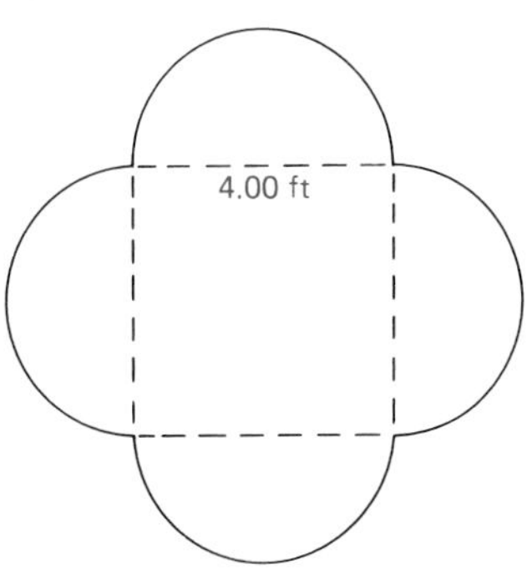

FIGURE 6.50

16. Find the area of the object shown in Figure 6.51. The length of the rectangle is 4 inches. The width of the rectangle and the diameter of the small half-circle are both 2 inches. The radius of the large quarter-circle is 2 inches.

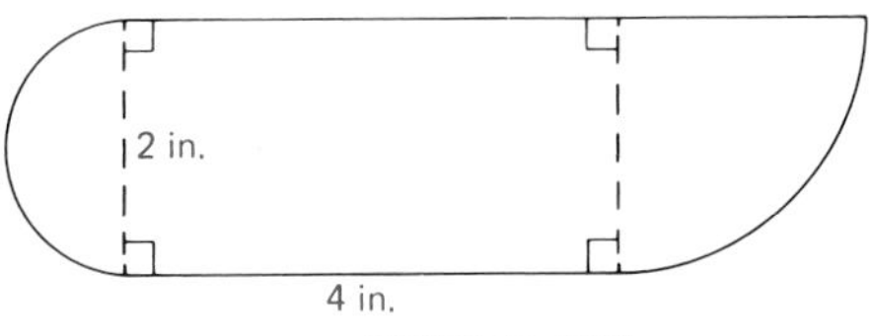

FIGURE 6.51

17. Figure 6.52 shows an oddly shaped field consisting of a triangular region and a quarter-circle. This field must be fertilized. Green Lawn, Inc. will do the job for $0.40 per square yard. How much will the job cost?

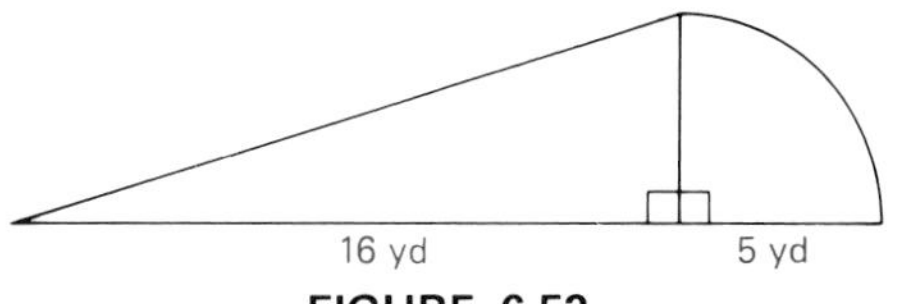

FIGURE 6.52

18. A sign constructed in the shape of a rainbow needs painting. The area to be painted is shaded in Figure 6.53. The sign was constructed from two semicircles. The radius of the larger semicircle is 8 feet while the radius of the smaller semicircle is 6 feet. If the painter charges \$0.20 per square foot to paint the sign, find the cost to paint the sign. (Use 3.14 as an approximation for π. Round your answer to the nearest cent.)

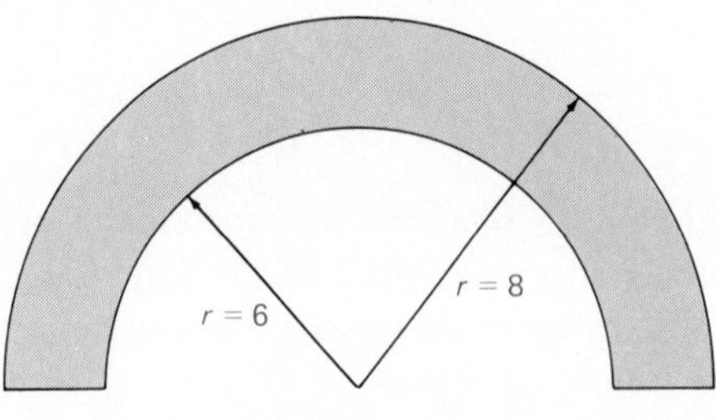

FIGURE 6.53

Challenge Problems

19. Three paper circles of radius 2 inches are cut from a rectangular piece of paper that measures 4 inches by 12 inches. They are cut in such a way as to minimize the leftover paper (Figure 6.54). The shaded region indicates the remaining paper. How many square inches of paper are left over? (Use $\pi = 3.14$.)

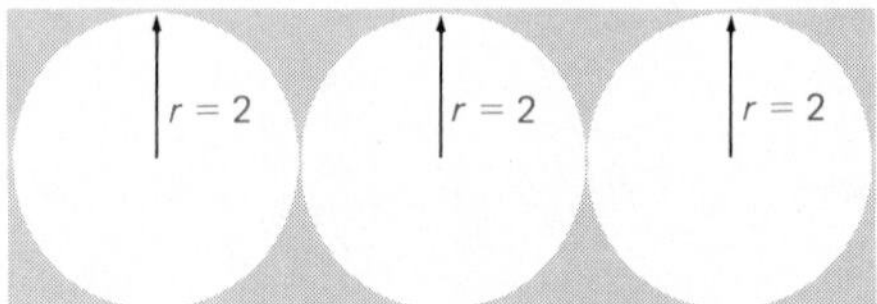

FIGURE 6.54

20. Assume that a very long rope is stretched around the moon supported by poles that are *3 feet tall.* (Neglect gravitational pull and assume that the rope takes the shape of a large circle, as shown in Figure 6.55). The radius of the moon is approximately 1080 miles. How much longer would you need to make the rope if you wanted the rope to be supported by poles that are *4 feet tall*?

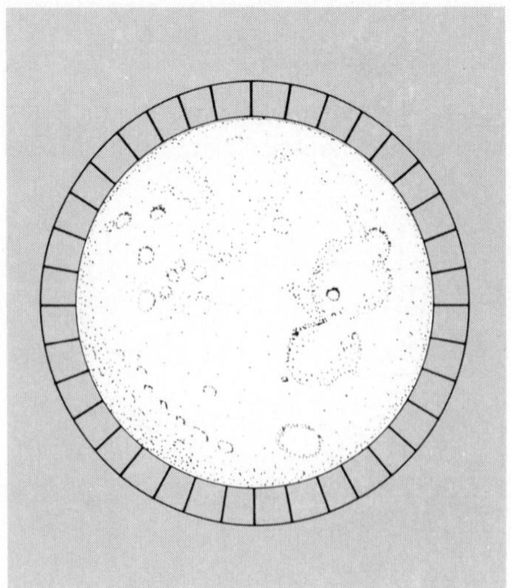

FIGURE 6.55

21. Find the total surface area of (a) a sphere of radius 3 inches and (b) a right circular cylinder with radius 2 inches and height 0.5 inches. Which object has a greater surface area?

Solutions to Practice Problems for Exercise 6.5

1. $V = \frac{4}{3} \cdot \pi r^3$, where $r = 4$ and $\pi = 3.14$ (approximately).

$V = \frac{4}{3}(3.14)(4)^3$	Substitute values for 4 and π.
$V = \frac{4}{3}(3.14)(64)$	Raise to a power first.
$V = \frac{(4)(3.14)(64)}{3}$	Write the product as one fraction to simplify the calculation.
$V = \frac{803.84}{3}$	Multiply numerators.
$V = 267.95$	Carry out division to two decimal places.

We have obtained an approximate value that is accurate to the nearest hundredth. The volume = 267.95 cubic centimeters or 267.95 cc.

2. Let c = unknown side, as shown in Figure 6.56.

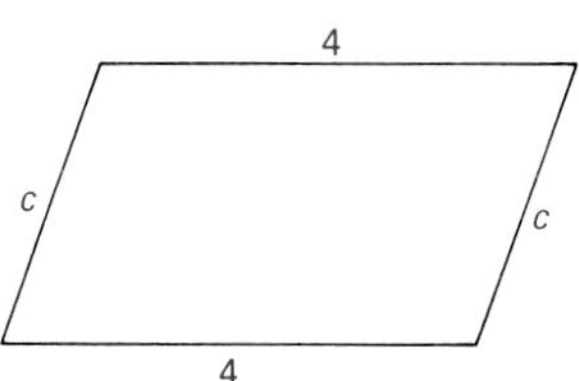

FIGURE 6.56

$$P = 2b + 2c$$
$$P = 2(4) + 2c$$
$$P = 8 + 2c$$

3. The circumference = 28.26 meters and $\pi = 3.14$ (approximately).

$C = 2\pi r$	
$28.26 = 2(3.14)r$	Substitute the given values for C and π.
$28.26 = 6.28r$	Simplify.
$\frac{28.26}{6.28} = r$	Divide both sides by 6.28.
$4.5 = r$	The radius is 4.5 meters.

4. The desired area is the sum of three areas, which we will call B, C, and D (Figure 6.57). Area B is rectangular in shape.

$$A = (l)(w) = (3.0)(1.5) = 4.5 \text{ square yards}$$

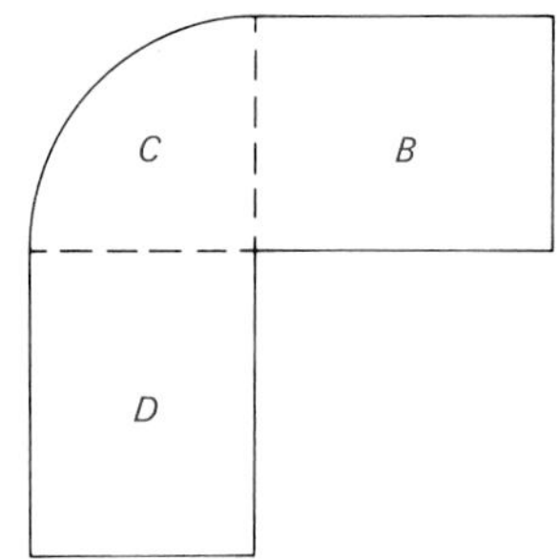

FIGURE 6.57

Area C is one quarter of a circle. The radius of the circle is 1.5 yards.

$$A = \frac{\pi r^2}{4} = \frac{(3.14)(1.5)^2}{4} = \frac{(3.14)(2.25)}{4}$$

$$= \frac{7.065}{4} = 1.76625$$

Round off to 1.77 square yards.

Area D is rectangular in shape.

$$A = (l)(w) = (5.0)(1.5) = 7.5 \text{ square yards.}$$

Total area $= 4.5 + 1.77 + 7.5 = 13.77$ square yards (approximation to nearest hundredth).

Note: This is the amount of carpeting to be *placed on the floor*. There would undoubtedly be some waste when the carpeting is cut to these dimensions.

Now do the problems in Exercise 6.5.

An Application for Chapter 6: Geometrical Formulas for More Than Three Dimensions

Warning

The following application is quite abstract. It may be hazardous to your mental well-being if you do not appreciate reflecting on unusual or farfetched ideas. The topic you are about to read about is literally "out of this world."

In this chapter we have discussed finding the area of two-dimensional objects and the volumes of three-dimensional objects. How do you suppose you could find the "volume" of an object in the fourth dimension? This is an interesting area of study in mathematics. One has to use more theoretical approaches because it is hard to visualize the fourth dimension. (In many areas of mathematics and physics time is considered to be the fourth dimension. Mathematical models of five, six, and higher dimensions are used in various advanced theoretical studies.)

Wayne Hoover of Cape Cod Community College in West Barnstable, Massachusetts, has developed the following interesting approach to the creation of formulas in higher dimensions.

Let us consider two-dimensional space.

For a circular object let us consider two measurements of interest.

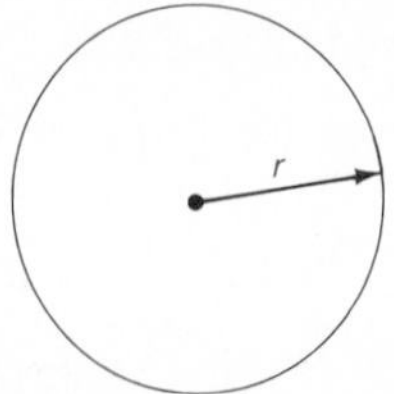

Area of the circle $= \pi r^2$

Distance around the circle (circumference) $= 2\pi r$

We will use the subscript 2 to denote these measurements in two-dimensional space. The formula for the amount contained inside the object is therefore

$$Area = V_2 = \pi r^2$$

The formula for the measurement of the boundary that contains the object is

$$Circumference = S_2 = 2\pi r$$

Now let us examine three-dimensional space.

We call a circular object in three dimensions a sphere. The volume inside the sphere is given by the formula

$$V = \frac{4}{3}\pi r^3$$

The surface area around the sphere is given by the formula

$$S = 4\pi r^2$$

We will now use the subscript 3 to denote these measurements in three-dimensional space.

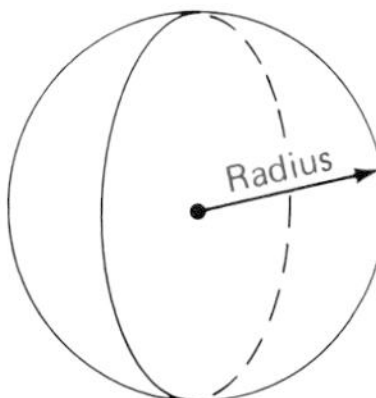

$$\text{Volume} = V_3 = \frac{4}{3}\pi r^3$$

$$\text{Surface area} = S_3 = 4\pi r^2$$

How do these formulas relate?

For any given dimensional space n, we an relate a "V" formula (the amount inside) to an "S" formula (the amount of the containing boundary) by the relationship

Formula A

$$nV_n = rS_n$$

where n represents the dimension of the space. (This formula can be proved, but it is beyond the scope of this book.)

For example, suppose we know the formula for the volume of a sphere, $V_3 = \frac{4}{3}\pi r^3$, but cannot remember the formula for the surface area S_3. We can use Formula A to find it. We substitute the equation above for V_3 and the value of 3 for n.

$nV_n = rS_n$	Formula A.
$(3)\left(\frac{4}{3}\pi r^3\right) = rS_3$	Substitution.
$4\pi r^3 = rS_3$	Simplification of the left side.
$\frac{4\pi r^3}{r} = S_3$	We divide both sides by r (the coefficient of S_3).
$4\pi r^2 = S_3$	We derive the formula for surface area.

There is another formula that allows us to find formulas in a higher dimension.

Formula B

$$S_n = 2\pi r V_{n-2}$$

(This formula can also be proved, but it also is well beyond the scope of this book.)

Suppose that you wanted to find the "surface area" of a "sphere" in the fourth dimension. You would want S_4. Now merely replace n by 4 in the equation above.

$S_n = 2\pi r V_{n-2}$	Formula B.
$S_4 = 2\pi r V_{4-2}$	Replace n by 4.
$S_4 = 2\pi r V_2$	Simplify.
$S_4 = 2\pi r(\pi r^2)$	Replace V_2 by πr^2.
$S_4 = 2\pi^2 r^3$	Simplify.

This formula gives us the measurement of the boundary that contains a fourth-dimensional "sphere."

To find V_4 we will use Formula A.

$nV_n = rS_n$	Formula A.
$4V_4 = rS_4$	Replace n by 4.
$4V_4 = r(2\pi^2 r^3)$	Replace S_4 by the formula we derived above.
$4V_4 = 2\pi^2 r^4$	Simplify the right side.
$V_4 = \dfrac{2\pi^2 r^4}{4}$	Divide both sides by 4.
$V_4 = \dfrac{1}{2}\pi^2 r^4$	

This formula gives us the measurement of the amount contained inside a fourth-dimensional "sphere."

A Challenge for You!

Find S_5 and V_5 using the techniques discussed above.

Interested in the idea of other dimensions? *You may want to read one or more of the following:*

Abbott, Edwin A. *Flatland: A Romance of Many Dimensions.* New York: Barnes & Noble Books, 1963.

Manning, Henry P., ed. *The Fourth Dimension Simply Explained.* New York: Dover Publications, Inc., 1960.

Rucker, Rudolf V. *Geometry, Relativity, and the Fourth Dimension.* New York: Dover Publications, Inc., 1977.

Supplementary Problems for Chapter 6

Graph the straight lines in problems 1 to 4.

1. $3y = 2x + 6$

2. $5y - x = 8$

3. $y + 3 = \dfrac{1}{4}x$

4. $-3x - 5y + 6 = 0$

5. Find the slope of a line passing through $(5, -3)$ and $(2, -\frac{1}{2})$.

6. Find the slope and y-intercept of the line $5x - 8y + 12 = 0$.

7. Find the slope and y-intercept of the line $3(x - 2) = \frac{1}{2}(4 - x + 2y)$.

8. Write the equation of a line passing through (5, 6) having a slope of -2.
9. Write the equation of a line passing through $(\frac{1}{2}, -3)$ having a slope of $\frac{1}{3}$.
10. Write the equation of a line passing through (0.3, 2) and (1.5, 6).
11. Write the equation of a line passing through (6, 1) and $(2, -3)$.
12. Write the equation of a line passing through $(3, -\frac{1}{2})$ that has a y-intercept of -6.
13. Write the equation of a line passing through $(-4, -2)$ that has a y-intercept of $\frac{1}{2}$.
14. Write the equation of a line with slope 0 and a y-intercept of 7.
15. Write the equation of a line with no slope and an x-intercept of -5.

Graph the regions in problems 16 to 22.

16. $y < \frac{1}{3}x + 2$
17. $3y + 2x > 8$
18. $7(x + 3) \geq 2(y - 3)$
19. $3x + 4y - 9 \leq 0$
20. $y + 2x > 3y - x$
21. $\frac{1}{3}(x + y) \geq \frac{5}{6}$
22. $5x + 2y + 6 < 6x - 8y + 4$
23. A rhombus has a perimeter of 37 inches. How long is each side?
24. Find the diameter of a circle whose circumference is 25.12 meters. (Use $\pi \doteq 3.14$.)
25. What is the perimeter of a rectangle whose length is 8 inches and whose width is $\frac{1}{4}$ inch?
26. Find the third angle of a triangle if one angle measures 62° and a second angle measures 47°.
27. Find the area of a triangle whose base is 8 miles and whose altitude is 10.5 miles.
28. Find the perimeter of a triangle whose smallest side is 8 inches and whose second side is double the first and whose third side is double the second.
29. Write an expression for the area of a parallelogram whose altitude is 8 centimeters and whose base is represented by the variable b.
30. Write an expression for the volume of a rectangular solid whose width is 8 feet, whose length is 12 feet, and whose height is represented by the variable h.
31. Find the volume of a sphere if the radius is 3 centimeters. (Use $\pi \doteq 3.14$.)
32. The perimeter of an isosceles triangle is 46 inches. The two equal sides are each 17 inches long. How long is the other side?
33. Find the area of the region shown in Figure 6.58 consisting of two semicircles with diameter 3 cm and one rectangle with dimensions labeled.

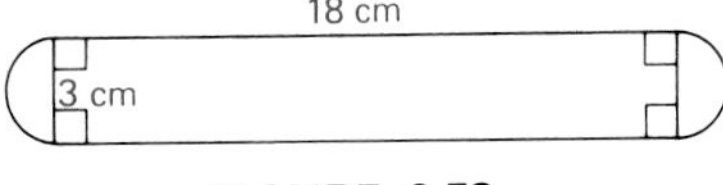

FIGURE 6.58

34. Find the area of the shaded region in Figure 6.59. All dimensions are labeled. (*Hint:* $A = \frac{1}{2}ab$ two times.)

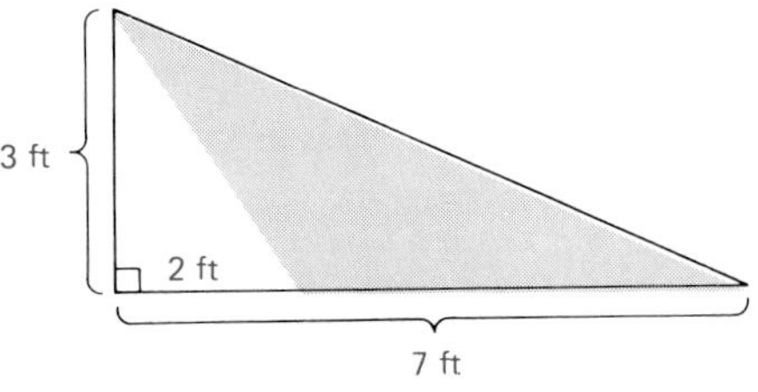

FIGURE 6.59

35. A carpet installer will charge \$2.00 per square yard to install carpet on the region sketched in Figure 6.60.

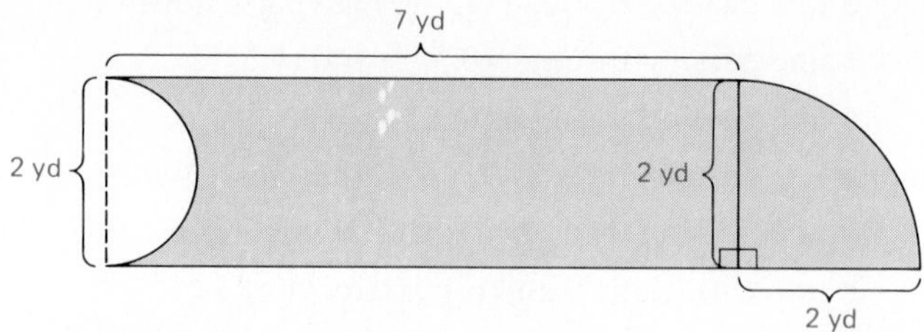

FIGURE 6.60

(The region consists of a rectangle increased by a quarter-circle and decreased by a semicircle.)

(a) What is the area in square yards of the shaded region?

(b) What is the installation cost for the carpet?

Chapter 6 Summary

Graphing Straight Lines

1. Plot three points that lie on the line and connect them by drawing a line through them.
2. Use the x-intercept and the y-intercept as two of the points unless the line passes through the origin or passes very close to it.

Properties of Straight Lines

1. The slope m of any nonvertical line passing through (x_1, y_1) and (x_2, y_2) is given by the formula

$$m = \frac{y_2 - y_1}{x_2 - x_1}$$

The slope of a vertical line is not defined.

2. A line of slope m that has a y-intercept at $(0, b)$ is described by the equation $y = mx + b$. This is called the slope-intercept equation of a line.

Graphing Linear Inequalities

1. Replace the inequality symbol by the equality symbol and draw the line represented by that equation. This line will be a boundary line.
2. The line itself will be a solid line and is part of the answer if the inequality uses the $\geq$ or the $\leq$ symbols. A dashed line is used if the inequality involves the $<$ or $>$ symbols.
3. To determine if you should shade above the line or below the line you should pick a test point that does not lie on the line. Substitute the x and y coordinates of that point into the inequality. If you obtain an inequality that is true, shade in the half-plane that includes that point. If you obtain a false inequality, shade in the opposite half-plane. Unless the line goes through or is very close to the origin, the point $(0, 0)$ is a good test point.

Geometrical Formulas

Be sure that you are familiar with the formulas to find the area, perimeter, circumference, and volume of common geometrical objects. If you are not clear on these, review these formulas in Section 6.5.

Practice Quiz for Chapter 6

Sections 6.1–6.3

1. Give the coordinates of points A, B, C, D, E, F, G, and H shown in Figure 6.61.

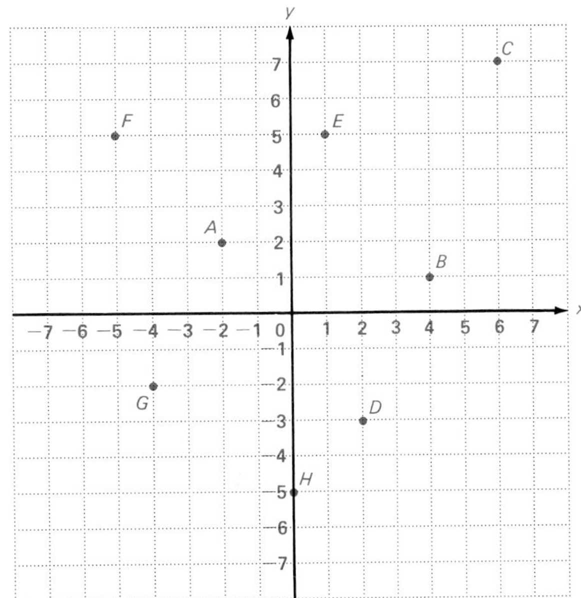

FIGURE 6.61

2. Plot and label on graph paper the following points:

$I = (-3, 2)$ $J = (-6, 0)$ $K = (5, -2)$
$L = (-6, -3)$ $M = (7, 1)$ $N = (0, 5)$

3. Plot the points $(-2, -6)$, $(1, 3)$, and $(-1, -4)$. Do they lie on a straight line?

In problems 4 to 8, graph the straight lines given by the following equations.

4. $y = 1 - 2x$
5. $-7y + 2x = 8$
6. $3x - 2 = 8 - 6x$
7. $y + 5 = \frac{2}{3}x$
8. $3x + 2y = 2x - (7 - x)$
9. Find the slope of a line passing through $(0, -4)$ and $(-2, 6)$.
10. What is the slope and the y-intercept of the line $3y + 2x = 6 + 3x$?
11. Write the equation of a line passing through $(-2, -3)$ and having a slope of $-\frac{2}{3}$.
12. Write the equation of a line passing through $(1, 4)$ and $(-2, 3)$.
13. Write the equation of a line passing through $(-5, 1)$ and having a y-intercept of $\frac{2}{3}$.
14. Write the equation of a line passing through $(-4, 8)$ having an x-intercept of -4.

Sections 6.4–6.5

In problems 1 to 4, graph the region described.

1. $y \geq -\frac{2}{3}x$
2. $3x + 2y < -4$
3. $5x - 6y - 12 > 0$
4. $2y - 7 \leq 8 + 3y$
5. Find the circumference of a circle with radius 15 inches. (Use $\pi = 3.14$ as an approximation.)
6. A rhombus has a perimeter of 29 meters. How long is one side?
7. Write an expression for the area of any triangle with an altitude of 64 centimeters.
8. A trapezoid has an area of 104 square meters. One base is 6 meters long. The altitude is 10 meters. How long is the other base?

9. Rug binding costs \$0.18 per linear foot. How much will it cost to bind a rectangular rug that is 14 feet wide and 18 feet long?
10. Find the area of the shaded region in Figure 6.62. The radius of the circle is 3 centimeters. The sides of the square measure 6 centimeters.

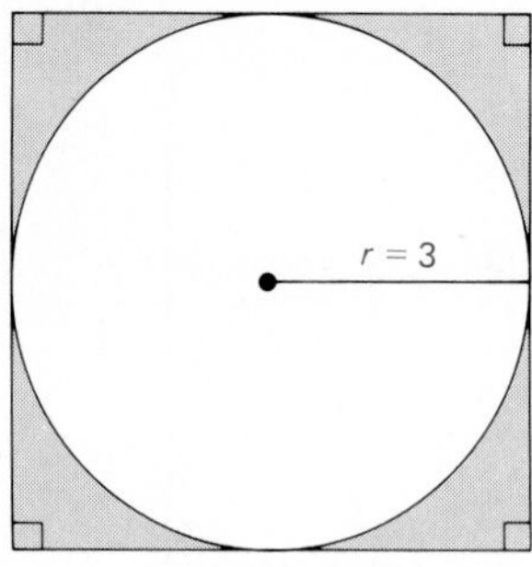

FIGURE 6.62

11. Write an expression for the volume of a rectangular solid whose width is 0.056 centimeter and whose length is 0.003 centimeter.

Practice Test for Chapter 6

1. Plot and label the following points:

$$A = (0, 5) \qquad B = (6, 1) \qquad C = (-4, -3)$$
$$D = (-3, 0) \qquad E = (5, -2)$$

2. Graph the line $6x - 3 = 5x - 2y$.
3. Graph the line $12x - 3y = 6$.
4. What is the slope and the y-intercept of the line $3x + 2y - 5 = 0$?
5. Write the equation of the line that passes through $(4, -2)$ and has a slope of $\frac{1}{2}$.
6. Find the slope of a line that passes through $(5, -2)$ and $(-3, -6)$.
7. Graph the region described by $4y \leq 3x$.
8. Graph the region described by $-3x - 2y > 10$.
9. Find the circumference of a circle with radius of 22.0 inches. Use $\pi = 3.14$ as an approximation.
10. Write an expression to find the area of any trapezoid if the two bases are 10 inches and 14 inches, respectively.
11. How much would it cost to carpet the area shown in Figure 6.63 if carpeting will cost \$12.00 per square yard?

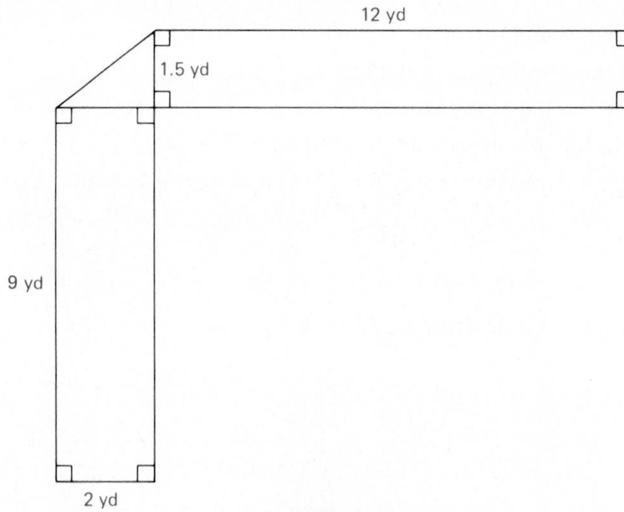

FIGURE 6.63

12. Find the volume of a sphere with radius 10 inches. (Use $\pi = 3.14$ as approximation and round your answer to the nearest cubic inch.)

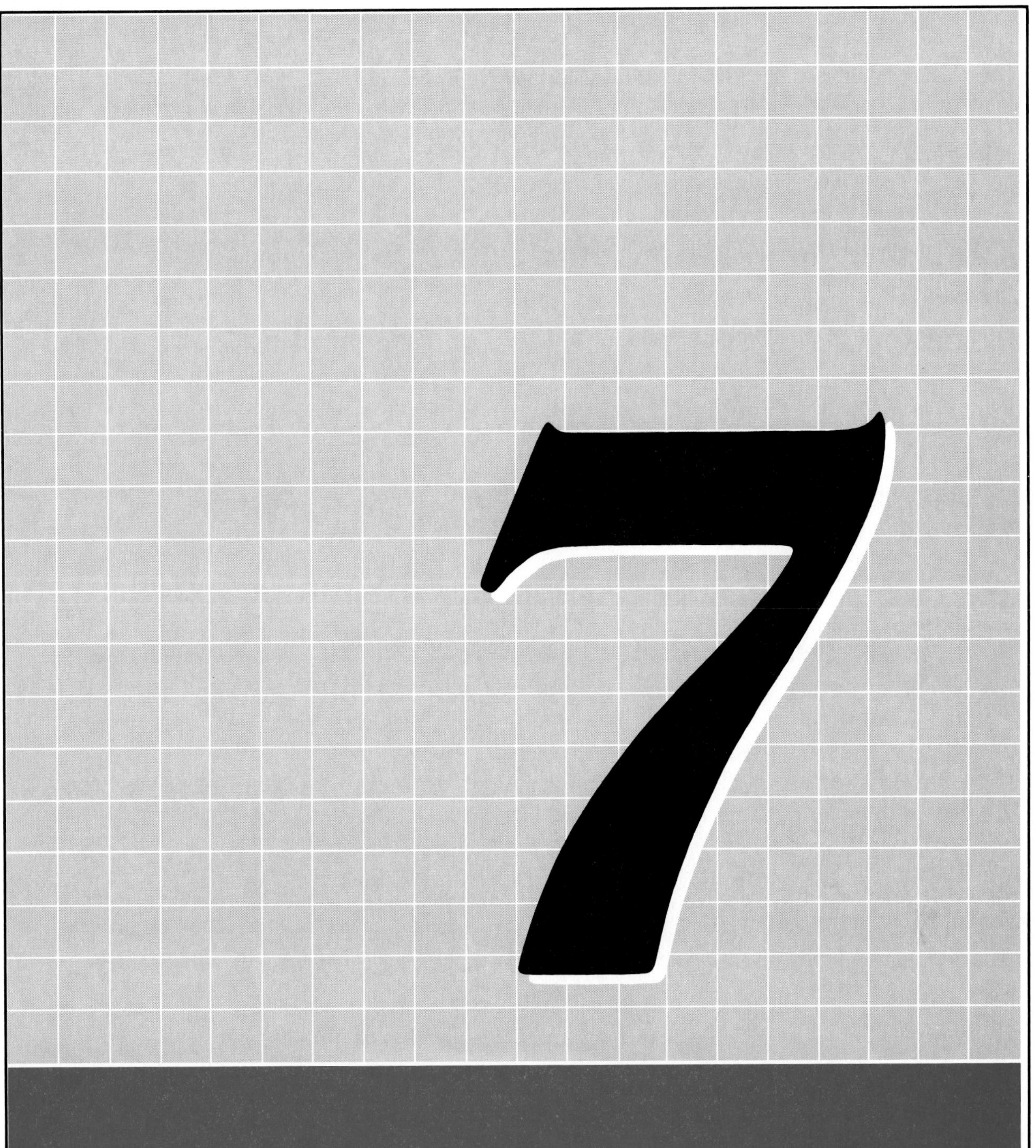

7

Systems of Equations

Diagnostic Pretest for Chapter 7

This diagnostic pretest is intended for those students who have previously studied algebra. Work as many problems as you can. Then, check your answers against those listed at the back of the book. Where you have obtained an incorrect answer or have been unable to work the problem, you will need to study the corresponding sections of the chapter.

If you have not previously studied algebra, do not take this test now. You should study the chapter and work the problems and exercises. After finishing the chapter, you may use this test as a review or as a practice test.

Follow the directions and solve each problem.

Section 7.1

Solve each system by the substitution method.

1. $3x + y = -11$
$4x - 3y = -6$

2. $5x + 2y = 3$
$6x - y = -10$

Section 7.2

Solve each system by the addition method.

3. $5x - 3y = 10$
$4x + 7y = 8$

4. $2x + 7y = -1$
$-5x - 6y = -9$

Section 7.3

Solve each system by the graphing method.

5. $4x + y = -10$
$-2x + 4y = 14$

6. $x - 2y = 2$
$x - y = 0$

Solve each system by any method to find the one solution (if possible). If there is not one solution for the system, state the reason.

7. $-10x + 8y = 5$
$5x - 4y = 3$

8. $4x - 2y = -6$
$-6x + 3y = +9$

Section 7.4

Solve problems 9 and 10 by any method.

9. $x + 8y = 7$
$5x - 2y = 14$

10. $3x - y = 5$
$-5x + 2y = -10$

11. James and Michael work sorting letters for the post office. Yesterday James sorted letters for 3 hours and Michael for 2 hours. Together they sorted 2250 letters. Today James sorted letters for 2 hours and Michael for 3 hours. Together they sorted 2650 letters. How many letters per hour can be sorted by James? by Michael?

12. A theatre sold 380 tickets for a musical production. Children's tickets were \$2 each and adults' tickets were \$3. The receipts for the day were \$970. How many tickets of each type were sold?

Student Learning Objectives for Chapter 7

After successfully learning the content of each section, you will be able to:

Section 7.1

A. Solve a system of two linear equations with two variables by the substitution method.

B. Check a solution for a system of two linear equations with two variables.

Section 7.2

C. Solve a system of two linear equations with two variables by the addition method.

Section 7.3

D. Solve a system of two linear equations with two variables by the graphing method.

E. Identify dependent linear equations by graphical or algebraic methods.

F. Identify an inconsistent system of linear equations by graphical or algebraic methods.

Section 7.4

G. Solve a variety of applied word problems that can be readily solved through the use of a system of two linear equations containing two variables.

Introduction: The Concept of a System of Linear Equations

We examined in Chapter 6 some properties of linear equations with two variables. An equation like $3x + 2y = 6$ contains the variables x and y. We found in graphing linear equations that an unlimited number of ordered pairs of (x, y) values satisfied any given linear equation. For example, (0, 3), (2, 0) and (−2, 6) all satisfy the equation $3x + 2y = 6$.

However, suppose that we desired to know if there was a solution that satisfied two linear equations with two variables at the same time. When two linear equations with two variables are considered together they are known as *a system of two equations in two variables.* Most systems have one solution. In other words, one value of x and one value of y satisfy each equation in the system simultaneously.

Let us state a more precise definition. *A solution to a system of two equations in two variables* is an ordered pair that can be substituted into each equation and an identity will be obtained in each case. A solution to a system of two equations in two variables is sometimes called a *simultaneous solution* since the ordered pair satisfies both equations simultaneously.

EXAMPLE A Is (0, 3) a solution for the system of equations?

$$3x + 2y = 6$$
$$x + y = 3$$

Solution We substitute $x = 0$ and $y = 3$ into each equation.

$$3x + 2y = 6$$
$$3(0) + 2(3) \stackrel{?}{=} 6 \qquad \text{if } x = 0 \text{ and } y = 3$$
$$0 + 6 \stackrel{?}{=} 6$$
$$6 = 6\checkmark$$

This is a true mathematical statement. We have obtained an identity.

$$x + y = 3$$
$$0 + 3 \stackrel{?}{=} 3 \qquad \text{if } x = 0 \text{ and } y = 3$$
$$3 = 3\checkmark$$

This is a true mathematical statement. Again, it is an identity.

Thus the values $x = 0$ and $y = 3$ are a solution to the system

$$3x + 2y = 6$$
$$x + y = 3$$

7.1 Finding a Solution to a System of Equations by the Substitution Method

In such cases as Example A we need to develop a method to find the solution to the system of equations. The difficulty is that we have two equations with two unknown values and we have only learned to solve equations with one unknown value. Therefore, we want to eliminate one unknown if possible. We can do this by replacing one variable by an expression equal in value.

EXAMPLE B Find the solution to the following system of equations:

$$y = 3x - 1$$
$$3x - 2y = -4$$

Solution We observe that the first equation is solved for the variable y in terms of x. This means that y is equivalent to the expression $3x - 1$. Let us eliminate the variable y in the second equation by substituting in its place the equivalent expression $3x - 1$.

$3x - 2y = -4$	The original second equation.
$3x - 2(3x - 1) = -4$	Replace the variable y by $3x - 1$. Note that we need parentheses here.
$3x - 6x + 2 = -4$	Remove the parentheses.
$-3x + 2 = -4$	Collect like terms.
$-3x = -6$	Subtract 2 from both sides.
$x = 2$	Divide both sides by -3.

We have now found the value for x. Now we substitute this value for x into one of the equations with two variables to obtain the corresponding value for y.

$y = 3x - 1$	The original first equation.
$y = 3(2) - 1$	Replace the variable x by the value 2.
$y = 6 - 1$	Simplify.
$y = 5$	

Thus we see that the solution to the system of equations is $x = 2$, $y = 5$.

Usually when we do this type of problem an extra step is needed to isolate one variable at the start of the procedure. (If you find this step of isolating the variable to be difficult you may want to review Section 2.3.) We will summarize the steps as follows:

Procedure to Find a Solution to a System of Equations by the Substitution Method

1. **Solve for one variable in terms of the other variable in one of the two equations.**
2. **Substitute the expression you obtain in step 1 for this variable in the *other* equation.**
3. **You now have one equation with one variable. Solve this equation to find the value for that one variable.**
4. **Substitute this value for the variable in one of the equations with two variables to obtain a value for the second variable.**
5. **Check each solution against each equation to verify your results.**

EXAMPLE C Find the solution to the following system of equations:

$$x - 2y = 7$$

$$-5x + 4y = -5$$

Solution Step 1. Solve for one variable.

$x - 2y = 7$ The first equation is the easiest one in which to isolate a variable.

$x = 7 + 2y$ Add $2y$ to both sides to solve for x.

Step 2. Substitute the expression into the other equation.

$-5x + 4y = -5$ This is the original second equation.

$-5(7 + 2y) + 4y = -5$ Substitute the value $7 + 2y$ for x in this equation.

Step 3. Solve this equation.

$-35 - 10y + 4y = -5$ Remove the parentheses.

$-35 - 6y = -5$ Collect like terms.

$-6y = +30$ Add 35 to both sides.

$y = -5$ Divide both sides by -6.

Step 4. Obtain the value of the second variable.
We will now use this value $y = -5$ to find the value for x in one of the equations that contains both variables.

$x = 7 + 2y$ The easiest equation to use is the one that is already solved for x.

$x = 7 + 2(-5)$ Replace y by -5.

$$x = 7 - 10$$

$$x = -3$$

The solution is $x = -3$ and $y = -5$.

Step 5. Check.
To be sure that we have the correct solution we will need to check that the obtained values of x and y can be substituted into *both original* equations to obtain true mathematical statements.

$x - 2y = 7$	$-5x + 4y = -5$
$-3 - 2(-5) \stackrel{?}{=} 7$	$-5(-3) + 4(-5) \stackrel{?}{=} -5$
$-3 + 10 \stackrel{?}{=} 7$	$15 - 20 \stackrel{?}{=} -5$
$7 = 7\checkmark$	$-5 = -5\checkmark$

Equations with Fractional Coefficients

If a system of equations contains fractions it is usually easier to obtain equivalent equations that do not have fractions before following the procedure to find a solution. This is done by multiplying each term of the equation containing fractions by the LCD of the denominators in that equation.

EXAMPLE D Find the solution to the following system of equations:

$$\frac{3}{2}x + y = \frac{5}{2}$$

$$-y + 2x = -1$$

Solution Before following Steps 1 to 5 we obtain an equivalent system of equations with no fractional coefficients.

$$\frac{3}{2}x + y = \frac{5}{2}$$

We observe that the LCD of the first equation is 2.

$$2\left(\frac{3}{2}x\right) + 2(y) = 2\left(\frac{5}{2}\right)$$

We multiply each term of the equation by the LCD.

$$3x + 2y = 5$$

We obtain an equivalent equation that does not contain a fraction.

Step 1. Solve for one variable.

$$3x + 2y = 5$$
$$-y + 2x = -1$$

We now follow the procedure to solve this equivalent system of equations.

$$-y = -1 - 2x$$

Isolate the $-y$ term in the second equation.

$$-1y = -1 - 2x$$

Since $-y$ is actually $-1y$, the coefficient of y is -1.

$$y = 1 + 2x$$

We divide each term of both sides of the equation by -1 to obtain an expression for y (or $+1y$).

Step 2. Substitute the expression into the other equation.

$$3x + 2(1 + 2x) = 5$$

Substitute this expression for y into the other equation.

$$3x + 2 + 4x = 5$$

Remove parentheses.

Step 3. Solve this equation.

$$7x + 2 = 5$$

Collect like terms.

$$7x = 3$$

Subtract 2 from both sides.

$$x = \frac{3}{7}$$

Divide both sides by 7.

Step 4. Obtain the value of the second variable.

$y = 1 + 2x$ — To obtain y, we use the equation that is already solved for y.

$y = 1 + 2\left(\frac{3}{7}\right)$ — Substitute $\frac{3}{7}$ for x.

$y = 1 + \frac{6}{7}$ — Simplify.

$y = \frac{7}{7} + \frac{6}{7}$ — Transform 1 to the equivalent fraction $\frac{7}{7}$ so that fractions may be added.

$y = \frac{13}{7}$

The solution is $x = \frac{3}{7}$ and $y = \frac{13}{7}$.

Step 5. Check.
Equations that have fractional solutions are more difficult to check, but it is an important skill. Be sure that you can check fractional answers. Remember: to have a complete check you must check the solution in *both* original equations.

Replace x by $\frac{3}{7}$, replace y by $\frac{13}{7}$.

$\frac{3}{2}x + y = \frac{5}{2}$	$-y + 2x = -1$
$\frac{3}{2}\left(\frac{3}{7}\right) + \frac{13}{7} \stackrel{?}{=} \frac{5}{2}$	$-\frac{13}{7} + 2\left(\frac{3}{7}\right) \stackrel{?}{=} -1$
$\frac{9}{14} + \frac{13}{7} \stackrel{?}{=} \frac{5}{2}$	$-\frac{13}{7} + \frac{6}{7} \stackrel{?}{=} -1$
$\frac{9}{14} + \frac{26}{14} \stackrel{?}{=} \frac{5}{2}$	$-\frac{7}{7} \stackrel{?}{=} -1$
$\frac{35}{14} \stackrel{?}{=} \frac{5}{2}$	$-1 = -1 \checkmark$
$\frac{5}{2} = \frac{5}{2} \checkmark$	

Review Examples A to D and then try these practice problems.

Practice Problems for Exercise 7.1

Find the solution for the following systems of equations and check your answers.

1. $y - 3x = 10$
$2x - 3y = -16$

2. $5x + 3y = 19$
$2x - y = 12$

3. $\frac{x}{3} - \frac{y}{2} = 1$
$x + 4y = -8$

Now turn to the answers following Exercise 7.1 and check your solutions for these practice problems.

Exercise 7.1

Find the solution for the following systems of equations. Check your answers for problems 1 to 8.

1. $y = 4x - 3$
$3x - y = 2$

2. $x = 6 + 2y$
$3x - 2y = 12$

3. $4x + 3y = 9$
$x - 3y = 6$

4. $x + 4y = 4$
$-x + 2y = 2$

5. $2x + y = 4$
$2x - y = 0$

6. $7x - 3y = -10$
$x + 3y = 2$

7. $3x - 4y = -8$
$y = 2x$

8. $2x - 7y = -3$
$x = 3y$

9. $x + 3y = 2$
$2x + 3y = 7$

10. $4x + 2y = 4$
$3x + y = 4$

11. $-x + 2y = -3$
$-2x + 3y = -6$

12. $6x - y = 6$
$-x + 3y = -1$

13. $3a - 5b = 2$
$3a + b = 32$

14. $a - 5b = 21$
$3a + 4b = -13$

15. $3x - 4y = 6$
$-2x - y = -4$

16. $-2x - 3y = -1$
$3x - y = 7$

17. $5c + 8d + 1 = 0$

$3c + d - 7 = 0$

18. $4s - 3t - 9 = 0$

$-2s + t + 3 = 0$

19. $3x - y - 9 = 0$

$8x + 5y - 1 = 0$

20. $8x + 2y - 7 = 0$

$-2x - y + 2 = 0$

21. $\frac{x}{3} + \frac{y}{2} = 2$

$-2x + 3y = 24$

22. $-x + y = -4$

$\frac{3x}{7} + \frac{2y}{3} = 5$

Challenge Problems

Be sure that you can do problems 1 to 22 before you attempt these!

23. $\frac{x}{2} + \frac{y}{3} = 14$

$\frac{x}{3} + \frac{y}{2} = 11$

24. $-\frac{3x}{2} - \frac{2y}{3} = +\frac{2}{3}$

$-\frac{5}{2}x + \frac{4}{3}y = -5$

25. $\frac{x}{4} - 3y = 0$

$\frac{3x}{4} + 6y = \frac{15}{4}$

26. $\frac{x}{4} + \frac{5y}{6} = 14$

$\frac{x}{8} + \frac{y}{6} = 4$

Solutions to Practice Problems for Exercise 7.1

1. $y - 3x = 10$

$2x - 3y = -16$ — We want to isolate y in first equation.

Step 1. Solve for one variable.

$y = 10 + 3x$ — Add $3x$ to each side to solve for y in first equation.

Step 2. Substitute the expression into the other equation.

$2x - 3(10 + 3x) = -16$ — Replace y by $10 + 3x$ in the second equation.

$2x - 30 - 9x = -16$ — Remove parentheses.

	Step 3. Solve this equation.
$-7x - 30 = -16$	Collect like terms.
$-7x = 14$	Add 30 to both sides.
$x = -2$	Divide both sides by -7.
	Step 4. Obtain the value of the second variable.
$y = 10 + 3(-2)$	Replace x by -2 in the equation $y = 10 + 3x$
$y = 10 - 6$	
$y = 4$	Simplify.

Step 5. Check.

Check the solution ($x = -2$ and $y = 4$) in both original equations.

$y - 3x = 10$	$2x - 3y = -16$
$4 - 3(-2) \stackrel{?}{=} 10$	$2(-2) - 3(4) \stackrel{?}{=} -16$
$4 + 6 \stackrel{?}{=} 10$	$-4 - 12 \stackrel{?}{=} -16$
$10 = 10\checkmark$	$-16 = -16\checkmark$

2. $5x + 3y = 19$

$2x - y = 12$	We want to solve for y in the second equation.
	Step 1. Solve for one variable.
$-y = 12 - 2x$	Isolate the $-y$ term.
$y = -12 + 2x$	Since $-y$ is $-1y$, we divide both sides of equation by -1. (See Example D.)
	Step 2. Substitute the expression into the other equation.
$5x + 3(-12 + 2x) = 19$	Replace y by $-12 + 2x$ in the first equation.
$5x - 36 + 6x = 19$	Remove parentheses.
	Step 3. Solve this equation.
$11x - 36 = 19$	Collect like terms.
$11x = 55$	Add 36 to both sides.
$x = 5$	Divide both sides by 11.
	Step 4. Obtain the value of the second variable.
$y = -12 + 2x$	Use the equation solved for y.
$y = -12 + 2(5)$	Replace x by 5.
$y = -12 + 10 = -2$	Simplify.

Step 5. Check.

Check the solution ($x = 5$ and $y = -2$) in both original equations.

$$\begin{aligned} 5x + 3y &= 19 \\ 5(5) + 3(-2) &\stackrel{?}{=} 19 \\ 25 - 6 &\stackrel{?}{=} 19 \\ 19 &= 19 \checkmark \end{aligned} \qquad \begin{aligned} 2x - y &= 12 \\ 2(5) - (-2) &\stackrel{?}{=} 12 \\ 10 + 2 &\stackrel{?}{=} 12 \\ 12 &= 12 \checkmark \end{aligned}$$

3. $\frac{x}{3} - \frac{y}{2} = 1$ — We observe that the first equation has fractions with a LCD of 6.

$6\left(\frac{x}{3}\right) - 6\left(\frac{y}{2}\right) = 6(1)$ — We multiply each term of the equation by the LCD.

$2x - 3y = 6$ — We obtain an equivalent equation with no fractions.

Step 1. Solve for one variable.

$2x - 3y = 6$

$x + 4y = -8$ — To solve the solution of this system we will want first to solve for x in the second equation.

$x = -8 - 4y$ — Subtract $4y$ from both sides of the second equation.

Step 2. Substitute the expression into the other equation.

$2(-8 - 4y) - 3y = 6$ — We substitute this expression for x in the equation $2x - 3y = 6$.

$-16 - 8y - 3y = 6$ — Remove parentheses.

Step 3. Solve this equation.

$-16 - 11y = 6$ — Collect like terms.

$-11y = 22$ — Add 16 to both sides.

$y = -2$ — Divide both sides by -11.

Step 4. Obtain the value of the second variable.

$x = -8 - 4y$ — Use the equation already solved for x.

$x = -8 - 4(-2)$ — Substitute -2 for y.

$x = -8 + 8$ — Simplify.

$x = 0$

Step 5. Check.

Check the solution ($x = 0$ and $y = -2$) in both original equations.

$$\begin{aligned} \frac{x}{3} - \frac{y}{2} &= 1 \\ \frac{0}{3} - \frac{-2}{2} &\stackrel{?}{=} 1 \\ 0 - (-1) &\stackrel{?}{=} 1 \\ 1 &= 1 \checkmark \end{aligned} \qquad \begin{aligned} x + 4y &= -8 \\ 0 + 4(-2) &\stackrel{?}{=} -8 \\ 0 - 8 &\stackrel{?}{=} -8 \\ -8 &= -8 \checkmark \end{aligned}$$

Now do the problems in Exercise 7.1.

7.2 Finding a Solution to a System of Equations by the Addition Method

The method of substitution is a very useful one in solving systems of equations at both elementary and advanced levels in mathematics. However the method of substitution is not convenient in solving certain systems of equations with two unknowns. The following example will illustrate.

EXAMPLE A Find the solution to the following system of equations by the *substitution method*:

$$2x + 5y = -2$$

$$-7x - 5y = 32$$

Solution

$2x = -2 - 5y$ Isolate the x term in the first equation.

$x = \dfrac{-2 - 5y}{2}$ Divide both sides by 2.

$-7\left(\dfrac{-2 - 5y}{2}\right) - 5y = 32$ Substitute $\dfrac{-2 - 5y}{2}$ for x in the second equation.

$\dfrac{14 + 35y}{2} - 5y = 32$ Remove parentheses by multiplying by -7.

$2\left(\dfrac{14 + 35y}{2}\right) - 2(5y) = 2(32)$ Multiply each term by the LCD of 2 to remove fractions.

$14 + 35y - 10y = 64$ Remove parentheses.

$14 + 25y = 64$ Collect like terms.

$25y = 50$ Subtract 14 from both sides.

$y = 2$ Divide both sides by 25.

$x = \dfrac{-2 - 5y}{2}$ Take the form of the equation already solved for x.

$x = \dfrac{-2 - 5(2)}{2}$ Substitute 2 for y.

$x = \dfrac{-2 - 10}{2} = \dfrac{-12}{2} = -6$ Simplify.

The solution is $x = -6$ and $y = 2$.

Probably most students would find this example not too difficult to understand. However, the chance for student error in solving such a system by the substitution method is high, largely due to the variable expression for x that involves a fraction. We will now develop a new technique to solve the system that is more convenient in such cases. (This new method is more efficient in solving more involved problems such as those with three equations and three unknowns. These more involved systems will be covered in a more advanced course.)

In developing the substitution method our goal was to eliminate one variable. This can also be done by adding the two equations together. When we do this we use the principle: If $a = b$ and $c = d$, then $a + c = b + d$. Let us add the first equation to the second equation in Example A and see if this will aid us in finding the solution to the system.

EXAMPLE B Find the solution to the following system of equations by the *addition method*:

$$\begin{array}{r} 2x + 5y = -2 \\ \underline{-7x - 5y = 32} \end{array}$$

Solution

$-5x = 30$	Add the two equations. Since $5y - 5y = 0y = 0$, we have eliminated the y variable.
$\dfrac{-5x}{-5} = \dfrac{30}{-5}$	Divide both sides by -5.
$x = -6$	Simplify.
$2(-6) + 5y = -2$	Substitute $x = -6$ into one of the original equations.
$-12 + 5y = -2$	Remove parentheses.
$5y = 10$	Add 12 to both sides.
$y = 2$	Divide both sides by 5.

We again obtain the solution of $x = -6$ and $y = 2$ for the system. Most students will agree that the chance for making a mistake in using the method of Example B will be much less than that in using the method of Example A.

Most systems of equations cannot be solved this easily. In many cases you will first need to multiply each term of one equation by a constant value. After that step, when you add the two equations together, one variable is eliminated. Study Examples C and D carefully to see how this is done.

EXAMPLE C Find the solution to the following system of equations by the addition method:

$$\begin{array}{r} 5x + 2y = 7 \\ 3x - y = 13 \end{array}$$

Solution If we add the equations together in this form a variable will not be eliminated. Here we need to multiply each item of one equation by the appropriate number.

$5x + 2y = 7$	The first equation is not changed.
$2(3x) - 2(y) = 2(13)$	Multiply each term of the second equation by 2.
$5x + 2y = 7$ $6x - 2y = 26$	We now have an equivalent system of equations.
$11x = 33$	Add the two equations. This will eliminate the y variable.
$x = 3$	Divide both sides by 11.
$5(3) + 2y = 7$	Substitute $x = 3$ into one of the original equations.
$15 + 2y = 7$	Remove parentheses.
$2y = -8$	Subtract 15 from both sides.
$y = -4$	Divide both sides by 2.

Check Replace x by 3 and y by -4 in *both* original equations.

$$5x + 2y = 7 \qquad\qquad 3x - y = 13$$

$$5(3) + 2(-4) \stackrel{?}{=} 7 \qquad\qquad 3(3) - (-4) \stackrel{?}{=} 13$$

$$15 - 8 \stackrel{?}{=} 7 \qquad\qquad 9 + 4 \stackrel{?}{=} 13$$

$$7 = 7 \checkmark \qquad\qquad 13 = 13 \checkmark$$

If the coefficients of the variable you wish to eliminate have the same sign, you will need to multiply each term of one equation by a negative number.

If the system of equations has fractional coefficients, you should first multiply each equation by the LCD of that equation.

EXAMPLE D Find the solution to the following system of equations by the addition method:

$$x - \frac{5}{2}y = \frac{5}{2}$$

$$\frac{4}{3}x + y = \frac{23}{3}$$

Solution

$$2(x) - 2\left(\frac{5}{2}y\right) = 2\left(\frac{5}{2}\right)$$ Multiply each term of the first equation by 2.

$$2x - 5y = 5$$ Simplify.

$$3\left(\frac{4}{3}x\right) + 3(y) = 3\left(\frac{23}{3}\right)$$ Multiply each term of the second equation by 3.

$$4x + 3y = 23$$ Simplify.

We now have an equivalent system of equations that does not contain fractions. We will now perform steps to solve the system.

$$2x - 5y = 5$$

$$4x + 3y = 23$$

Let us eliminate the x variable.

$$(-2)(2x) - (-2)5y = (-2)(5)$$ Multiply each term of the first equation by -2.

$$4x + 3y = 23$$ The second equation is not changed.

$$\begin{array}{r} -4x + 10y = -10 \\ 4x + 3y = 23 \\ \hline \end{array}$$ We now have an equivalent system of equations.

$$13y = 13$$ Add the two equations. This will eliminate the variable x.

$$y = 1$$ Divide both sides by 13.

$$4x + 3(1) = 23$$ Substitute $y = 1$ into one of the equations with two variables.

$$4x + 3 = 23$$

$$4x = 20$$

$$x = 5$$ Solve for x.

Check See if you can verify the solution.

In Example C, only one equation was multiplied by a constant value. Often it is necessary to multiply each equation by a constant in order that one variable will be eliminated when the equations are combined. Study the following examples carefully to see how this is done.

EXAMPLE E Find the solution to the following system of equations by the addition method:

$$3x + 4y = 7$$

$$2x + 7y = 9$$

Solution

$(2)(3x) + (2)(4y) = (2)(7)$ — We multiply each term of the first equation by 2.

$(-3)(2x) + (-3)(7y) = (-3)(9)$ — We multiply each term of the second equation by -3.

$$\begin{array}{r} 6x + 8y = 14 \\ -6x - 21y = -27 \\ \hline \end{array}$$

We now have an equivalent system of equations.

$-13y = -13$ — Add the two equations. This will eliminate the variable x.

$y = 1$ — Divide both sides by -13.

Substitute $y = 1$ into one of the original equations.

$$3x + 4(1) = 7$$

$$3x + 4 = 7$$

$$3x = 3$$

$x = 1$ — Solve for x.

The solution is $x = 1$ and $y = 1$.

Alternative Solution You could also have eliminated the y variable. For example, if you multiply the first equation by 7 and the second equation by -4, you would obtain the equivalent system.

$$21x + 28y = 49$$

$$-8x - 28y = -36$$

This system can be added to eliminate the y variable. Since the numbers involved in this approach are somewhat larger, it is probably wiser to eliminate the x variable in Example E.

Check Can you verify that $x = 1$ and $y = 1$ will check in both original equations?

The problem of decimal coefficients

It is best to have the linear equations in two variables in a form with integer coefficients for each variable. If an equation contains decimal numbers, you are advised to multiply each term of the equation by the appropriate power of 10 to obtain an equivalent equation with integer coefficients. Observe how this is done in Example F.

EXAMPLE F Find the solution to the following system of equations by the addition method:

$$0.12x + 0.05y = -0.02$$

$$0.08x - 0.03y = -0.14$$

Solution

$$100(0.12x) + 100(0.05y) = 100(-0.02)$$

$$100(0.08x) + 100(-0.03y) = 100(-0.14)$$

We will multiply each term of both equations by 100 to obtain integer coefficients in each equation.

equation 1 $12x + 5y = -2$

equation 2 $8x - 3y = -14$

We now have an equivalent system of equations.

$$3(12x) + 3(5y) = 3(-2)$$

$$5(8x) + 5(-3y) = 5(-14)$$

To eliminate the variable y we multiply the first equation by 3 and the second equation by 5.

$$\begin{aligned} 36x + 15y &= -6 \\ 40x - 15y &= -70 \\ \hline 76x \qquad &= -76 \end{aligned}$$

We now have an equivalent system of equations.

Add the equations.

$$x = -1$$

Solve for x.

$$12(-1) + 5y = -2$$

$$-12 + 5y = -2$$

$$5y = -2 + 12$$

$$5y = 10$$

$$y = 2$$

Now we replace x by -1 in equation 1, and solve for y.
(We would prefer not to use the original equations because of the chance of error in decimal operations.)

We have obtained the solution $x = -1$ and $y = 2$.

Check We substitute $x = -1$ and $y = 2$ into the original equations. (I'm sure most students would rather use equation 1 and equation 2!) However, *we must check the solutions into the original equations* if we want to be sure of our answers. It is possible to miscopy a number and obtain an error in equation 1 and equation 2. Therefore, you could obtain a solution that would check when you use those equations, but would not satisfy the original equations.

$$0.12x + 0.05y = -0.02$$

$$0.12(-1) + 0.05(2) \stackrel{?}{=} -0.02$$

$$-0.12 + 0.10 \stackrel{?}{=} -0.02$$

$$-0.02 = -0.02 \checkmark$$

$$0.08x - 0.03y = -0.14$$

$$0.08(-1) - 0.03(2) \stackrel{?}{=} -0.14$$

$$-0.08 - 0.06 \stackrel{?}{=} -0.14$$

$$-0.14 = -0.14 \checkmark$$

Review Examples B to F and then try to use the addition method on these practice problems.

Practice Problems for Exercise 7.2

Find the solution to each of the following systems of equations by the *addition method*, and check your answers.

1. $3x + y = 7$
$5x - 2y = 8$

2. $4x + 5y = 7$
$3x + 7y = 2$

3. $0.2x + 0.3y = -0.1$
$0.5x - 0.1y = -1.1$

Now turn to the answers following Exercise 7.2 and check your solutions for these practice problems.

Exercise 7.2

Find the solution to each of the following systems of equations by the *addition method.* Check your answers for problems 1 to 8.

1. $2x + y = -1$
$3x - y = -9$

2. $2x + y = 2$
$4x - y = -8$

3. $2x + 3y = 1$
$x - 2y = 4$

4. $2x + y = 4$
$3x - 2y = -1$

5. $x + 5y = 2$
$2x + 3y = -3$

6. $x + y = 5$
$2x - y = -5$

7. $2x + 5y = 1$
$x - 2y = -4$

8. $2x + 3y = 6$
$-3x + y = 13$

9. $2x + 3y = 1$
$3x + 2y = -6$

10. $2x + 5y = 11$
$3x + 8y = 16$

11. $a - 2b = 0$
$6a - 13b = 5$

12. $a + 3b = 0$
$-3a - 10b = -2$

13. $8x + 6y = -2$
$10x - 9y = -8$

14. $4x + 9y = 0$
$8x - 5y = -23$

15. $x + \frac{5}{4}y = \frac{9}{4}$
$\frac{3}{2}x - y = \frac{1}{2}$

16. $\frac{2}{3}x + y = \frac{1}{3}$
$x + \frac{3}{5}y = \frac{7}{5}$

17. $\begin{aligned} 5x + 6y &= -2 \\ -8x + 9y &= 28 \end{aligned}$

18. $\begin{aligned} 2x - 3y &= 9 \\ 4x + 9y &= -2 \end{aligned}$

19. $\begin{aligned} 0.2x - 0.3y &= 0.4 \\ 0.3x - 0.4y &= 0.9 \end{aligned}$

20. $\begin{aligned} 0.5x - 0.2y &= 0.5 \\ 0.4x + 0.7y &= 0.4 \end{aligned}$

21. $\begin{aligned} 0.02x - 0.04y &= 0.26 \\ 0.07x - 0.09y &= 0.66 \end{aligned}$

22. $\begin{aligned} 0.04x - 0.03y &= 0.05 \\ 0.05x + 0.08y &= -0.76 \end{aligned}$

23. $\begin{aligned} 4x - 5y &= -18 \\ -3x - 2y &= 2 \end{aligned}$

24. $\begin{aligned} -6x - 7y &= -4 \\ -5x - 8y &= 1 \end{aligned}$

Challenge Problems

Let a and b represent constants. Find the solution to each system of equations by the addition method.

25. $\begin{aligned} ax + by &= b^2 + a^2 \\ -bx + ay &= b^2 + a^2 \end{aligned}$

26. $\begin{aligned} -\frac{x}{a} + \frac{y}{b} &= -2 \\ \frac{x}{a} + \frac{y}{b} &= 6 \end{aligned}$

Solutions to Practice Problems for Exercise 7.2

1. $3x + y = 7$
$5x - 2y = 8$

In this problem it is best to eliminate the y variable.

$6x + 2y = 14$ — Multiply each term in the first equation by 2.

$5x - 2y = 8$ — The second equation remains unchanged.

$11x = 22$ — Add the two equations.

$x = 2$ — Solve for x.

$3(2) + y = 7$ — Substitute 2 for x in an original equation.

$6 + y = 7$

$y = 1$

Check

$$3(2) + (1) \stackrel{?}{=} 7 \qquad 5(2) - 2(1) \stackrel{?}{=} 8$$
$$6 + 1 \stackrel{?}{=} 7 \qquad 10 - 2 \stackrel{?}{=} 8$$
$$7 = 7\checkmark \qquad 8 = 8\checkmark$$

2. $4x + 5y = 7$
$3x + 7y = 2$

We will be working with smaller numbers if we eliminate the x variable.

$(-3)(4x) + (-3)(5y) = (-3)(7)$ — Multiply the first equation by -3.

$(4)(3x) + (4)(7y) = (4)(2)$ — Multiply the second equation by 4.

$$-12x - 15y = -21$$
$$+12x + 28y = 8$$

$13y = -13$ — Add the equations.

$y = -1$ — Solve for y.

$4x + 5(-1) = 7$ — Substitute -1 for y in an original equation.

$$4x - 5 = 7$$
$$4x = 12$$

$x = 3$ — Solve for x.

Check

$$4(3) + 5(-1) \stackrel{?}{=} 7 \qquad 3(3) + 7(-1) \stackrel{?}{=} 2$$
$$12 - 5 \stackrel{?}{=} 7 \qquad 9 - 7 \stackrel{?}{=} 2$$
$$7 = 7\checkmark \qquad 2 = 2\checkmark$$

3. $0.2x + 0.3y = -0.1$
$0.5x - 0.1y = -1.1$

We observe that the decimals are tenths, so we will multiply each term of each equation by 10.

$2x + 3y = -1$
$5x - y = -11$

An equivalent system of equations. We observe that it is best to eliminate the variable y.

$2x + 3y = -1$ — The first equation is unchanged.

$15x - 3y = -33$ — The second equation has had each term multiplied by 3.

$17x = -34$ — Add equations.

$x = -2$ — Solve for x.

$2(-2) + 3y = -1$ — Substitute -2 for x in one of the equations that do not have decimals.

$-4 + 3y = -1$ — Simplify.

$3y = -1 + 4$ — Add 4 to both sides.

$3y = 3$ — Collect like terms.

$y = 1$ — Solve for y.

Check $x = -2$ and $y = 1$ are substituted into both *original* equations. If you do not check the original equations, you do not have a valid verification of your answers.

$$0.2x + 0.3y = -0.1 \qquad 0.5x - 0.1y = -1.1$$
$$0.2(-2) + 0.3(1) \stackrel{?}{=} -0.1 \qquad 0.5(-2) - 0.1(1) \stackrel{?}{=} -1.1$$
$$-0.4 + 0.3 \stackrel{?}{=} -0.1 \qquad -1.0 - 0.1 \stackrel{?}{=} -1.1$$
$$-0.1 = -0.1\checkmark \qquad -1.1 = -1.1\checkmark$$

Study these solutions and then do the problems in Exercise 7.2.

7.3 The Graphing Method and General Solutions for Systems of Equations

The graphing method

If there is a solution to a system of two linear equations with two variables, we have learned how to obtain that solution by the substitution method and the addition method. These are algebraic methods to obtain the answer. The solution may also be obtained graphically. We will use the techniques of graphing a linear equation discussed in Chapter 6. We will need to graph each line on one set of axes.

EXAMPLE A Graph the two linear equations in the following system and determine the solution of the system from the graph:

$$-2x + 3y = 6$$

$$+2x + 3y = 18$$

Solution

$-2x + 3y = 6$ — We will deal with each equation separately. We will start with the first equation.

if $x = 0$, then $y = 2$ — First we find the intercepts.

if $y = 0$, then $x = -3$

$-2(6) + 3y = 6$ — To find a third point, we pick an arbitrary value of x (say $x = 6$) and find y.

$3y = 18$

$y = 6$

So we have the 3 points $(0, 2)$, $(-3, 0)$, $(6, 6)$ to plot to obtain the graph of $-2x + 3y = 6$.

$2x + 3y = 18$ — Now we work with the second equation.

if $x = 0$, then $y = 6$ — First we find the intercepts.

if $y = 0$, then $x = 9$

if $x = -3$, then — To find a third point we pick an arbitrary value of x (say $x = -3$).

$2(-3) + 3y = 18$

$3y = 24$

$y = 8$

So we have the three points $(0, 6)$, $(9, 0)$, and $(-3, 8)$ to plot for the graph of $2x + 3y = 18$.

When we draw each line we label the line so that we know which line corresponds to a given equation (Figure 7.1). We do not label the coordinates of the six points so that our graph does not become cluttered.

When drawing a graph, we will usually label each unit of each axis. However, this practice is not always continued in higher level mathematics and science courses. It is not absolutely necessary to label each axis with the units $-3, -2, -1, 0, 1, 2, 3, \ldots$ since it is a common convention in graphing that *each square represents one unit unless the axis is labeled otherwise*.

We observe that the lines intersect at (3, 4); therefore, the solution to the system is $x = 3$ and $y = 4$.

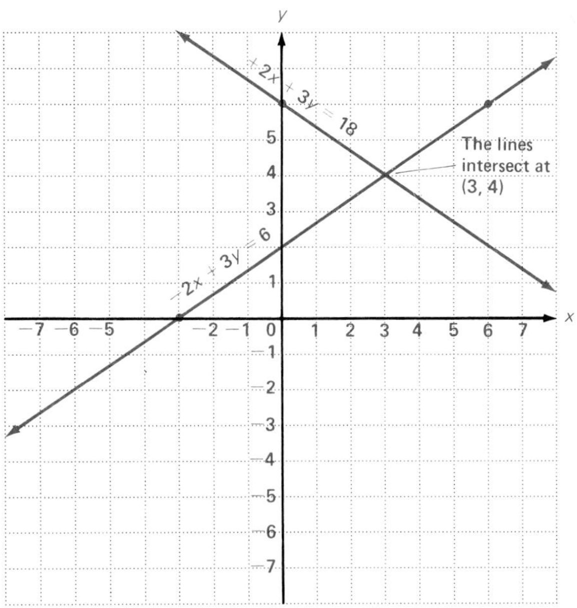

FIGURE 7.1

When we obtain a solution by the graphing method there is some margin of error. Unless the solution contains integer values for x and y we will probably only be able to approximate the answer.

EXAMPLE B Determine the solution graphically for the following system:

$$x + 2y = 3$$
$$-3x + 2y = -7$$

Solution

$x + 2y = 3$ — We first examine the first equation.

if $x = 0$, then $y = \frac{3}{2}$ or $1\frac{1}{2}$ — We find the intercepts of the first line.

if $y = 0$, then $x = 3$

if $x = -3$, then $2y = 6$ and $y = 3$ — We find a third point by letting $x = -3$.

We have the points $(0, 1\frac{1}{2})$, $(3, 0)$, and $(-3, 3)$ for $x + 2y = 3$.

$-3x + 2y = -7$ — We now examine the second equation.

if $x = 0$, then $y = -\frac{7}{2}$ or $-3\frac{1}{2}$ — We find the intercepts of the second line.

if $y = 0$, then $x = \frac{7}{3}$ or $2\frac{1}{3}$

if $x = -1$, then we have — We find a third point by letting $x = -1$.

$2y = -10$ and $y = -5$

We now have the points $(0, -3\frac{1}{2})$, $(2\frac{1}{3}, 0)$, and $(-1, -5)$ for $-3x + 2y = -7$.

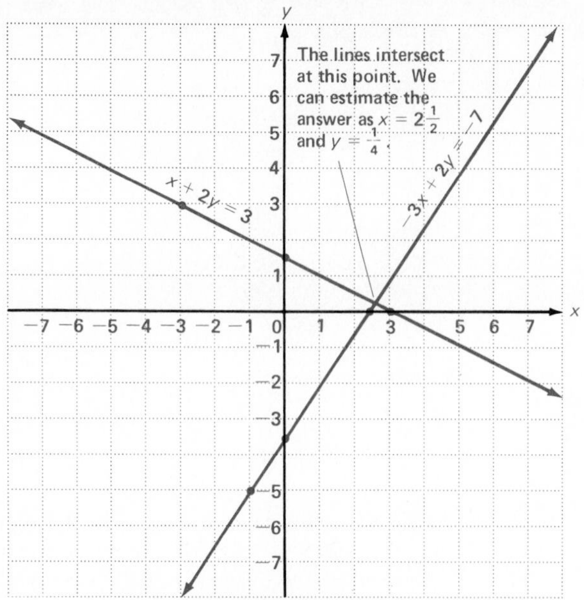

FIGURE 7.2

Now we draw each line and label it with the appropriate equation (Figure 7.2). Our estimate above was correct. The solution to the system is exactly

$$x = \frac{5}{2}\left(\text{or } 2\frac{1}{2}\right) \quad \text{and} \quad y = \frac{1}{4}$$

We can prove this by checking the solution in each equation as follows:

$$x + 2y = 3 \qquad\qquad -3x + 2y = -7$$

$$\frac{5}{2} + 2\left(\frac{1}{4}\right) \stackrel{?}{=} 3 \qquad\qquad -3\left(\frac{5}{2}\right) + 2\left(\frac{1}{4}\right) \stackrel{?}{=} -7$$

$$\frac{5}{2} + \frac{1}{2} \stackrel{?}{=} 3 \qquad\qquad \frac{-15}{2} + \frac{1}{2} \stackrel{?}{=} -7$$

$$\frac{6}{2} = 3 \checkmark \text{ It checks.} \qquad\qquad \frac{-14}{2} = -7 \checkmark \text{ It checks.}$$

However, if you look carefully at the graph you can see it involved some luck that we did not pick a value such as $x = 2\frac{1}{3}$ and $y = \frac{1}{3}$. Thus it is evident that if you seek a solution to a linear system by graphing, you will have to accept the fact that your solution is only an approximation, especially if your lines do not cross exactly at grid squares.

We have found that if a system of linear equations has a solution, we can find it by the substitution method, the addition method, or the graphing method. The graphing method usually takes somewhat longer and is not as accurate as the other two methods. However, the graphing method is especially helpful in assisting us to answer the question: Does a linear system of equations always have a solution? The answer, we will discover, is no!

The inconsistent linear system

EXAMPLE C Try to find the solution by the graphing method for the following system:

$$3x - y = -1$$

$$3x - y = -7$$

Solution $3x - y = -1$ Three convenient points that satisfy this equation are the two intercepts and (1, 4).

x intercept $\left(-\frac{1}{3}, 0\right)$

y intercept (0, 1)

$3x - y = -7$ Three convenient points that satisfy this equation are the two intercepts and (−2, 1).

x intercept $\left(-\frac{7}{3}, 0\right)$

y intercept (0, 7)

We will graph each line and label each with the appropriate equation (Figure 7.3). We observe that the two lines do not intersect. Lines that never meet are called *parallel lines.* Obviously, in such cases there will be no solution to the system. The two lines have no ordered pair (x, y) in common. There is no solution to this system of equations.

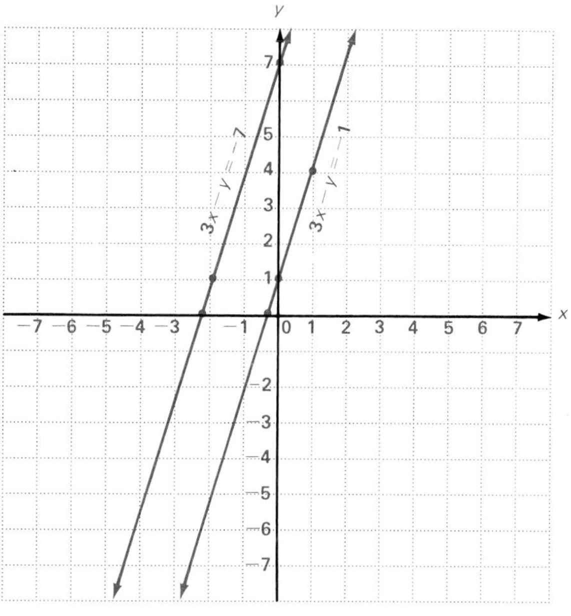

FIGURE 7.3

In mathematics a linear system that has no solution is called *inconsistent.* When a linear system has a solution it is called *consistent.* If we graph a linear system and obtain *parallel lines* we know that the system is inconsistent and there is *no solution.*

What would happen if we had used algebraic methods to find the solution of the inconsistent system in Example C?

EXAMPLE D Try to find the solution by an algebraic method for the following system:

$$3x - y = -1$$

$$3x - y = -7$$

Solution Clearly, the addition method would be very convenient in this case.

$3x - y = -1$	Keep the first equation unchanged.
$-3x + y = +7$	Multiply each term in the second equation by -1.
$0 = 6$	Add the two equations.
	We obtain $0 = 6$, which we know is not true.

The statement $0 = 6$ is *inconsistent* with known mathematical facts. No possible x and y values can make that equation true. There is *no solution* to this system of equations.

The system of dependent equations

There is one additional case that is important. In this case one equation is dependent on another and it is impossible to find just one solution. The graphical method is most helpful in understanding the situation.

EXAMPLE E Try to find the solution by the graphing method for the following system:

$$x + y = 4$$

$$3x + 3y = 12$$

Solution

$$x + y = 4$$

Three points that satisfy the equation are (0, 4), (4, 0), and (2, 2).

$$3x + 3y = 12$$

Three points that satisfy the equation are (1, 3), (3, 1), and $(-1, 5)$.

We plot the points and attempt to graph the two lines (Figure 7.4). Every point on the line satisfies both $x + y = 4$ and $3x + 3y = 12$. Both equations can be represented by one line.

We observe that any solution that satisfies one equation will satisfy the other equation. There are an infinite number of solutions to the system. These equations are said to be dependent. *Dependent equations* form a system of equations where every set of values that satisfies one equation will satisfy all equations in the system. There is no *one* solution to dependent equations,

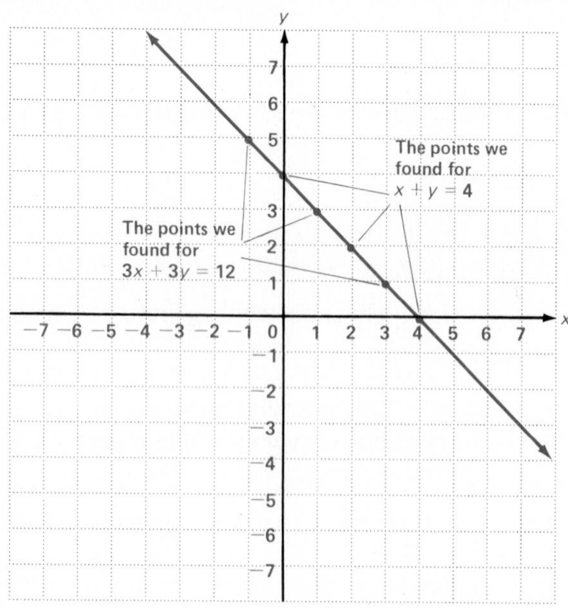

FIGURE 7.4

but rather an unlimited number of solutions. If a system of equations is not dependent, it is said to be *independent*.

What would happen if we try to solve dependent equations by algebraic methods?

EXAMPLE F Try to find the one solution by an algebraic method for the following system:

$$x + y = 4$$
$$3x + 3y = 12$$

Solution

$y = 4 - x$ — Let us use the substitution method. We will solve the first equation for y.

$3x + 3(4 - x) = 12$ — We substitute $4 - x$ for y in the second equation.

$3x + 12 - 3x = 12$ — Remove parentheses.

$12 = 12$ — Since $3x - 3x = 0$ we have $12 = 12$, which is always true.

Thus we obtain an equation that is true for any value for x. An equation that is *always* true is called an *identity*. To find the corresponding y value we could substitute the x value into one of the original equations. (0, 4), (1, 3), (2, 2), (3, 1), (4, 0), (5, −1), and (6, −2) are all ordered pairs that satisfy both equations. There are an *unlimited number of solutions* to the system that has *dependent equations*.

SUMMARY OF POSSIBLE RESULTS WHEN SOLVING TWO EQUATIONS WITH TWO VARIABLES

	Type of System		
	Independent Equations: Consistent System	*Independent Equations: Inconsistent System*	*Dependent Equations*
Examples of each type in Section 7.3	A, B	C, D	E, F
Graph of the system	Two lines intersect in one point:	Parallel lines:	One line:
Number of solutions	One	None	Unlimited
What happens when you try to solve the system algebraically?	You obtain one value for x and one value for y.	You obtain a mathematical equation that does not contain variables and is *not* true (inconsistent with known facts).	You obtain a mathematical equation that does not contain variables and is always true (an identity).

Review Examples A to F and the summary chart. Now try these practice problems.

Practice Problems for Exercise 7.3

Solve the following three systems by the *graphing method* to find a solution if possible. If there is not *one* solution to a system, state the reason.

1. $x + y = 12$
$-x + y = 4$

2. $5x + 4y = 7$
$2x - 3y = 12$

3. $4x + 2y = 2$
$-6x - 3y = 6$

Try to solve the following system by an *algebraic method* (without the aid of graphing). If there is not *one* solution to the system, state the reason.

4. $3x - 9y = 18$

$-4x + 12y = -24$

Now turn to the answers following Exercise 7.3 and check your solutions for these practice problems.

Exercise 7.3

Solve the following systems by the *graphing* method to find one solution, if possible. If there is not *one* solution to a system, state the reason. (Extra sheets of graph paper are to be found at the end of this book.)

1. $x + y = 8$

$x - y = 2$

2. $x + y = 6$

$-x + y = -2$

3. $2x + y = 6$

$x - 3y = 10$

4. $2x + 3y = 4$

$-3x + y = 5$

5. $2x + y = 6$

$-2x + y = 2$

6. $x - 3y = 6$

$4x + 3y = 9$

7. $2x - y + 6 = 0$

$3x + 2y - 1 = 0$

8. $x - 2y - 10 = 0$

$2x + 3y - 6 = 0$

9. $3x - 2y = -18$

$2x + 3y = 14$

10. $3x + 2y = -10$

$-2x + 3y = 24$

11. $y = \frac{2}{3}x - 3$

$y = -2x + 5$

12. $y = \frac{3}{4}x + \frac{4}{3}$

$y = -\frac{3}{2}x + 3$

13. $4x - 6y = 8$

$-2x + 3y = -4$

14. $3x - 2y = -4$

$-9x + 6y = -8$

15. $2y + x - 6 = 0$

$y + \frac{1}{2}x = 4$

16. $y - 2x - 6 = 0$

$\frac{1}{2}y - 3 = x$

If possible, solve the following systems by an algebraic method (without the use of graphing) to find the one solution of the system. If there is not one solution to a system, state the reason.

17. $x - 4y = 6$
$-x + 2y = 4$

18. $-2x + 2y = 7$
$4x - 2y = 1$

19. $4x - 6y = 10$
$-10x + 15y = -25$

20. $3x + 2y = -2$
$-4x + 5y = 18$

21. $-2x - 3y = 15$
$5x + 2y = 1$

22. $2x - 4y = 5$
$-4x + 8y = 9$

23. $5x + 10y = 15$
$-2x - 4y = -1$

24. $3x - y = 4$
$-9x + 3y = -12$

25. $0.3x - 0.5y = 0.4$
$0.6x + 1.0y = 2.8$

26. $-0.05x + 0.02y = -0.01$
$0.10x + 0.04y = -0.22$

Challenge Problems

27. Solve by the graphing method:

$$\frac{1}{7}x - \frac{3}{7}y = 1$$

$$\frac{1}{2}x - \frac{1}{4}y = -1$$

28. Let c and d represent constants in the system:

$$cx + dy = -3$$
$$dx - cy = -3$$

Find the values for c and d so that the solution to the system is $x = 1$ and $y = 2$.

Solutions to Practice Problems for Exercise 7.3

1. $x + y = 12$ First equation.

Intercepts: if $x = 0$, $y = 12$
if $y = 0$, $x = 12$

Pick any other third point that you wish. A convenient third point is obtained if we let $x = 6$; then $y = 6$. So three points for $x + y = 12$ are (0, 12), (12, 0), and (6, 6).

$-x + y = 4$ Second equation.

Intercepts: if $x = 0$, $y = 4$
if $y = 0$, $x = -4$

Another convenient point is if $x = 2$, then $y = 6$. So three points for $-x + y = 4$ are (0, 4), (−4, 0), and (2, 6).

We graph each line and label each line by its equation (Figure 7.5). From this graph we see that the solution to the system is $x = 4$ and $y = 8$.

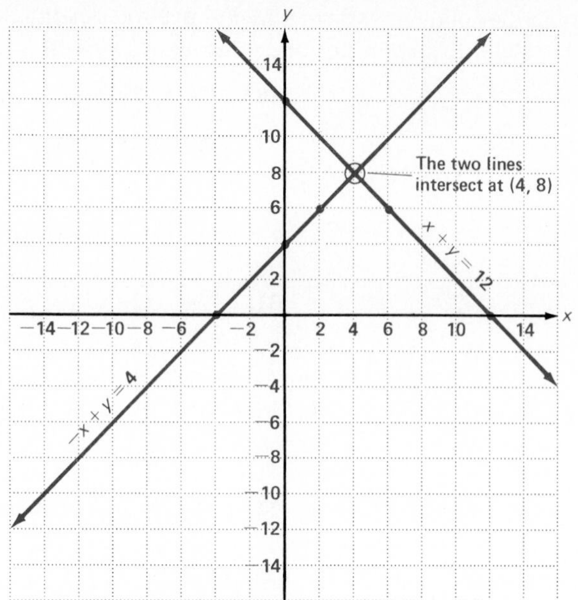

FIGURE 7.5

2. $5x + 4y = 7$ First equation.

Intercepts: if $x = 0$, $4y = 7$ $\quad y = \frac{7}{4}$ or $1\frac{3}{4}$

if $y = 0$, $5x = 7$ $\quad x = \frac{7}{5}$ or $1\frac{2}{5}$

With fractions like this, plotting accurately is very difficult. In such cases the accuracy of your graph is much improved if you plot two additional points that come out integer values. Pick out another couple of points of your choice. Some convenient values are:

If $x = 3$, then

$$\begin{aligned} 5(3) + 4y &= 7 \\ 4y &= 7 - 15 \\ 4y &= -8 \\ y &= -2 \end{aligned}$$

If $x = -1$, then

$$\begin{aligned} 5(-1) + 4y &= 7 \\ 4y &= 7 + 5 \\ 4y &= 12 \\ y &= 3 \end{aligned}$$

Thus four points that satisfy $5x + 4y = 7$ are $\left(0, 1\frac{3}{4}\right)$, $\left(1\frac{2}{5}, 0\right)$, $(3, -2)$, and $(-1, 3)$.

$2x - 3y = 12$ Second equation.

Intercepts: if $x = 0$, $y = -4$

if $y = 0$, $x = 6$

Pick out another point of your choice. A convenient value is obtained by letting $x = -3$. Then

$$-6 - 3y = 12$$
$$-3y = 6 + 12$$
$$-3y = 18$$
$$y = -6$$

Thus three points that satisfy $2x - 3y = 12$ are $(0, -4)$, $(6, 0)$, and $(-3, -6)$.

We graph each line and label each line by its equation (Figure 7.6). By coincidence we plotted $(3, -2)$ as a point that lies on $5x + 4y = 7$. Sometimes you will happen to pick the point of intersection when you graph a given line. Thus the solution to the system is $x = 3$ and $y = -2$.

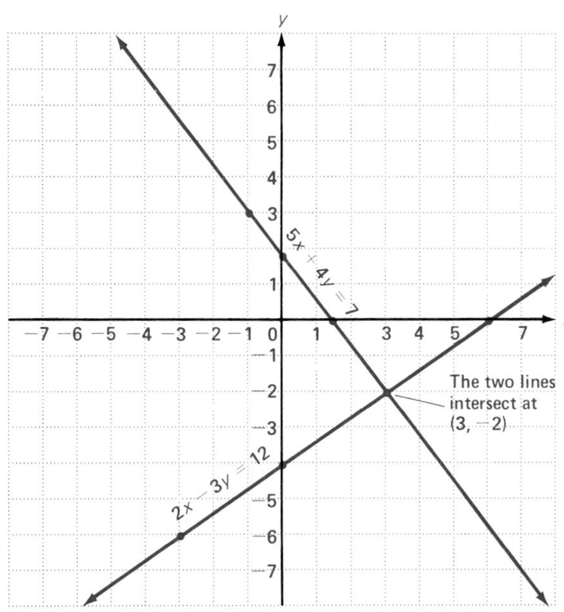

FIGURE 7.6

3. $4x + 2y = 2$ First equation.

Intercepts: if $x = 0$, $y = 1$

if $y = 0$, $x = \frac{1}{2}$

Pick another arbitrary value of x to find out a third point on the line. A convenient choice is to let $x = 2$. Then

$$4(2) + 2y = 2$$
$$2y = -6$$
$$y = -3$$

So three points on the line are $(0, 1)$, $\left(\frac{1}{2}, 0\right)$, and $(2, -3)$.

$-6x - 3y = 6$ Second equation.

Intercepts: if $x = 0$, $-3y = 6$, $y = -2$

if $y = 0$, $-6x = 6$, $x = -1$

Pick another arbitrary value of x to find out a third point on the line. A convenient choice is to let $x = -3$. Then

$$-6(-3) - 3y = 6$$
$$18 - 3y = 6$$
$$-3y = -12$$
$$y = 4$$

We graph each line and label each line by its equation (Figure 7.7). There is no point of intersection since the lines are parallel. There is no solution to the system. We have an *inconsistent* system of equations.

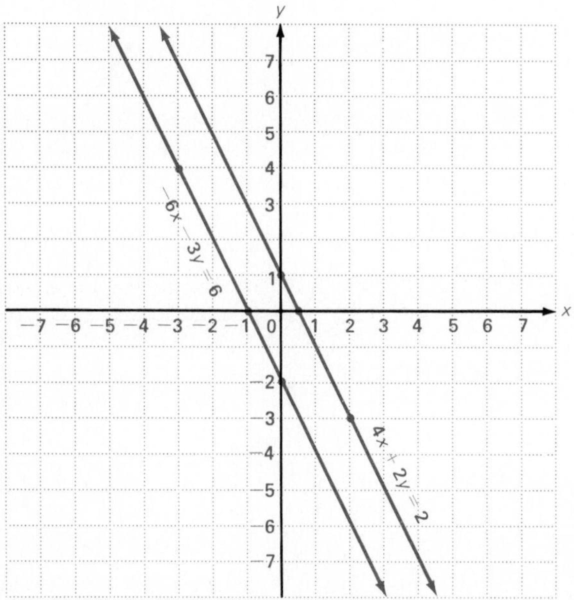

FIGURE 7.7

4. If we try to solve the system algebraically we would probably use the addition method.

$$3x - 9y = 18$$
$$-4x + 12y = -24$$

We will find it easier to eliminate the x variables.

$$4(3x) - (4)(9y) = (4)(18)$$

Multiply each term of first equation by 4.

$$3(-4x) + 3(12y) = (3)(-24)$$

Multiply each term of second equation by 3.

$$\begin{array}{r} 12x - 36y = 72 \\ -12x + 36y = -72 \\ \hline 0 = 0 \end{array}$$

Now we add the two equations and observe the variables drop out.

We obtain an equation that is always true: $0 = 0$. Thus the *equations are dependent*. Any solution to the first equation is also a solution to the second equation [such as $(0, -2)$, $(6, 0)$, $(3, -1)$, $(-3, -3)$]. Thus there are unlimited ordered pairs of solutions. The equations are dependent.

Alternative solution—a short cut

Perhaps you observed that we can transform these equations to equivalent equations with smaller coefficients.

$$3x - 9y = 18 \longrightarrow \text{Divide each term by 3} \longrightarrow x - 3y = 6$$

$$-4x + 12y = -24 \longrightarrow \text{Divide each term by 4} \longrightarrow -x + 3y = -6$$

Now add the equations together $\quad 0 = 0$

We again conclude that the equations are dependent.

Now do the problems in Exercise 7.3.

7.4 Word Problems Involving Systems of Equations

Many word problems can be solved in a variety of ways. In particular, several problems can be solved using one linear equation or two simultaneous linear equations. In such cases, students should try to be aware of all methods but use the method that seems most obvious and reasonable.

In this section we examine a variety of problems that can readily be solved by representing the variables in simultaneous linear equations. A typical problem that can be readily solved by using two variables in two linear equations is the following:

EXAMPLE A The student concert committee on campus sold tickets for a concert given by a popular music group. Tickets purchased in advance were \$4. Tickets purchased at the door were \$5. A total of 480 students bought tickets and attended the concert. The revenue from the ticket sales was \$2100. How many students bought tickets in advance and how many bought tickets at the door?

Solution Let

x = the number of students who purchased tickets in advance

y = the number of students who purchased tickets at the door

$x + y = 480$ — Since 480 students bought tickets and attended the concert.

The amount of money received from advance ticket sales was \$4 for every advance ticket, or $4x$. The amount of money received from sales at the door was \$5 for each ticket, or $5y$.

$4x + 5y = 2100$	We can write this equation since the total revenue from each type of ticket came to \$2100.
(1) $x + y = 480$ (2) $4x + 5y = 2100$	We have a system of two equations with two unknowns. We will solve the systems by the substitution method.
(3) $y = 480 - x$	Solve equation (1) for variable y.
$4x + 5(480 - x) = 2100$	Use equation (2) and replace y by the expression in (3).
$4x + 2400 - 5x = 2100$	Remove parentheses.
$-1x + 2400 = 2100$	Collect like terms.
$-1x = -300$	Subtract 2400 from each side.
$x = 300$	Solve for x by dividing each side by -1.
$y = 480 - x$	Substitute the value 300 obtained for x into (3) to obtain y.
$y = 480 - 300$	
$y = 180$	

Thus 300 tickets were sold in advance and 180 tickets were sold at the door.

Check Does this answer seem reasonable? Yes. Does the number of tickets obtained total 480?

$$300 + 180 \stackrel{?}{=} 480$$

$$480 = 480\checkmark$$

Does the number of each type of ticket obtained yield a sales revenue of \$2100?

$$4(300) + 5(180) \stackrel{?}{=} 2100$$

$$1200 + 900 \stackrel{?}{=} 2100$$

$$2100 = 2100\checkmark$$

EXAMPLE B A technician in a data processing office is trying to verify the rate at which two electronic card-sorting machines operate. Yesterday the first machine sorted for 3 minutes and the second machine sorted for 4 minutes. The total work load both machines processed during that time period was 10,300 cards. Two days ago the first machine sorted for 2 minutes and the second machine for 3 minutes. The total work load both machines processed during that time period was 7400 cards. Can you determine how many cards per minute each machine sorts?

Solution Let

$x =$ the number of cards per minute sorted by the first machine

$y =$ the number of cards per minute sorted by the second machine

We can visualize the problem this way:

Number of cards sorted by first machine for specified number of minutes	+	number of cards sorted by second machine for specified number of minutes	=	number of cards processed for a given day

The situation yesterday:

(3)	(x)	+	(4)	(y)	=	10,300
Number of minutes first machine is running	rate of first machine		number of minutes second machine is running	rate of second machine		total number of cards yesterday

The situation two days ago:

(2)	(x)	+	(3)	(y)	=	7400
Number of minutes first machine is running	rate of first machine		number of minutes second machine is running	rate of second machine		total number of cards two days ago

(1) $3x + 4y = 10{,}300$

(2) $2x + 3y = 7400$

We can now solve these simultaneous linear equations.

We will solve by the method of addition of equations.

$6x + 8y = 20{,}600$	Multiply equation (1) by 2.
$-6x - 9y = -22{,}200$	Multiply equation (2) by -3.
$-1y = -1600$	Since we want $+y$, not $-y$, we divide (or multiply) both sides of equation by -1.
$y = 1600$	
$2x + 3(1600) = 7400$	Substitute this value of y into equation (2).
$2x + 4800 = 7400$	Simplify.
$2x = 2600$	Subtract 4800 from both sides.
$x = 1300$	Divide both sides of the equation by 2.

Therefore, the first machine sorts at the rate of 1300 cards per minute and the second machine at the rate of 1600 cards per minute.

Check

1. The numbers appear to be reasonable. Just by guessing, we would probably estimate that each machine would sort somewhere between 1000 and 2000 cards per minute.
2. We can verify each statement in the problem:

(a) Yesterday: Were 10,300 cards sorted?

$$3(1300) + 4(1600) \stackrel{?}{=} 10{,}300$$

$$3900 + 6400 \stackrel{?}{=} 10{,}300$$

$$10{,}300 = 10{,}300 \checkmark$$

(b) Two days ago: Were 7400 cards sorted?

$$2(1300) + 3(1600) \stackrel{?}{=} 7400$$

$$2600 + 4800 \stackrel{?}{=} 7400$$

$$7400 = 7400 \checkmark$$

In Example B we were trying to verify some information on the rate of card sorting of two data processing machines. Many times the use of mathematics in technology, business, and industrial production is to verify numbers or inspect for quality control by checking numbers.

Another major use is to find a value that cannot be directly measured or that represents information you cannot obtain. Consider the following example.

EXAMPLE C Fred is considering working for a company that sells encyclopedias. During his job interview the company was not willing to release the specific salary structure for beginning sales representatives. Fred found out that all starting representatives receive the same annual base salary and a standard commission of a certain percentage of the sales they make during the first year. He has been told that one representative sold \$50,000 worth of encyclopedias her first year and that she earned \$14,000. He was able to find out that another representative sold \$80,000 worth of encyclopedias and that he earned \$17,600. Determine the base salary and the commission rate of a beginning sales representative.

Solution Let

$$b = \text{the amount of base salary}$$

$$c = \text{the commission rate}$$

The salary structure is defined by: Base salary + additional amount from commission = total salary

$b + 50{,}000c = 14{,}000$ For the first sales representative.

$b + 80{,}000c = 17{,}600$ For the second sales representative.

(1) $b + 50{,}000c = 14{,}000$
(2) $b + 80{,}000c = 17{,}600$

We have two simultaneous linear equations in two variables. We will solve by the addition method.

$-b - 50{,}000c = -14{,}000$ Multiply equation (1) by -1.
$b + 80{,}000c = 17{,}600$ Equation (2) is unchanged.
$30{,}000c = 3600$ Add the two equations.

$c = \dfrac{3600}{30{,}000}$ Divide both sides of equation by 30,000.

$c = 0.12$ The commission rate is 12% or 0.12.

$b + (50{,}000)(0.12) = 14{,}000$ Substitute this value into equation (1).

$b + 6000 = 14{,}000$ Simplify.

$b = 8000$ Solve for b.

Thus the base salary is $8000 per year, and the commission rate is 12%.

Check The student should verify that these values are reasonable and that they satisfy the statements in the original problem.

EXAMPLE D Mike has a boat on Lazy River. He would like to find the speed of his boat and the speed of the current on the river. He took a 48-mile trip up the river traveling against the current in exactly 3 hours. He refueled and made the return trip in exactly 2 hours. Can you find the speed of his boat and the speed of the current on the river?

Solution Let

$$x = \text{the speed of the boat in mph}$$

$$y = \text{the speed of the river current in mph}$$

Now to go upstream against the current, we can find the rate of travel from the equation $D = R \cdot T$ by solving for R.

$$R = \frac{D}{T} = \frac{48}{3} = 16$$

(1) $x - y = 16$ Equation for speed traveling against the current (see Figure 7.8).

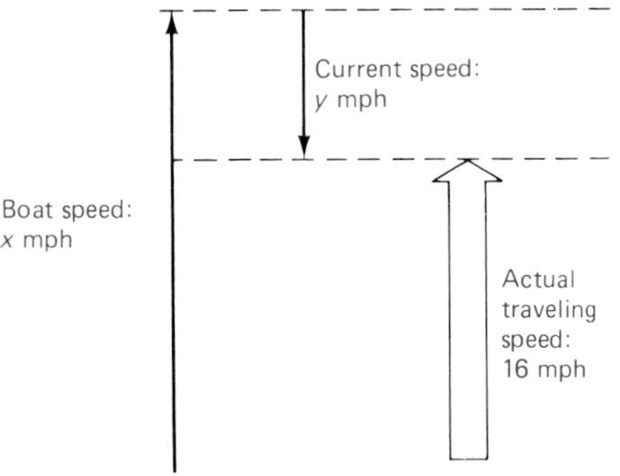

FIGURE 7.8

To go down stream with the current, we have

$$R = \frac{D}{T} = \frac{48}{2} = 24$$

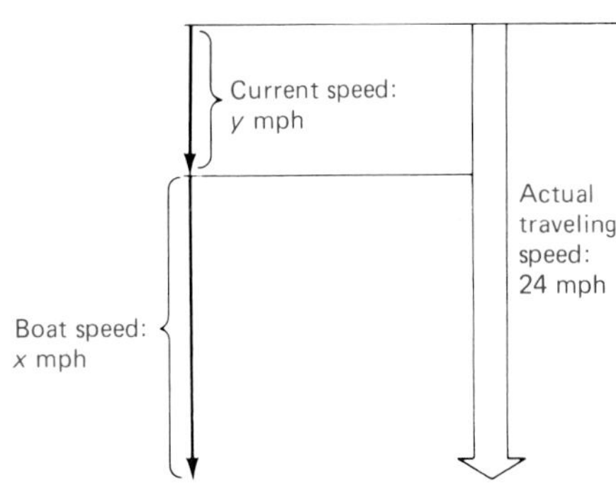

FIGURE 7.9

(2) $x + y = 24$	Equation for speed traveling with the current (see Figure 7.9).
(1) $x - y = 16$ (2) $x + y = 24$	This system is solved most easily by the addition method.
$2x = 40$	Add the two equations.
$x = 20$	Solve for x by dividing both sides by 2.
(2) $x + y = 24$	Substitute the value of $x = 20$ into one of the original equations and solve for y.
$20 + y = 24$	Here we substitute $x = 20$ into equation (2).
$y = 4$	

Thus the speed of Mike's boat is 20 mph and the current speed of Lazy River is 4 mph.

Check Can you verify these answers?

The use of two variables is very helpful in solving a class of problems often referred to as mixture problems.

EXAMPLE E A lab technician is required to prepare 200 liters of a fungicide. The prepared solution must contain 42% fungicide. The technician wishes to use a solution that contains 30% fungicide with a solution that contains 50% fungicide. How much of each solution should he use (see Figure 7.10)?

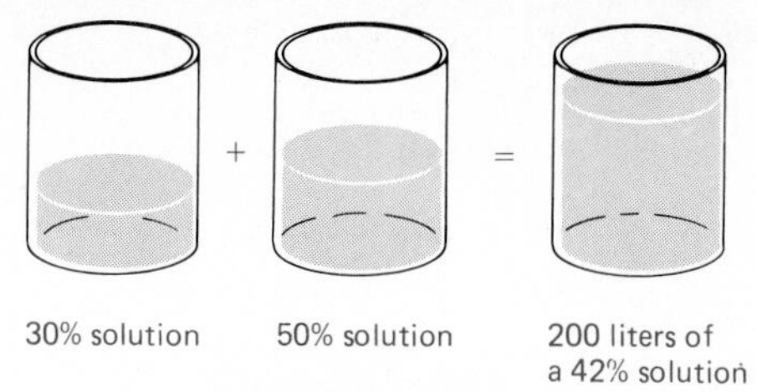

FIGURE 7.10

Solution Let

x = the number of liters of the 30% solution to be used

y = the number of liters of the 50% solution to be used

He will obtain a final solution of 200 liters, so

$$x + y = 200$$

How much *fungicide* is in the final solution?

$$42\% \text{ of } 200 \text{ or } (0.42)(200) = 84$$

The *first solution* has a strength of 30%, so the amount of *fungicide* in the first amount to be used is $0.30x$. The *second solution* has a strength of 50%, so the amount of *fungicide* in the second amount to be used is $0.50y$.

Amount of fungicide in the first solution to be used	+	amount of fungicide in the second solution to be used	=	amount of fungicide in the final solution

$0.3x + 0.5y = 84$ — We write the equation omitting the extra zero in the decimals.

(1) $x + y = 200$

(2) $0.3x + 0.5y = 84$

We will solve the system of equations by the addition method.

$$\begin{aligned} 1.0x + 1.0y &= 200 \\ -0.6x - 1.0y &= -168 \end{aligned}$$

Equation (1) remains the same, but we write the coefficients with decimals. Multiply equation (2) by -2.

$0.4x = 32$ — Add the equations.

$\dfrac{0.4x}{0.4} = \dfrac{32}{0.4}$ — Divide both sides by 0.4.

$x = 80$ — Solve for x.

$x + y = 200$ — Substitute the value $x = 80$ into one of the original equations to obtain y.

$80 + y = 200$

$y = 120$

The technician will use 80 liters of the 30% fungicide and 120 liters of the 50% fungicide to obtain the required solution.

Check Do the amounts total 200 liters?

$$80 + 120 = 200\checkmark$$

Do the amounts yield a 42% strength mixture?

$$0.30x + 0.50y \stackrel{?}{=} 0.42(200)$$

$$0.30(80) + 0.50(120) \stackrel{?}{=} 84$$

$$24 + 60 = 84\checkmark$$

Now try these practice problems.

Practice Problems for Exercise 7.4

Solve each of the following problems by using two equations involving two variables.

1. During the first week of college Alicia took math and verbal placement tests. She just received her test scores. The combined score of the math test and the verbal test was 1153 points. She shared that figure with her friends, but she did not reveal the score for the separate tests. Later, she reluctantly admitted that the difference of the two scores was 409. Can you determine her actual score on each test if you know her math test score was the higher of the two?
2. Robert has an older car that requires leaded premium gasoline. In his town, no gas station still sells leaded premium gasoline. A mechanic advised Robert to use some leaded regular gasoline mixed with some unleaded premium. Last week Robert bought 7 gallons of unleaded premium and 8 gallons of leaded regular gasoline and paid $27.22. This week he purchased 8 gallons of unleaded premium and 4 gallons of leaded regular and paid $22.16. He forgot to record how much each type of fuel cost per gallon. Can you determine these values?
3. Mr. Caminetti traveled downstream in his boat a distance of 72 miles in a time of 3 hours. His return trip up the river against the current took 4 hours. Find the speed of the boat in still water. Find the speed of the current in the river.

Now turn to the answers following Exercise 7.4 and check your solutions for these practice problems.

Exercise 7.4

Solve each of the following problems by using two equations involving two variables.

1. The sum of two numbers is -4. The difference between the two numbers is 30. Find the two numbers.

2. The sum of two numbers is 24. The difference between the two numbers is 14. Find the two numbers.

3. A new compact car has an EPA estimated highway mileage that is 12 miles per gallon more than the estimated city mileage. If you average the city and highway mileage rating for the car, you obtain 36 miles per gallon. Find the city and the highway mileage ratings.

4. In a right triangle the sum of the two smaller interior angles is 90°. In a certain right triangle the difference between those two angles is 28°. Find each of these two angles.

5. There are 46 managers in a large company. The majority of the managers have attended college. The difference between the number of managers who have attended college and those who have not is 18. How many managers attended college? How many managers did not attend college?

6. A large department store starts its salespersons at a salary of $3.50 per hour. The starting assistant managers earn $5.20 per hour. Two roommates work at the same store. The person who was an assistant manager worked 4 more hours than the salesperson. The first 7 days of the job they earned a total of $421.00. How many hours did each work?

7. John's father is a theater manager. A recent matinee performance of a new musical was attended by 420 paying customers. In this theatre mezzanine tickets cost $8 and orchestra tickets are $13 for matinee showings. The sales receipts for this performance were $4610. How many tickets of each type were sold?

8. A temporary office employment agency pays secretaries $35 per day and receptionists $25 per day. Yesterday, 46 people were employed by the agency. Yesterday's payroll was $1470. How many secretaries and how many receptionists were employed yesterday?

9. Alice planted a rooftop garden for her apartment in the city. Last year she needed 48 feet of fencing to enclose the rectangular garden. This year her neighbor gave her permission to use her adjoining roof space. This allowed Alice to triple the width of her garden and to double the length. Now she needs 118 feet of fencing. What was the length and the width of her original garden last year?

10. Abdul is selling cars and trucks at a local dealership. He is considered the top salesman. Last week he sold 14 vehicles. The manager said the rest of the entire dealership sales staff (eight other salespersons) sold four times as many cars as Abdul and three times as many trucks as Abdul. This amounted to a sale of 51 vehicles. How many cars and how many trucks did Abdul sell?

11. An airplane travels between two cities that are 630 miles apart. The trip with a tailwind was completed in 3 hours. The return trip traveling against the wind took 3.5 hours. Find the speed of the wind. Find the speed of the airplane in still air.

12. An airplane travels between two cities that are 3000 kilometers apart. The trip against the wind took 6 hours. The return trip with the benefit of the wind was 5 hours. What is the wind speed in kilometers per hour? What is the speed of the plane in still air?

13. Alicia took her boat on a 45-mile trip. The trip against the current took 4 hours. Her return trip with the current took only 3 hours. Find the speed of the boat in still water. Find the speed of the current.

14. Bob's leaking radiator holds 16 quarts of water. It is now partially full. Bob temporarily patched the leak. He added a mixture from his workbench of 80% antifreeze. The radiator is now full and Bob tested it and found it is 65% antifreeze. Before he added anything he had tested it and found it was 50% antifreeze. How many quarts of solution were in the radiator just before Bob added antifreeze? How many quarts of 80% solution antifreeze did Bob add?

15. A manufacturing company wants to utilize 20 kilograms of a special 45% aluminum alloy. To obtain this, a 50% aluminum alloy is combined with a 30% aluminum alloy in a special heating process. How many kilograms of the 50% alloy and how many kilograms of the 30% alloy are used to make the required 20 kilograms?

16. The hiking club wants to sell 50 pounds of an energy snack food that will cost them only \$1.80 per pound. The mixture will be made up of nuts that cost \$2.00 per pound and raisins that cost \$1.50 per pound. How much of each should they buy to obtain 50 pounds of the desired mixture?

17. How much flour that costs \$0.85 per pound should be mixed with an unrefined flour that costs \$0.90 per pound to obtain 10 pounds of a mixture that will cost \$0.88 per pound?

Challenge Problem

18. Margaret's commission is a given percent of the total sales cost of the books she sells. Margaret receives an 8% commission for selling encyclopedia sets. She receives a 6% commission for selling collegiate dictionaries. Last week she earned \$350 in commissions. This week she sold the same quantity of encyclopedias and the same quantity of dictionaries. However, her company gave her a raise and she now earns 10% commission for selling encyclopedia sets and an 8% commission for selling dictionaries. This week she earned \$440. What is the value of the encyclopedia sets she sold last week? What is the value of the collegiate dictionaries she sold last week?

1. During the first week of college, Alicia took math and verbal placement tests. She just received her test scores. The combined score of the math test and the verbal test was 1153 points. She shared that figure with her friends, but she did not reveal the score for the separate tests. Later, she reluctantly admitted that the difference of the two scores was 409. Can you determine her actual score on each test if you know her math test score was the higher of the two?

Solution: Let

$$m = \text{Alicia's score on the math test}$$

$$v = \text{Alicia's score on the verbal test}$$

Then

(1) $m + v = 1153$	We know the combined score was 1153.
(2) $m - v = 409$	We know that the difference of her scores was 409. We subtract v from m because the math score is higher.
$2m = 1562$	If we add equation (1) to equation (2) we have
$m = 781$	Divide both sides by 2.

Her score on the math test was 781.

$781 + v = 1153$	Replace m by 781.
$v = 1153 - 781$	Subtract 781 from both sides.
$v = 372$	Simplify.

Her score on the verbal test was 372.

Check:

$372 + 781 = 1153\checkmark$	The total score is 1153.
$781 - 372 = 409\checkmark$	The difference in her scores is 409.

2. Robert has an older car that requires leaded premium gasoline. In his town, no gas station still sells leaded premium gasoline. A mechanic advised Robert to use some leaded regular gasoline mixed with some unleaded premium. Last week Robert bought 7 gallons of unleaded premium and 8 gallons of leaded regular gasoline and paid $27.22. This week he purchased 8 gallons of unleaded premium and 4 gallons of leaded regular and paid $22.16. He forgot to record how much each type of fuel cost per gallon. Can you determine these values?

Solution: Let

$$x = \text{the cost per gallon of unleaded premium gasoline}$$

$$y = \text{the cost per gallon of leaded regular gasoline}$$

For each purchase we have:

$$\begin{pmatrix}\text{Number of}\\\text{gallons}\\\text{premium}\end{pmatrix} \times \begin{pmatrix}\text{cost per}\\\text{gallon}\\\text{premium}\end{pmatrix} + \begin{pmatrix}\text{number of}\\\text{gallons}\\\text{regular}\end{pmatrix} \times \begin{pmatrix}\text{cost per}\\\text{gallon}\\\text{regular}\end{pmatrix} = \begin{matrix}\text{purchase}\\\text{price}\end{matrix}$$

$(7)(x) + (8)(y) = 27.22$ Last week's purchase.

$(8)(x) + (4)(y) = 22.16$ This week's purchase.

(1) $7x + 8y = 27.22$

(2) $8x + 4y = 22.16$

This pair of equations can be readily solved by eliminating y.

$7x + 8y = 27.22$

$-16x - 8y = -44.32$ Multiply (2) by -2.

$-9x = -17.10$ Add the two equations.

$\dfrac{-9x}{-9} = \dfrac{-17.10}{-9}$ Divide both sides by -9.

$x = 1.90$

Therefore, unleaded premium gasoline cost \$1.90 per gallon.

$7(1.90) + 8y = 27.22$ Replace x by 1.90 in equation (1).

$13.30 + 8y = 27.22$ Simplify.

$8y = 13.92$ Subtract 13.30 from both sides.

$\dfrac{8y}{8} = \dfrac{13.92}{8}$ Divide both sides by 8.

$y = 1.74$

Thus leaded regular gasoline cost \$1.74 per gallon.

Check:

$7(\$1.90) + 8(\$1.74) \stackrel{?}{=} \$27.22$

$\$13.30 + \$13.92 \stackrel{?}{=} \$27.22$

$\$27.22 = \$27.22\checkmark$

Last week's purchase: 7 gallons at \$1.90 added to 8 gallons at \$1.74 does total \$27.22.

$8(\$1.90) + 4(\$1.74) \stackrel{?}{=} \$22.16$

$\$15.20 + \$6.96 \stackrel{?}{=} \$22.16$

$\$22.16 = \$22.16\checkmark$

This week's purchase: 8 gallons at \$1.90 added to 4 gallons at \$1.74 does total \$22.16.

3. Mr. Caminetti traveled downstream in his boat a distance of 72 miles in a time of 3 hours. His return trip up the river against the current took 4 hours. Find the speed of the boat in still water. Find the speed of the current in the river.

Solution: To go downstream with the current we have as the rate of travel

$$R = \frac{D}{T} = \frac{72}{3} = 24 \text{ mph}$$

To go upstream against the current we have as the rate of travel

$$R = \frac{D}{T} = \frac{72}{4} = 18 \text{ mph}$$

Let

$$x = \text{the speed of the boat in mph}$$

$$y = \text{the speed of the river current in mph}$$

$x + y = 24$ Equation for speed traveling with the current.

$x - y = 18$ Equation for speed traveling against the current.

$$\begin{aligned} x + y &= 24 \\ x - y &= 18 \\ \hline 2x \quad &= 42 \\ x &= 21 \end{aligned}$$

We solve this system by the addition method.

$x + y = 24$ We substitute into the original equation and solve for y.

$$21 + y = 24$$

$$y = 3$$

Thus the boat speed is 21 mph and the speed of the current is 3 mph.

Check:

Do the numbers check on the trip downstream with the current?

$$\text{Rate} = \frac{\text{distance}}{\text{time}} = \frac{72 \text{ miles}}{3 \text{ hours}} = x + y$$

$$24 = 21 + 3\checkmark$$

Do the numbers check on the trip upstream against the current?

$$\text{Rate} = \frac{\text{distance}}{\text{time}} = \frac{72 \text{ miles}}{4 \text{ hours}} = x - y$$

$$18 = 21 - 3\checkmark$$

Now do the problems in Exercise 7.4.

An Application for Chapter 7: Break-Even Analysis

One way that simultaneous systems of equations are used in the business world is in the area known as break-even analysis. The following simple business illustration will help you to see the basic aspects of this approach.

Dave's father has a family business manufacturing candy and supplying the candy to retailers. Each year Dave's father figures out an annual profit and loss statement. When Dave returned home from his first year of college his father asked him to analyze the *daily profit and loss status* of the candy business.

Dave was given the following information:

1. The company has fixed costs from salaries, rent, light, heat, and telephone of \$300 per day.
2. Each pound of candy produced costs \$2 per pound for ingredients.
3. Each pound of candy sold yields \$3 in revenue.

Dave was asked by his father for the following:

1. Make some type of chart that shows how I can visualize the daily profit and loss of my company.
2. Assume that the company makes the same number of pounds of candy each day as it will sell on that same day.
   (a) How many pounds of candy must be sold each day to break even?
   (b) What is the loss for the day if only 200 pounds of candy are sold?
   (c) How many pounds of candy must be sold in order to make a profit of $200 on a given day?

Dave has just finished a college course in beginning algebra using this text. He analyzed the problem in this way.

The major variable in this situation is the amount of candy. Dave let

x = number of pounds of candy produced and sold each day

The cost to daily operate the business is given by a fixed cost of $300 and a cost of $2 for every pound of candy made, so let

Cost = fixed cost + (cost per pound of candy)(number of pounds of candy)

$C = 300 + 2x$ since x is the number of pounds of candy and each pound costs $2 in ingredients.

The daily revenue is calculated as $3 for every pound of candy sold. Let

Revenue = (revenue per pound of candy)(number of pounds of candy)

$R = 3x$ since x is the number of pounds of candy and each pound sold yields $3.

Dave now has two equations:

$$R = 3x$$
$$C = 2x + 300$$

where R and C are always an amount of money.

He now graphs each equation (Figure 7.11), where x is the number of pounds of candy and the values of R and C are amounts of money in dollars. This provides a visual chart to show his father the daily profit and loss of the company.

If the R line crosses the C line, then revenues are equal to costs and the business exactly breaks even.

If the R line is below the C line, then revenues are less than costs and the business is losing money on a daily basis. This region is shaded with horizontal lines:

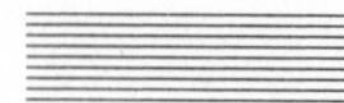

If the R line is higher than the C line, then revenues are greater than costs and a profit is achieved. This region is shaded with vertical lines:

(a) To break even, R must equal C. On the graph this takes place if $x = 300$. The company must make and sell 300 pounds of candy per day to break even for that day.

(b) If 200 pounds of candy are made and sold, the revenue value is \$600 but the cost value is \$700. The daily loss would be \$100. The amount by which the *R* is beneath the *C* line is 100, as shown on the graph.

(c) If the *R* line is above the *C* line by 200 units, a profit of \$200 is achieved. This occurs on the graph when $x = 500$. Thus 500 pounds of candy must be made and sold to yield a daily profit of \$200.

Today many small business computers have software to create charts similar to Figure 7.11. This can assist managers or owners of a small business in making timely decisions that help to maximize profit.

A Challenge for You!

1. Given the information in this problem, if you manufacture and sell daily 625 pounds of candy, how much is the profit or loss?
2. Campus T-Shirts, Inc. is a local student-run enterprise. It has daily fixed costs of \$400. Every T-shirt sold yields \$8 in revenue. Every T-shirt sold costs Campus T-Shirts, Inc. a sum of \$5. Make a graph that illustrates the profit or loss if 320 shirts are sold each day.

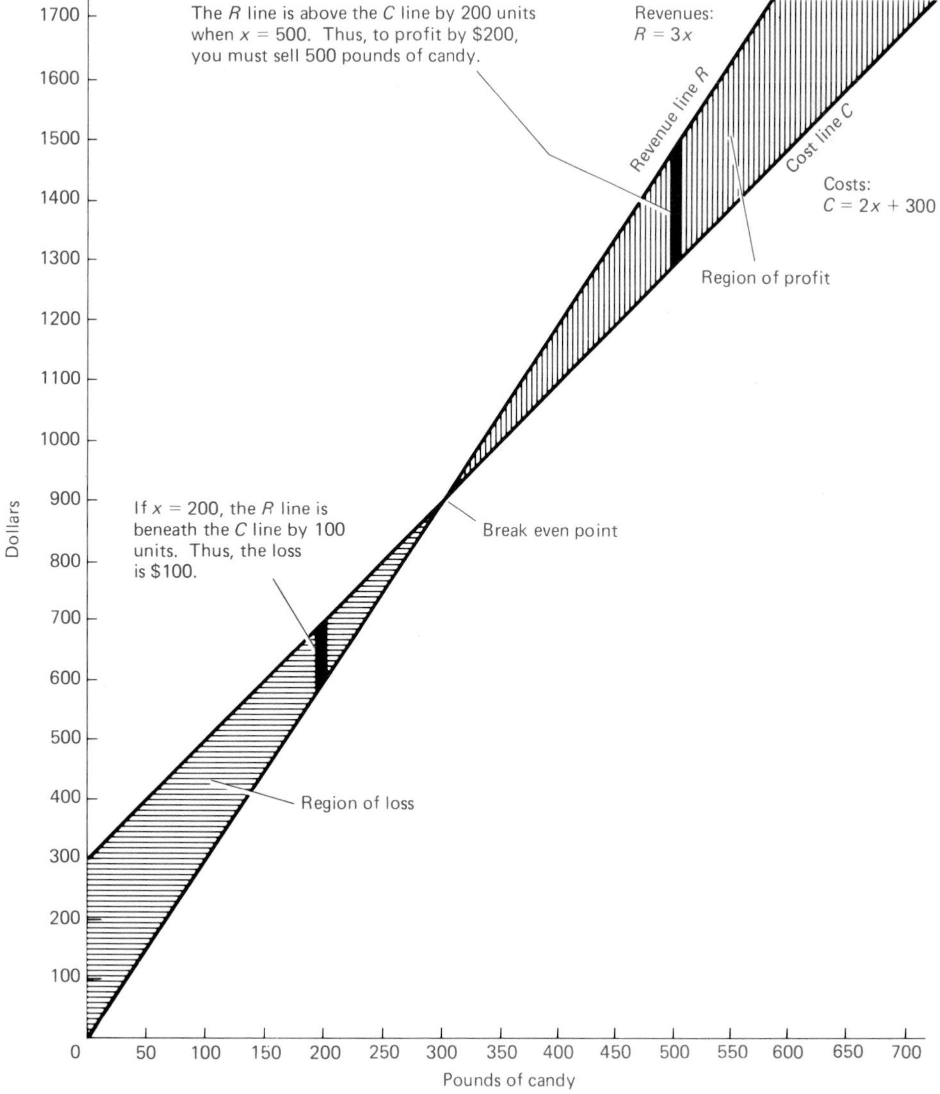

FIGURE 7.11

Supplementary Problems for Chapter 7

If possible, solve for x and y by any appropriate method to obtain the one solution for x and y. If it is not possible to obtain one solution, state the reason.

1. $3x - 2y = 17$
$x + y = 9$

2. $x - 2y = 2$
$2x + 3y = 11$

3. $11x + 3y = 12$
$4x + y = 5$

4. $7x - 5y = 8$
$-x + y = -2$

5. $5x - 3y = 32$
$5x + 7y = -8$

6. $4x + 3y = 46$
$-2x + 3y = -14$

7. $7x + 3y = 2$
$-8x - 7y = 2$

8. $x + 5y = 7$
$-3x - 4y = 1$

9. $5x - 11y = -4$
$6x - 8y = -10$

10. $3x + 5y = -9$
$4x - 3y = 17$

11. $2x - 4y = 6$
$-3x + 6y = 7$

12. $4x - 7y = 8$
$5x + 9y = 81$

13. $2x - 9y = 0$
$3x + 5 = 6y$

14. $2x + 10y = 1$
$-4x - 20y = -2$

15. $5x - 8y = 3x + 12$
$7x + y = 6y - 4$

16. $1 + x - y = y + 4$
$4(x - y) = 3 - x$

17. $3x + y = 9$
$x - 2y = 10$

18. $x + 12y = 0$
$3x - 5y = -2$

19. $2(x + 3) = y + 4$
$4x - 2y = -4$

20. $x + y = 3000$
$x - 2y = -120$

21. $4x - 3y + 1 = 6$
$5x + 8y + 2 = -74$

22. $5x + 4y + 3 = 23$
$8x - 3y - 4 = 75$

23. $\frac{2x}{3} - \frac{3y}{4} = \frac{7}{12}$
$8x + 5y = 9$

24. $\frac{x}{2} + \frac{y}{5} = 4$
$\frac{x}{3} + \frac{y}{5} = \frac{10}{3}$

If possible, solve for the two variables by any appropriate method to find the one solution for each variable. If it is not possible to obtain one solution, state the reason.

25. $\frac{1}{5}a + \frac{1}{2}b = 6$
$\frac{3}{5}a - \frac{1}{2}b = 2$

26. $\frac{2}{3}a + \frac{3}{5}b = -17$
$\frac{1}{2}a - \frac{1}{3}b = -1$

27. $0.2s - 0.3t = 0.3$
$0.4s + 0.6t = -0.2$

28. $0.3s + 0.2t = 0.9$
$0.2s - 0.3t = -0.6$

29. $3m + 2n = 5$
$4n = 10 - 6m$

30. $12n - 8m = 6$
$4m + 3 = 6n$

31. $3(x + 2) = -2 - (x + 3y)$
$3(x + y) = 3 - 2(y - 1)$

32. $13 - x = 3(x + y) + 1$
$14 + 2x = 5(x + y) + 3x$

33. $0.2b = 1.4 - 0.3a$
$0.1b + 0.6 = 0.5a$

34. $0.3a = 1.1 - 0.2b$
$0.3b = 0.4a - 0.9$

35. $\frac{b}{5} = \frac{2}{5} - \frac{a - 3}{2}$
$4(a - b) = 3b - 2(a - 2)$

36. $9(b + 4) = 2(2a + 5b)$
$\frac{b}{5} + \frac{a}{2} = \frac{18}{5}$

In problems 37 and 38, solve for x and y. Assume that a and b are constants.

37. $ax - by = ab + b^2$
$ax + by = 2a^2 + ab - b^2$

38. $5 + ax + by = 2a + 5$
$2b + by = ax$

39. During a concert \$3280 was received. A total of 760 students attended the concert. Reserved seats were \$6 and general admission seats were \$4. How many of each kind of ticket were sold?

40. In a local fruit market apples are on sale for \$0.25 per pound. Oranges are on sale for \$0.39 per pound. Coach Smith purchased 12 pounds of a mixture of apples and oranges before the basketball team left on a bus trip to play at Center City. If the coach paid \$3.70, how many pounds of each fruit did he get?

41. A plane travels 1500 miles in 5 hours with the benefit of a tailwind. On the return trip it requires 6 hours to fly against the wind. Can you find the speed of the plane in still air and the wind speed?

42. A chemist needs to combine two solutions of a certain acid. How many liters of a 30% acid solution should be added to 40 liters of a 12% acid solution to obtain a final mixture that is 20% acid?

Chapter 7 Summary

If there is one solution to a system of two linear equations in two variables it may be obtained by three methods. Such a system is called a consistent system of equations.

The Substitution Method

The substitution method is most appropriate when *at least one variable has a coefficient of 1 or −1.*

Procedure

1. Solve for one variable in one of the equations.
2. In the other equation replace that variable with the expression you obtained in step 1.
3. Solve the resulting equation.
4. Substitute the numerical value you obtain for a variable into the equation you found in step 1.
5. Solve this equation to find the other variable.

The Addition Method

The addition method is most appropriate when the variables *all have coefficients other than 1 or −1.*

Procedure

1. Multiply one or both equations by appropriate numerical values so that when the two resulting equations are added one variable is eliminated.
2. Solve the resulting equation.
3. Substitute the numerical value you obtain for the variable in one of the original equations.
4. Solve this equation to find the other variable.

The Graphing Method

The graphing method is an approximate method to find the solution to a system of linear equations. It is appropriate if a graphical representation is desired.

Procedure

1. Determine three points that satisfy each linear equation.
2. Draw a graph for each linear equation.
3. Determine the point where the lines intersect.
4. The x and y coordinates of that point are the x and y values that satisfy (approximately) the system.

If there is not one solution to a system of two linear equations:

1. There may be no solution. In this case the *system is inconsistent.* This is represented graphically by parallel lines. Algebraically, if you try to solve this system you get a math equation that is not true (like $0 = 3$).

2. There may be an infinite number of solutions. In this case the *equations are dependent*. If you graph each equation you will get the same line. Algebraically, if you solve this system you get a math equation that is always true (such as $0 = 0$).

Checking solutions to a system of linear equations

To check the solutions obtained by solving a system of two linear equations you must substitute the obtained values for each variable into *both* original equations.

Practice Quiz for Chapter 7

Sections 7.1–7.2

1. Solve by the substitution method:

$$2x - y = 5$$
$$5x + 2y = 8$$

2. Solve by the addition method:

$$5x + 6y = 14$$
$$3x - 2y = -14$$

Solve by either the substitution method or the addition method.

3. $2x + y = 5$
   $-x + 4y = 11$

4. $-6x + 15y = -27$
   $6x + 8y = 4$

5. $5x - 3y = 13$
   $-2x + 9y = -13$

6. $5x - 2y = 17$
   $2x - 3y = 9$

7. $\frac{3}{4}x - \frac{1}{2}y = -2$
   $\frac{1}{3}x - \frac{7}{6}y = 1$

Sections 7.3–7.4

Solve by the graphing method.

1. $x + 2y = 7$
   $2x - y = -1$

2. $2x - y = 3$
   $x + 3y = 5$

Solve by any method.

3. $x + 3y = 7$
   $x - 5y = -1$

4. $2x + 3y = 0$
   $-x + y = -5$

Solve problems 5 to 7 by any method to find one solution if possible. If it is not possible, state why.

5. $3x - 4y = 8$
   $-6x + 8y = -16$

6. $4x + 2y = 9$
   $x + \frac{1}{2}y = 2$

7. $4x - 5y = -2$
   $7x + 3y = 20$

8. The girls and boys sponsoring the Scouts Car Wash had 86 customers on Saturday. They charged \$2 to wash regular size cars and \$1.50 to wash compact cars. The gross receipts for the day were \$151. How many cars of each type were washed?
9. Carl manages a highway department garage in Maine. Salt and sand mixtures are stored in a garage for use during the winter. Carl needs 24 tons of a salt/sand mixture that is 25% salt. He will combine shipments of 15% salt and shipments of 30% to achieve the desired 24 tons. How much of each type should he require?
10. A water ski boat traveled 23 kilometers per hour going with the current. It returned and went back in the opposite direction against the current and only made 15 kilometers per hour. What was the speed of the boat? What was the speed of the current?

11. Fred is analyzing the cost to produce two different items at an electronics company. An electrical sensing device requires 5 grams of copper and requires 3 hours to assemble. A smaller sensing device made by the same company requires 4 grams of copper but requires 5 hours to assemble. The first device has a production cost of \$27. The second device has a production cost of \$32. How much does it cost the company for a gram of this type of copper? What is the hourly labor cost at this company? (Assume that production cost is obtained by adding copper cost and labor cost.)

Practice Test for Chapter 7

Solve by the method specified.

1. Substitution method:

$$3x - y = -5$$
$$-2x + 5y = -14$$

2. Addition method:

$$3x + 4y = 7$$
$$2x + 3y = 6$$

3. Graphing method:

$$x + y = 5$$
$$x - 4y = 5$$

4. Any method:

$$x - 3y = 0$$
$$2x - 3y = -6$$

Solve problems 5 to 10 by any method to find one solution if this is possible. If there is not one solution to a system, state why.

5. $2x - y = 5$
$-x + 3y = 5$

6. $2x + y = 4$
$3x + 7y = 17$

7. $\frac{2}{3}x - \frac{1}{5}y = 2$
$\frac{4}{3}x + 4y = 4$

8. $-2x + 4y = 14$
$3x - 6y = 5$

9. $5x - 2 = y$
$10x = 4 + 2y$

10. $0.3x + 0.2y = 0$
$1.0x + 0.5y = -0.5$

11. Two machines place address labels on business envelopes. The first machine operated for 8 hours and the second machine for 3 hours. That day 5200 envelopes were affixed with address labels. The next day 4400 envelopes were addressed when the first machine was operated for 4 hours and the second machine for 6 hours. How many envelopes are addressed per hour by each machine?

12. A jet plane flew 2000 kilometers against the wind in a time of 5 hours. It refueled and flew back with the wind in 4 hours. Find the wind speed in kilometers per hour. Find the speed of the jet in still air in kilometers per hour.

8

Radicals

Diagnostic Pretest for Chapter 8

This diagnostic pretest is intended for those students who have previously studied algebra. Work as many problems as you can. Then, check your answers against those listed at the back of the book. Where you have obtained an incorrect answer or have been unable to work the problem, you will need to study the corresponding sections of the chapter.

If you have not previously studied algebra, do not take this test now. You should study the chapter and work the problems and exercises. After finishing the chapter, you may use this test as a review or as a practice test.

Section 8.1

Find the square roots if possible. (If it is impossible to obtain a real number for an answer, so state.)

1. $\sqrt{121}$

2. $\sqrt{\dfrac{25}{64}}$

3. $-\sqrt{100}$

4. $\sqrt{-81}$

Section 8.2

Simplify.

5. $\sqrt{12x^3y^5}$

6. $\sqrt{36x^6y^3}$

Section 8.3

Combine if possible.

7. $3\sqrt{2} - \sqrt{8} + \sqrt{18}$

8. $2\sqrt{28} + 3\sqrt{63} - \sqrt{49}$

Section 8.4

Multiply. Simplify your answer.

9. $2\sqrt{5}\,(3\sqrt{5} - \sqrt{15})$

10. $(2\sqrt{6} - \sqrt{2})(\sqrt{6} + 2\sqrt{2})$

Section 8.5

Simplify. Rationalize all denominators.

11. $\dfrac{5}{\sqrt{3}}$

12. $\dfrac{3}{\sqrt{5} + 2\sqrt{3}}$

Section 8.6

13. Using the Table of Square Roots in Appendix A, approximate:

$$\frac{\sqrt{3}}{\sqrt{5}}$$

14. Simplify the following higher-order radical:

$$\sqrt[3]{125x^6y^9}$$

15. Multiply the following two radicals and simplify your answer:

$$\sqrt[4]{2x^3y}\;\sqrt[4]{8x^5y^3}$$

Section 8.7

16. The air resistance of a car traveling on the highway varies directly as the square of the car's velocity. The air resistance is 300 pounds when the car is traveling at 30 miles per hour. Find the air resistance when the car is traveling at 50 miles per hour.

Student Learning Objectives for Chapter 8

After successfully learning the content of each section, you will be able to:

Section 8.1

A. Identify and explain the concept of a square root.

B. Evaluate square roots of numbers that are perfect squares.

C. Identify cases where no real number exists when the radicand contains a negative number.

Section 8.2

D. Simplify radicals when the radicand is a number that is not a perfect square.

E. Simplify radicals when the radicand is a product of a numerical coefficient and one or more variable quantities.

Section 8.3

F. Add or subtract similar radical expressions.

Section 8.4

G. Multiply radical expressions and simplify the results.

Section 8.5

H. Rationalize the denominator of radical expressions.

Section 8.6

I. Use a square root table to approximate the value of a square root radical.

J. Simplify or evaluate a higher-order radical.

Section 8.7

K. Solve a variety of word problems that involve direct and inverse variation of two variable quantities.

Introduction

In Sections 1.1, 2.3, 2.4 we briefly mentioned irrational numbers. We now begin to study irrational numbers more extensively. In algebra, we need to be able to perform the operations of addition, subtraction, multiplication, and division on all types of irrational numbers as well as variable expressions that represent irrational numbers. The most common type of irrational number is expressed by means of a radical. In this chapter we will study how to evaluate, simplify, and perform elementary operations on radical expressions.

We will be considering those types of irrational numbers that are obtained by taking roots of rational numbers. Here is a specific real world example of how we might encounter such a number.

Suppose we ask this question: How long is the side of a square whose area is 4? Using our previously learned formula for the area of a square we could state $s^2 = 4$ and therefore conclude that the side $s = 2$. We see in this case the rational number 2 is the answer.

Let's consider a more difficult question. How long is the side of a square whose area is 2? We could express this situation by the formula $s^2 = 2$. However we cannot find a rational number for the value of s. It turns out that $s = \sqrt{2}$ (this is read "the square root of 2"). This irrational value cannot be written as a fraction (it cannot be expressed in the form p/q where p, q are integers) and it cannot be written as a finite decimal. We can use an approximate value (such as 1.414) to indicate the value accurate to three decimal places and can use that approximation to indicate the relative position on the number line.

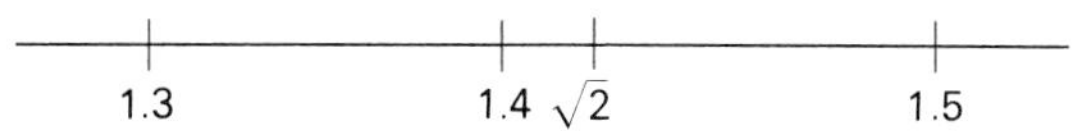

This raises some questions. Mathematically speaking, what are "square roots?" Exactly what do we mean by $\sqrt{2}$? We will begin chapter 8 by explaining precisely what we mean by the idea of a square root.

8.1 Square Roots

In a variety of areas such as geometry and statistics we encounter the idea of a square root. This concept is not difficult; it has been used by mathematicians for thousands of years.

Suppose that we square several positive numbers as follows:

Let $x = 3$; then x^2 is $(3)^2 = 3 \cdot 3 = 9$.
Let $x = 8$; then x^2 is $(8)^2 = 8 \cdot 8 = 64$.
Let $x = 12$; then x^2 is $(12)^2 = 12 \cdot 12 = 144$.

Now let us reverse the process assuming that *x is a positive number.*

If $x^2 = 16$, what is x?

Since $4^2 = 16$ and since we want a positive value for x, we select $x = 4$.

If $x^2 = 81$, what is x?

Since $9^2 = 81$ and since we want a positive value for x, we select $x = 9$.

In these last two cases we found the *square root* of a number. We were given a number and we wanted to find a number which, when multiplied by itself, would result in that given number.

EXAMPLE A Find a positive number that is the square root of 121.

Solution The answer is 11 since $(11)^2 = (11)(11) = 121$. When we square 11 we obtain 121. Thus the positive number that is the square root of 121 is 11.

Are all square roots positive numbers?

No, although there are many times in mathematics when we are only interested in the positive square roots.

EXAMPLE B Find a negative number that is the square root of 64.

Solution The answer is -8. We know that $(-8)^2 = 64$. Thus the negative number that is the square root of 64 is -8.

The radical sign

The symbol $\sqrt{\ }$ is used in mathematics to indicate a square root of a number. The symbol itself $\sqrt{\ }$ is defined as the *radical sign.* It is understood to be a nonnegative square root. The nonnegative square root of a number is called principal square root of that number. If we desire to find the negative square root of a number, we use the symbol $-\sqrt{\ }$. The following definition is helpful.

> ***Definition of Square Root***
>
> **For all nonnegative numbers N the square root of N (written $\sqrt{N}$) is defined to be the nonnegative number a if and only if $a^2 = N$.**

Notice in the definition that we did not use the word "all positive numbers N" because we did not want to exclude the fact that $\sqrt{0} = 0$ since $0^2 = 0$. The working part of the definition of a square root is that $\sqrt{N} = a$ if and only if $a^2 = N$. An understanding of that statement will enable you to find square roots.

Let us now examine the use of this definition with a few examples.

EXAMPLE C Find: $\sqrt{144}$.

Solution $\sqrt{144} = 12$ since $(12)^2 = (12)(12) = 144$.

EXAMPLE D Find: $-\sqrt{9}$.

Solution $-\sqrt{9} = -3$ since $(-3)^2 = (-3)(-3) = 9$

The $-\sqrt{\ }$ symbol (called the negative square root of a number) indicates that we want to find a negative number that, when squared, is equal to 9. When we write the symbol $-\sqrt{9}$ we are indicating the problem "find a negative number that is the square root of 9." The symbol $-\sqrt{9}$ is read "the negative square root of 9."

The number beneath the radical sign is called the *radicand.* It is not necessary that the radicand be an integer for us to evaluate the square root of a number.

EXAMPLE E Find:

$$\sqrt{\frac{1}{4}}$$

Solution $\sqrt{\frac{1}{4}} = \frac{1}{2}$ since $\left(\frac{1}{2}\right)^2 = \left(\frac{1}{2}\right)\left(\frac{1}{2}\right) = \frac{1}{4}$.

In this case the radicand is the rational number $\frac{1}{4}$.

EXAMPLE F Find:

$$-\sqrt{\frac{4}{9}}$$

Solution First we observe that we are being asked to find the negative square root of $\frac{4}{9}$. We are seeking a negative number that can be squared to obtain $\frac{4}{9}$. That number is $-\frac{2}{3}$ since $(-\frac{2}{3})^2 = (-\frac{2}{3})(-\frac{2}{3}) = \frac{4}{9}$. Thus

$$-\sqrt{\frac{4}{9}} = -\frac{2}{3}$$

The radicand can be a rational number expressed in decimal form.

EXAMPLE G Find: $\sqrt{0.09}$.

Solution We know that $(0.3)^2 = (0.3)(0.3) = 0.09$. Thus $\sqrt{0.09} = 0.3$.

Alternate solution Since $0.09 = \frac{9}{100}$ we could say that $\sqrt{0.09} = \sqrt{\frac{9}{100}}$.

$$\sqrt{\frac{9}{100}} = \frac{3}{10} \quad \text{since} \quad \left(\frac{3}{10}\right)^2 = \left(\frac{3}{10}\right)\left(\frac{3}{10}\right) = \frac{9}{100}$$

Thus

$$\sqrt{0.09} = \sqrt{\frac{9}{100}} = \frac{3}{10} = 0.3$$

Does the radicand need to be a nonnegative number? What if we write $\sqrt{-4}$? Obviously, there is *no real number* that you can square to get -4. We know that $(2)^2 = 4$ and $(-2)^2 = 4$. Any number that is squared will be nonnegative. We therefore conclude that $\sqrt{-4}$ does not represent a real number. Our definition of the square root $\sqrt{b}$ requires that b be nonnegative.

An Interesting Sidelight for Future Study

In mathematics there are useful things that can be done with numbers like $\sqrt{-4}$. Such numbers are called *imaginary numbers*. Even though such numbers were introduced in the 1500s to solve certain equations, most mathematicians would not accept them as valid numbers. Great arguments were fought concerning what to do with numbers such as $\sqrt{-4}$. Today they are a part of the study of complex numbers and are very useful in the study of electronic circuit theory and in a variety of areas in science. You will probably encounter them in a later course. The name *imaginary number* continues in use today because for hundreds of years mathematicians claimed that such numbers did not exist!

EXAMPLE H Compare: $-\sqrt{16}$ and $\sqrt{-16}$.

Solution $-\sqrt{16}$ is the negative square root of 16. Since $(-4)^2 = (-4)(-4) = 16$, $-\sqrt{16} = -4$. $\sqrt{-16}$ is not a real number. $\sqrt{-16}$ means the square root of -16. There is no real number that you can square to obtain a negative number. Thus $-\sqrt{16} = -4$, but $\sqrt{-16}$ is not a real number.

Now try these practice problems.

Practice Problems for Exercise 8.1

Find, if possible, the following square roots.

1. $\sqrt{64}$

2. $\sqrt{\frac{9}{25}}$

3. $-\sqrt{121}$

4. $\sqrt{-81}$

Now turn to the answers following Exercise 8.1 and check your solutions for these practice problems.

Exercise 8.1

Find, if possible, the following square roots. If it is impossible to obtain a real number for an answer, so state. Do *not* use a calculator or a table of square roots.

1. $\sqrt{25}$

2. $\sqrt{36}$

3. $\sqrt{0}$

4. $\sqrt{1}$

5. $\sqrt{49}$

6. $\sqrt{16}$

7. $\sqrt{100}$

8. $\sqrt{9}$

9. $-\sqrt{36}$

10. $-\sqrt{25}$

11. $\sqrt{0.81}$

12. $\sqrt{0.64}$

13. $\sqrt{0.01}$

14. $\sqrt{0.16}$

15. $\sqrt{\frac{16}{25}}$

16. $\sqrt{\frac{9}{64}}$

17. $\sqrt{\frac{49}{64}}$

18. $\sqrt{\frac{49}{100}}$

19. $\sqrt{-\frac{9}{25}}$

20. $\sqrt{-\frac{1}{49}}$

21. $-\sqrt{-16}$

22. $-\sqrt{-25}$

23. $\sqrt{169}$

24. $\sqrt{225}$

25. $-\sqrt{\frac{1}{64}}$

26. $\sqrt{\frac{81}{100}}$

27. $\sqrt{\frac{25}{49}}$

28. $-\sqrt{\dfrac{9}{64}}$

29. $\sqrt{0.36}$

30. $\sqrt{0.25}$

31. $\sqrt{-\dfrac{36}{121}}$

32. $\sqrt{\dfrac{1}{196}}$

33. $\sqrt{\dfrac{49}{100}}$

34. $\sqrt{-\dfrac{16}{25}}$

35. $\sqrt{0.0064}$

36. $\sqrt{0.000081}$

In problems 37 to 42, use the following table of values.

$16^2 = 256$	$22^2 = 484$	$28^2 = 784$
$17^2 = 289$	$23^2 = 529$	$29^2 = 841$
$18^2 = 324$	$24^2 = 576$	$30^2 = 900$
$19^2 = 361$	$25^2 = 625$	$31^2 = 961$
$20^2 = 400$	$26^2 = 676$	$32^2 = 1024$
$21^2 = 441$	$27^2 = 729$	$33^2 = 1089$

37. $\sqrt{576}$

38. $-\sqrt{361}$

39. $-\sqrt{841}$

40. $\sqrt{529}$

41. $\sqrt{\dfrac{289}{900}}$

42. $\sqrt{\dfrac{400}{729}}$

Challenge Problems

Without the aid of a calculator see if you can evaluate the following. (You may need to make several educated guesses!)

43. $\sqrt{3249}$

44. $\sqrt{4356}$

Solutions to Practice Problems for Exercise 8.1

1. $\sqrt{64} = 8$ since $(8)^2 = (8)(8) = 64$

2. $\sqrt{\dfrac{9}{25}} = \dfrac{3}{5}$ since $\left(\dfrac{3}{5}\right)^2 = \left(\dfrac{3}{5}\right)\left(\dfrac{3}{5}\right) = \dfrac{9}{25}$

3. Here we want the negative square root of 121.

$$-\sqrt{121} = -11 \quad \text{since} \quad (-11)^2 = (-11)(-11) = 121$$

4. By our definition of square root we only take the square root of a nonnegative number. If the radicand is negative, the square root is not a real number. Thus $\sqrt{-81}$ is *not a real number.*

Now do the problems in Exercise 8.1.

8.2 Simplification of Radicals

We can find the square roots of certain expressions in exponent form. Since we know that $\sqrt{25} = 5$, we also know that $\sqrt{5^2} = 5$. This concept can be extended to any positive number raised to an even power.

EXAMPLE A Find: $\sqrt{9^4}$.

Solution We would certainly not want to evaluate 9^4 since that would be rather cumbersome. However, we know that $\sqrt{9^4} = 9^2$ since $(9^2)^2 = (9^2)(9^2) = 9^4$ by the laws of exponents.

EXAMPLE B Find: $\sqrt{18^{12}}$.

Solution $\sqrt{18^{12}} = 18^6$ since we know that $18^6 \cdot 18^6 = 18^{12}$ by the laws of exponents.

This same concept can be used with variable expressions. We will restrict the values of each variable to be nonnegative if the variable is under the radical sign. Thus *all variable radicands in this chapter* are assumed to *represent positive numbers.*

EXAMPLE C Find: $\sqrt{x^6}$.

Solution $\sqrt{x^6} = x^3$ since we know that $x^3 \cdot x^3 = x^6$ by the laws of exponents.

EXAMPLE D Find: $\sqrt{y^{10}}$.

Solution $\sqrt{y^{10}} = y^5$ since we know that $y^5 \cdot y^5 = y^{10}$ by the laws of exponents.

Most of the time when we encounter square roots the radicand is not a perfect square. Thus we cannot remove the radical sign and find the square root. However, in these cases we can often simplify the radical so that the radicand is a smaller value. To perform this simplification, we will utilize the following:

> ***Multiplication Law for Square Roots***
>
> **For all nonnegative numbers a, b,**
>
> $$\sqrt{a} \cdot \sqrt{b} = \sqrt{ab}$$

When we encounter radicals that can be simplified we will use this property to express the radicand as the product of two factors. One of these factors will be the square of a nonnegative real number or variable quantity. In other words, the first radicand will contain a perfect square. (You recall from our discussion in Section 3.4 that the numbers 1, 4, 9, 16, 25, 36, $\cdots$ and the variables $x^2, x^4, x^6, x^8, x^{10}, x^{12}, \cdots$ are called **perfect squares**.)

EXAMPLE E Simplify: $\sqrt{20}$.

Solution $\sqrt{20} = \sqrt{4 \cdot 5}$ since we can write 20 as the product of 4 and 5.

Therefore $\sqrt{20} = \sqrt{4}\sqrt{5}$ by using the multiplication law for square roots. Since we know $\sqrt{4} = 2$, we can write $\sqrt{20} = \sqrt{4}\sqrt{5} = 2\sqrt{5}$.

In general, to simplify a square root that has a numerical radicand, first write it as a product of the form $\sqrt{a}\sqrt{b}$ where a is the largest perfect square factor of the radicand.

Before doing Example F see if you can mentally finish finding the square roots of the following perfect squares:

$$\sqrt{1} = 1, \sqrt{4} = 2, \sqrt{9} = 3, \sqrt{16} = 4, \sqrt{25} = 5, \cdots, \sqrt{100} = 10$$

$$\sqrt{x^2} = x, \sqrt{x^4} = x^2, \sqrt{x^6} = x^3, \sqrt{x^8} = x^4, \sqrt{x^{10}} = x^5, \sqrt{x^{12}} = x^6, \cdots, \sqrt{x^{20}} = x^{10}$$

EXAMPLE F Simplify: $\sqrt{12}$.

Solution

$$\sqrt{12} = \sqrt{4 \cdot 3} = \sqrt{4}\sqrt{3} = 2\sqrt{3}$$

Note that 4 is the largest perfect square that is a factor of 12

EXAMPLE G Simplify:

1. $\sqrt{75}$. **2.** $\sqrt{72}$

Solution **1.**

$$\sqrt{75} = \sqrt{25 \cdot 3} = \sqrt{25}\sqrt{3} = 5\sqrt{3}$$

Note that 25 is the largest perfect square that is a factor of 75.

2.

$$\sqrt{72} = \sqrt{36 \cdot 2} = \sqrt{36} \cdot \sqrt{2} = 6\sqrt{2}$$

Note that 36 is the largest perfect square that is a factor of 72. (Do you see that we do *not* use $4 \cdot 18 = 72$ or $9 \cdot 8 = 72$ since neither 4 nor 9 is the largest perfect square factors of 72?)

If variable expressions representing nonnegative real numbers appear in the radicand, we can employ a similar procedure.

EXAMPLE H Simplify: $\sqrt{x^3}$.

Solution

$$\sqrt{x^3} = \sqrt{x^2}\sqrt{x} = x\sqrt{x}$$

Note that x^2 is the largest perfect square quantity that is a factor of x^3.

EXAMPLE I Simplify: $\sqrt{y^5}$.

Solution

$$\sqrt{y^5} = \sqrt{y^4}\sqrt{y} = y^2\sqrt{y}$$

You can observe the pattern that we always try to have the first factor have the largest possible even exponent. So we *did not write*

$$\sqrt{y^5} = \sqrt{y^2}\sqrt{y^3}$$

Do you see why we did this?

In general, to simplify a square root that has a variable with an odd exponent in the radicand, first write it as a product of the form $\sqrt{x^n}\sqrt{x}$ where n is the largest possible even exponent.

EXAMPLE J Simplify: $\sqrt{w^{17}}$.

Solution

$$\sqrt{w^{17}} = \sqrt{w^{16}}\sqrt{w} = w^8\sqrt{w}$$

We can also use the multiplication law for square roots to simplify square roots that contain nonnegative numbers multiplied by variables.

EXAMPLE K Simplify: $\sqrt{20x^3}$.

Solution We see that this problem is a combination of Examples E and H. We want the first factor to be a square root that can be evaluated exactly. We write

$$\sqrt{20x^3} = \sqrt{4x^2}\sqrt{5x} \qquad = 2x\sqrt{5x}$$

EXAMPLE L Simplify: $\sqrt{12y^5}$.

Solution This problem is a combination of Examples F and I. Remember that we want the first factor to be the largest square root that can be evaluated exactly. In other words the first radicand will be the largest possible perfect square that is a factor of the original radicand.

$$\sqrt{12y^5} = \sqrt{4y^4}\sqrt{3y}$$
$$= 2y^2\sqrt{3y}$$

EXAMPLE M Simplify: $\sqrt{48x^{11}}$.

Solution

$$\sqrt{48x^{11}} = \sqrt{16x^{10}}\sqrt{3x}$$
$$= 4x^5\sqrt{3x}$$

(Remember that the first factor is the largest square root that can be evaluated exactly.)

Finally, we can do problems of this type with two or more variables in the radicand.

EXAMPLE N Simplify: $\sqrt{27x^5y^6}$.

Solution

$$\sqrt{27x^5y^6} = \sqrt{9x^4y^6}\sqrt{3x}$$

Notice that since the exponent of y is even, we place y^6 in the first radical.

$$= 3x^2y^3\sqrt{3x}$$

EXAMPLE O Simplify: $\sqrt{18x^3y^7w^{10}}$.

Solution

$$\sqrt{18x^3y^7w^{10}} = \sqrt{9x^2y^6w^{10}}\sqrt{2xy}$$

Since the exponent of w is even, the entire expression w^{10} is placed is the first radical.

$$= 3xy^3w^5\sqrt{2xy}$$

EXAMPLE P Simplify: $\sqrt{81x^5y^{11}}$.

Solution

$$\sqrt{81x^5y^{11}} = \sqrt{81x^4y^{10}}\sqrt{xy}$$

Note that the complete numerical coefficient 81 is placed in the first radical since 81 is a perfect square.

$$= 9x^2y^5\sqrt{xy}$$

We will be using this method of simplifying radicals throughout Chapter 8. Carefully review Examples A to P and then try these practice problems.

Practice Problems for Exercise 8.2

Simplify. Do not use a calculator or a table of square roots.

1. $\sqrt{98}$

2. $\sqrt{w^{13}}$

3. $\sqrt{50x^5}$

4. $\sqrt{72x^6y^9}$

Now turn to the answers following Exercise 8.2 and check your solutions for these practice problems.

Exercise 8.2

Simplify.

1. $\sqrt{27}$

2. $\sqrt{18}$

3. $\sqrt{32}$

4. $\sqrt{50}$

5. $\sqrt{48}$

6. $\sqrt{72}$

7. $\sqrt{8^4}$

8. $\sqrt{5^{10}}$

9. $\sqrt{56^6}$

10. $\sqrt{39^8}$

11. $\sqrt{w^{12}}$

12. $\sqrt{y^6}$

13. $\sqrt{x^2y^6}$

14. $\sqrt{x^4y^{12}}$

15. $\sqrt{25x^8y^4}$

16. $\sqrt{16x^2y^{20}}$

17. $\sqrt{8x^3}$

18. $\sqrt{12y^5}$

19. $\sqrt{27w^5}$

20. $\sqrt{18y^{13}}$

21. $\sqrt{16x^7}$

22. $\sqrt{49x^5}$

23. $\sqrt{75x^3y^5}$

24. $\sqrt{50x^7y}$

25. $\sqrt{48y^3w}$

26. $\sqrt{12x^2y^3}$

27. $\sqrt{75x^2y^3}$

28. $\sqrt{27a^8y^7}$

29. $\sqrt{45ab^3c^4}$

30. $\sqrt{32a^3bc^4}$

31. $\sqrt{36a^2b^4c^8d^{12}}$

32. $\sqrt{63a^4b^6c^3}$

33. $\sqrt{28a^2b^5c^8}$

34. $\sqrt{81x^{12}y^{11}w^5}$

35. $\sqrt{64x^8y^9w^{11}}$

36. $\sqrt{135x^5y^7}$

37. $\sqrt{140x^7y^{11}}$

Challenge Problem

38. Simplify the following:

$$\sqrt{3645(a+b)^8c^{11}d^{12}e^{17}}$$

Solutions to Practice Problems for Exercise 8.2

1. $\sqrt{98} = \sqrt{49}\sqrt{2}$

$= 7\sqrt{2}$

Notice that we want the first factor to contain the *largest possible value* that represents a *number squared.* Do you see that the number 49 is the only perfect square that is a factor of 98? Therefore, we select $\sqrt{49}\sqrt{2}$ instead of other products such as $\sqrt{14}\sqrt{7}$ or $\sqrt{7}\sqrt{14}$.

2. $\sqrt{w^{13}} = \sqrt{w^{12}}\sqrt{w}$

$= w^6\sqrt{w}$

Note the first radicand is w^{12} (not w^{10} or w^8, etc.). We want the first radicand to have the largest possible *even* exponent.

3. $\sqrt{50x^5} = \sqrt{25x^4}\sqrt{2x}$

$= 5x^2\sqrt{2x}$

4. $\sqrt{72x^6y^9} = \sqrt{36x^6y^8}\sqrt{2y}$

$= 6x^3y^4\sqrt{2y}$

Note that since x^6 has an even exponent, we place x^6 in the first radical.

Now do the problems in Exercise 8.2.

8.3 Addition and Subtraction of Radicals

We have previously seen that we can combine like terms such as $8b + 5b$ to obtain $13b$. Suppose that $b = \sqrt{3}$; then we would have

$$8\sqrt{3} + 5\sqrt{3} = 13\sqrt{3}$$

This is an example of combining *like square roots. Like square roots* are square roots that have exactly the same radicand. Only like square roots can be added or subtracted.

EXAMPLE A Combine: $5\sqrt{2} - 8\sqrt{2}$.

Solution

$$5\sqrt{2} - 8\sqrt{2} = -3\sqrt{2}$$

(Note that this is similar to the problems in Chapter 1, such as $5x - 8x = -3x$.)

EXAMPLE B Combine: $7\sqrt{a} + 3\sqrt{a} - 5\sqrt{a}$.

Solution

$$7\sqrt{a} + 3\sqrt{a} - 5\sqrt{a} = 5\sqrt{a}$$

Here we have the terms $7\sqrt{a}$, $3\sqrt{a}$, and $-5\sqrt{a}$, which all contain like square roots.

? Why does this work? What are we really doing when we combine like square roots?

The addition of like square roots is an application of the distributive property. Look at the following situation.

$$5\sqrt{2a} + 4\sqrt{2a} = (5 + 4)\sqrt{2a} = 9\sqrt{2a}$$

$$ac + bc = (a + b)c$$

Here the a and b quantities are integers and the c quantity is a square root. Specifically $a = 5$, $b = 4$, and $c = \sqrt{2a}$.

Note: Only like square roots can be combined.

EXAMPLE C Combine: $5\sqrt{2a} + 3\sqrt{a} - 7\sqrt{2} + 3\sqrt{2a}$.

Solution The only like square roots are in the terms $5\sqrt{2a}$ and $3\sqrt{2a}$, so these are the only terms that may be combined. Therefore,

$$5\sqrt{2a} + 3\sqrt{a} - 7\sqrt{2} + 3\sqrt{2a} = 8\sqrt{2a} + 3\sqrt{a} - 7\sqrt{2}$$

EXAMPLE D Combine: $3\sqrt{x} + 2\sqrt{xy} + 5\sqrt{y}$.

Solution There are no like square roots. None of these terms can be combined.

Sometimes it is necessary to simplify one or more radicals before the terms can be combined.

EXAMPLE E Combine: $2\sqrt{3} + \sqrt{12}$.

Solution

$$2\sqrt{3} + \sqrt{12} = 2\sqrt{3} + \sqrt{4}\sqrt{3} \qquad \text{We simplify } \sqrt{12}.$$

$$= 2\sqrt{3} + 2\sqrt{3} \qquad \text{Now we collect like terms.}$$

$$= 4\sqrt{3}$$

EXAMPLE F Combine: $\sqrt{50} - \sqrt{18} + \sqrt{98}$.

Solution

$$\sqrt{50} - \sqrt{18} + \sqrt{98} = \sqrt{25}\sqrt{2} - \sqrt{9}\sqrt{2} + \sqrt{49}\sqrt{2}$$

$$= 5\sqrt{2} - 3\sqrt{2} + 7\sqrt{2}$$

$$= 9\sqrt{2}$$

Be alert to add only like square roots after the radicals have been simplified.

EXAMPLE G Combine: $\sqrt{12} - \sqrt{27} + \sqrt{50}$.

Solution

$$\sqrt{12} - \sqrt{27} + \sqrt{50} = \sqrt{4}\sqrt{3} - \sqrt{9}\sqrt{3} + \sqrt{25}\sqrt{2}$$
$$= 2\sqrt{3} - 3\sqrt{3} + 5\sqrt{2}$$
$$= -\sqrt{3} + 5\sqrt{2}$$
$$\text{(or)} \quad 5\sqrt{2} - \sqrt{3} \qquad \text{(either answer)}$$

EXAMPLE H Combine: $\sqrt{2a} + \sqrt{8a} + \sqrt{27a}$.

Solution

$$\sqrt{2a} + \sqrt{8a} + \sqrt{27a} = \sqrt{2a} + \sqrt{4}\sqrt{2a} + \sqrt{9}\sqrt{3a}$$
$$= \sqrt{2a} + 2\sqrt{2a} + 3\sqrt{3a}$$
$$= 3\sqrt{2a} + 3\sqrt{3a}$$

(Note that we *cannot add* these last two terms since they are not like square roots. We could factor out a 3 to obtain an equivalent answer $3(\sqrt{2a} + \sqrt{3a})$. This factoring step is not usually necessary, but it is helpful in a few situations and the procedure should be understood by the student.)

Special care should be taken if the radical has a numerical coefficient. If the radical can be simplified further, the two resulting numerical coefficients should be multiplied.

EXAMPLE I Combine: $3\sqrt{18} - 5\sqrt{2}$.

Solution

$$3\sqrt{18} - 5\sqrt{2} = 3 \cdot \sqrt{9} \cdot \sqrt{2} - 5\sqrt{2}$$
$$= 3 \cdot 3\sqrt{2} - 5\sqrt{2}$$

Remember to multiply $3 \cdot 3$ here.

$$= 9\sqrt{2} - 5\sqrt{2}$$
$$= 4\sqrt{2}$$

EXAMPLE J Combine: $2\sqrt{18} + 3\sqrt{8}$.

Solution

$$2\sqrt{18} + 3\sqrt{8} = 2 \cdot \sqrt{9} \cdot \sqrt{2} + 3 \cdot \sqrt{4} \cdot \sqrt{2}$$
$$= 2 \cdot 3 \cdot \sqrt{2} + 3 \cdot 2 \cdot \sqrt{2}$$
$$= 6\sqrt{2} + 6\sqrt{2}$$
$$= 12\sqrt{2}$$

EXAMPLE K Combine: $2\sqrt{20} + 3\sqrt{45} - 4\sqrt{80}$.

Solution

$$2\sqrt{20} + 3\sqrt{45} - 4\sqrt{80} = 2 \cdot \sqrt{4} \cdot \sqrt{5} + 3 \cdot \sqrt{9} \cdot \sqrt{5} - 4 \cdot \sqrt{16} \cdot \sqrt{5}$$
$$= 2 \cdot 2 \cdot \sqrt{5} + 3 \cdot 3 \cdot \sqrt{5} - 4 \cdot 4 \cdot \sqrt{5}$$
$$= 4\sqrt{5} + 9\sqrt{5} - 16\sqrt{5}$$
$$= -3\sqrt{5}$$

If the radicands contain both a numerical coefficient and a variable, it may take several steps of simplification before it is evident that we have like square roots.

EXAMPLE L Combine: $3a\sqrt{8a} + 2\sqrt{50a^3}$.

Solution

$$3a\sqrt{8a} + 2\sqrt{50a^3}$$ We simplify each radical.

$$= 3a\sqrt{4}\sqrt{2a} + 2\sqrt{25a^2}\sqrt{2a}$$

$$= 3a \cdot 2 \cdot \sqrt{2a} + 2 \cdot 5a \cdot \sqrt{2a}$$

$$= 6a\sqrt{2a} + 10a\sqrt{2a}$$

$$= 16a\sqrt{2a}$$

(*Note*: If you are unsure of the last step, show the use of the distributive property in performing the addition.)

$$6a\sqrt{2a} + 10a\sqrt{2a} = (6 + 10)a\sqrt{2a}$$

$$= 16a\sqrt{2a}$$

Review Examples A to L and then try these practice problems.

Practice Problems for Exercise 8.3

Combine. Do not use a calculator or a table of square roots.

1. $3\sqrt{5} + 5\sqrt{2} - 8\sqrt{5} - 2\sqrt{2}$

2. $\sqrt{27} - 4\sqrt{3} + 2\sqrt{75}$

3. $2\sqrt{12x} - 3\sqrt{45x} - 3\sqrt{27x} + \sqrt{20}$

Now turn to the answers following Exercise 8.3 and check your solutions for these practice problems.

Exercise 8.3

Simplify each term if necessary. Then combine each of the following, if possible. Do not use a calculator or a table of square roots.

1. $\sqrt{3} - 5\sqrt{3} + 2\sqrt{3}$

2. $\sqrt{6} + 4\sqrt{6} - 7\sqrt{6}$

3. $\sqrt{2} + 8\sqrt{3} - 5\sqrt{3} + 4\sqrt{2}$

4. $\sqrt{5} - \sqrt{6} + 3\sqrt{5} - 2\sqrt{6}$

5. $3\sqrt{2x} + 5\sqrt{x} - 7\sqrt{2x}$

6. $\sqrt{ab} - 2\sqrt{b} + 3\sqrt{a}$

7. $\sqrt{2} - \sqrt{12}$

8. $\sqrt{5} - \sqrt{20}$

9. $\sqrt{50} + 3\sqrt{32}$

10. $2\sqrt{12} + \sqrt{48}$

11. $2\sqrt{8} - 3\sqrt{2}$

12. $2\sqrt{27} - 4\sqrt{3}$

13. $\sqrt{75} + \sqrt{3} - 2\sqrt{27}$

14. $\sqrt{28} - 3\sqrt{7} + 2\sqrt{63}$

15. $2\sqrt{12} + \sqrt{20} + \sqrt{36}$

16. $\sqrt{24} - 2\sqrt{54} + 2\sqrt{18}$

17. $\sqrt{20} - \sqrt{80} + 3\sqrt{48}$

18. $\sqrt{50} + \sqrt{28} - \sqrt{9}$

19. $3\sqrt{3x} + \sqrt{12x}$

20. $5\sqrt{2y} - \sqrt{18y}$

21. $3\sqrt{8xy} - \sqrt{50xy}$

22. $2\sqrt{45ab} - \sqrt{20ab}$

23. $1.2\sqrt{3x} - 0.5\sqrt{12x}$

24. $-1.5\sqrt{2a} + 0.2\sqrt{8a}$

25. $\sqrt{20y} + 2\sqrt{45y} - \sqrt{5y}$

26. $\sqrt{72w} - 3\sqrt{2w} - \sqrt{50w}$

27. $3\sqrt{y^3} + 2y\sqrt{y}$

28. $4x\sqrt{x} - \sqrt{16x^3}$

29. $2x\sqrt{8x} + 5\sqrt{18x}$

30. $-3a\sqrt{12a} + \sqrt{75a}$

31. $5\sqrt{8x^3} - 3x\sqrt{50x}$

32. $-2\sqrt{27y^3} + y\sqrt{12y}$

33. $3\sqrt{27x^2} - 2\sqrt{48x^2}$

34. $-5\sqrt{8y^2} + 2\sqrt{32y^2}$

35. $2\sqrt{6y^3} - 2y\sqrt{54}$

36. $3\sqrt{24x^4} - 2\sqrt{54x^4}$

37. $-5\sqrt{72x^6} - 2\sqrt{50x^6}$

38. $3\sqrt{20x^3} - 4x\sqrt{5x^2}$

Challenge Problem

39. Simplify and combine:

$$8\sqrt{99a^3b^3} - 2a\sqrt{44ab^3} + 3b\sqrt{176a^3b} - ab\sqrt{275ab}$$

Solutions to Practice Problems for Exercise 8.3

1. $3\sqrt{5} + 5\sqrt{2} - 8\sqrt{5} - 2\sqrt{2}$

Now since

$$3\sqrt{5} - 8\sqrt{5} = -5\sqrt{5}$$

and also

$$5\sqrt{2} - 2\sqrt{2} = 3\sqrt{2}$$

The final answer is

$$3\sqrt{2} - 5\sqrt{5} \quad (\text{or } -5\sqrt{5} + 3\sqrt{2})$$

2.
$$\begin{aligned}\sqrt{27} - 4\sqrt{3} + 2\sqrt{75} &= \sqrt{9}\cdot\sqrt{3} - 4\cdot\sqrt{3} + 2\sqrt{25}\cdot\sqrt{3}\\ &= 3\sqrt{3} - 4\sqrt{3} + 2\cdot 5\sqrt{3}\\ &= 3\sqrt{3} - 4\sqrt{3} + 10\sqrt{3}\\ &= 9\sqrt{3}\end{aligned}$$

3. $2\sqrt{12x} - 3\sqrt{45x} - 3\sqrt{27x} + \sqrt{20}$

$$= 2\cdot\sqrt{4}\cdot\sqrt{3x} - 3\cdot\sqrt{9}\cdot\sqrt{5x} - 3\cdot\sqrt{9}\cdot\sqrt{3x} + \sqrt{4}\cdot\sqrt{5}$$

$$= 2\cdot 2\cdot\sqrt{3x} - 3\cdot 3\cdot\sqrt{5x} - 3\cdot 3\cdot\sqrt{3x} + 2\cdot\sqrt{5}$$

$$= 4\sqrt{3x} - 9\sqrt{5x} - 9\sqrt{3x} + 2\sqrt{5}$$

Note that these two terms have like square roots.

$\sqrt{5x} \neq \sqrt{5}$
These terms **do not have like** square roots.

$$= -5\sqrt{3x} - 9\sqrt{5x} + 2\sqrt{5}$$

Be sure you see that we cannot add $\sqrt{5x} + \sqrt{5}$!

Now do the problems in Exercise 8.3.

8.4 Multiplication of Radicals

The basic property of multiplication of radicals is $\sqrt{a}\sqrt{b} = \sqrt{ab}$. We have used this concept extensively in order to simplify radicals. In this section we use the concept to multiply radical expressions.

EXAMPLE A Multiply: $\sqrt{5}\sqrt{2}$.

Solution

$$\sqrt{5}\sqrt{2} = \sqrt{10}$$

EXAMPLE B Multiply: $\sqrt{7}\sqrt{14x}$ and then simplify your answer.

Solution

$$\sqrt{7}\sqrt{14x} = \sqrt{98x}$$

We do *not* stop here, because the radical $\sqrt{98x}$ can be simplified.

$$\sqrt{98x} = \sqrt{49\cdot 2x} = \sqrt{49}\sqrt{2x} = 7\sqrt{2x}$$

(Note: We are using the factoring procedures of Chapter 3 to write $\sqrt{98x}$ as $\sqrt{49\cdot 2x}$ so that we can simplify the radical.)

If one or more of the radicals has a numerical coefficient, you should multiply these coefficients separately.

EXAMPLE C Multiply: $(2\sqrt{3})(5\sqrt{7})$.

Solution

$$(2\sqrt{3})(5\sqrt{7}) = 10\sqrt{21}$$

We multiply (2)(5) to obtain 10 as the coefficient for the radical.

EXAMPLE D Multiply: $(x\sqrt{3})(4\sqrt{5})$.

Solution

$$(x\sqrt{3})(4\sqrt{5}) = 4x\sqrt{15}$$

Here we multiply the two coefficients x and 4 to obtain $4x$. The coefficient of $\sqrt{15}$ is $4x$.

Why can't we simplify $\sqrt{15}$? Do you see that the only factors of 15 are $15 \cdot 1$ and $3 \cdot 5$ and that neither of these will lead to a perfect square as a radicand?

EXAMPLE E Multiply: $(2a\sqrt{3})(3a^2\sqrt{6})$.

Solution

$$\begin{aligned} &(2a\sqrt{3})(3a^2\sqrt{6}) \\ &= 6a^3\sqrt{18} && \text{We can simplify the } \sqrt{18} \text{ radical.} \\ &= 6a^3\sqrt{9}\sqrt{2} \\ &= 6a^3(3)\sqrt{2} \\ &= 18a^3\sqrt{2} \end{aligned}$$

EXAMPLE F Multiply: $(\sqrt{3x})(2\sqrt{5x})$.

Solution

$$\begin{aligned} &(\sqrt{3x})(2\sqrt{5x}) \\ &= 2\sqrt{15x^2} && \text{We can simplify } \sqrt{15x^2}. \\ &= 2\sqrt{x^2}\sqrt{15} \\ &= 2(x)\sqrt{15} \\ &= 2x\sqrt{15} \end{aligned}$$

If one expression being multiplied has two or more terms, we will need to use the distributive property. In $a(b + c) = ab + ac$, suppose that $a = \sqrt{3}$, $b = \sqrt{2}$ and $c = \sqrt{5}$. Then we would have

$$\begin{aligned} \sqrt{3}(\sqrt{2} + \sqrt{5}) &= (\sqrt{3})(\sqrt{2}) + (\sqrt{3})(\sqrt{5}) \\ &= \sqrt{6} + \sqrt{15} \end{aligned}$$

Let us examine some examples of this.

EXAMPLE G Multiply: $\sqrt{5}(\sqrt{2} + 3\sqrt{7})$.

Solution

$$\begin{aligned} &\sqrt{5}(\sqrt{2} + 3\sqrt{7}) \\ &= \sqrt{5}\sqrt{2} + 3\sqrt{5}\sqrt{7} \\ &= \sqrt{10} + 3\sqrt{35} \end{aligned}$$

Simplify wherever it is possible after multiplying.

EXAMPLE H Multiply: $\sqrt{5}(\sqrt{10} + 2\sqrt{3})$.

Solution

$$\sqrt{5}(\sqrt{10} + 2\sqrt{3})$$

$= \sqrt{50} + 2\sqrt{15}$ Using the distributive property.

$= \sqrt{25}\sqrt{2} + 2\sqrt{15}$ We simplify $\sqrt{50}$.

$= 5\sqrt{2} + 2\sqrt{15}$

A useful pattern can be observed from multiplication of a radical by itself: $\sqrt{2}\cdot\sqrt{2} = 2$, $\sqrt{3}\cdot\sqrt{3} = 3$, and $\sqrt{11}\cdot\sqrt{11} = 11$. Thus we state the following (which is a specific case of the *multiplication law of radicals*):

> **For any nonnegative real number a,**
>
> $$\sqrt{a}\cdot\sqrt{a} = a$$

EXAMPLE I Multiply: $\sqrt{a}(3\sqrt{a} - 2\sqrt{5})$.

Solution

$$\sqrt{a}(3\sqrt{a} - 2\sqrt{5})$$

$= 3a - 2\sqrt{5a}$ Using the distributive property and the property that $\sqrt{a}\sqrt{a} = a$.

We must be alert to simplify all radicals after multiplication. Be sure that you understand the next example.

EXAMPLE J Multiply: $2\sqrt{2x}(3\sqrt{10x} - 2\sqrt{6})$.

Solution

$$2\sqrt{2x}(3\sqrt{10x} - 2\sqrt{6})$$

$= 6\sqrt{20x^2} - 4\sqrt{12x}$ Using the distributive property.

$= 6\sqrt{4x^2}\sqrt{5} - 4\sqrt{4}\sqrt{3x}$ We simplify $\sqrt{20x^2}$ and $\sqrt{12x}$.

$= 6(2x)\sqrt{5} - 4(2)\sqrt{3x}$ Since $\sqrt{4x^2} = 2x$ and $\sqrt{4} = 2$.

$= 12x\sqrt{5} - 8\sqrt{3x}$

EXAMPLE K Multiply: $\sqrt{2}(\sqrt{6} + 3\sqrt{2} - \sqrt{14})$.

Solution

$$\sqrt{2}(\sqrt{6} + 3\sqrt{2} - \sqrt{14})$$

$= \sqrt{12} + 3\sqrt{4} - \sqrt{28}$ Observe that all three radicals can be simplified.

$= \sqrt{4}\sqrt{3} + 3(2) - \sqrt{4}\sqrt{7}$

$= 2\sqrt{3} + 6 - 2\sqrt{7}$

The two expressions being multiplied may each contain two or more terms. In such cases you will find it helpful to realize that the multiplication of two binomials in Chapter 3 is very similar to this operation. Compare the following two problems.

$$\begin{aligned}(2x + y)(x - 2y) &= 2x^2 - 4xy + xy - 2y^2 \\ &= 2x^2 - 3xy - 2y^2 \\ (2\sqrt{3} + \sqrt{5})(\sqrt{3} - 2\sqrt{5}) &= 2\sqrt{9} - 4\sqrt{15} + \sqrt{15} - 2\sqrt{25} \\ &= 2(3) - 3\sqrt{15} - 2(5) \\ &= 6 - 3\sqrt{15} - 10 \\ &= -4 - 3\sqrt{15}\end{aligned}$$

Although the problem involving radicals is a bit longer, the first two steps are very similar. In fact, the two problems are really the same if you let $x = \sqrt{3}$ and $y = \sqrt{5}$ in the first problem. (This is another example of using the method of substitution we discussed in Section 1.6).

Now study the following example.

EXAMPLE L Multiply: $(\sqrt{2} + 5)(\sqrt{2} - 3)$.

Solution $(\sqrt{2} + 5)(\sqrt{2} - 3)$

$= \sqrt{4} - 3\sqrt{2} + 5\sqrt{2} - 15$ We multiply the four products.

$= 2 + 2\sqrt{2} - 15$ We simplify $\sqrt{4}$ and collect like terms.

$= -13 + 2\sqrt{2}$ We combine $2 - 15$ to obtain -13.

EXAMPLE M Multiply: $(\sqrt{6} + \sqrt{5})(\sqrt{2} + 2\sqrt{5})$.

Solution $(\sqrt{6} + \sqrt{5})(\sqrt{2} + 2\sqrt{5})$

$= \sqrt{12} + 2\sqrt{30} + \sqrt{10} + 2\sqrt{25}$ We need to simplify the first and last terms.

$= \sqrt{4}\sqrt{3} + 2\sqrt{30} + \sqrt{10} + 2(5)$

$= 2\sqrt{3} + 2\sqrt{30} + \sqrt{10} + 10$ Notice that we cannot add together any of the four terms. None of the radicals are like square roots.

EXAMPLE N Multiply: $(\sqrt{2} - 3\sqrt{6})(\sqrt{2} + \sqrt{6})$.

Solution $(\sqrt{2} - 3\sqrt{6})(\sqrt{2} + \sqrt{6})$

$= \sqrt{4} + \sqrt{12} - 3\sqrt{12} - 3\sqrt{36}$ We multiply the four products.

$= 2 - 2\sqrt{12} - 3(6)$ We collect like terms and simplify $\sqrt{4} = 2$ and $\sqrt{36} = 6$.

$= 2 - 2\sqrt{12} - 18$ Multiply $-3(6) = -18$.

$= -16 - 2\sqrt{12}$ Collect like terms.

$= -16 - 2\sqrt{4}\sqrt{3}$ We write $\sqrt{12} = \sqrt{4}\sqrt{3}$.

$= -16 - 2(2)\sqrt{3}$ Use the fact that $\sqrt{4} = 2$.

$= -16 - 4\sqrt{3}$ Multiply the coefficients of $\sqrt{3}$.

It is not necessary to show every step as was done in Example N. It is more likely that you will combine some steps as you do these problems. We will now do a similar problem with fewer steps shown.

EXAMPLE O Multiply: $(\sqrt{5} + \sqrt{3})(\sqrt{5} + 2\sqrt{3})$.

Solution

$$
\begin{aligned}
&(\sqrt{5} + \sqrt{3})(\sqrt{5} + 2\sqrt{3}) \\
&= \sqrt{25} + 2\sqrt{15} + \sqrt{15} + 2\sqrt{9} \\
&= 5 + 3\sqrt{15} + 6 \\
&= 11 + 3\sqrt{15}
\end{aligned}
$$

This same procedure can be used if the expression is squared.

EXAMPLE P Multiply: $(2 + 3\sqrt{5})^2$.

Solution

$$
\begin{aligned}
(2 + 3\sqrt{5})^2 &= (2 + 3\sqrt{5})(2 + 3\sqrt{5}) \\
&= 4 + 6\sqrt{5} + 6\sqrt{5} + 9\sqrt{25} \\
&= 4 + 12\sqrt{5} + 45 \\
&= 49 + 12\sqrt{5}
\end{aligned}
$$

EXAMPLE Q Multiply: $(2\sqrt{3} - \sqrt{6})^2$.

Solution

$$
\begin{aligned}
(2\sqrt{3} - \sqrt{6})^2 &= (2\sqrt{3} - \sqrt{6})(2\sqrt{3} - \sqrt{6}) \\
&= 4\sqrt{9} - 2\sqrt{18} - 2\sqrt{18} + \sqrt{36} \\
&= 12 - 4\sqrt{18} + 6 \\
&= 18 - 4(3)\sqrt{2} \\
&= 18 - 12\sqrt{2}
\end{aligned}
$$

Do you see a shorter way that you could do Examples P and Q? You can use Formula B for squaring a binomial from Chapter 3.

$$
\begin{aligned}
(a + b)^2 &= a^2 + 2ab + b^2 \\
(a - b)^2 &= a^2 - 2ab + b^2
\end{aligned}
$$

Some students find this to be an effective short cut. Take a minute and study the following example.

EXAMPLE P (Alternate Method) Multiply: $(2 + 3\sqrt{5})^2$ by using Formula B for squaring a binomial.

Solution

$$
\begin{aligned}
(a + b)^2 &= a^2 + 2ab + b^2 \\
(2 + 3\sqrt{5})^2 &= (2)^2 + 2(2)(3\sqrt{5}) + (3\sqrt{5})^2 \\
&= 4 + 12\sqrt{5} + 45 \\
&= 49 + 12\sqrt{5}
\end{aligned}
$$

See if you can use this approach to do example Q by the alternate method. Carefully review Examples A to Q and then try these practice problems.

Practice Problems for Exercise 8.4

Multiply. Be sure to simplify any radicals in your answer. Do not use a calculator or a table of square roots.

1. $(2\sqrt{3x})(5\sqrt{6x})$

2. $2\sqrt{3}(\sqrt{3} + \sqrt{5} - \sqrt{12})$

3. $(2 - \sqrt{5})^2$

4. $(\sqrt{2} + \sqrt{6})(2\sqrt{2} - \sqrt{6})$

Now turn to the answers following Exercise 8.4 and check your solutions for these practice problems.

Exercise 8.4

Multiply. Be sure to simplify any radicals in your answer. Do not use a calculator or a table of square roots.

1. $\sqrt{7}\sqrt{5}$

2. $\sqrt{3}\sqrt{11}$

3. $\sqrt{2}\sqrt{22}$

4. $\sqrt{5}\sqrt{15}$

5. $\sqrt{3a}\sqrt{5a}$

6. $\sqrt{6x}\sqrt{3}$

7. $\sqrt{5x}\sqrt{10}$

8. $\sqrt{12b}\sqrt{2b}$

9. $(3\sqrt{5})(2\sqrt{6})$

10. $(-3a\sqrt{2})(5a\sqrt{3a})$

11. $(2x\sqrt{x})(3x\sqrt{5x})$

12. $(4\sqrt{12})(2\sqrt{6})$

13. $(-3\sqrt{ab})(2\sqrt{b})$

14. $(2\sqrt{x})(5\sqrt{xy})$

15. $\sqrt{3}(\sqrt{2} + 2\sqrt{5})$

16. $\sqrt{2}(\sqrt{5} - 2\sqrt{2})$

17. $\sqrt{2}(\sqrt{10} - 3\sqrt{6})$

18. $\sqrt{5}(\sqrt{15} + 3\sqrt{5})$

19. $2\sqrt{x}(\sqrt{x} - 8\sqrt{5})$

20. $-3\sqrt{b}(2\sqrt{a} + 3\sqrt{b})$

21. $\sqrt{6}(\sqrt{2} - 3\sqrt{6} + 2\sqrt{10})$

22. $\sqrt{10}(\sqrt{5} - 3\sqrt{10} + 5\sqrt{2})$

23. $2\sqrt{a}(3\sqrt{b} + \sqrt{ab} - 2\sqrt{a})$

24. $3\sqrt{x}(\sqrt{y} + 2\sqrt{xy} - 4\sqrt{x})$

25. $(\sqrt{7} + \sqrt{2})(2\sqrt{7} + \sqrt{2})$

26. $(3\sqrt{5} + \sqrt{3})(\sqrt{5} + \sqrt{3})$

27. $(8 + \sqrt{3})(2 + 2\sqrt{3})$

28. $(5 + \sqrt{2})(6 + \sqrt{2})$

29. $(2\sqrt{7} - 3\sqrt{3})(\sqrt{7} + \sqrt{3})$

30. $(5\sqrt{6} - \sqrt{2})(\sqrt{6} + 3\sqrt{2})$

31. $(\sqrt{3} + 2\sqrt{6})(2\sqrt{3} - \sqrt{6})$

32. $(\sqrt{2} + 3\sqrt{10})(2\sqrt{2} - \sqrt{10})$

33. $(3\sqrt{7} - \sqrt{8})(\sqrt{8} + 2\sqrt{7})$

34. $(\sqrt{12} - \sqrt{5})(\sqrt{5} + 2\sqrt{12})$

35. $(2\sqrt{5} - 3)^2$

36. $(3\sqrt{2} + 5)^2$

37. $(\sqrt{3} + 5\sqrt{2})^2$

38. $(\sqrt{5} - 2\sqrt{6})^2$

39. $(\sqrt{6} - 2\sqrt{3})^2$

40. $(2\sqrt{3} + 5\sqrt{5})^2$

41. $(\sqrt{a} - \sqrt{ab})^2$

42. $(\sqrt{xy} - 2\sqrt{x})^2$

43. $(2a\sqrt{3} - \sqrt{a})(\sqrt{3} + \sqrt{a})$

44. $(5x\sqrt{y} + \sqrt{2})(\sqrt{y} - 2\sqrt{2})$

45. $(5x\sqrt{x} - 2\sqrt{5x})(\sqrt{5} - \sqrt{x})$

46. $(4a\sqrt{a} - 2\sqrt{3a})(\sqrt{a} - 6\sqrt{3})$

Challenge Problems

Multiply. Simplify your answer.

47. $(3\sqrt{x} + 2\sqrt{y})(\sqrt{x} - 2\sqrt{y})(\sqrt{x} + 4\sqrt{y})$

48. $(5\sqrt{2} + \sqrt{6})(3\sqrt{2} - \sqrt{3})(\sqrt{6} + 3\sqrt{2})$

Solutions to Practice Problems for Exercise 8.4

1.
$$\begin{aligned}(2\sqrt{3x})(5\sqrt{6x}) &= 10\sqrt{18x^2}\\ &= 10\sqrt{9x^2}\sqrt{2}\\ &= 10(3x)\sqrt{2}\\ &= 30x\sqrt{2}\end{aligned}$$

2. $2\sqrt{3}(\sqrt{3} + \sqrt{5} - \sqrt{12}) = 2\sqrt{9} + 2\sqrt{15} - 2\sqrt{36}$
$= 2(3) + 2\sqrt{15} - 2(6)$
$= 6 + 2\sqrt{15} - 12$
$= -6 + 2\sqrt{15}$

3. $(2 - \sqrt{5})^2 = (2 - \sqrt{5})(2 - \sqrt{5})$
$= 4 - 2\sqrt{5} - 2\sqrt{5} + \sqrt{25}$
$= 4 - 4\sqrt{5} + 5$
$= 9 - 4\sqrt{5}$

4. $(\sqrt{2} + \sqrt{6})(2\sqrt{2} - \sqrt{6}) = 2\sqrt{4} - \sqrt{12} + 2\sqrt{12} - \sqrt{36}$
$= 2(2) + \sqrt{12} - 6$
$= 4 + \sqrt{12} - 6$
$= -2 + \sqrt{12}$ Don't forget to simplify $\sqrt{12}$!
$= -2 + 2\sqrt{3}$

Now do the problems in Exercise 8.4.

8.5 Division of Radicals

Sometimes a square root is divided by a square root. In examining this situation it is helpful to use a property known as the *quotient rule for square roots*:

> ***Quotient Rule for Square Roots***
>
> **For all positive numbers *a* and *b*,**
>
> $$\frac{\sqrt{a}}{\sqrt{b}} = \sqrt{\frac{a}{b}}$$

There are some cases where using this rule will enable you to simplify the quotient of two square roots very easily.

EXAMPLE A Simplify:

$$\frac{\sqrt{75}}{\sqrt{3}}$$

Solution

$$\frac{\sqrt{75}}{\sqrt{3}} = \sqrt{\frac{75}{3}} = \sqrt{25} = 5$$

EXAMPLE B Simplify:

$$\frac{\sqrt{x^3}}{\sqrt{x}}$$

Solution

$$\frac{\sqrt{x^3}}{\sqrt{x}} = \sqrt{\frac{x^3}{x}} = \sqrt{x^2} = x$$

This rule gives us an alternative way to find the square root of fractions. We merely use the rule in the reverse direction.

EXAMPLE C Simplify:

$$\sqrt{\frac{25}{36}}$$

Solution

$$\sqrt{\frac{25}{36}} = \frac{\sqrt{25}}{\sqrt{36}} = \frac{5}{6}$$

EXAMPLE D Simplify:

$$\sqrt{\frac{49}{x^2}}$$

Solution

$$\sqrt{\frac{49}{x^2}} = \frac{\sqrt{49}}{\sqrt{x^2}} = \frac{7}{x}$$

Sometimes, using the rule will help to simplify a radical but not all the square roots can be evaluated exactly.

EXAMPLE E Simplify:

$$\sqrt{\frac{20}{x^6}}$$

Solution

$$\sqrt{\frac{20}{x^6}} = \frac{\sqrt{20}}{\sqrt{x^6}} = \frac{\sqrt{4}\sqrt{5}}{x^3} = \frac{2\sqrt{5}}{x^3}$$

Don't forget to simplify $\sqrt{20}$ to $2\sqrt{5}$.

Frequently in mathematics we desire to remove any square roots from the denominator. The process of changing a fraction to an equivalent fraction that does not have a square root in the denominator is called *rationalizing* the denominator. (In more advanced courses, sometimes we need to rationalize the numerator. We will not discuss that procedure in this text.) In some cases the denominator can be simplified by multiplying numerator and denominator by the radical in the denominator.

EXAMPLE F Rationalize the denominator:

$$\frac{3}{\sqrt{2}}$$

Solution

$$\frac{3}{\sqrt{2}} \times 1 = \frac{3}{\sqrt{2}} \times \frac{\sqrt{2}}{\sqrt{2}} = \frac{3\sqrt{2}}{\sqrt{4}} = \frac{3\sqrt{2}}{2}$$

Do you see that these fractions are equivalent since we are multiplying by 1?

EXAMPLE G Rationalize the denominator:

$$\frac{a}{\sqrt{5}}$$

Solution

$$\frac{a}{\sqrt{5}} \times \frac{\sqrt{5}}{\sqrt{5}} = \frac{a\sqrt{5}}{5}$$

We want to multiply the numerator and denominator by the smallest possible radical that will yield a square root in the denominator that can be evaluated exactly. Very often we do *not* multiply the denominator by the same square root that is in the denominator because multiplying by a smaller square root is sufficient to achieve the desired result.

EXAMPLE H Rationalize the denominator:

$$\frac{\sqrt{7}}{\sqrt{8}}$$

Solution We want to multiply numerator and denominator by a square root so that the square root in the denominator can be evaluated exactly. What is the smallest radical that will achieve this goal? Do you see that it is $\sqrt{2}$? (If you did not see it, remember that $\sqrt{8} = 2\sqrt{2}$. It is then more obvious that you should multiply by $\sqrt{2}$.)

$$\frac{\sqrt{7}}{\sqrt{8}} = \frac{\sqrt{7}}{\sqrt{8}} \times \frac{\sqrt{2}}{\sqrt{2}} = \frac{\sqrt{14}}{\sqrt{16}} = \frac{\sqrt{14}}{4}$$

(*Note*: We actually could have multiplied by $\sqrt{8}$, and obtained an equivalent answer. This would have involved larger radicands, however, and extra steps to simplify the resulting expression.)

EXAMPLE I Rationalize the denominator:

$$\frac{3}{\sqrt{x^3}}$$

Solution We would like the denominator to be $\sqrt{x^4}$. Then we could find the square root exactly. Do you see that we should multiply numerator and denominator by $\sqrt{x}$?

$$\frac{3}{\sqrt{x^3}} \times \frac{\sqrt{x}}{\sqrt{x}} = \frac{3\sqrt{x}}{\sqrt{x^4}} = \frac{3\sqrt{x}}{x^2}$$

It may be necessary to multiply by a radical that contains a variable as well as a numerical value.

EXAMPLE J Rationalize the denominator:

$$\frac{\sqrt{2}}{\sqrt{27x}}$$

Solution We would like the denominator to be $\sqrt{81x^2}$ so that we could find the square root exactly. Therefore, we will multiply numerator and denominator by $\sqrt{3x}$.

$$\frac{\sqrt{2}}{\sqrt{27x}} \times \frac{\sqrt{3x}}{\sqrt{3x}} = \frac{\sqrt{6x}}{\sqrt{81x^2}} = \frac{\sqrt{6x}}{9x}$$

A fraction inside a square root sign is sometimes difficult to work with. If you cannot evaluate the square root exactly, it is best to change it to the quotient of two radicals and rationalize.

EXAMPLE K Simplify:

$$\sqrt{\frac{2}{5}}$$

Solution $$\sqrt{\frac{2}{5}} = \frac{\sqrt{2}}{\sqrt{5}} \cdot \frac{\sqrt{5}}{\sqrt{5}} = \frac{\sqrt{10}}{\sqrt{25}} = \frac{\sqrt{10}}{5}$$

EXAMPLE L Simplify:

$$\sqrt{\frac{12}{x}}$$

Solution $$\sqrt{\frac{12}{x}} = \frac{\sqrt{12}}{\sqrt{x}} = \frac{2\sqrt{3}}{\sqrt{x}} \cdot \frac{\sqrt{x}}{\sqrt{x}} = \frac{2\sqrt{3x}}{\sqrt{x^2}} = \frac{2\sqrt{3x}}{x}$$

There is one more type of problem that involves slightly more work. The denominator of the fraction may be a binomial. One or both terms of the binomial may contain a radical. In such cases we multiply the fraction by the conjugate of the denominator. The *conjugate* of a binomial that contains square roots is the same binomial except that the second term contains the opposite sign. The conjugate of $\sqrt{3} + 2\sqrt{2}$ is $\sqrt{3} - 2\sqrt{2}$. The conjugate of $5 - 6\sqrt{3}$ is $5 + 6\sqrt{3}$.

EXAMPLE M Rationalize the denominator:

$$\frac{2}{\sqrt{3} - 4}$$

Solution The conjugate of $\sqrt{3} - 4$ is $\sqrt{3} + 4$.

$$\frac{2}{(\sqrt{3} - 4)} \cdot \frac{(\sqrt{3} + 4)}{(\sqrt{3} + 4)} = \frac{2\sqrt{3} + 8}{\sqrt{9} + 4\sqrt{3} - 4\sqrt{3} - 16}$$

$$= \frac{2\sqrt{3} + 8}{3 - 16}$$

$$= \frac{2\sqrt{3} + 8}{-13} \quad \text{or} \quad -\frac{2\sqrt{3} + 8}{13}$$

EXAMPLE N Rationalize the denominator:

$$\frac{\sqrt{x}}{\sqrt{5} + \sqrt{3}}$$

Solution The conjugate of $\sqrt{5} + \sqrt{3}$ is $\sqrt{5} - \sqrt{3}$.

$$\frac{\sqrt{x}}{(\sqrt{5} + \sqrt{3})} \cdot \frac{(\sqrt{5} - \sqrt{3})}{(\sqrt{5} - \sqrt{3})} = \frac{\sqrt{5x} - \sqrt{3x}}{\sqrt{25} - \sqrt{15} + \sqrt{15} - \sqrt{9}}$$

$$= \frac{\sqrt{5x} - \sqrt{3x}}{5 - 3} = \frac{\sqrt{5x} - \sqrt{3x}}{2}$$

Be careful not to combine $\sqrt{5x}$ and $\sqrt{3x}$ in the numerator. They are not like square roots.

Why does this work? Why does the denominator never have a radical remaining after we have multiplied by the conjugate?

This method is based on Formula A we developed in Chapter 3. That formula is $(a + b)(a - b) = a^2 - b^2$. In such cases there is no middle term. If the two binomials contain radicals, we see that when we multiply $(\sqrt{a}+\sqrt{b})(\sqrt{a}-\sqrt{b})$ we obtain $(\sqrt{a})^2-(\sqrt{b})^2=a-b$. There are no terms containing $\sqrt{a}$ or $\sqrt{b}$. Since this is true, it is not necessary to multiply out all four terms in the denominator. We can just write the step that shows that

$$(\sqrt{a}+\sqrt{b})(\sqrt{a}-\sqrt{b})=(\sqrt{a})^2-(\sqrt{b})^2=a-b$$

EXAMPLE O Rationalize the denominator:

$$\frac{\sqrt{3}+\sqrt{2}}{\sqrt{3}-\sqrt{2}}$$

Solution The conjugate of $\sqrt{3}-\sqrt{2}$ is $\sqrt{3}+\sqrt{2}$.

$$\frac{(\sqrt{3}+\sqrt{2})}{(\sqrt{3}-\sqrt{2})}\cdot\frac{(\sqrt{3}+\sqrt{2})}{(\sqrt{3}+\sqrt{2})}$$

$$=\frac{\sqrt{9}+\sqrt{6}+\sqrt{6}+\sqrt{4}}{(\sqrt{3})^2-(\sqrt{2})^2} \longleftarrow$$ Notice that we are multiplying two binomials in the numerator.

$$=\frac{3+2\sqrt{6}+2}{3-2}$$ We simplify and collect like terms.

$$=\frac{5+2\sqrt{6}}{1}=5+2\sqrt{6}$$

Review Examples A to O, making sure that you see the various steps that can be used to rationalize the denominator. Then try these practice problems.

Practice Problems for Exercise 8.5

Simplify. Be sure to rationalize all denominators. Do not use a calculator or a table of square roots.

1. $\dfrac{9}{\sqrt{7}}$ **2.** $\sqrt{\dfrac{5}{6}}$

3. $\dfrac{2x}{\sqrt{8x}}$ **4.** $\dfrac{4}{\sqrt{3}+\sqrt{5}}$

Now turn to the answers following Exercise 8.5 and check your solutions for these practice problems.

Exercise 8.5

Simplify. Be sure to rationalize all denominators. Do not use a calculator or a table of square roots.

1. $\dfrac{\sqrt{32}}{\sqrt{2}}$ **2.** $\dfrac{\sqrt{7}}{\sqrt{28}}$ **3.** $\dfrac{\sqrt{5}}{\sqrt{20}}$

4. $\frac{\sqrt{26}}{\sqrt{6}}$

5. $\frac{\sqrt{27}}{\sqrt{3}}$

6. $\frac{\sqrt{15}}{\sqrt{5}}$

7. $\frac{\sqrt{6}}{\sqrt{x^4}}$

8. $\frac{\sqrt{12}}{\sqrt{x^2}}$

9. $\frac{5}{\sqrt{6}}$

10. $\frac{3}{\sqrt{5}}$

11. $\frac{x}{\sqrt{3}}$

12. $\frac{\sqrt{y}}{\sqrt{6}}$

13. $\frac{\sqrt{8}}{\sqrt{x}}$

14. $\frac{\sqrt{18}}{\sqrt{y}}$

15. $\frac{3}{\sqrt{12}}$

16. $\frac{4}{\sqrt{20}}$

17. $\frac{7}{\sqrt{a^3}}$

18. $\frac{5}{\sqrt{b^5}}$

19. $\frac{x}{\sqrt{2x^5}}$

20. $\frac{y}{\sqrt{5x^3}}$

21. $\frac{\sqrt{3}}{\sqrt{12x^7}}$

22. $\frac{\sqrt{5}}{\sqrt{18x^5}}$

23. $\sqrt{\frac{2}{7}}$

24. $\sqrt{\frac{3}{10}}$

25. $\sqrt{\frac{3}{ab^2}}$

26. $\sqrt{\frac{5}{2a^2b}}$

27. $\frac{9}{\sqrt{32x}}$

28. $\dfrac{3}{\sqrt{50x}}$

29. $\dfrac{3}{\sqrt{2} - 1}$

30. $\dfrac{2}{\sqrt{3} + 1}$

31. $\dfrac{3}{\sqrt{2} + \sqrt{5}}$

32. $\dfrac{4}{\sqrt{7} - \sqrt{2}}$

33. $\dfrac{x}{\sqrt{7} + 2\sqrt{2}}$

34. $\dfrac{x}{\sqrt{6} + 3\sqrt{5}}$

35. $\dfrac{x}{\sqrt{3} - 2x}$

36. $\dfrac{y}{\sqrt{5} - 2y}$

37. $\dfrac{\sqrt{7}}{\sqrt{8} + \sqrt{7}}$

38. $\dfrac{\sqrt{5}}{\sqrt{5} + \sqrt{6}}$

39. $\dfrac{\sqrt{5} - \sqrt{2}}{\sqrt{5} + \sqrt{2}}$

40. $\dfrac{\sqrt{7} + \sqrt{3}}{\sqrt{7} - \sqrt{3}}$

41. $\dfrac{3\sqrt{5} + 6}{\sqrt{7} - \sqrt{8}}$

42. $\dfrac{\sqrt{6} + 2\sqrt{2}}{\sqrt{5} - 4\sqrt{7}}$

43. $\dfrac{x - 4}{\sqrt{x} + 2}$

44. $\dfrac{x - 9}{\sqrt{x} - 3}$

Challenge Problems

Rationalize the following and simplify your answer.

45. $\dfrac{3h}{\sqrt{x + h} + \sqrt{x}}$

46. $\dfrac{5\sqrt{4x} + 2\sqrt{12xy}}{3\sqrt{9x} - 2\sqrt{18y}}$

Solutions to Practice Problems for Exercise 8.5

1. $\dfrac{9}{\sqrt{7}} \cdot \dfrac{\sqrt{7}}{\sqrt{7}} = \dfrac{9\sqrt{7}}{\sqrt{49}} = \dfrac{9\sqrt{7}}{7}$

2. $\sqrt{\dfrac{5}{6}} = \dfrac{\sqrt{5}}{\sqrt{6}} \cdot \dfrac{\sqrt{6}}{\sqrt{6}} = \dfrac{\sqrt{30}}{\sqrt{36}} = \dfrac{\sqrt{30}}{6}$

3. $\dfrac{2x}{\sqrt{8x}} \cdot \dfrac{\sqrt{2x}}{\sqrt{2x}} = \dfrac{2x\sqrt{2x}}{\sqrt{16x^2}} = \dfrac{2x\sqrt{2x}}{4x} = \dfrac{\sqrt{2x}}{2}$

Notice here that we did *not* multiply both numerator and denominator by $\sqrt{8x}$. We chose the smaller radicand when we multiplied by $\sqrt{2x}$. Do not forget to reduce the fraction in the last two steps.

4. $\dfrac{4}{(\sqrt{3}+\sqrt{5})} \cdot \dfrac{(\sqrt{3}-\sqrt{5})}{(\sqrt{3}-\sqrt{5})} = \dfrac{4\sqrt{3}-4\sqrt{5}}{(\sqrt{3})^2-(\sqrt{5})^2} = \dfrac{4\sqrt{3}-4\sqrt{5}}{3-5}$

$= \dfrac{4(\sqrt{3}-\sqrt{5})}{-2}$ Observe that we can remove a common factor in the numerator. Then it is possible to cancel and reduce the fraction. We can leave the answer in any of several forms.

$= \dfrac{2(\sqrt{3}-\sqrt{5})}{-1}$ or $-2(\sqrt{3}-\sqrt{5})$ or $-2\sqrt{3}+2\sqrt{5}$

Now do the problems in Exercise 8.5.

8.6 Irrational Numbers and Decimal Approximation

At this point in the chapter, we need to pause to review and clarify the concepts of rational and irrational numbers. It is important that you see how these two types of numbers relate to the subject of radicals.

A *rational number* is a number that can be expressed in the form a/b, where a and b are integers and $b \neq 0$. Some square roots can be evaluated exactly and we obtain a rational number. We found that $\sqrt{4} = 2$ is rational number.

An *irrational number* is a real number than *cannot* be expressed in the form a/b, where a and b are integers and $b \neq 0$. We could show that $\sqrt{3}$, $\sqrt{5}$, and $\sqrt{12}$ are irrational numbers, since they cannot be expressed as the quotient of the two integers. A formal proof of this fact is usually covered in a more advanced math course. We will not prove the statement here.

A **real number** is a number that is rational or irrational. The set of real numbers consists of all the rational numbers and all the irrational numbers. Obviously, we often need to know "how large" certain irrational numbers are. For example, is a rope $\sqrt{2}$ meters long going to be longer or shorter than a rope exactly 1.410 meters long? To answer such questions we will use the fact that every irrational number can be *approximated* by a rational number in decimal form. We will also use the properties of inequalities developed in Section 2.4.

The irrational number $\sqrt{2}$ can be approximated by the decimal 1.414213562.... More commonly we would find it useful to have an answer involving fewer decimals. We might prefer the following (depending on how accurate we wanted our answer):

$\sqrt{2} = 1.41421$ (an approximation accurate to five decimal places)

$\sqrt{2} = 1.414$ (an approximation accurate to three decimal places)

$\sqrt{2} = 1.4$ (an approximation accurate to one decimal place)

To answer the question "Is a rope $\sqrt{2}$ meters long going to be longer or shorter than a rope exactly 1.410 meters long?" we would use the approximation 1.414. Since $1.414 > 1.410$ we would say that the rope $\sqrt{2}$ meters in length would be longer than a rope 1.410 meters in length. We could also state that $1.410 < \sqrt{2} < 1.415$. A rope $\sqrt{2}$ meters long is more than 1.410 meters in length but less than the 1.415 meters.

Decimal approximations for square roots can be found from computers, from a calculator with a $\sqrt{\ }$ key, and from square root tables. Appendix A is a table of square root values that represent approximations accurate to three decimal places.

EXAMPLE A Using the table in Appendix A, evaluate $\sqrt{3}$.

Solution $\sqrt{3} = 1.732$ accurate to three decimal places.

To avoid making the statement "the square root of 3 is equal to 1.732," mathematicians have developed a notation to stress than 1.732 is only an approximation of $\sqrt{3}$. The symbol $\doteq$ means "is approximately equal to." Thus when we want to say "the square root of 3 is approximately equal to 1.732" we write

$$\sqrt{3} \doteq 1.732$$

(Note: Not all mathematics books use this notation for an approximation. Sometimes the symbol $\approx$ is used for "is approximately equal to.")

EXAMPLE B Using the table in Appendix A, evaluate $5 + \sqrt{5}$.

Solution From the table we see that $\sqrt{5} \doteq 2.236$.

Therefore we have $5 + \sqrt{5} \doteq 5.000 + 2.236 = 7.236$

Thus $5 + \sqrt{5} \doteq 7.236$

EXAMPLE C Using the table in Appendix A, evaluate $3\sqrt{87}$.

Solution

$$3\sqrt{87} \doteq 3(9.327) = 27.981$$

Therefore $3\sqrt{87} \doteq 27.981$

Our work in these three examples helps us to see one advantage in rationalizing denominators that contain radicals. Consider the following example.

EXAMPLE D Using the table in Appendix A, evaluate:

$$\frac{\sqrt{3}}{\sqrt{7}}$$

Solution We will not do the problem directly by looking up each table value and dividing. It would take us longer to divide 1.732 by 2.646, so we *first* rationalize the denominator.

$$\frac{\sqrt{3}}{\sqrt{7}} \times \frac{\sqrt{7}}{\sqrt{7}} = \frac{\sqrt{21}}{7}$$

Now we use the table approximation for $\sqrt{21}$ and divide by 7.

$$\frac{\sqrt{21}}{7} \doteq \frac{4.583}{7} \doteq 0.655 \quad \text{(to three significant digits)}$$

Example D gives evidence for the wisdom of rationalizing *before* using decimal approximations for square roots. The resulting division is shorter and less prone to student error.

If you need to find the square root of a number that is too large for a square root table, try to simplify the radical. In many cases, the remaining radicand will be small enough to be found in the table.

EXAMPLE E Evaluate: $\sqrt{925}$.

Solution

$$\sqrt{925} = \sqrt{25}\sqrt{37} = 5\sqrt{37}$$

$$5\sqrt{37} \doteq 5(6.083) = 30.415$$

How accurate are these approximations?

They are close, but there is some error. As a calculation with approximate numbers becomes more involved, the amount of error in the final answer usually increases.

Our answers to Examples A, B, and D are correct to the nearest thousandth (three decimal places). The answers to Examples C and E are in error by one digit in the thousandth place. The correct answer to Example C to the nearest thousandth is 27.982. We obtained 27.981 using the table. The correct answer to Example E to the nearest thousandth is 30.414 and we obtained 30.415 using the table.

If you have a calculator, you will obtain more accurate answers than if you use a table. However, even the calculator values for square roots are an approximation (usually accurate to eight decimal places in a scientific calculator.)

Other irrational numbers are often encountered in mathematics. The most common is *pi*, which we represent by the symbol π. An approximation for π can be obtained by dividing the circumference of a circle by the diameter of the same circle. (See Example D of Section 2.3). An approximation for π accurate to eight decimal places is $\pi \doteq 3.14159265$. Modern computers have calculated *pi* to over a hundred decimal places.

A second irrational number frequently encountered is the number e. This number is used extensively in calculus and physics and is encountered in the sciences as we study the growth and decay of organic and radioactive objects. An approximation for e is $e \doteq 2.71828183$ (accurate to eight decimal places).

Higher-order radicals

Not all radicals are square roots, although these are the most common. We also have cube roots, fourth roots, etc. For example, the cube root of a number is a number we cube to get the original number. The cube root of 27 is 3 since $3^3 = 27$. We use this notation

$$\sqrt[3]{27} = 3$$

Some other common cube roots are

$$\sqrt[3]{1} = 1 \qquad \text{since } 1^3 = 1$$

$$\sqrt[3]{8} = 2 \qquad \text{since } 2^3 = 8$$

$$\sqrt[3]{64} = 4 \qquad \text{since } 4^3 = 64$$

$$\sqrt[3]{125} = 5 \qquad \text{since } 5^3 = 125$$

The small 3 written in the radical sign is called the *index*:

$$\sqrt[3]{27} = 3$$

How many irrational numbers are there?

The set of all rational numbers is infinite. The set of irrational numbers is also infinite. However, in more advanced math courses we can prove that there are infinitely many more irrational numbers than rational. Let us make a brief analysis of square and cube roots of integers from 0 to 16.

Square Roots			*Cube Roots*		
Number	*Square Root*	*Check If It Is Irrational*	*Number*	*Cube Root*	*Check If It Is Irrational*
0	$\sqrt{0} = 0$		0	$\sqrt[3]{0} = 0$	
1	$\sqrt{1} = 1$		1	$\sqrt[3]{1} = 1$	
2	$\sqrt{2}$	✓	2	$\sqrt[3]{2}$	✓
3	$\sqrt{3}$	✓	3	$\sqrt[3]{3}$	✓
4	$\sqrt{4} = 2$		4	$\sqrt[3]{4}$	✓
5	$\sqrt{5}$	✓	5	$\sqrt[3]{5}$	✓
6	$\sqrt{6}$	✓	6	$\sqrt[3]{6}$	✓
7	$\sqrt{7}$	✓	7	$\sqrt[3]{7}$	✓
8	$\sqrt{8}$	✓	8	$\sqrt[3]{8} = 2$	
9	$\sqrt{9} = 3$		9	$\sqrt[3]{9}$	✓
10	$\sqrt{10}$	✓	10	$\sqrt[3]{10}$	✓
11	$\sqrt{11}$	✓	11	$\sqrt[3]{11}$	✓
12	$\sqrt{12}$	✓	12	$\sqrt[3]{12}$	✓
13	$\sqrt{13}$	✓	13	$\sqrt[3]{13}$	✓
14	$\sqrt{14}$	✓	14	$\sqrt[3]{14}$	✓
15	$\sqrt{15}$	✓	15	$\sqrt[3]{15}$	✓
16	$\sqrt{16} = 4$		16	$\sqrt[3]{16}$	✓
Results for Square Roots: 17 Numbers of which { 5 are rational, 12 are irrational }			Results for Cube Roots: 17 Numbers of which { 3 are rational, 14 are irrational }		

Thus we see two trends. First, as you continue to take the roots of larger numbers you observe that more of the results are irrational. Second, the higher the order of the root, the more likely you are to obtain an irrational number.

Let us now consider how to take the cube root of variable quantities. For example,

$$\sqrt[3]{x^3} = x \qquad \text{since } (x^1)^3 = x^3$$

$$\sqrt[3]{x^6} = x^2 \qquad \text{since } (x^2)^3 = x^6$$

$$\sqrt[3]{x^9} = x^3 \qquad \text{since } (x^3)^3 = x^9$$

If the exponent of the variable is divisible by 3, you can find the cube root exactly.

EXAMPLE F Find: $\sqrt[3]{y^{15}}$.

Solution

$$\sqrt[3]{y^{15}} = y^5 \qquad \text{since } (y^5)^3 = y^{15}$$

We see that $\sqrt[3]{y^{15}} = y^{15/3} = y^5$ and therefore 15 is divisible by the index 3.

An Interesting Sidelight For Future Study

The procedure we used in Example F suggests that $\sqrt[3]{y^{15}}$ may be written as $y^{15/3}$. In general we can show also that

$$\sqrt[3]{x^2} = x^{2/3}$$

or

$$\sqrt[3]{x^7} = x^{7/3}$$

or

$$\sqrt{x} = x^{1/2}$$

In general for any *n*th root $\sqrt[n]{x^a} = x^{a/n}$.
The writing of radicals as fractional exponents will be covered more extensively in higher level math courses.
The use of fractional exponents is a means of evaluating radicals on computers and some scientific calculators.
If you wanted to evaluate $\sqrt[3]{15}$ on a computer or a scientific calculator you could enter the expression $15^{1/3}$ on the machine and obtain the result.

We have examined cube roots. Working in similar fashion, we can study fourth roots, fifth roots, etc.

EXAMPLE G Find the fourth root of 16.

Solution

$$\sqrt[4]{16} = 2 \qquad \text{since } 2^4 = 16$$

EXAMPLE H Find the fifth root of 32.

Solution

$$\sqrt[5]{32} = 2 \qquad \text{since } 2^5 = 32$$

EXAMPLE I Find: $\sqrt[5]{x^5}$.

Solution

$$\sqrt[5]{x^5} = x \qquad \text{since } (x^1)^5 = x^5$$

Note that the exponent of the radicand is divisible by 5.

EXAMPLE J Find: $\sqrt[6]{x^{12}}$.

Solution

$$\sqrt[6]{x^{12}} = x^2 \qquad \text{since } (x^2)^6 = x^{12}$$

What about negative numbers in the radicand?

1. *We cannot have even roots of negative numbers and obtain a real number.* So $\sqrt{-4}$, $\sqrt[4]{-16}$, and $\sqrt[6]{-64}$ are not real numbers. If the index of the root is a positive even number, you cannot have a negative number in the radicand and obtain a real number. (A square root is considered to have an index of 2!) As we mentioned in Section 8.1 $\sqrt{-4}$ is called an imaginary number. $\sqrt[4]{-16}$ and $\sqrt[6]{-64}$ are also imaginary numbers. You will study them more extensively in higher level math courses.
2. *It is possible to have odd roots of negative numbers.*

EXAMPLE K Find: $\sqrt[3]{-64}$.

Solution

$$\sqrt[3]{-64} = -4 \qquad \text{since } (-4)^3 = -64$$

EXAMPLE L Find: $\sqrt[5]{-32}$.

Solution

$$\sqrt[5]{-32} = -2 \qquad \text{since } (-2)^5 = -32$$

> **Thus we see that**
>
> **1. if the *index* of the root is a *positive odd number*, then you may take the root of a negative radicand and obtain a real number.**
>
> **2. if the *index* of the root is a *positive even number*, then you may *not* take the root of a negative radicand and obtain a real number.**

EXAMPLE M Find: $\sqrt[7]{-1}$ and $\sqrt[8]{-1}$.

Solution

$$\sqrt[7]{-1} = -1 \qquad \text{since } (-1)^7 = -1$$

Here the index is an odd number and we can take an odd root of a negative radicand.

$\sqrt[8]{-1}$ is not a real number. 8 is an *even* index. We cannot take an even root of a negative radicand and obtain a real number.

Both the product and the quotient rule for square roots hold for cube roots, fourth roots, etc., just as they do for square roots. However the index of each radical must be the same.

> **For any positive integer n, and for real numbers a and b,**
>
> $$\sqrt[n]{ab} = \sqrt[n]{a} \cdot \sqrt[n]{b}$$
>
> $$\sqrt[n]{\frac{a}{b}} = \frac{\sqrt[n]{a}}{\sqrt[n]{b}} \qquad \text{if } b \neq 0$$

An Interesting Sidelight For Future Study

Do you see that if you can express radicals as fractional exponents that this rule follows from the laws of exponents? You can show that $\sqrt[3]{x}\sqrt[3]{x} = \sqrt[3]{x^2}$ since $x^{1/3} \cdot x^{1/3} = x^{2/3}$ and $x^{2/3} = \sqrt[3]{x^2}$. This use of fractional exponents will be studied more extensively in a more advanced mathematics course.

EXAMPLE N Evaluate:

$$\sqrt[4]{\frac{16}{y^{20}}}$$

Solution

$$\sqrt[4]{\frac{16}{y^{20}}} = \frac{\sqrt[4]{16}}{\sqrt[4]{y^{20}}} \qquad \text{We use the quotient rule.}$$

$$= \frac{2}{y^5} \qquad \text{We evaluate each fourth root.}$$

EXAMPLE O Use the product rule to multiply:

$$(\sqrt[4]{15})(\sqrt[3]{3})$$

Solution The product rule does *not* apply since each index is *not* the same. Therefore we do *not* perform the multiplication.

EXAMPLE P Use the product rule to multiply:

$$(2\sqrt[5]{2})(3\sqrt[5]{5})$$

Solution $(2\sqrt[5]{2})(3\sqrt[5]{5}) = 6\sqrt[5]{10}$ The index is 5 for each radical, so we can use the product rule. It cannot be further simplified.

EXAMPLE Q Evaluate: $\sqrt[3]{8x^9}$

Solution $\sqrt[3]{8x^9} = \sqrt[3]{8}\sqrt[3]{x^9}$ We use the product rule.

$= (2)(x^3)$ We evaluate each cube root.

$= 2x^3$

Now try these practice problems.

Practice Problems for Exercise 8.6.

1. Evaluate using the table $4 + \sqrt{6}$
2. Evaluate using the table $\dfrac{\sqrt{2}}{\sqrt{5}}$
3. Find the indicated root $\sqrt[3]{27x^{12}}$
4. Find the indicated root $\sqrt[4]{\dfrac{x^8}{16}}$

Now turn to the answers following Exercise 8.6 and check your solutions for these practice problems.

Exercise 8.6

Approximate the expression using the square root table in Appendix A. Perform all necessary steps. Express your answer to three decimal places. (If you use a calculator your answers will vary slightly from those listed in the answer key.)

1. $\sqrt{51}$ **2.** $\sqrt{65}$ **3.** $2\sqrt{23}$

4. $3\sqrt{5}$ **5.** $\sqrt{11} + 6$ **6.** $2 + \sqrt{14}$

7. $\sqrt{46} + \sqrt{47}$ **8.** $\sqrt{37} + \sqrt{38}$ **9.** $\dfrac{\sqrt{132}}{2}$

10. $\dfrac{\sqrt{53}}{3}$ **11.** $\sqrt{348}$ **12.** $\sqrt{432}$

13. $2\sqrt{47} - 3$

14. $12 - 3\sqrt{31}$

15. $\dfrac{\sqrt{5}}{\sqrt{11}}$

16. $\dfrac{\sqrt{3}}{\sqrt{13}}$

17. $\sqrt{\dfrac{3}{2}}$

18. $\sqrt{\dfrac{5}{7}}$

19. $\sqrt{2}(\sqrt{3} - \sqrt{2})$

20. $\sqrt{5}(\sqrt{2} - \sqrt{5})$

See if you can evaluate exactly each of the following higher-order radicals. If there is no real number for some, state the reason.

21. $\sqrt[3]{27}$

22. $\sqrt[3]{8}$

23. $\sqrt[4]{16}$

24. $\sqrt[4]{81}$

25. $\sqrt[5]{243}$

26. $\sqrt[5]{32}$

27. $\sqrt[9]{1}$

28. $\sqrt[18]{0}$

29. $\sqrt[3]{-125}$

30. $\sqrt[3]{-27}$

31. $\sqrt[4]{-81}$

32. $\sqrt[4]{256}$

33. $\sqrt[5]{-32}$

34. $\sqrt[6]{-64}$

35. $-\sqrt[3]{-27}$

36. $-\sqrt[4]{16}$

37. $\sqrt[3]{w^6}$

38. $\sqrt[3]{y^9}$

39. $\sqrt[5]{x^{15}}$

40. $\sqrt[5]{x^{20}}$

41. $\sqrt[4]{x^{12}}$

42. $\sqrt[4]{y^{28}}$

43. $\sqrt[8]{y^{16}}$

44. $\sqrt[7]{x^{21}}$

45. $\sqrt[3]{8x^6}$

46. $\sqrt[3]{27y^3}$

47. $\sqrt[3]{64x^{12}y^9}$

48. $\sqrt[3]{125y^9w^{15}}$

49. $(2\sqrt[3]{x})(3\sqrt[3]{x^2})$

50. $(5\sqrt[4]{x})(2\sqrt[4]{x^7})$

51. $(\sqrt[4]{16w^8})(\sqrt[4]{16w^4})$

52. $(2\sqrt[3]{27x^5})(-3\sqrt[3]{3x^4})$

53. $\dfrac{\sqrt[5]{64x^3}}{\sqrt[5]{2x^8}}$

54. $\dfrac{\sqrt[4]{625x}}{\sqrt[4]{81x^{13}}}$

Challenge Problems

Let n be any positive integer. Simplify the following.

55. $(\sqrt[n]{2^4x^{4n}})(\sqrt[n]{2^{n-4}x^{3n}})$

56. $\dfrac{\sqrt[n]{3^{6+n}x^{2n}}}{\sqrt[n]{3^6x^{4n}}}$

Solutions to Practice Problems for Exercise 8.6

1. The table value gives us $\sqrt{6} \doteq 2.449$. Thus $4 + \sqrt{6} \doteq 4.000 + 2.449 = 6.449$.

2. $\dfrac{\sqrt{2}}{\sqrt{5}}$ The denominator should first be rationalized.

$$\frac{\sqrt{2}}{\sqrt{5}} \cdot \frac{\sqrt{5}}{\sqrt{5}} = \frac{\sqrt{10}}{5} \doteq \frac{3.162}{5} \doteq 0.632$$

3. $\sqrt[3]{27x^{12}} = \sqrt[3]{27}\sqrt[3]{x^{12}}$ by the product rule.

$\sqrt[3]{27} = 3$ since $3^3 = 27$

$\sqrt[3]{x^{12}} = x^4$ since $(x^4)^3 = x^{12}$

Note that the exponent of x^{12} is divisible by 3.

Thus

$$\sqrt[3]{27x^{12}} = 3 \cdot x^4 = 3x^4$$

4. $\sqrt[4]{\dfrac{x^8}{16}} = \dfrac{\sqrt[4]{x^8}}{\sqrt[4]{16}}$ by the quotient rule.

Now $\sqrt[4]{x^8} = x^2$ since $(x^2)^4 = x^8$. Note that the exponent of x^8 is divisible by 4. Also, $\sqrt[4]{16} = 2$ since $2^4 = 16$. Thus

$$\sqrt[4]{\frac{x^8}{16}} = \frac{x^2}{2}$$

Now do the problems in Exercise 8.6.

8.7 Word Problems Involving Radicals: Direct and Inverse Variation

In many instances in daily life we observe that a change in one variable quantity produces a change in another variable quantity. For example, we turn up our thermostat and our heating bill increases, or the faster we drive our car the lower our gas mileage becomes. Frequently, we are able to write a formula that relates the two variables and accurately predicts the amount of change in the second variable produced by a change in the first variable. In some cases the formulas are derived through scientific observation and experimental evidence. In other cases they can be derived from theoretical concepts or geometrical evidence. We will now study two important types of variation.

Direct variation

If one variable is a constant multiple of the other, then the two variables are said to *vary directly*. We usually write this with an equation. If y varies directly as x, then $y = kx$, where k is a constant. The constant is often called the constant of variation. Consider the following example.

EXAMPLE A Cliff works part-time in a local supermarket while going to college. His salary varies directly as the number of hours worked. Last week he earned \$26.60 for working seven hours. This week he earned \$41.80. How many hours did he work?

Solution Let

S = his salary

h = the number of hours he worked

k = the constant of variation

Since his salary varies directly as the number of hours worked, we write

$$S = k \cdot h$$

We can find the constant k by substituting the known values of $S = 26.60$ when $h = 7$.

$$26.60 = k \cdot 7 = 7k$$

$$\frac{26.60}{7} = \frac{7k}{7}$$ Divide both sides by 7.

$$3.80 = k$$ The constant of variation is 3.80.

$$S = 3.80h$$ We replace k in the variation equation by 3.80.

How many hours did he work to earn \$41.80?

$S = 3.80h$	The direct variation equation with the obtained value for k.
$41.80 = 3.80h$	Substitute 41.80 for S.
$\frac{41.80}{3.80} = \frac{3.80h}{3.80}$	Divide both sides by 3.80.
$11 = h$	Simplify.

Cliff worked 11 hours.

Are all direct variation equations this simple? I can do this problem another way. Why do I need to learn direct variation?

No, not all direct variation equations are this simple. Direct variation equations apply to a number of cases, some of which are more difficult to solve. Direct variation is particularly useful in a number of scientific and technical problems. The following chart shows some of the more common forms of direct variation. In each case k = the constant of variation.

Sample Direct Variation Situations

Verbal Description	*Variation Equation*
y varies directly as x	$y = kx$
b varies directly as the square of c	$b = kc^2$
l varies directly as the cube of m	$l = km^3$
d varies directly as the square root of e	$d = k\sqrt{e}$

EXAMPLE B In a certain class of racing cars the maximum speed varies directly as the square root of the horsepower of the engine. If a car with 225 horsepower is able to achieve a maximum speed of 120 mph, what speed could it achieve if the engine developed 256 horsepower?

Solution Let

V = the maximum speed

h = the horsepower of the engine

k = the constant of variation

Since the maximum speed (V) varies directly as the square root of the horsepower of the engine,

$V = k\sqrt{h}$	
$120 = k\sqrt{225}$	Substitute known values of V and h.
$120 = k \cdot 15$	Since $\sqrt{225} = 15$.
$8 = k$	Divide both sides by 15.

Now we can write the direct variation equation with a known constant of variation.

$$V = 8\sqrt{h}$$

$V = 8\sqrt{256}$ Substitute value of $h = 256$.

$V = (8)(16)$ Since $\sqrt{256} = 16$.

$$V = 128$$

Thus a maximum speed of 128 mph could be achieved if the engine developed 256 horsepower.

Inverse variation

If one variable is a constant multiple of the reciprocal of the other, the two variables are said to *vary inversely*. If y varies inversely as x, we express this by the equation $y = k/x$, where k is the constant of variation.

EXAMPLE C A car manufacturer is thinking of reducing the size of the wheel used in a subcompact car. The number of times a car wheel must turn to cover a given distance varies inversely as the radius of the wheel. A wheel with a radius of 0.35 meter must turn 400 times to cover a specified distance on a test track. How many times would it have to turn if the radius is reduced to 0.30 meter (see Figure 8.1)?

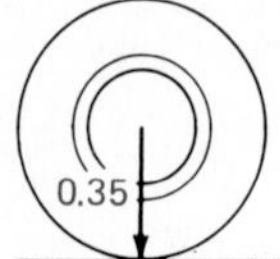

Test distance

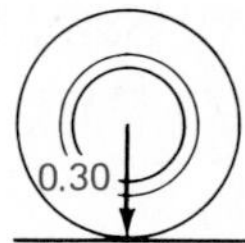

Test distance

FIGURE 8.1

Solution Let

n = the number of times the car wheel will turn

r = the radius of the wheel

k = the constant of variation

Since the number of turns varies inversely as the radius, we can write

$n = \dfrac{k}{r}$ The variation equation.

$400 = \dfrac{k}{0.35}$ We substitute known values of n and r.

$(0.35)(400) = (0.35)\left(\dfrac{k}{0.35}\right)$ Multiply both sides of the equation by the LCD = 0.35.

$140 = k$ Simplify to find k, the constant of variation.

How many times will the wheel turn if the radius is 0.30 meter?

$n = \dfrac{140}{r}$	We use the variation equation where k is known.
$n = \dfrac{140}{0.30}$	We substitute the radius value of 0.30.
$n = 466\dfrac{2}{3}$	Simplify.

The wheel would have to turn $466\frac{2}{3}$ times to cover the same distance if the radius were only 0.30 meter.

The following chart demonstrates some situations involving inverse variation. In each case k = the constant of variation.

Sample Inverse Variation Situations

Verbal Description	*Variation Equation*
y varies inversely as x	$y = \dfrac{k}{x}$
b varies inversely as the square of c	$b = \dfrac{k}{c^2}$
l varies inversely as the cube of m	$l = \dfrac{k}{m^3}$
d varies inversely as the square root of e	$d = \dfrac{k}{\sqrt{e}}$

EXAMPLE D The illuminance of a light source varies inversely as the square of the distance from the source. The illuminance measures 25 units when a certain light is 4 meters away. Find the illuminance when the light is 8 meters away (see Figure 8.2).

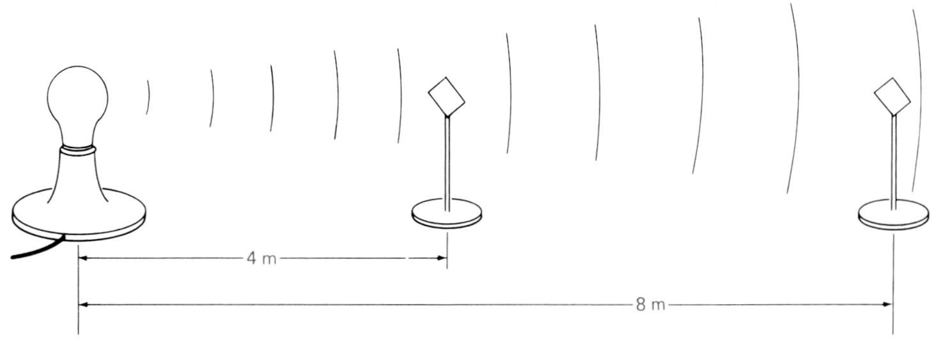

FIGURE 8.2

Solution Let

I = the measurement of illuminance

d = the distance from the light source

k = the constant of variation

Since the illuminance varies as the square of the distance,

$$I = \frac{k}{d^2}$$

We evaluate the constant by substituting the given values.

$$25 = \frac{k}{4^2}$$ Where $I = 25$ and distance $= 4$.

$$25 = \frac{k}{16}$$ Simplify.

$$400 = k$$ Multiply both sides by 16 to obtain the constant k.

We may now write the variation equation with the constant evaluated.

$$I = \frac{400}{d^2}$$ Variation equation.

$$I = \frac{400}{8^2}$$ Substitute distance of 8 meters.

$$I = \frac{400}{64}$$ Square 8.

$$I = \frac{25}{4} \quad \text{or} \quad 6\frac{1}{4}$$

The illuminance is $6\frac{1}{4}$ units when the light source is 8 meters away.

Review Examples A to D, making sure that you understand the difference between direct and inverse variation. Then try these practice problems.

Practice Problems for Exercise 8.7

1. For a certain class of automobiles the distance to stop the car varies directly as the square of its speed. On ice-covered roads a car that is traveling at a speed of 20 mph could stop in 60 feet. If the car had been traveling at 40 mph, how many feet would be necessary for it to have come to a stop?
2. Over the last three years the market research division of a calculator company found the volume of sales of scientific calculators varies inversely as the price of the calculator. One year 120,000 calculators were sold at $30 each. How many calculators were sold the next year when the price was $24 for each calculator?
3. If the amount of power in an electrical circuit is held constant, the resistance in the circuit varies inversely as the square of the amount of current. If the amount of current is 0.01 ampere, then the resistance is 800 ohms. What is the resistance if the amount of current is 0.02 ampere?

Now turn to the answers following Exercise 8.7 and check your solutions for these three practice problems.

Exercise 8.7

1. If y varies directly as x, and $y = 7$ when $x = 4$, find y when $x = 9$.

2. If y varies directly as x, and $y = 9$ when $x = 12$, find y when $x = 5$.

3. If y varies directly as the cube of x and $y = 12$ when $x = 2$, find y when $x = 7$.

4. If y varies directly as the square root of x and $y = 7$ when $x = 4$, find y when $x = 25$.

5. The pressure of water on an object submerged beneath the surface varies directly as the distance beneath the surface. A submarine experiences a pressure of 26 pounds per square inch at 60 feet below the surface. How much pressure will the submarine experience at 390 feet below the surface?

6. The weight of an object on the surface of the moon varies directly as the weight of an object on the surface of the earth. An astronaut with his protective suit weighs 80 kilograms on the earth surface, on the moon and wearing the same suit he will weigh 12.8 kilograms. If his moon rover vehicle weighs 900 kilograms on earth, how much will it weigh on the moon?

7. The time it takes to fill a storage cube with sand varies directly as the cube of the side of the box. A storage cube that is 2.0 meters (Figure 8.3) on each side can be filled in 7 minutes by a sand loader. How long will it take to fill a storage cube that is 4.0 meters on each side?

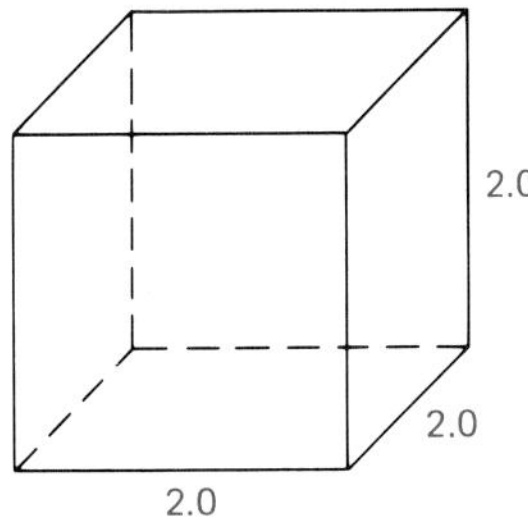

FIGURE 8.3

8. The distance that an object falls under the influence of gravity varies directly as the square of the time of fall. An object falls 64 feet in 2 seconds. How far will it fall in 7 seconds?

9. If y varies inversely as x, and $y = 12$ when $x = 4$, find y when $x = 7$.

10. If y varies inversely as x, and $y = 39$ when $x = 3$, find y when $x = 11$.

11. If y varies inversely as the square of x, and $y = 100$ when $x = 5$, find y when $x = 3$.

12. If y varies inversely as the cube of x, and $y = 12$ when $x = 3$, find y when $x = 6$.

13. The amount of current in an electric circuit varies inversely as the amount of resistance. In a certain circuit when the resistance is 5 ohms the amount of current is 42 amperes. If the resistance is 4 ohms, what is the amount of current?

14. The thickness of fiberglass insulation is an indicator of its thermal effectiveness. The heat loss through a certain type of fiberglass insulation varies inversely as the thickness of the fiberglass. The heat loss through 6 inches of this type of fiberglass is 3000 BTU/hour. Find the heat loss through 14 inches of fiberglass.

15. The weight of an object on the earth's surface varies inversely as the square of its distance from the center of the earth. An object weighs 1000 pounds at the earth's surface. This is approximately 4000 miles from the center of the earth. How much will the object weigh 6000 miles from the center of the earth?

16. The pressure exerted by a given amount of gas on a cubical storage container varies inversely as the cube of the side of the container. A certain amount of gas exerts a pressure of 60 pounds per square inch when placed in a cubical storage container that measures 0.2 meter in each dimension (Figure 8.4). If it is stored in a box 0.4 meter in each dimension, what will be the pressure?

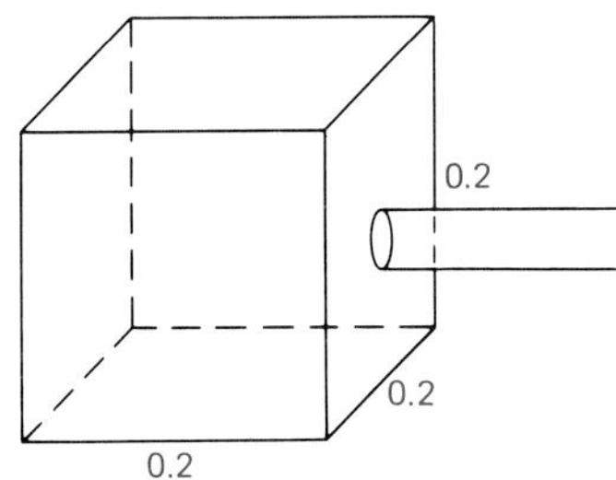

FIGURE 8.4

Challenge Problems

17. The amount of heat that passes through a wall can be related to the factors of the temperature difference on the two sides of the wall, the area of the wall, and the thickness of the wall. The heat transfer varies directly as the product of the area of the wall and the temperature difference between the two sides of the wall and inversely as the thickness of the wall. In a certain house 6000 BTU/hour were transferred by a wall measuring 80 square feet. The wall is 0.5 foot thick. The outside temperature was 40°F, the inside temperature was 70°F. Suppose that the wall was exactly 1.0 foot thick and the outside temperature was 20°F. Assuming the same wall area and inside temperature, how much heat would be transferred?

18. An important element of sailboat design is a calculation of the force of wind on the sail. The force varies directly as the square of the wind speed and directly as the area of the sail. On a certain class of racing boat the force on 100 square feet of sail is 280 pounds when the wind speed is 6 mph. What would be the force on the sail if it is rigged for a storm (the amount of sail exposed to the wind is reduced by lashing part of the sail to the boom) so that only 60 square feet of sail are exposed but the windspeed is 40 mph?

Solutions to Practice Problems for Exercise 8.7

1. For a certain class of automobiles the distance to stop the car varies directly as the square of its speed. On ice covered roads a car that is traveling at a speed of 20 mph could stop in 60 feet. If the car had been traveling at 40 mph how many feet would be necessary for it to have come to a stop? Let

$d =$ the distance to stop the car

$s =$ the speed of the car

$k =$ the constant of variation

Since the distance varies directly as the square of the speed, we have

$$d = ks^2$$

To evaluate k we substitute the known distance and speed.

$$60 = k(20)^2$$

$$60 = k(400) \qquad \text{We square 20 to obtain 400.}$$

$$\frac{60}{400} = k \qquad \text{Divide both sides by 400.}$$

$$\frac{3}{20} = k \qquad \text{We simplify to obtain } k.$$

Now we write the variation equation with the constant evaluated.

$$d = \left(\frac{3}{20}\right)s^2$$

$$d = \left(\frac{3}{20}\right)(40)^2 \qquad \text{Substitute } s = 40.$$

$$d = \left(\frac{3}{20}\right)(1600) \qquad \text{Square 40.}$$

$$d = 240 \qquad \text{Simplify.}$$

The distance required to stop the car going 40 mph on an ice-covered road is 240 feet!

2. Over the last three years the market research division of a calculator company found the volume of sales of scientific calculators varies inversely as the price of the calculator. One year 120,000 calculators were sold at $30 each. How many calculators were sold the next year when the price was $24 for each calculator? Let

V = the volume of sales

p = the price of the calculator

k = the constant of variation

Since the volume varies inversely as the price, we can write

$$V = \frac{k}{p}$$

To evaluate k we substitute the known V and p quantities.

$$120{,}000 = \frac{k}{30} \qquad \text{Multiply both sides by 30.}$$

$$3{,}600{,}000 = k \qquad \text{The constant of variation.}$$

$$V = \frac{3{,}600{,}000}{p} \qquad \text{The variation equation with the evaluated constant.}$$

$$V = \frac{3{,}600{,}000}{24} \qquad \text{Substitute the price of \$24.}$$

$$V = 150{,}000 \qquad \text{Simplify.}$$

Thus the volume increased to 150,000 sales per year when the price was reduced to $24 per calculator.

3. If the amount of power in an electrical circuit is held constant, the resistance in the circuit varies inversely as to the square of the amount of current. If the amount of current is 0.01 ampere, then the resistance is 800 ohms. What is the resistance if the amount of current is 0.02 ampere? Let

R = the resistance in the circuit

i = the amount of current

k = the constant of variation

Since the resistance in the circuit varies inversely as the square of the amount of current, we have

$$R = \frac{k}{i^2}$$

To evaluate k we substitute the known values of R and i.

$$800 = \frac{k}{(0.01)^2}$$

$$800 = \frac{k}{0.0001} \qquad (0.01)^2 = 0.0001.$$

$$0.08 = k \qquad \text{Multiply both sides by 0.0001.}$$

$$R = \frac{0.08}{i^2} \qquad \text{The variation equation with the evaluated constant of } k = 0.08.$$

$$R = \frac{0.08}{(0.02)^2} \qquad \text{Substitute } i = 0.02.$$

$$R = \frac{0.08}{0.0004} \qquad \text{Square } (0.02)^2 \text{ to obtain } 0.0004.$$

$$R = 200 \qquad \text{Simplify.}$$

Thus the resistance is 200 ohms if the current is increased to 0.02 ampere.

Now do the problems in Exercise 8.7.

An Application for Chapter 8: The Use of Radicals in Estimating the Speed of an Object that Orbits the Earth

A variety of human-made satellites now orbit the earth. Many of them orbit the earth in a circular pattern (or an orbit that is very, very close to circular; see Figure 8.5).

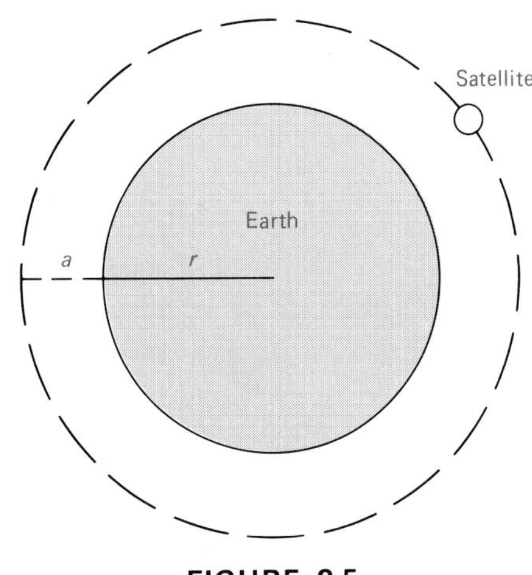

FIGURE 8.5

Have you ever wondered how fast these human-made satellites have to travel in order to stay in orbit?

An equation has been determined that approximates the required speed of a satellite orbiting the earth in a circular orbit. The speed S in miles per hour is given by

$$S \doteq \sqrt{\frac{1.24 \times 10^{12}}{r + a}}$$

where r = the radius of the earth

a = the altitude of the orbiting body

A Challenge for You!

Suppose that we want to know the speed of a satellite that will circle the earth at an altitude of $a = 140$ miles. Assume that the radius of the earth is approximately 3960 miles. Find the orbiting speed of the satellite.

Solution

$$S \doteq \sqrt{\frac{1.24 \times 10^{12}}{r + a}}$$

The equation for the speed of a satellite orbiting the earth.

$$S \doteq \sqrt{\frac{1.24 \times 10^{12}}{3960 + 140}}$$

Substitute $r \doteq 3960$ miles and $a = 140$ miles into the equation.

$$S \doteq \sqrt{\frac{1.24 \times 10^{12}}{4100}}$$

Add the two values in the denominator.

Now this expression can be evaluated immediately on a scientific calculator. However, even if you have access only to the square root table in Appendix A, you may still evaluate the radical if you do the following steps.

$$S \doteq \sqrt{\frac{1.24 \times 10^{12}}{4.1 \times 10^3}}$$

Since $4100 = 4.1 \times 1000 = 4.1 \times 10^3$.

$$= \sqrt{\frac{1.24}{4.1} \times \frac{10^{12}}{10^3}}$$

$$S \doteq \sqrt{\frac{1.24}{4.1} \times 10^9}$$

Since $\frac{10^{12}}{10^3} = 10^9$ by the laws of exponents.

$$S \doteq \sqrt{0.30 \times 10^9}$$

Dividing $1.24 \div 4.1$ and rounding the answer to the nearest hundredth.

$$S \doteq \sqrt{(0.30 \times 10^1) \times 10^8}$$

Since $10^9 = 10^1 \cdot 10^8$.

$$S \doteq \sqrt{(3) \times 10^8} = (\sqrt{3})(\sqrt{10^8})$$

Since $\sqrt{ab} = \sqrt{a}\sqrt{b}$.

$$S \doteq \sqrt{3} \times 10^4$$

Since we know that $\sqrt{10^8} = 10^4$.

$$S \doteq \sqrt{3} \times 10{,}000$$

Since $10^4 = 10{,}000$.

$$S \doteq 1.732 \times 10{,}000$$

Obtain $\sqrt{3} \doteq 1.732$ from the table in Appendix A.

$$S \doteq 17{,}320 \text{ mph}$$

Since we rounded the result of $1.24 \div 41$ to only two significant digits, our answer is accurate to only two significant digits. Therefore, the speed is approximately 17,000 miles per hour. (If we had used a calculator to evaluate

$$\sqrt{\frac{1.24 \times 10^{12}}{3960 + 140}}$$

we would obtain 17,390.774.)

Supplementary Problems for Chapter 8

Simplify or evaluate, if possible.

1. $\sqrt{225}$

2. $\sqrt{169}$

3. $\sqrt{0.81}$

4. $\sqrt{0.36}$

5. $\sqrt{\frac{1}{25}}$

6. $\sqrt{\frac{36}{49}}$

7. $-\sqrt{0.0004}$

8. $\sqrt{0.0009}$

9. $\sqrt{-0.0016}$

10. $-\sqrt{0.0081}$

11. $\sqrt{50}$

12. $\sqrt{48}$

13. $\sqrt{x^8}$

14. $\sqrt{y^{10}}$

15. $\sqrt{x^5y^6}$

16. $\sqrt{a^3b^4}$

17. $\sqrt{16x^3y^5}$

18. $\sqrt{98x^4y^6}$

19. $\sqrt{120a^3b^4c^5}$

20. $\sqrt{121a^6b^4c}$

21. $\sqrt{56x^7y^9}$

22. $\sqrt{99x^{13}y^7}$

23. $\sqrt{2} - \sqrt{8} + \sqrt{32}$

24. $3\sqrt{5} - 2\sqrt{20} - \sqrt{5}$

25. $x\sqrt{3} + 3x\sqrt{3} + \sqrt{27x^2}$

26. $a\sqrt{2} + \sqrt{12a^2} + a\sqrt{98}$

27. $5\sqrt{5} - 6\sqrt{20} + 2\sqrt{10}$

28. $3\sqrt{6} - 5\sqrt{18} + 3\sqrt{24}$

29. $2\sqrt{28} - 3\sqrt{63} + 2x\sqrt{7}$

30. $5a\sqrt{3} - 12\sqrt{27} + 2\sqrt{75}$

31. $(2\sqrt{x})(3\sqrt{x^3})$

32. $(-5\sqrt{a})(2\sqrt{ab})$

33. $(\sqrt{2a^3})(\sqrt{8b^2})$

34. $(5x\sqrt{x})(-3x^2\sqrt{x})$

35. $\sqrt{5}(3\sqrt{5} - \sqrt{20})$

36. $2\sqrt{3}(\sqrt{27} - 6\sqrt{3})$

37. $\sqrt{2}(\sqrt{5} - \sqrt{3} - 2\sqrt{2})$

38. $\sqrt{5}(\sqrt{6} - 2\sqrt{5} + \sqrt{10})$

39. $(\sqrt{11} + 2)(2\sqrt{11} - 1)$

40. $(\sqrt{10} + 3)(3\sqrt{10} - 1)$

41. $(2 + 3\sqrt{6})(4 - 2\sqrt{3})$

42. $(5 - \sqrt{2})(3 - \sqrt{12})$

43. $(2\sqrt{3} + 3\sqrt{6})^2$

44. $(5\sqrt{2} - 2\sqrt{6})^2$

45. $(a\sqrt{b} - 2\sqrt{a})(\sqrt{b} - 3\sqrt{a})$

46. $(x\sqrt{y} - 2\sqrt{x})(3\sqrt{y} + \sqrt{x})$

47. $\frac{1}{\sqrt{3x}}$

48. $\frac{2y}{\sqrt{5}}$

49. $\frac{x^2y}{\sqrt{8}}$

50. $\frac{3ab}{\sqrt{2b}}$

51. $\sqrt{\frac{3}{7}}$

52. $\sqrt{\frac{2}{9}}$

53. $\dfrac{15\sqrt{60}}{5\sqrt{15}}$

54. $\dfrac{\sqrt{120}}{2\sqrt{5}}$

55. $\dfrac{\sqrt{a^5}}{\sqrt{2a}}$

56. $\dfrac{\sqrt{x^3}}{\sqrt{3x}}$

57. $\dfrac{a\sqrt{a^2b}}{ab\sqrt{ab^2}}$

58. $\dfrac{2xy\sqrt{x^3}}{2y\sqrt{y^3}}$

59. $\dfrac{3}{\sqrt{5}+\sqrt{2}}$

60. $\dfrac{2}{\sqrt{6}-\sqrt{3}}$

61. $\dfrac{2\sqrt{6}+\sqrt{3}}{5\sqrt{2}-3\sqrt{3}}$

62. $\dfrac{4\sqrt{5}-\sqrt{7}}{2\sqrt{5}+\sqrt{2}}$

Evaluate, using the square root table in Appendix A.

63. $\sqrt{\dfrac{3}{8}}$

64. $\sqrt{\dfrac{2}{7}}$

65. $\sqrt{684}$

66. $\sqrt{549}$

Simplify problems 67 to 82.

67. $\sqrt[3]{-27}$

68. $\sqrt[4]{16}$

69. $\sqrt[4]{81}$

70. $\sqrt[5]{32}$

71. $\sqrt[5]{x^{10}y^{25}}$

72. $\sqrt[4]{a^{20}b^8}$

73. $-\sqrt[4]{16x^{12}y^{24}}$

74. $\sqrt[3]{-8x^6y^{21}}$

75. $(\sqrt[3]{-2x})(\sqrt[3]{4x^2y^6})$

76. $(\sqrt[3]{-3ab^5})(\sqrt[3]{-9a^5b^4})$

77. $(\sqrt[4]{4x^2y^3})(\sqrt[4]{4x^2y^{17}})$

78. $(\sqrt[5]{2x^3y})(\sqrt[5]{16x^2y^{24}})$

79. $\dfrac{\sqrt[3]{27x^5y^7}}{\sqrt[3]{x^3y^{13}}}$

80. $\dfrac{\sqrt[4]{2a^5b^6}}{\sqrt[4]{32ab^2}}$

81. $\sqrt[5]{\dfrac{32a^6b^{12}}{ab^{17}}}$

82. $\sqrt[3]{\dfrac{-81x^5yz^6}{3x^{14}y}}$

83. The insect population in a potato field varies inversely as the amount of pesticide used. When 40 pounds of pesticide were used, the insect population of the field was estimated to be 1000 bugs. How many pounds of pesticide should be required to reduce the number of bugs in the field to 100?

84. The length of skid marks on the road when a car slams on the brakes varies directly as the square of the speed of the car. At 30 mph a certain car had skid marks 40 feet long. How long would the skid marks be if the car were traveling at 55 mph?

85. The horsepower that is needed to drive a racing boat through water varies directly as the cube of the speed of the boat. What will happen to the horsepower requirement if someone wants to double the maximum speed of a given boat?

Chapter 8 Summary

The Square Root

1. For all nonnegative numbers N, the *square root of* N (written $\sqrt{N}$) is defined to be the nonnegative number a if and only if $a^2 = N$.
2. The number beneath the radical sign is called the *radicand*.
3. There is *no real number* obtained in a given square root if the radicand is negative.

Simplification of Square Roots

1. For all nonnegative numbers a, b, it is true that $(\sqrt{a})(\sqrt{b}) = \sqrt{ab}$.
2. For any nonnegative number a, $(\sqrt{a})(\sqrt{a}) = a$.
3. The properties above can be used to simplify indicated square roots by writing the radicand as a product of two factors where the first factor is a number that is a perfect square or a variable raised to an even power.

Addition and Subtraction of Square Roots

1. Like square roots are square roots that have the same radicand. Only like square roots can be added or subtracted.
2. To combine square roots, simplify each radical expression. Then add or subtract the numerical coefficients of like square roots.

Multiplication of Square Roots

1. *Monomial radical expressions*
   - **(a)** Multiply the coefficients of the radicals.
   - **(b)** Multiply the radicands and write the product under the radical sign.
   - **(c)** Simplify the resulting square root if possible.
2. *Radical expressions of more than one term*
   - **(a)** Use the procedure above for multiplying each term of the expression.
   - **(b)** Perform all steps of multiplying that are required in multiplying polynomials.
   - **(c)** Simplify resulting radicals and collect like terms.

Division of Square Roots

1. For all nonnegative numbers a, b with $b \neq 0$, it is true that

$$\sqrt{\frac{a}{b}} = \frac{\sqrt{a}}{\sqrt{b}}$$

2. Expressions that contain radicals are not considered to be simplified if:
   - **(a)** The radicand contains a fraction.
   - **(b)** The denominator of an expression contains a square root.
   - **(c)** If either a or b occurs, the denominator should be rationalized.

Rationalizing the Denominator Containing a Square Root

1. If the denominator is a monomial, use the following technique, where $\sqrt{b}$ is in simplified form:

$$\frac{a}{\sqrt{b}} = \frac{a}{\sqrt{b}} \cdot \frac{\sqrt{b}}{\sqrt{b}} = \frac{a\sqrt{b}}{b}$$

   Simplify your answer.

2. If the denominator is a polynomial, use the following technique, where $\sqrt{c}$ and $\sqrt{d}$ are in simplified form:

$$\frac{b}{\sqrt{c} + \sqrt{d}} = \frac{b}{\sqrt{c} + \sqrt{d}} \cdot \frac{\sqrt{c} - \sqrt{d}}{\sqrt{c} - \sqrt{d}} = \frac{b(\sqrt{c} - \sqrt{d})}{c - d}$$

   Simplify your answer.

Approximation of Square Roots Using a Table

1. Simplify the radical expression.
2. Find the approximate value of the square root from a table and substitute it for the radical.
3. Simplify the result.

Higher-Order Radicals

1. Higher-order radicals are radicals with an index greater than 2.
2. The symbol $\sqrt[n]{x}$ indicates the nth root of x. If x is positive, the root is positive. If x is negative, then:
   (a) The root is negative if the index n is an odd number.
   (b) The root is not a real number if the index n is an even number.
3. For all radicals with the same index number n (where n is a positive integer and where a and b are real numbers):

$$\sqrt[n]{ab} = (\sqrt[n]{a})(\sqrt[n]{b})$$

$$\sqrt[n]{\frac{a}{b}} = \frac{\sqrt[n]{a}}{\sqrt[n]{b}} \qquad \text{where } b \neq 0$$

Practice Quiz for Chapter 8

Sections 8.1–8.3

Find the following square roots, if possible.

1. $\sqrt{144}$ **2.** $\sqrt{-36}$ **3.** $\sqrt{\frac{25}{36}}$

4. $\sqrt{0.49}$ **5.** $-\sqrt{\frac{64}{81}}$ **6.** $\sqrt{\frac{1}{169}}$

Simplify.

7. $\sqrt{48x^7}$ **8.** $\sqrt{98x^6y^3}$ **9.** $\sqrt{81a^3b^8}$

10. $\sqrt{2^{20}x^5y^6z^7}$

Combine each of the following expressions, if possible.

11. $\sqrt{20} + \sqrt{45} - \sqrt{25}$ **12.** $2\sqrt{50} - 2\sqrt{18} + \sqrt{8}$ **13.** $3\sqrt{27} - 2\sqrt{75} + \sqrt{48}$

14. $\sqrt{24x} + \sqrt{150x} - \sqrt{96x}$ **15.** $3a\sqrt{8a} + \sqrt{18a^3} - 5a\sqrt{2a}$

Sections 8.4–8.7

Multiply. Simplify your answer.

1. $(\sqrt{6x})(2\sqrt{4x})$ **2.** $\sqrt{5x}(\sqrt{2x} - \sqrt{10})$ **3.** $(\sqrt{3} - 2\sqrt{2})^2$

4. $(2\sqrt{3} + \sqrt{5})(3\sqrt{3} - \sqrt{10})$ **5.** $(2a\sqrt{3} + \sqrt{b})(\sqrt{5} - 2\sqrt{a})$

Simplify. Rationalize all denominators.

6. $\dfrac{x}{\sqrt{y}}$

7. $\dfrac{5}{\sqrt{12}}$

8. $\dfrac{2}{\sqrt{5} - 2\sqrt{3}}$

9. $\dfrac{\sqrt{3} + \sqrt{6}}{2\sqrt{3} + \sqrt{2}}$

10. Use the table of square roots in Appendix A to approximate the expression:

$$\sqrt{3}(\sqrt{2} + \sqrt{3})$$

Simplify problems 11 to 14, if possible.

11. $\sqrt[5]{-32x^{10}}$

12. $\sqrt[3]{5xy^2}\,\sqrt[3]{25x^8y}$

13. $\sqrt[6]{-128}$

14. $\dfrac{\sqrt[4]{32x^5}}{\sqrt[4]{2xy^8}}$

15. The surface area of a sphere varies directly as the square of the radius. When the radius of a sphere is 2 meters, the surface area is 1256 square meters. Find the surface area if the radius is 0.5 meter.

Practice Test for Chapter 8

1. Evaluate:

$$\sqrt{81}$$

2. Evaluate:

$$\sqrt{\frac{9}{100}}$$

3. Simplify:

$$\sqrt{48x^2y^7}$$

4. Simplify:

$$\sqrt{100x^3yz^4}$$

5. Combine and simplify:

$$\sqrt{5} + 3\sqrt{20} - 2\sqrt{45}$$

6. Combine and simplify:

$$\sqrt{4a} + \sqrt{8a} + \sqrt{36a} + \sqrt{18a}$$

7. Multiply:

$$(2\sqrt{a})(3\sqrt{b})(2\sqrt{ab})$$

8. Multiply:

$$\sqrt{3}(\sqrt{6} - \sqrt{2} + 5\sqrt{27})$$

9. Multiply:

$$(2 - \sqrt{6})^2$$

10. Multiply:

$$(4\sqrt{2} - \sqrt{5})(3\sqrt{2} + \sqrt{5})$$

Simplify.

11. $\sqrt{\dfrac{x}{5}}$

12. $\dfrac{3}{\sqrt{12}}$

13. $\dfrac{\sqrt{3} - \sqrt{2}}{2\sqrt{5} + \sqrt{2}}$

14. $\dfrac{3a}{\sqrt{5} + \sqrt{2}}$

15. Use the table of square roots in Appendix A to approximate the expression:

$$\frac{2\sqrt{3}}{\sqrt{5}}$$

16. Simplify:

$$\sqrt[4]{81x^8y^{12}}$$

17. Simplify:

$$\sqrt[5]{4x^2}\sqrt[5]{-8x^8}$$

18. The intensity of illumination from a light source varies inversely as the square of the distance from the light source. A photoelectric cell is placed 8 inches from a light source. Then it is moved so that it only receives one-fourth as much illumination. How far is it from the light source?

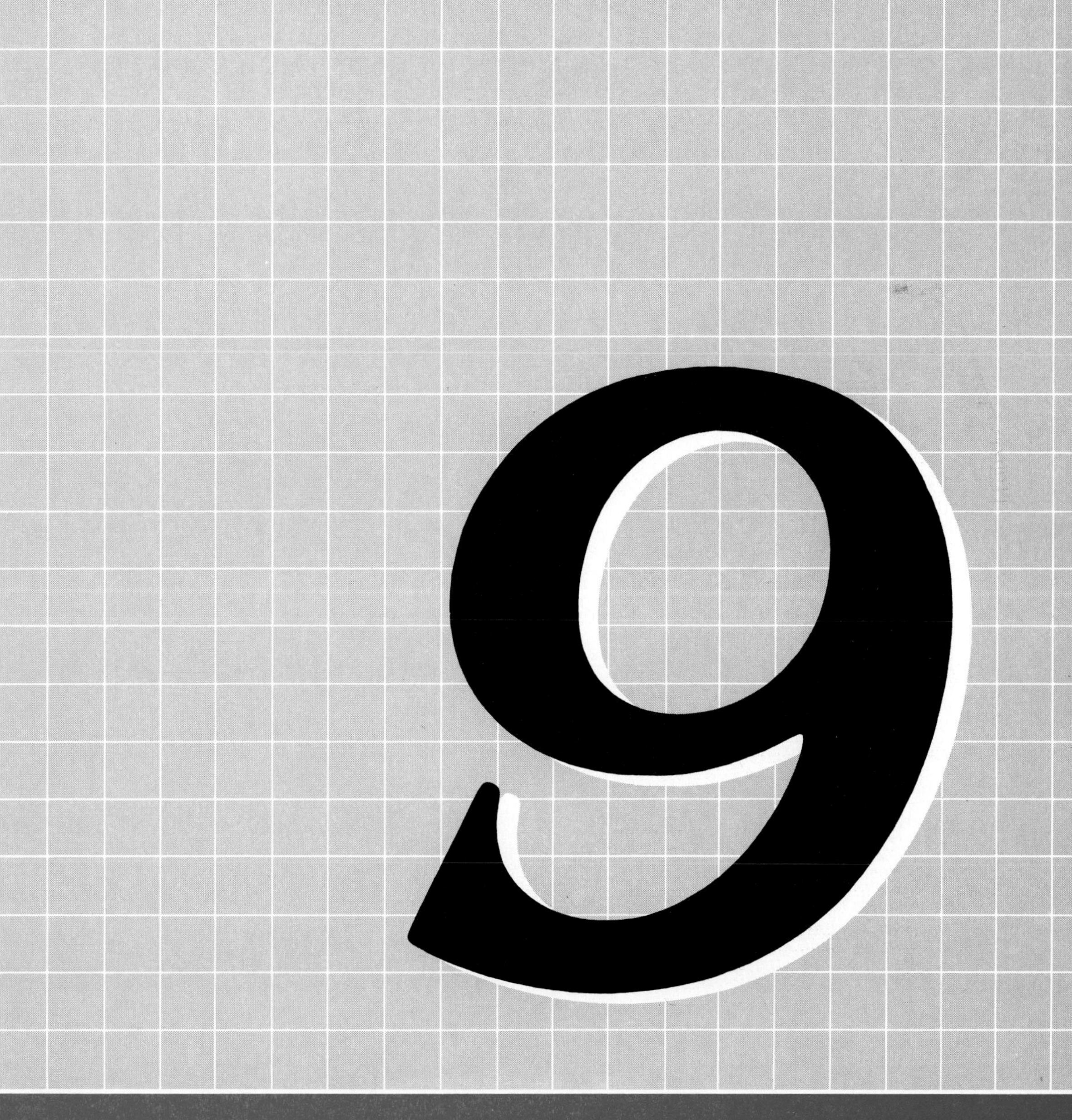

9 Quadratic Equations

Diagnostic Pretest for Chapter 9

This diagnostic pretest is intended for those students who have previously studied algebra. Work as many problems as you can. Then, check your answers against those listed at the back of the book. Where you have obtained an incorrect answer or have been unable to work the problem, you will need to study the corresponding sections of the chapter.

If you have not previously studied algebra, do not take this test now. You should study the chapter and work the problems and exercises. After finishing the chapter, you may use this test as a review or as a practice test.

Follow the directions and solve each problem.

Section 9.1

Solve each quadratic equation by the *factoring method.*

1. $x^2 - 12x - 28 = 0$

2. $3x^2 + 2x = 6x$

3. $5x^2 = 22x - 8$

Section 9.2

Solve each quadratic equation by the *completing the square method.*

4. $x^2 + 8x + 5 = 0$

5. $2x^2 + 3x - 7 = 0$

Section 9.3

Solve each quadratic equation by using the *quadratic formula.*

6. $3x^2 - 2x - 5 = 0$

7. $4x^2 = -7x + 2$

Section 9.4

Use the *Pythagorean theorem* to find the length of the third side of the right triangle labeled x.

8.

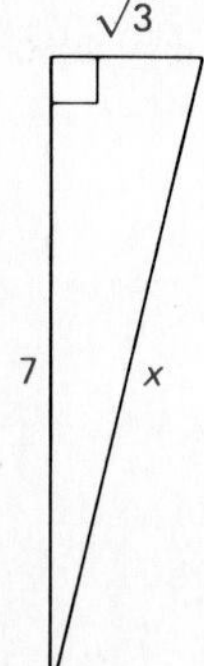

9.

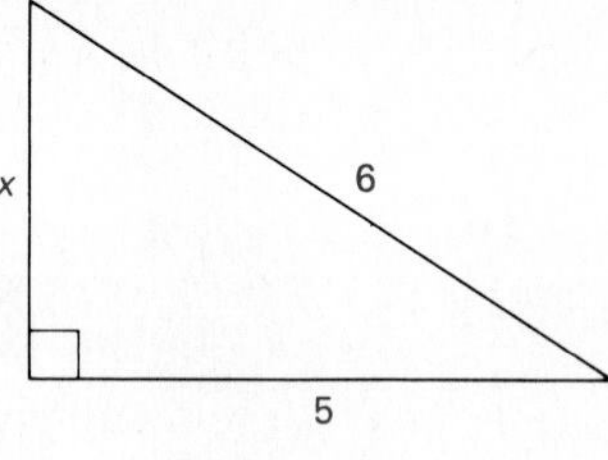

10. Solve the radical equation and verify your answer:

$$2 + \sqrt{3x + 4} = x$$

Section 9.5

Graph the quadratic equations. Plot at least six points for each graph.

11. $y = 5 - 3x^2$

12. $y = 2x^2 - 3x - 1$

Section 9.6

The two problems in this section are more rapidly solved using one variable and a quadratic equation.

13. A triangle has an area of 39 square centimeters. The altitude of the triangle is 1 centimeter longer than double the length of the base. Find the dimensions of the triangle.

14. A number is added to twice its reciprocal. The result is $\frac{17}{6}$. Find all the numbers that satisfy these properties.

Student Learning Objectives for Chapter 9

After successfully learning the content of each section, you will be able to:

Section 9.1

A. Perform necessary algebraic operations to place a quadratic equation in the standard form $ax^2 + bx + c = 0$.

B. Solve a quadratic equation by the *factoring method*.

C. Verify that the apparent solutions to a quadratic equation are in fact correct.

Section 9.2

D. Solve a quadratic equation by the method of *completing the square*.

Section 9.3

E. Memorize the *quadratic formula* and understand how it was obtained.

F. Solve a quadratic equation by the *quadratic formula*.

G. Identify when there are no real number solutions to a quadratic equation.

Section 9.4

H. Find the length of a third side of a right triangle given any two sides through the use of the Pythagorean theorem.

I. Analyze stated word problems involving a right triangle and apply the Pythagorean theorem to solve for the required unknown quantity.

J. Solve an equation involving a square root radical and verify any apparent solutions.

Section 9.5

K. Identify the general shape and properties of a *parabola*.

L. Graph any given quadratic equation of the form $y = ax^2 + bx + c$.

Section 9.6

M. Solve a variety of applied word problems that can be solved through the use of one quadratic equation containing one variable.

9.1 Solutions by Factoring

In Chapter 2 we learned how to solve linear equations such as $3x + 5 = 0$ to find the root (or value of x) that satisfied the equation. Now we turn to the question of how to solve equations like $3x^2 + 5x + 2 = 0$. Such equations are called *quadratic equations*. A quadratic equation is a polynomial equation that contains a second power of a variable as the highest power in the equation. It is often easiest to write the terms of the quadratic equation with the highest power of the variable first, followed by the first-degree term, and then the constant term.

The *standard form* of a quadratic equation is $ax^2 + bx + c = 0$, where a, b, and c are real numbers and $a \neq 0$.

We use quadratic equations in standard form in solving by the factoring method in this section. Furthermore, it is a common practice to write the quadratic equation in standard form,

$$ax^2 + bx + c = 0$$

with a positive value for a and with a, b, and c all being integers. We follow this generally accepted approach in this chapter. (Quadratic equations can be solved if a, b, and c are not integers, but we will not study that type in this chapter.)

Usually, we can find two real roots that satisfy any quadratic equation. The equation $3x^2 + 5x + 2 = 0$ has two roots. They are $x = -\frac{2}{3}$ and $x = -1$. We can verify this by substituting those values into the original equation to see if we can obtain a true mathematical statement. In other words we want to verify that $x = -\frac{2}{3}$ and $x = -1$ are solutions to the equation $3x^2 + 5x + 2 = 0$, by using our definition of *solution of an equation* that we discussed in Section 2.1.3.

$$3x^2 + 5x + 2 = 0$$

$3\left(-\frac{2}{3}\right)^2 + 5\left(-\frac{2}{3}\right) + 2 \stackrel{?}{=} 0$ Replace x by $-\frac{2}{3}$.

$3\left(\frac{4}{9}\right) + 5\left(-\frac{2}{3}\right) + 2 \stackrel{?}{=} 0$ Square $-\frac{2}{3}$.

$\frac{4}{3} - \frac{10}{3} + 2 \stackrel{?}{=} 0$ Multiply

$\frac{4}{3} - \frac{10}{3} + \frac{6}{3} = \frac{0}{3} \stackrel{?}{=} 0$ Collect like terms. It checks. An identity is obtained.

$0 = 0 \checkmark$

Thus $x = -\frac{2}{3}$ is a root, or a solution to the equation $3x^2 + 5x + 2 = 0$.

$3(-1)^2 + 5(-1) + 2 \stackrel{?}{=} 0$ Replace x by -1.

$3(1) + 5(-1) + 2 \stackrel{?}{=} 0$ Square.

$3 - 5 + 2 \stackrel{?}{=} 0$ Multiply.

$-2 + 2 \stackrel{?}{=} 0$ Collect like terms. It checks.

$0 = 0 \checkmark$ An identity is obtained.

Thus $x = -1$ is also a root, or a solution to the equation $3x^2 + 5x + 2 = 0$.

If there are two roots to a given quadratic equation, how do we find them? The most direct approach is often the factoring method. (If your knowledge of factoring is weak you should review Chapter 3 before studying this section.) It is dependent on a mathematical principle that if the product of two or more factors is 0, then one or both of the factors is 0. We can state this as follows:

Zero Factor Property

If $a \cdot b = 0$ then $a = 0$ or $b = 0$ or $a = b = 0$

We can implement this principle by:

1. Factoring, if possible, a quadratic expression.
2. Set each factor equal to 0.
3. Solve the resulting equations to find each root.
4. Checking each root.

EXAMPLE A Find the two roots: $3x^2 + 5x + 2 = 0$.

Solution

$3x^2 + 5x + 2 = 0$	The equation is in standard form.
$(3x + 2)(x + 1) = 0$	Factor the trinomial.
$3x + 2 = 0 \qquad x + 1 = 0$	Set each factor equal to 0.
$3x = -2 \qquad x = -1$	Solve each equation to find the two roots.
$x = -\frac{2}{3}$	

Check We have already checked to be sure that these two roots satisfy the equation.

EXAMPLE B Find the two roots: $2x^2 + 13x - 7 = 0$.

Solution

$2x^2 + 13x - 7 = 0$	The equation is in standard form.
$(2x - 1)(x + 7) = 0$	Factor the trinomial.
$2x - 1 = 0 \qquad x + 7 = 0$	Set each factor equal to 0.
$2x = 1 \qquad x = -7$	Solve each equation to find the two roots.
$x = \frac{1}{2}$	

Check If $x = \frac{1}{2}$, then

$$2\left(\frac{1}{2}\right)^2 + 13\left(\frac{1}{2}\right) - 7 = 2\left(\frac{1}{4}\right) + 13\left(\frac{1}{2}\right) - 7$$

$$= \frac{1}{2} + \frac{13}{2} - \frac{14}{2} = 0\checkmark$$

If $x = -7$, then

$$2(-7)^2 + 13(-7) - 7 = 2(49) + 13(-7) - 7$$

$$= 98 - 91 - 7 = 0\checkmark$$

Thus $x = \frac{1}{2}$ and $x = -7$ are both roots for the quadratic equation $2x^2 + 13x - 7 = 0$.

What if there is no constant term?

If the quadratic equation $ax^2 + bx + c = 0$ has $c = 0$, there will not be a constant term. All such quadratic equations can be factored to obtain two solutions that are real numbers.

EXAMPLE C Find the two roots: $7x^2 - 3x = 0$.

Solution

$7x^2 - 3x = 0$	The equation is in standard form.
$x(7x - 3) = 0$	Factor by removing the common factor.
$x = 0 \qquad 7x - 3 = 0$	Set each factor equal to 0 by the zero factor property.
$7x = 3$	
$x = \frac{3}{7}$	Solve each equation to find the two roots.

Check Can you verify that $x = 0$ and $x = \frac{3}{7}$ are the two roots of $7x^2 - 3x = 0$?

Quadratic equations not in standard form

If the quadratic equation is not in standard form, we use the same basic algebraic methods we studied in Section 2.1 to obtain the form $ax^2 + bx + c = 0$. The purpose of obtaining the standard form is to have everything equal to zero so that we can use the zero factor property.

EXAMPLE D Find the two roots: $x^2 = 12 - x$.

Solution

$x^2 = 12 - x$	The equation is not in standard form.
$x^2 + x - 12 = 0$	Add $(+x - 12)$ to both sides of the equation. The equation is now in standard form.
$(x - 3)(x + 4) = 0$	The left hand side is equal to zero. We can now factor.
$x - 3 = 0 \qquad x + 4 = 0$	Set each factor equal to 0 by the zero factor property.
$x = 3 \qquad x = -4$	Solve each equation for x.

Check

If $x = 3$: $(3)^2 \stackrel{?}{=} 12 - 3$	If $x = -4$: $(-4)^2 \stackrel{?}{=} 12 - (-4)$
$9 \stackrel{?}{=} 12 - 3$	$16 \stackrel{?}{=} 12 + 4$
$9 = 9 \checkmark$	$16 = 16 \checkmark$

Both roots check.

If the equation has fractional expressions, multiply both sides by the LCD to obtain an equivalent equation without fractions.

EXAMPLE E Find the two roots: $8x - 6 + \frac{1}{x} = 0$.

Solution

$$8x - 6 + \frac{1}{x} = 0$$

$$x(8x) - (x)(6) + x\left(\frac{1}{x}\right) = x(0)$$ We will need to multiply each term by the LCD $= x$.

$$8x^2 - 6x + 1 = 0$$ Simplify.

$$(4x - 1)(2x - 1) = 0$$ Factor.

$$4x - 1 = 0 \qquad 2x - 1 = 0$$ Set each factor equal to 0.

$$4x = 1 \qquad 2x = 1$$ Solve each equation for x and obtain the two roots.

$$x = \frac{1}{4} \qquad x = \frac{1}{2}$$

Check Checking fractional roots is more difficult, but you should be able to do it if you work carefully.

$$\text{If } x = \frac{1}{4}: \quad 8\left(\frac{1}{4}\right) - 6 + \frac{1}{\frac{1}{4}} = 2 - 6 + 4 = 0$$

$$0 = 0 \checkmark$$

$$\left(\text{since } 1 \div \frac{1}{4} = 1 \cdot \frac{4}{1} = 4\right)$$

$$\text{If } x = \frac{1}{2}: \quad 8\left(\frac{1}{2}\right) - 6 + \frac{1}{\frac{1}{2}} = 4 - 6 + 2 = 0$$

$$0 = 0 \checkmark$$

Both roots check, so $x = \frac{1}{2}$ and $x = \frac{1}{4}$ are the two roots that satisfy

$$8x - 6 + \frac{1}{x} = 0.$$

Sometimes you will obtain the same value for x twice. This is known as a *double root*.

EXAMPLE F Find the two roots: $x(x + 12) = -36$.

Solution

$$x(x + 12) = -36$$ The equation is not in standard form.

$$x^2 + 12x = -36$$ Remove parentheses.

$$x^2 + 12x + 36 = 0$$ Add 36 to both sides.

$$(x + 6)^2 = 0$$ Factor. (Notice that this is a perfect square trinomial like those studied in Section 3.4.)

$$x + 6 = 0 \qquad x + 6 = 0$$ Set each factor $= 0$.

$$x = -6 \qquad x = -6$$ We obtain a double root of $x = -6$.

Usually, we list the answer only once as $x = -6$, but mention that it is a double root.

Check Can you verify that $x = -6$ is the root of $x(x + 12) = -36$?

Review Examples A to F and then try these practice problems.

Practice Problem for Exercise 9.1

Using the factoring method, solve for the two roots of each quadratic equation. Check your answers.

1. $10x^2 - x - 2 = 0$

2. $x^2 - 6x + 4 = -8 + x$

3. $x - 6 + \dfrac{9}{x} = 0$

Now turn to the answers following Exercise 9.1 and check your solutions for these practice problems.

Exercise 9.1

Using the factoring method, solve for the two roots of each quadratic equation. Check your answers for problems 1 to 12.

1. $x^2 - 3x - 10 = 0$

2. $x^2 + 2x - 24 = 0$

3. $x^2 + 13x + 40 = 0$

4. $x^2 + 16x + 63 = 0$

5. $2x^2 - 5x - 3 = 0$

6. $3x^2 - 5x - 2 = 0$

7. $4x^2 - 11x - 3 = 0$

8. $2x^2 + 15x - 8 = 0$

9. $x^2 - 6x = 16$

10. $x^2 - 21 = 4x$

11. $10 = 7x - x^2$

12. $6 = 7x - x^2$

13. $6x^2 - 13x = -6$

14. $10x^2 + 19x = 15$

15. $x^2 - 8x = 0$

16. $5x^2 - x = 0$

17. $3x^2 = 2x$

18. $8x = 6x^2$

19. $5x^2 + 3x = 8x$

20. $6x^2 - 4x = 3x$

21. $x(9 + x) = 4(2x + 5)$

22. $(x - 2)(x - 3) = 6(x - 3)$

23. $x^2 - 12x + 36 = 0$

24. $x^2 - 14x + 49 = 0$

25. $4x^2 - 3x + 1 = -7x$

26. $9x^2 - 2x + 4 = 10x$

27. $x + \frac{12}{x} = 8$

28. $x - 8 = -\frac{15}{x}$

29. $\frac{x}{2} + \frac{4}{x} = -3$

30. $\frac{x}{2} + \frac{5}{2} = -\frac{3}{x}$

31. $\frac{x^2 - 3x}{2} = 27$

32. $\frac{x^2 - 9x}{2} = -9$

33. $6x - 2x^2 = -8$

34. $6x - 3x^2 = 3$

35. $60 - 2x^2 = 2x$

36. $4x^2 = 28 + 24x$

37. $\frac{x^2}{2} - 8 + x = -8$

38. $4 + \frac{x^2}{3} = 2x + 4$

Challenge Problem

39. Let a, b, d, and f represent constant values. Solve for x by using the factoring method.

$$abx^2 + adx - bfx - fd = 0$$

Solutions for Practice Problems for Exercise 9.1

1.

$$10x^2 - x - 2 = 0$$

$$(5x + 2)(2x - 1) = 0 \qquad \text{Factor.}$$

$$5x + 2 = 0 \qquad 2x - 1 = 0 \qquad \text{Set factors equal to 0.}$$

$$5x = -2 \qquad 2x = 1$$

$$x = -\frac{2}{5} \qquad x = \frac{1}{2} \qquad \text{Solve for each root.}$$

Check

$$10\left(-\frac{2}{5}\right)^2 - \left(-\frac{2}{5}\right) - 2 = 10\left(\frac{4}{25}\right) + \frac{2}{5} - 2$$

$$= \frac{8}{5} + \frac{2}{5} - \frac{10}{5} = 0 \checkmark$$

$$10\left(\frac{1}{2}\right)^2 - \left(\frac{1}{2}\right) - 2 = 10\left(\frac{1}{4}\right) - \frac{1}{2} - 2$$

$$= \frac{5}{2} - \frac{1}{2} - \frac{4}{2} = 0 \checkmark$$

2.

$$x^2 - 6x + 4 = -8 + x \qquad \text{It is not in standard form.}$$

$$x^2 - 7x + 12 = 0 \qquad \text{Add } (8 - x) \text{ to both sides. It is now in standard form.}$$

$$(x - 4)(x - 3) = 0 \qquad \text{Set each factor equal to 0.}$$

$$x = 4 \qquad x = 3 \qquad \text{Solve each equation.}$$

Check

$$(4)^2 - 6(4) + 4 \stackrel{?}{=} -8 + 4 \qquad (3)^2 - 6(3) + 4 \stackrel{?}{=} -8 + 3$$

$$16 - 24 + 4 \stackrel{?}{=} -4 \qquad 9 - 18 + 4 \stackrel{?}{=} -5$$

$$-4 = -4 \checkmark \qquad -5 = -5 \checkmark$$

3.

$$x - 6 + \frac{9}{x} = 0 \qquad \text{It is not in standard form.}$$

$$x(x) - x(6) + x\left(\frac{9}{x}\right) = x(0) \qquad \text{Multiply each term by the LCD} = x.$$

$$x^2 - 6x + 9 = 0 \qquad \text{Simplify. The equation is now in standard form.}$$

$$(x - 3)^2 = 0 \qquad \text{Factor.}$$

$$x - 3 = 0 \qquad x - 3 = 0$$

$$x = 3 \qquad x = 3 \qquad \text{We obtain a double root of } x = 3.$$

Check

$$3 - 6 + \frac{9}{3} = 3 - 6 + 3 = 0 \checkmark$$

Now do the problems in Exercise 9.1.

9.2 Completing the Square

In many cases, we are not able to obtain the solution of a quadratic equation by the factoring method. In cases where we cannot factor the trinomial by the methods of Chapter 3, we must develop an approach that will yield the two real roots for the quadratic equation (if such roots exist). Before we learn the method of completing the square there is one basic property that is very useful for certain special quadratic equations. We can state it as follows.

> **For all nonnegative real numbers a, if $x^2 = a$, then $x = \sqrt{a}$ or $x = -\sqrt{a}$.**

EXAMPLE A Solve: $x^2 = 49$.

Solution From the property above, we see that

$$x = \sqrt{49} \quad \text{or} \quad x = -\sqrt{49} \qquad \text{Simplify each radical.}$$

$$x = 7 \quad \text{or} \quad x = -7$$

If the value of a is not a perfect square, the roots of the quadratic equation will thus be irrational numbers.

EXAMPLE B Solve: $x^2 = 13$.

Solution We obtain two irrational numbers for our solution.

$$x = \sqrt{13} \quad \text{or} \quad x = -\sqrt{13}$$

If we wish to have a decimal approximation for these numbers, we can use the square root table in Appendix A to obtain

$$x \doteq 3.606 \quad \text{or} \quad x \doteq -3.606$$

EXAMPLE C Solve: $x^2 = 20$. Leave the solutions as irrational numbers in simplified form.

Solution

$$x = \sqrt{20} \quad \text{or} \quad x = -\sqrt{20}$$

However, we can simplify $\sqrt{20}$ since $\sqrt{20} = \sqrt{4}\sqrt{5} = 2\sqrt{5}$. Therefore, we have

$$x = 2\sqrt{5} \quad \text{or} \quad x = -2\sqrt{5}$$

Mathematicians usually abbreviate these two answers by using the $\pm$ symbol, which is read as "plus or minus." It means that there are two values, one with the plus sign and the other with the minus sign.

EXAMPLE D Solve: $5x^2 = 125$.

Solution

$$5x^2 = 125 \qquad \text{We want to obtain the form } x^2 = a.$$

$$x^2 = 25 \qquad \text{We divide both sides by 5.}$$

$$x = \pm 5$$

We express the solutions $x = +5$ or $x = -5$ by the new notation $x = \pm 5$. This is read "x equals plus or minus 5."

When we are doing this type of problem, aren't we just taking the square root of both sides of the equation?

Essentially, yes. But you must remember that there are two possibilities for the sign.

If you have an equation such as $x^2 = a$ and you wish to find the square root of each side of the equation, you must include the two signs at the step of taking a square root.

EXAMPLE E Solve: $x^2 - 40 = 0$. Leave the solutions as irrational numbers in simplified form.

Solution

$x^2 - 40 = 0$ — We want to obtain the form $x^2 = a$.

$x^2 = 40$ — Add 40 to both sides.

$\sqrt{x^2} = \pm\sqrt{40}$ — Take the square root of both sides.

↑ Two sign possibilities are necessary.

$x = \pm\sqrt{40}$ — We observe that $\sqrt{40}$ can be simplified.

$x = \pm 2\sqrt{10}$ — Since $\sqrt{40} = \sqrt{4}\sqrt{10} = 2\sqrt{10}$.

The left side of the equation may be a binomial.

EXAMPLE F Solve: $(x - 2)^2 = 36$.

Solution

$(x - 2)^2 = 36$

$\sqrt{(x - 2)^2} = \pm\sqrt{36}$ — Take the square root of both sides. Remember both sign possibilities.

$x - 2 = \pm 6$ — Simplify.

Note that we now have two equations since $\pm$ indicates plus or minus. We will solve each equation.

$x - 2 = +6 \qquad x - 2 = -6$

$x = 8 \qquad x = -4$

The two roots of the equation are $x = 8$ and $x = -4$.

Using this method we may obtain an irrational number. This will occur if the number on the right side of the equation is not a perfect square.

EXAMPLE G Solve: $(3x + 1)^2 = 8$.

Solution

$(3x + 1)^2 = 8$

$\sqrt{(3x + 1)^2} = \pm\sqrt{8}$ — Take the square root of both sides. Remember both sign possibilities.

$3x + 1 = \pm 2\sqrt{2}$ — We simplify as much as possible.

Now we must solve the two equations expressed by the plus or minus statement.

$$3x + 1 = +2\sqrt{2} \qquad 3x + 1 = -2\sqrt{2}$$

$$3x = 2\sqrt{2} - 1 \qquad 3x = -2\sqrt{2} - 1$$ Add -1 to each side.

$$x = \frac{2\sqrt{2} - 1}{3} \qquad x = \frac{-2\sqrt{2} - 1}{3}$$ Divide both sides by coefficient of x.

The roots of this quadratic equation are irrational numbers. They are

$$x = \frac{2\sqrt{2} - 1}{3} \quad \text{and} \quad x = \frac{-2\sqrt{2} - 1}{3}$$

If a quadratic equation is not in a form such as those in Examples F and G, we can usually transform it to an equivalent form by adding a constant to both sides of the equation.

EXAMPLE H Solve: $x^2 - 2x = 8$.

Solution

$$x^2 - 2x = 8$$ We observe that the left side is not a binomial squared.

$$x^2 - 2x + 1 = 8 + 1$$ Add 1 to both sides. (We will explain why shortly!)

$$(x - 1)^2 = 9$$ Since $x^2 - 2x + 1 = (x - 1)^2$, we can write the left side as a binomial squared.

$$\sqrt{(x - 1)^2} = \pm\sqrt{9}$$ Take the square root of both sides.

$$x - 1 = \pm 3$$

$$x - 1 = 3 \qquad x - 1 = -3$$ Solve each equuation for x.

$$x = 4 \qquad x = -2$$

Why do we add 1 to both sides of the equation? How do we know what number should be added?

We want the left-hand side of the equation to be in the form of a perfect square trinomial. Consider this list of perfect square trinomials and their corresponding factored forms (for a further review, look over Section 3.4):

$$x^2 + 4x + 4 = (x + 2)^2$$

$$x^2 + 10x + 25 = (x + 5)^2$$

$$x^2 - 6x + 9 = (x - 3)^2$$

$$x^2 - 8x + 16 = (x - 4)^2$$

The major pattern we want to observe is how to predict the last term of the trinomial if we know the numerical coefficient of the middle term.

1. In the trinomial $x^2 + 4x + 4$ the last term, 4, is the square of $\frac{1}{2}$ of the numerical coefficient of the middle term, thus:

$$\left[\frac{1}{2} \cdot 4\right]^2 = [2]^2 = 4$$

2. In the trinomial $x^2 + 10x + 25$ the last term 25 is (10/2) squared. That is,

$$\left[\frac{1}{2} \cdot 10\right]^2 = (5)^2 = 25$$

3. In the trinomial $x^2 - 6x + 9$ we find $\frac{1}{2}$ of (-6) and square the result to obtain the last term 9:

$$\left[\left(\frac{1}{2}\right)(-6)\right]^2 = (-3)^2 = 9$$

It is this last term that we wish to add to both sides so that the left side of the equation is a perfect square trinomial.

EXAMPLE I Solve: $x^2 + 12x = 4$.

Solution

$x^2 + 12x + \_ = 4 + \_$ — We want to add a number to both sides so that the left side is a perfect square trinomial.

$x^2 + 12x + 36 = 4 + 36$ — We take $\left[\frac{1}{2} \cdot 12\right]^2 = (6)^2 = 36$ as the number to add.

$x^2 + 12x + 36 = 40$ — Simplify.

$(x + 6)^2 = 40$ — Factor the left side.

$\sqrt{(x + 6)^2} = \pm\sqrt{40}$ — Take the square root of both sides.

$x + 6 = \pm 2\sqrt{10}$ — We simplify $\sqrt{40}$.

$x + 6 = +2\sqrt{10}$ $\quad$ $x + 6 = -2\sqrt{10}$ — We write the two equations.

$x = -6 + 2\sqrt{10}$ $\quad$ $x = -6 - 2\sqrt{10}$ — Solve to obtain the two roots.

Look over Examples H and I and see if you see the pattern of steps needed to obtain the solutions.

Here is a summary of the steps to be used for any quadratic equation. In some cases you will not need to use each step.

Procedure to Solve a Quadratic Equation by Completing the Square

1. **Perform any necessary steps to transform the equation to the form $ax^2 + bx = c$.**
2. **If $a \neq 1$, divide each term of the equation by a.**
3. **Square half of the numerical coefficient of the x term. Add the result to both sides of the equation. (In other words, if the equation is in the form $x^2 + dx = e$ then add $(d/2)^2$ to both sides.)**
4. **The left side is now a perfect trinomial square so you can factor the left side. Then take the square root of both sides of the equation. (If the radicand is *negative* there will be no real number solution.)**
5. **Solve each resulting equation for x.**

The number you add to both sides may be a fraction. We will now employ the 5 step procedure to solve Example J.

EXAMPLE J Solve: $x^2 - 3x = 1$.

Solution

Steps 1, 2 are not needed.

$$x^2 - 3x + \_\_ = 1 + \_\_$$

Step 3. We calculate $\left[\frac{1}{2}(-3)\right]^2 = \left(-\frac{3}{2}\right)^2 = \frac{9}{4}$ as the number to add to both sides.

$$x^2 - 3x + \frac{9}{4} = \frac{4}{4} + \frac{9}{4}$$

We write 1 as $\frac{4}{4}$.

$$\left(x - \frac{3}{2}\right)^2 = \frac{13}{4}$$

Step 4. Factor the left side and collect like terms on the right side.

$$\sqrt{\left(x - \frac{3}{2}\right)^2} = \pm\sqrt{\frac{13}{4}}$$

Take the square root of both sides.

$$x - \frac{3}{2} = \pm\frac{\sqrt{13}}{2}$$

Step 5. Simplify, using the properties of radicals.

$$x - \frac{3}{2} = +\frac{\sqrt{13}}{2} \qquad x - \frac{3}{2} = -\frac{\sqrt{13}}{2}$$

Write the two equations.

$$x = \frac{3}{2} + \frac{\sqrt{13}}{2} \qquad x = \frac{3}{2} - \frac{\sqrt{13}}{2}$$

Solve each equation for x.

$$x = \frac{3 + \sqrt{13}}{2} \qquad x = \frac{3 - \sqrt{13}}{2}$$

Since the two fractions have a common denominator, we write them as one fraction.

If the numerical coefficient of the squared variable is not 1, we divide all terms of the equation by that numerical coefficient.

EXAMPLE K Solve: $4y^2 + 4y - 3 = 0$

Solution

$$4y^2 + 4y - 3 = 0$$

$$4y^2 + 4y = 3$$

Step 1. First add 3 to both sides.

$$y^2 + y = \frac{3}{4}$$

Step 2. Divide all terms by the numerical coefficient of the y^2 term.

$$y^2 + y + \frac{1}{4} = \frac{3}{4} + \frac{1}{4}$$

Step 3. The coefficient of y is 1: $\left[\left(\frac{1}{2}\right)1\right]^2 = \left(\frac{1}{2}\right)^2 = \frac{1}{4}$. Add $\frac{1}{4}$ to both sides.

$$\left(y + \frac{1}{2}\right)^2 = 1$$

Step 4. Factor and collect like terms.

$$\sqrt{\left(y + \frac{1}{2}\right)^2} = \pm\sqrt{1}$$

Take the square root of both sides.

$$y + \frac{1}{2} = \pm 1$$ *Step 5.* Simplify.

$$y + \frac{1}{2} = 1 \qquad y + \frac{1}{2} = -1$$ Write the two equations.

$$y = \frac{2}{2} - \frac{1}{2} \qquad y = -\frac{2}{2} - \frac{1}{2}$$ Subtract $\frac{1}{2}$ from each side.

$$y = \frac{1}{2} \qquad y = -\frac{3}{2}$$ Solve each equation for y.

This method will enable us to solve any quadratic equation that has real roots. It is usually faster, however, to factor the equadratic equation if the polynomial is factorable.

Now try these practice problems.

Practice Problems for Exercise 9.2

Solve for x by taking the square root of each side of the equation. Complete the square, if necessary.

1. $x^2 - 50 = 0$

2. $x^2 - 8x = 20$

3. $6x^2 - 6x - 3 = 0$

Now turn to the answers following Exercise 9.2 and check your solutions for these practice problems.

Exercise 9.2

Solve each quadratic equation by taking the square root of each side.

1. $x^2 = 49$

2. $x^2 = 100$

3. $x^2 - 75 = 0$

4. $x^2 - 40 = 0$

5. $3x^2 = 75$

6. $4x^2 = 64$

7. $4x^2 - 12 = 0$

8. $5x^2 - 15 = 0$

9. $2x^2 - 10 = 62$

10. $3x^2 - 18 = 57$

11. $(x - 3)^2 = 5$

12. $(x + 2)^2 = 3$

13. $(2x + 5)^2 = 2$

14. $(3x - 4)^2 = 6$

Solve each quadratic equation by completing the square.

15. $x^2 + 14x = 15$

16. $x^2 - 2x = 5$

17. $x^2 - 4x = 11$

18. $x^2 + 4x = 12$

19. $x^2 + 6x + 7 = 0$

20. $x^2 - 12x - 4 = 0$

21. $x^2 - 10x - 4 = 0$

22. $x^2 + 20x + 72 = 0$

23. $x^2 + 3x = 0$

24. $x^2 + 5x = 3$

25. $x^2 - x - 6 = 0$

26. $x^2 + x - 20 = 0$

27. $2x^2 - 3x = 9$

28. $2x^2 + 10x + 11 = 0$

29. $4x^2 - 8x + 3 = 0$

30. $3x^2 - 6x - 2 = 0$

31. $2y^2 + 4y = 7$

32. $2y^2 - 7y = 9$

33. $3y^2 + 5y - 2 = 0$

34. $3y^2 - 8y + 1 = 0$

Challenge Problem

35. Let b represent a constant value. Solve for x by completing the square.

$$x^2 + bx - 7 = 0$$

Solutions to Practice Problems for Exercise 9.2

1. $x^2 - 50 = 0$

$x^2 = 50$ — Add 50 to both sides.

$\sqrt{x^2} = \pm\sqrt{50}$ — Take square root of both sides. Remember both sign possibilities.

$x = \pm 5\sqrt{2}$ — Simplify using properties of radicals.

2. $x^2 - 8x = 20$	Take one half of the coefficient of x and square it.
$x^2 - 8x + 16 = 20 + 16$	$\left[\left(\frac{1}{2}\right)(-8)\right]^2 = (-4)^2 = 16$. Add this value of 16 to both sides.
$(x - 4)^2 = 36$	Factor and collect like terms.
$\sqrt{(x - 4)^2} = \pm\sqrt{36}$	Take the square root of each side.
$x - 4 = \pm 6$	Simplify.
$x - 4 = 6 \qquad x - 4 = -6$	Solve each equation for x.
$x = 10 \qquad x = -2$	
3. $6x^2 - 6x - 3 = 0$	
$6x^3 - 6x = +3$	Add 3 to both sides.
$x^2 - x = +\frac{1}{2}$	Divide all terms by 6 since the x^2 term coefficient is not 1.
	Take one half of the coefficient of x and square it,
$x^2 - x + \frac{1}{4} = +\frac{2}{4} + \frac{1}{4}$	$\left[(-1)\left(\frac{1}{2}\right)\right]^2 = \left(-\frac{1}{2}\right)^2 = \frac{1}{4}$. So add $\frac{1}{4}$ to both sides.
$\left(x - \frac{1}{2}\right)^2 = \frac{3}{4}$	Factor and collect like terms.
$\sqrt{\left(x - \frac{1}{2}\right)^2} = \pm\sqrt{\frac{3}{4}}$	Take square root of each side.
$x - \frac{1}{2} = \pm\frac{\sqrt{3}}{2}$	Simplify.
$x - \frac{1}{2} = \frac{\sqrt{3}}{2} \qquad x - \frac{1}{2} = -\frac{\sqrt{3}}{2}$	Write two equations.
$x = \frac{1}{2} + \frac{\sqrt{3}}{2} \qquad x = \frac{1}{2} - \frac{\sqrt{3}}{2}$	Add $\frac{1}{2}$ to each side.
$x = \frac{1 + \sqrt{3}}{2} \qquad x = \frac{1 - \sqrt{3}}{2}$	Write each answer in the form of one fraction.

Now do the problems in Exercise 9.2.

9.3 The Quadratic Formula

If a quadratic equation can be factored, it is usually best to use the factoring method. If it does not factor, we can use the method of completing the square to find any real number roots that may exist for the equation. However, the method of completing the square is not a rapid method for many students. Often, students find that quadratic equations can be solved more quickly by using the quadratic formula.

The quadratic formula is a procedure to find the roots of a quadratic equation. The formula utilizes the method of completing the square for any general quadratic equation $ax^2 + bx + c = 0$, where a, b, and c are real numbers and $a \neq 0$.

Quadratic Formula

The roots of any quadratic equation in the form $ax^2 + bx + c = 0$, where a, b, and c are real numbers and $a \neq 0$, are

$$x = \frac{-b \pm \sqrt{b^2 - 4ac}}{2a}$$

Where does this formula come from? By using the steps of completing the square on the equation $ax^2 + bx + c = 0$. So that you can follow the steps more easily, we will solve the specific equation $2x^2 + 3x - 1 = 0$ at the same time. You can compare the two solutions at any step. One is general, the other specific.

General Quadratic Equation	*Specific Quadratic Equation*
$ax^2 + bx + c = 0$	$2x^2 + 3x - 1 = 0$

Add to obtain the constant on the right side of the equation.

$ax^2 + bx = -c$	$2x^2 + 3x = 1$

Divide each term by the coefficient of x^2.

$x^2 + \frac{b}{a}x = -\frac{c}{a}$	$x^2 + \frac{3}{2}x = \frac{1}{2}$

Take $\frac{1}{2}$ of the coefficient of the x term.

$\frac{1}{2}\left(\frac{b}{a}\right) = \frac{b}{2a}$	$\frac{1}{2}\left(\frac{3}{2}\right) = \frac{3}{4}$

Add the square of this value to both sides.

$x^2 + \frac{b}{a}x + \frac{b^2}{4a^2} = \frac{b^2}{4a^2} - \frac{c}{a}$	$x^2 + \frac{3}{2}x + \frac{9}{16} = \frac{9}{16} + \frac{1}{2}$

In each case the left side of the equation is a perfect trinomial. Therefore we can write the left side in factored form.

$\left(x + \frac{b}{2a}\right)^2 = \frac{b^2}{4a^2} - \frac{c}{a}$	$\left(x + \frac{3}{4}\right)^2 = \frac{9}{16} + \frac{1}{2}$

Transform the two fractions on the right to fractions with a common denominator.

$\left(x + \frac{b}{2a}\right)^2 = \frac{b^2}{4a^2} - \frac{4ac}{4a^2}$	$\left(x + \frac{3}{4}\right)^2 = \frac{9}{16} + \frac{8}{16}$

Combine the two fractions on the right side of the equation.

$\left(x + \frac{b}{2a}\right)^2 = \frac{b^2 - 4ac}{4a^2}$	$\left(x + \frac{3}{4}\right)^2 = \frac{17}{16}$

Take the square root of each side.

$x + \frac{b}{2a} = \pm\sqrt{\frac{b^2 - 4ac}{4a^2}}$	$x + \frac{3}{4} = \pm\sqrt{\frac{17}{16}}$

Simplify each radical.

$x + \frac{b}{2a} = \frac{\pm\sqrt{b^2 - 4ac}}{2a}$	$x + \frac{3}{4} = \frac{\pm\sqrt{17}}{4}$

Isolate the x by adding the appropriate expression to each side of the equation.

$x = -\frac{b}{2a} \pm \frac{\sqrt{b^2 - 4ac}}{2a}$	$x = -\frac{3}{4} \pm \frac{\sqrt{17}}{4}$

Since the two fractions have a common denominator, write them as one fraction.

$\underbrace{x = \frac{-b \pm \sqrt{b^2 - 4ac}}{2a}}$	$x = \frac{-3 \pm \sqrt{17}}{4}$

This equation is in fact the quadratic formula.

Now we wish to use the quadratic formula.

EXAMPLE A Solve by the quadratic formula: $3x^2 + 10x + 7 = 0$.

Solution In $3x^2 + 10x + 7 = 0$ we have

$$a = 3 \quad \text{(the coefficient of } x^2\text{)}$$

$$b = 10 \quad \text{(the coefficient of } x\text{)}$$

$$c = 7 \quad \text{(the constant term)}$$

$$x = \frac{-b \pm \sqrt{b^2 - 4ac}}{2a} = \frac{-10 \pm \sqrt{(10)^2 - 4(3)(7)}}{2(3)}$$

$$= \frac{-10 \pm \sqrt{100 - 84}}{6}$$

$$= \frac{-10 \pm \sqrt{16}}{6}$$

$$= \frac{-10 \pm 4}{6}$$

$$x = \frac{-10 + 4}{6} = \frac{-6}{6} = -1 \qquad \text{Using the positive sign.}$$

$$x = \frac{-10 - 4}{6} = \frac{-14}{6} = -\frac{7}{3} \qquad \text{Using the negative sign.}$$

Thus the two solutions are $x = -1$ and $x = -\frac{7}{3}$.

[*Note*: Here we obtain rational roots. We would obtain the same answer by factoring $3x^2 + 10x + 7 = 0$ into $(3x + 7)(x + 1)$ and setting each factor $= 0$.]

EXAMPLE B Solve by the quadratic formula: $x^2 + 5x - 3 = 0$.

Solution

$$a = 1 \qquad b = 5 \qquad c = -3$$

$$x = \frac{-b \pm \sqrt{b^2 - 4ac}}{2a} = \frac{-5 \pm \sqrt{(5)^2 - 4(1)(-3)}}{2(1)} = \frac{-5 \pm \sqrt{25 + 12}}{2} = \frac{-5 \pm \sqrt{37}}{2}$$

Thus the two solutions are

$$x = \frac{-5 + \sqrt{37}}{2} \quad \text{and} \quad x = \frac{-5 - \sqrt{37}}{2}$$

Sometimes the roots can be simplified even though we obtain irrational numbers. You should always leave your answer in simplest form.

EXAMPLE C Solve by the quadratic formula: $2x^2 - 4x - 9 = 0$.

Solution

$$a = 2 \qquad b = -4 \qquad c = -9$$

$$x = \frac{-b \pm \sqrt{b^2 - 4ac}}{2a} = \frac{-(-4) \pm \sqrt{(-4)^2 - 4(2)(-9)}}{2(2)}$$

$$= \frac{4 \pm \sqrt{16 + 72}}{4} = \frac{4 \pm \sqrt{88}}{4}$$

Notice that we can simplify $\sqrt{88}$, so

$$= \frac{4 \pm 2\sqrt{22}}{4} = \frac{2(2 \pm \sqrt{22})}{4} = \frac{\overset{1}{\not{2}}(2 \pm \sqrt{2})}{\underset{2}{\not{4}}}$$

$$x = \frac{2 \pm \sqrt{22}}{2}$$

> **WARNING: Students often make a mistake here. Be sure that you do *not* cancel the 4's! It is *not* correct to do this:**
>
> $$\frac{4 \pm 2\sqrt{22}}{4} \neq \frac{\overset{1}{\not{4}} \pm 2\sqrt{22}}{\underset{1}{\not{4}}}$$
>
> **Never cancel a value that is not a *factor* of the numerator. Please factor *before* you try to reduce the fraction! You cannot cancel terms. You may only cancel factors.**

? What if we want to know an approximate value for these roots?

Then we would use the square root table. From the table in Appendix A we have $\sqrt{22} \doteq 4.690$. If we wanted to know a decimal approximation for our answer to Example C, we would substitute that value. The results are

$$x \doteq \frac{2 + 4.690}{2} = \frac{6.690}{2} = 3.345$$

and

$$x \doteq \frac{2 - 4.690}{2} = \frac{-2.690}{2} = -1.345$$

> **The quadratic equation *must* be written in the general form $ax^2 + bx + c = 0$ *before* the quadratic equation can be used. Several algebraic steps may be needed to accomplish that objective.**

EXAMPLE D Solve by the quadratic formula:

$$2x^2 = 1 - \frac{7}{3}x$$

Solution

$2x^2 = 1 - \frac{7}{3}x$	The equation is not in general form
$3(2x^2) = 3(1) - 3\left(\frac{7}{3}x\right)$	Multiply each term by the LCD of 3.
$6x^2 = 3 - 7x$	Simplify
$6x^2 + 7x - 3 = 0$	Add $7x - 3$ to both sides.
$a = 6 \quad b = 7 \quad c = -3$	The equation is now in the proper form to determine a, b, and c.

Note: It is actually possible to use the quadratic formula if a, b, and c are fractions, that is, rational numbers. However the resulting calculations are usually very difficult. Therefore we stress the first step of Example D as a very important procedure that students should always follow.

$$x = \frac{-7 \pm \sqrt{(7)^2 - 4(6)(-3)}}{2(6)}$$

We substitute $a = 6$, $b = 7$, $c = -3$ into the quadratic formula.

$$= \frac{-7 \pm \sqrt{49 + 72}}{12}$$

$$= \frac{-7 \pm \sqrt{121}}{12} = \frac{-7 \pm 11}{12}$$

Thus we have $x = -\frac{18}{12} = -\frac{3}{2}$ and $x = \frac{4}{12} = \frac{1}{3}$.

The expression $b^2 - 4ac$ is often called the *discriminant*. 1. If $b^2 - 4ac$ is <0, then the radicand will be negative. In such a case there is no real number solution. 2. If $b^2 - 4ac = 0$ then the radicand will be zero. A double root will be obtained. 3. If $b^2 - 4ac > 0$, then two real roots will be obtained.

EXAMPLE E Solve by the quadratic formula: $2x^2 + 5 = -3x$.

Solution

$2x^2 + 5 = -3x$ — The equation is not in the proper form.

$2x^2 + 3x + 5 = 0$ — Add $3x$ to both sides.

$a = 2 \quad b = 3 \quad c = 5$ — The equation is now in proper form. We can find a, b, and c.

$$x = \frac{-3 \pm \sqrt{(3)^2 - 4(2)(5)}}{2(2)} = \frac{-3 \pm \sqrt{9 - 40}}{4} = \frac{-3 \pm \sqrt{-31}}{4}$$

There is no real number indicated by $\sqrt{-31}$. Since we are using only *real numbers* in this text, there is *no solution* to the problem. (*Note*: In more advanced math courses, these types of roots, called *complex numbers*, will be studied.)

Can irrational roots be checked by substituting back in the original equation?

Yes, but the operations are involved and it takes several steps. You will have to work carefully as you work through the substitution.

EXAMPLE F Solve by the quadratic formula: $x^2 + 3x - 1 = 0$. Check one of your answers by substitution.

Solution

$$a = 1 \quad b = 3 \quad c = -1$$

$$x = \frac{-3 \pm \sqrt{(3)^2 - 4(1)(-1)}}{2(1)} = \frac{-3 \pm \sqrt{9 + 4}}{2} = \frac{-3 \pm \sqrt{13}}{2}$$

Thus the two roots are

$$x = \frac{-3 + \sqrt{13}}{2} \quad \text{and} \quad x = \frac{-3 - \sqrt{13}}{2}$$

We will verify that the first root is correct by substitution.

$$x^2 + 3x - 1 = 0$$ Replace each x by $\dfrac{-3 + \sqrt{13}}{2}$.

$$\left(\frac{-3 + \sqrt{13}}{2}\right)^2 + 3\left(\frac{-3 + \sqrt{13}}{2}\right) - 1 \stackrel{?}{=} 0$$

$$\frac{9 - 6\sqrt{13} + 13}{4} + \frac{-9 + 3\sqrt{13}}{2} - 1 \stackrel{?}{=} 0$$ Square the first expression (Formula B of Chapter 3). Multiply the second expression by 3.

$$\frac{22 - 6\sqrt{13}}{4} + \frac{-9 + 3\sqrt{13}}{2} - 1 \stackrel{?}{=} 0$$ Collect like terms in the numerator of the first fraction.

$$\frac{{}^{1}\cancel{2}(11 - 3\sqrt{13})}{\cancel{4}\,2} + \frac{-9 + 3\sqrt{13}}{2} - 1 \stackrel{?}{=} 0$$ Factor the numerator of the first fraction, then reduce.

$$\frac{11 - 3\sqrt{13} - 9 + 3\sqrt{13}}{2} - 1 \stackrel{?}{=} 0$$ Since both fractions have a common denominator, we may write them as one fraction.

$$\frac{2}{2} - 1 \stackrel{?}{=} 0$$

$$0 = 0 \checkmark$$

Can you check $x = \dfrac{-3 - \sqrt{13}}{2}$ to verify that it is a root of $x^2 + 3x - 1 = 0$?

You should stop at this point and *memorize* the quadratic formula. Then try these practice problems.

Practice Problems for Exercise 9.3

Solve by the quadratic formula. If no real roots are possible, so state.

1. $x^2 - 7x + 6 = 0$ **2.** $2x^2 = 3x + 1$.

3. $5x^2 + 2x = -3$

Now turn to the answers following Exercise 9.3 and check your solutions for these practice problems.

Exercise 9.3

Solve by the quadratic formula. If no real roots are possible, so state.

1. $x^2 + 4x + 1 = 0$ **2.** $x^2 - 2x - 4 = 0$ **3.** $x^2 - 5x - 2 = 0$

4. $x^2 + 5x + 3 = 0$ **5.** $2x^2 - 7x - 9 = 0$ **6.** $2x^2 - 6x + 1 = 0$

7. $5x^2 + x - 1 = 0$ **8.** $2x^2 + 12x - 5 = 0$ **9.** $5x - 1 = 2x^2$

10. $5x^2 = -3x + 2$

11. $6x^2 - 3x = 1$

12. $3 = 7x - 4x^2$

13. $x - 1 = -\frac{2}{5}x^2$

14. $2x^2 + 3x = \frac{3}{4}$

15. $9x^2 - 6x = 2$

16. $6x^2 = 13x + 5$

17. $4x^2 + 3x + 2 = 0$

18. $\frac{x}{2} + \frac{2}{x} = \frac{5}{2}$

19. $\frac{x}{3} + \frac{2}{x} = \frac{7}{3}$

20. $3x^2 = 2x - 5$

21. $5y^2 = 3 - 7y$

22. $5y - 6 = -2y^2$

23. $\frac{1}{3}m^2 = \frac{1}{2}m + \frac{3}{2}$

24. $\frac{3}{5}m^2 - m = \frac{2}{5}$

25. $t^2 + 1 = \frac{13}{6}t$

26. $2t^2 - 1 = \frac{7}{3}t$

27. $3(s^2 + 1) = 10s$

28. $3 + p(p + 2) = 18$

29. $(p - 2)(p + 1) = 3$

30. $(s - 8)(s + 7) = 5$

31. $\frac{7}{3}y^2 - \frac{8}{3} = y$

32. $y = \frac{1}{3} - \frac{1}{9}y^2$

33. $3x^2 - 7 = 0$

34. $7x^2 + 5x = 0$

Challenge Problem

35. Let d, e, and f represent constant values. Solve for x by the quadratic formula.

$$4dx^2 + 5ex - 7f = 0$$

Solutions to Practice Problems for Exercise 9.3

1. $x^2 - 7x + 6 = 0$

$a = 1 \qquad b = -7 \qquad c = 6$

$$x = \frac{-b \pm \sqrt{b^2 - 4ac}}{2a} \qquad x = \frac{-(-7) \pm \sqrt{(-7)^2 - 4(1)(6)}}{2(1)}$$

$$= \frac{7 \pm \sqrt{49 - 24}}{2} \qquad x = \frac{7 \pm \sqrt{25}}{2} = \frac{7 \pm 5}{2}$$

The two roots are:

$$x = \frac{7 + 5}{2} = \frac{12}{2} = 6 = x$$

$$x = \frac{7 - 5}{2} = \frac{2}{2} = 1 = x$$

2. $2x^2 = 3x + 1$

The equation is not in standard form. Subtract $3x + 1$ from each side.

$2x^2 - 3x - 1 = 0$

$a = 2 \quad b = -3 \quad c = -1$

$$x = \frac{-(-3) \pm \sqrt{(-3)^2 - 4(2)(-1)}}{2(2)}$$

$$= \frac{3 \pm \sqrt{9 + 8}}{4}$$

$$x = \frac{3 \pm \sqrt{17}}{4}$$

3. $5x^2 + 2x = -3$

The equation is not in standard form. Add 3 to each side.

$5x^2 + 2x + 3 = 0$

$a = 5 \quad b = 2 \quad c = 3$

$$x = \frac{-2 \pm \sqrt{2^2 - 4(5)(3)}}{2(5)}$$

$$= \frac{-2 + \sqrt{4 - 60}}{10}$$

$$x = \frac{-2 \pm \sqrt{-56}}{10}$$

There is no real number that is the square root of a negative number. There are *no real roots* for this quadratic equation.

No real solution.

Now do the problems in Exercise 9.3.

9.4 The Pythagorean Theorem and Radical Equations

Certain equations that are frequently encountered involve a radical expression. We study two types in this section.

The Pythagorean theorem

In ancient Greece, mathematicians studied a number of properties of a right triangle (Figure 9.1). They were able to prove that the square of the longest side of a right triangle is equal to the sum of the squares of the other two sides. This property is

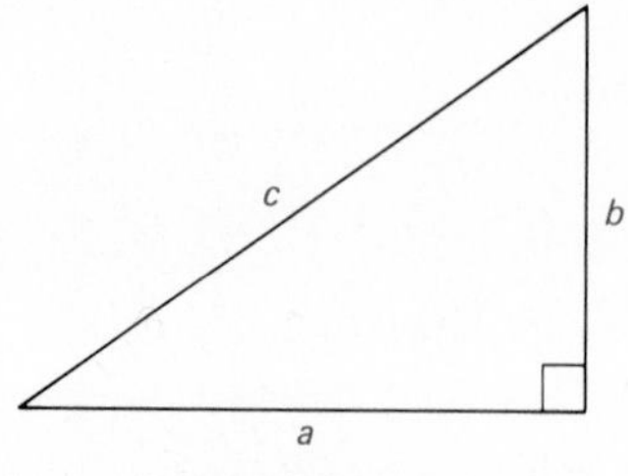

FIGURE 9.1

called the Pythagorean theorem, in honor of the Greek mathematician Pythagoras (ca. 590 BC). The shorter sides, a, b, are referred to as the *legs* of a right triangle. The longest side c is called the *hypotenuse* of a right triangle. We state the theorem thus:

> ***Pythagorean Theorem***
>
> **In any right triangle, if c is the length of the hypotenuse and a and b are the lengths of the two legs, then**
>
> $$c^2 = a^2 + b^2$$

If we know any two sides of a right triangle, we can find the third side using this theorem.

EXAMPLE A Find the hypotenuse of a right triangle whose legs are 5 and 12 (Figure 9.2).

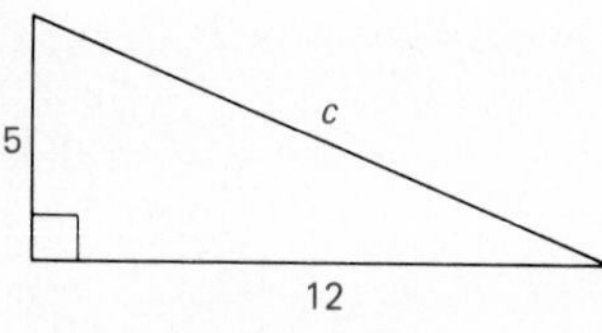

FIGURE 9.2

Solution

$$c^2 = a^2 + b^2$$

$$c^2 = 5^2 + 12^2$$

$$c^2 = 25 + 144$$

$$c^2 = 169$$

$$\sqrt{c^2} = \pm\sqrt{169} \qquad \text{Take the square root of each side.}$$

$$c = \pm 13$$

Only $c = 13$ is a measurement of length. The hypotenuse cannot be a negative number; thus $c = 13$.

One or more parts of the triangle may be an irrational number.

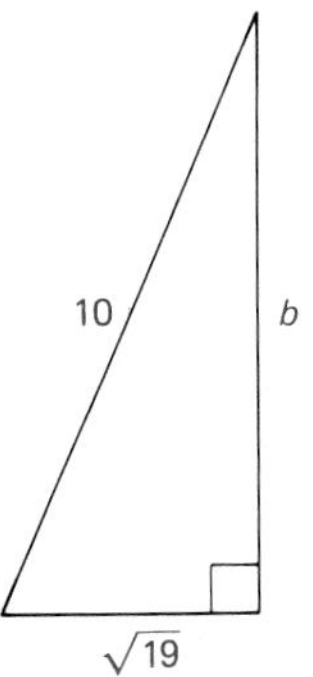

FIGURE 9.3

EXAMPLE B Find the other leg of a right triangle that has a hypotenuse of 10 and one leg of $\sqrt{19}$ (Figure 9.3).

Solution

$c^2 = a^2 + b^2$	
$10^2 = (\sqrt{19})^2 + b^2$	Replace c by 10 and a by $\sqrt{19}$.
$100 = 19 + b^2$	Square 10, square $\sqrt{19}$.
$81 = b^2$	Subtract 19 from each side.
$\pm\sqrt{81} = \sqrt{b^2}$	Take square root of each side.
$\pm 9 = b$	

The leg cannot be a negative number; therefore, $b = 9$.

Frequently, the answer will be an irrational number.

EXAMPLE C Find the approximate value of the other leg of a right triangle that has a hypotenuse of 7 and a leg of 4 (Figure 9.4).

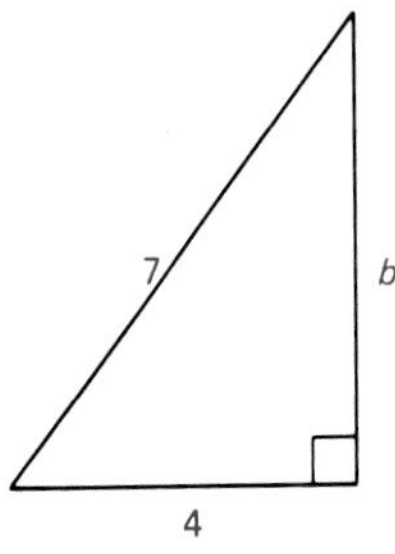

FIGURE 9.4

Solution

$$c^2 = a^2 + b^2$$

$$7^2 = 4^2 + b^2$$

$$49 = 16 + b^2$$

$$33 = b^2$$

$$\pm\sqrt{33} = \sqrt{b^2}$$

$$\pm\sqrt{33} = b$$

We only want a positive leg, so $b = \sqrt{33}$.

We can also express the answer as a decimal approximation by referring to the square root table in Appendix A.

$$b \doteq 5.745$$

Often the Pythagorean theorem is used to find a needed value in some practical situation involving a right triangle.

EXAMPLE D A 25-foot ladder is placed against a building (Figure 9.5). The foot of the ladder is 8 feet from the wall. How high on the building will the ladder reach?

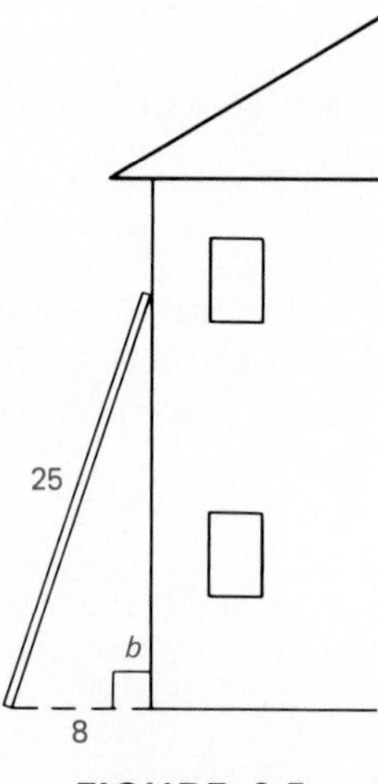

FIGURE 9.5

Solution

$$c^2 = a^2 + b^2$$

$$25^2 = 8^2 + b^2$$

$$625 = 64 + b^2$$

$$561 = b^2$$

$$\pm\sqrt{561} = \sqrt{b^2}$$

$$\pm\sqrt{561} = b$$

We only want positive value for the distance $b = \sqrt{561}$.

Our square root table only goes to the number 200! To approximate from our square root table we will have to write the radical in a different fashion. (Or use a calculator with a $\sqrt{x}$ key).

$b = \sqrt{3}\sqrt{187}$ Since $\sqrt{ab} = \sqrt{a}\sqrt{b}$ and $3 \cdot 187 = 561$.

$b \doteq (1.732)(13.675)$ Replacing each radical by a decimal approximation from the table.

$b \doteq 23.685$ (Multiply and round to the nearest thousandth.)

In actuality, with the accuracy of the original measurements of 25 feet and 8 feet we would probably use an answer of 24 feet or perhaps 23.7 feet. Unless the measurements of 25 feet and 8 feet were made with great precision, we would not be sure of an answer with three decimal places. Therefore, we would state the ladder reaches up approximately 23.7 feet on the building.

Radical equations

Some equations involve a variable underneath the square root sign. *A radical equation* is an equation with an unknown letter in one or more of the radicands. They can be simplified by squaring each side of the equation. The apparent solution to a radical equation *must* be checked.

EXAMPLE E Solve for x: $\sqrt{x+4} = 11$. Check your answer.

Solution

$$\sqrt{x+4} = 11$$

$$(\sqrt{x+4})^2 = (11)^2 \qquad \text{Square each side of the equation.}$$

$$x + 4 = 121 \qquad \text{Simplify.}$$

$$x = 117 \qquad \text{Solve for } x.$$

Check

$$\sqrt{117+4} \stackrel{?}{=} 11$$

$$\sqrt{121} \stackrel{?}{=} 11$$

$$11 = 11 \checkmark$$

Thus $x = 117$ is the solution.

If the equation contains one radical you must isolate the radical *before* you square each side of the equation.

EXAMPLE F Solve for x: $1 + \sqrt{5x-4} = 5$. Check your answer.

Solution

$$1 + \sqrt{5x-4} = 5 \qquad \text{We want to isolate the radical first.}$$

$$\sqrt{5x-4} = 4 \qquad \text{Subtract 1 from each side. Now the radical is isolated.}$$

$$(\sqrt{5x-4})^2 = (4)^2 \qquad \text{Square each side.}$$

$$5x - 4 = 16 \qquad \text{Simplify.}$$

$$5x = 20 \qquad \text{Add 4 to both sides.}$$

$$x = 4 \qquad \text{Solve for } x.$$

Check

$$1 + \sqrt{5(4)-4} \stackrel{?}{=} 5$$

$$1 + \sqrt{20-4} \stackrel{?}{=} 5$$

$$1 + \sqrt{16} \stackrel{?}{=} 5$$

$$1 + 4 = 5 \checkmark$$

Thus $x = 4$ is the solution.

Some problems will not have a solution. The apparent answer does not always check. An obtained solution that does not satisfy the original equation is called an **extraneous root**.

EXAMPLE G Solve for x: $\sqrt{x + 3} = -7$. Check your answer.

Solution

$$\sqrt{x+3} = -7$$

$$(\sqrt{x+3})^2 = (-7)^2 \qquad \text{Square each side.}$$

$$x + 3 = 49 \qquad \text{Simplify.}$$

$$x = 46 \qquad \text{Solve for } x.$$

Check

$$\sqrt{46+3} \neq -7$$

$$\sqrt{49} \neq -7$$

$$7 \neq -7!$$

Does not check! The apparent solution cannot be verified; there is no solution.

Wait a minute! Is that original equation possible in Example G?

No. $\sqrt{x+3} = -7$ is an impossible statement. By our definition of the $\sqrt{}$ symbol, we defined it to mean the positive square root of the radicand. If you noticed this in the first step, you could write down "no solution" immediately.

When you square both sides of the equation you may obtain a quadratic equation. In such cases you must check both apparent solutions.

EXAMPLE H Solve for x: $\sqrt{3x + 1} = x + 1$. Check your answer.

Solution

$$\sqrt{3x+1} = x + 1$$

$$(\sqrt{3x+1})^2 = (x+1)^2 \qquad \text{Square each side.}$$

$$3x + 1 = x^2 + 2x + 1$$

Simplify. Remember to square the binomial on the right side. (This is Formula B from Chapter 3)

$$0 = x^2 - x \qquad \text{Add } -3x - 1 \text{ to each side.}$$

$$0 = x(x-1) \qquad \text{Factor.}$$

$$x = 0 \qquad x - 1 = 0 \qquad \text{Set each factor equal to 0 and solve.}$$

$$x = 1$$

Check

$$x = 0: \quad \sqrt{3(0)+1} \stackrel{?}{=} 0 + 1$$

$$\sqrt{0+1} \stackrel{?}{=} 0 + 1$$

$$\sqrt{1} \stackrel{?}{=} 1$$

$$1 = 1\checkmark$$

$$x = 1: \quad \sqrt{3(1)+1} \stackrel{?}{=} 1 + 1$$

$$\sqrt{3+1} \stackrel{?}{=} 2$$

$$\sqrt{4} \stackrel{?}{=} 2$$

$$2 = 2\checkmark$$

Thus $x = 0$ and $x = 1$ are both solutions.

EXAMPLE I Solve for x: $\sqrt{2x-1} = x - 2$. Check your answer.

Solution

$$\sqrt{2x-1} = x - 2$$

$$(\sqrt{2x-1})^2 = (x-2)^2 \qquad \text{Square each side.}$$

$$2x - 1 = x^2 - 4x + 4 \qquad \text{Simplify. Remember to square the binomial carefully.}$$

$$0 = x^2 - 6x + 5 \qquad \text{Add } -2x + 1 \text{ to both sides.}$$

$$0 = (x-5)(x-1) \qquad \text{Factor.}$$

$$x - 5 = 0 \qquad x - 1 = 0 \qquad \text{Set each factor equal to zero and solve.}$$

$$x = 5 \qquad x = 1$$

Check

$$x = 5: \quad \sqrt{2(5)-1} \stackrel{?}{=} 5 - 2$$

$$\sqrt{10-1} \stackrel{?}{=} 3$$

$$\sqrt{9} \stackrel{?}{=} 3$$

$$3 = 3 \checkmark$$

$$x = 1: \quad \sqrt{2(1)-1} \neq 1 - 2$$

$$\sqrt{2-1} \neq -1$$

$$\sqrt{1} \neq -1$$

$1 \neq -1$ It does *not* check. In this case $x = 1$ is called an *extraneous root*.

Thus only $x = 5$ is a solution to this equation.

EXAMPLE J Solve for x: $-2 + \sqrt{6x-1} = 3x - 2$. Check your answer.

Solution

$$-2 + \sqrt{6x-1} = 3x - 2$$

As a first step we want to isolate the radical. This is always done before squaring each side.

$$\sqrt{6x-1} = 3x \qquad \text{Add 2 to each side.}$$

$$(\sqrt{6x-1})^2 = (3x)^2 \qquad \text{Square each side.}$$

$$6x - 1 = 9x^2 \qquad \text{Simplify.}$$

$$0 = 9x^2 - 6x + 1 \qquad \text{Add } -6x + 1 \text{ to each side.}$$

$$0 = (3x-1)^2 \qquad \text{Factor.}$$

$$\text{or} \quad (3x-1)(3x-1)$$

$$3x - 1 = 0 \qquad \text{Set factor equal to 0. Since we have a double root, we only need to solve the equation once.}$$

$$3x = 1$$

$$x = \frac{1}{3}$$

We have a double root of $\frac{1}{3}$ for an apparent solution.

Check

$$-2+\sqrt{6\left(\frac{1}{3}\right)-1}\stackrel{?}{=}3\left(\frac{1}{3}\right)-2$$

$$-2+\sqrt{2-1}\stackrel{?}{=}1-2$$

$$-2+\sqrt{1}\stackrel{?}{=}1-2$$

$$-2+1\stackrel{?}{=}1-2$$

$$-1=-1\checkmark$$

Thus $x=\frac{1}{3}$ is the solution to the equation.

EXAMPLE K Solve for x: $\sqrt{3x+3}=\sqrt{5x-1}$. Check your answer.

Solution

$\sqrt{3x+3}=\sqrt{5x-1}$	Each radical is already isolated.
$(\sqrt{3x+3})^2=(\sqrt{5x-1})^2$	Square each side.
$3x+3=5x-1$	Simplify.
$3=2x-1$	Add $-3x$ to each side.
$4=2x$	Add $+1$ to each side.
$2=x$	Solve for x.

Check

$$\sqrt{3(2)+3}\stackrel{?}{=}\sqrt{5(2)-1}$$

$$\sqrt{6+3}\stackrel{?}{=}\sqrt{10-1}$$

$$\sqrt{9}=\sqrt{9}\checkmark$$

Thus $x=2$ is a solution.

You may find the following summary helpful as a reference in solving radical equations.

> ***Procedure To Solve a Radical Equation***
>
> 1. **Perform algebraic operations to obtain one radical by itself on one side of the equation.**
> 2. **Square each side of the equation.**
> 3. **Simplify if possible.**
> 4. **Solve the resulting equation.**
> 5. **Check all apparent solutions. Solutions to radical equations *must* be verified.**

Review Examples A to K and then try to solve these practice problems.

Practice Problems for Exercise 9.4

1. In the right triangle in Figure 9.6, $c=8$ and $b=6$. Find a. Leave your answer in simplest radical form.
2. A support line is placed 6 meters away from the base of an 8-meter pole (Figure 9.7). If the support line is attached to the top of the pole and pulled tight (assume that it is a straight line), how long is the support line from the ground to the pole?

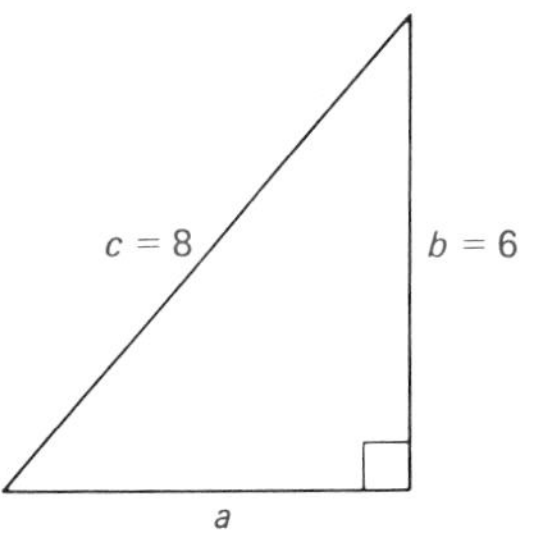

FIGURE 9.6

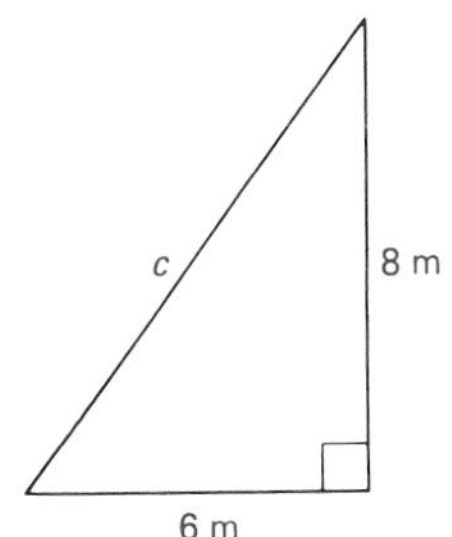

FIGURE 9.7

3. Solve for x and check your answer.

$$\sqrt{2x - 5} = 6$$

4. Solve for x and check your answer.

$$2 - x + \sqrt{x + 4} = 0$$

Now turn to the answers following Exercise 9.4 and check your solutions for these practice problems.

Exercise 9.4

Use the Pythagorean theorem to find the length of the third side of each right triangle in problems 1 to 10. Leave any irrational answers in radical form.

1.

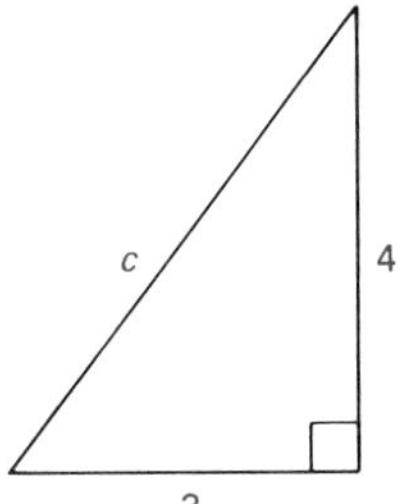

2.

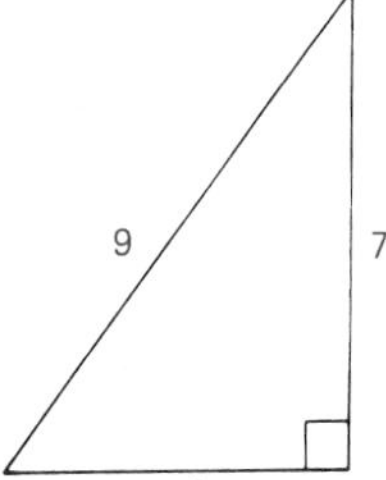

3.

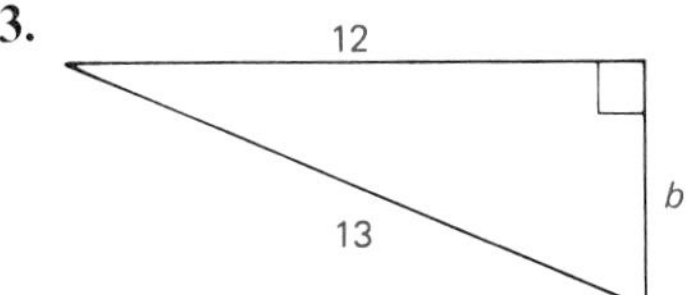

4.

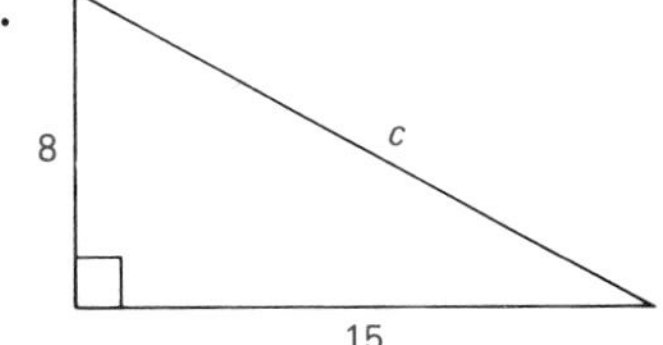

5.

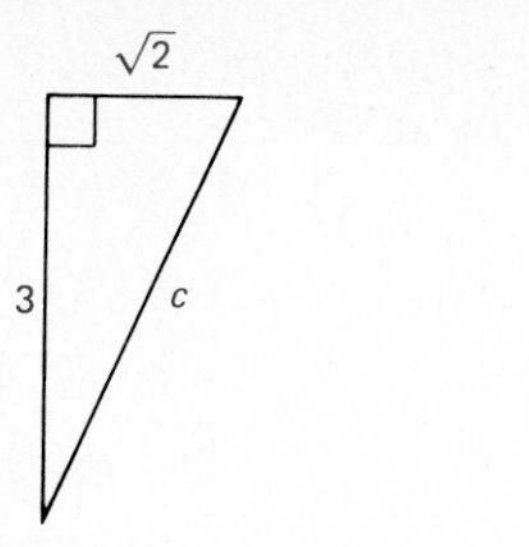

6.

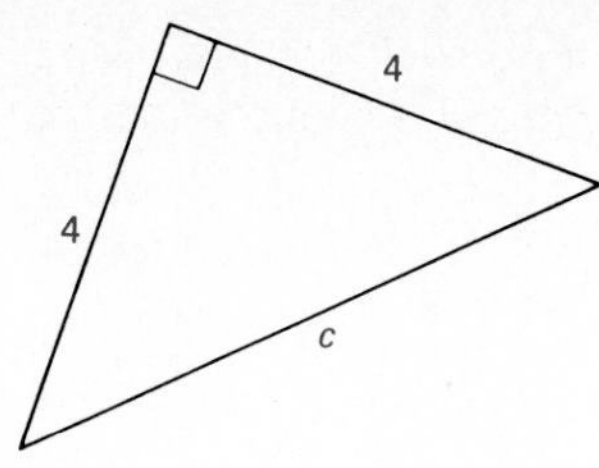

7.

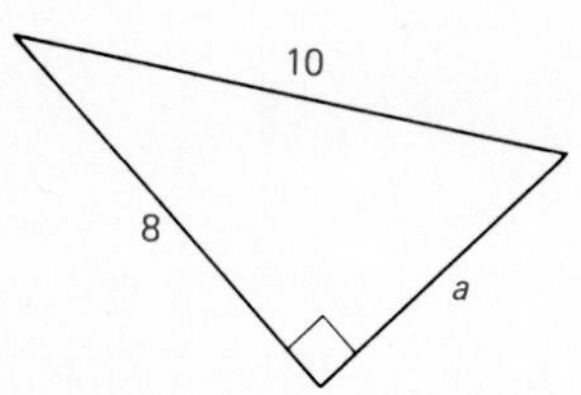

8.

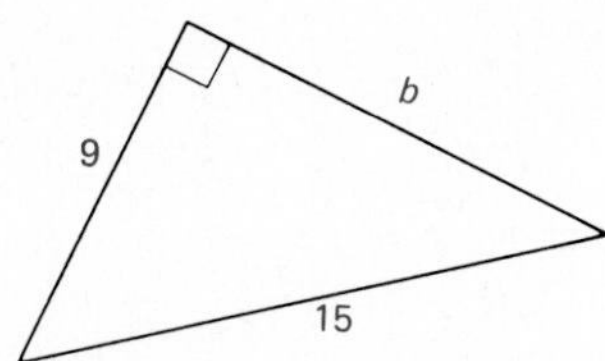

9.

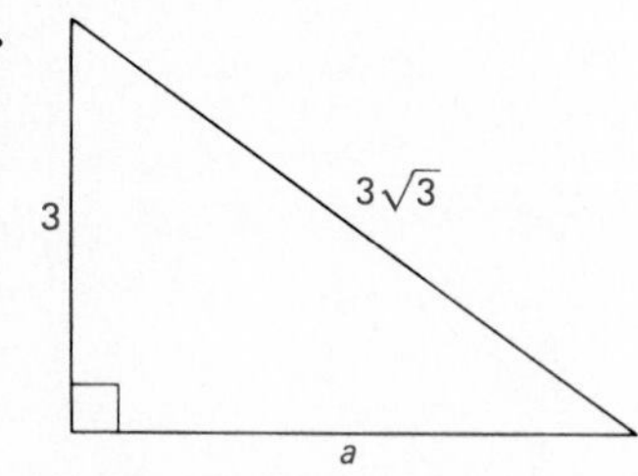

10.

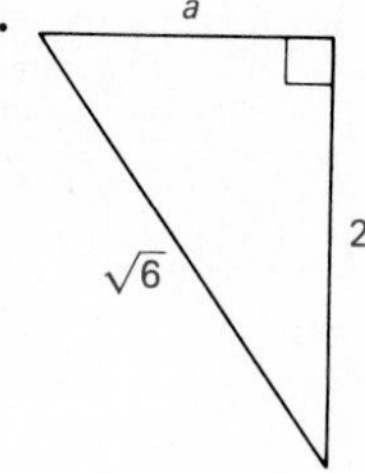

Draw a sketch of the appropriate right triangle in problems 11 to 14. Find the missing distance by using the Pythagorean theorem. Approximate to the nearest tenth any irrational numbers using the square root table in Appendix A.

11. A ladder is 18 feet long. It reaches 15 feet up on a wall. How far is the base of the ladder from the wall?

12. A kite is flying on a string 100 feet long. Assume that the string is a straight line. How high is the kite if it is flying above a point 20 feet away along the ground?

13. A baseball diamond is really constructed from a square. Each side of the square is 90 feet long. How far is it from home plate to second base? (*Hint*: Draw the diagonal.)

14. A boat must be moored 20 feet away from a dock. The boat will be 10 feet below the level of the dock at low tide. What is the minimum length of rope needed to go from the boat to the dock at low tide?

Solve for the variable. Check your solutions.

15. $\sqrt{x} = 11$

16. $\sqrt{x} = 3.5$

17. $\sqrt{x + 8} = 5$

18. $\sqrt{x + 5} = 7$

19. $\sqrt{3x + 6} = 2$

20. $\sqrt{3x - 8} = 4$

21. $\sqrt{2x + 2} = \sqrt{3x - 5}$

22. $\sqrt{5x - 5} = \sqrt{4x + 1}$

23. $\sqrt{2x} - 5 = 4$

24. $\sqrt{3x} + 5 = 8$

25. $\sqrt{3x + 10} = x$

26. $\sqrt{5x - 6} = x$

27. $\sqrt{5y + 1} = y + 1$

28. $\sqrt{2y + 9} = y + 3$

29. $\sqrt{3y + 1} - y = 1$

30. $\sqrt{3y - 8} + 2 - y = 0$

31. $y + \sqrt{3y - 14} = 6$

32. $3y - \sqrt{6y - 1} = 0$

33. $2\sqrt{2x - 3} + \sqrt{3x - 2} = 0$

34. $2\sqrt{x - 1} - \sqrt{2x - 1} = 0$

Challenge Problem

35. Solve for x. (*Hint*: Square each side first. Isolate the radical that remains. Then square each side again.)

$$\sqrt{2x + 5} = 2\sqrt{2x} + 1$$

Solutions to Practice Problems for Exercise 9.4

1\.

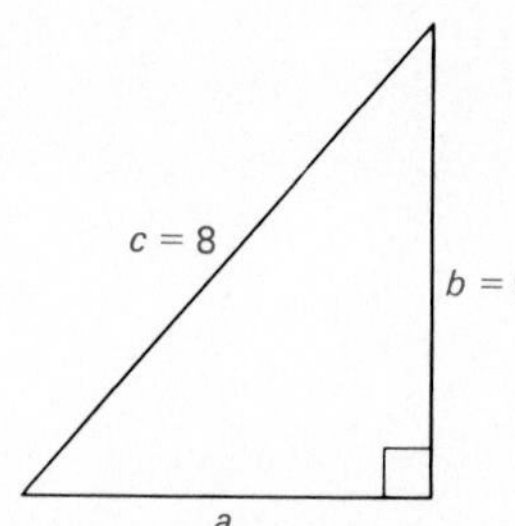

$$c = a^2 + b^2$$
$$8^2 = a^2 + 6^2$$
$$64 = a^2 + 36$$
$$28 = a^2$$
$$\pm\sqrt{28} = a$$ But we choose only the positive value.

So $a = \sqrt{28}$. Simplifying $a = 2\sqrt{7}$

2\.

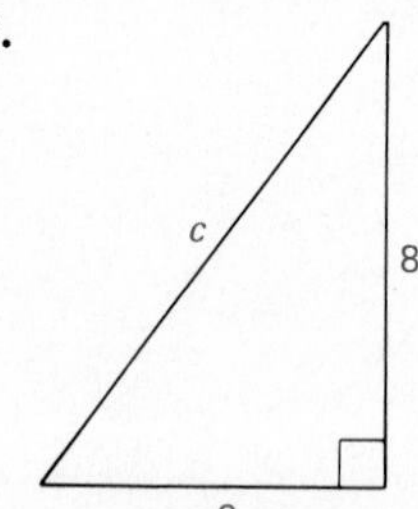

$$c^2 = 6^2 + 8^2$$
$$c^2 = 36 + 64$$
$$c^2 = 100$$
$$c = \pm\sqrt{100}$$

Choosing only the positive value, $c = 10$. The support line is 10 meters.

3\. $\sqrt{2x - 5} = 6$

$(\sqrt{2x - 5})^2 = (6)^2$ Square each side.

$2x - 5 = 36$ Simplify.

$2x = 41$ Add 5 to each side.

$x = \frac{41}{2}$ Solve for x.

Check

$$\sqrt{2\left(\frac{41}{2}\right) - 5} \stackrel{?}{=} 6$$
$$\sqrt{41 - 5} \stackrel{?}{=} 6$$
$$\sqrt{36} \stackrel{?}{=} 6$$
$$6 = 6 \checkmark$$

Thus the solution is $x = \frac{41}{2}$.

4. $2 - x + \sqrt{x + 4} = 0$ — We need to isolate the radical.

$\sqrt{x + 4} = x - 2$ — Add $+x - 2$ to both sides.

$(\sqrt{x + 4})^2 = (x - 2)^2$ — Square each side. Remember to carefully square the binomial.

$x + 4 = x^2 - 4x + 4$ — Remove parentheses.

$0 = x^2 - 5x$ — Add $-x - 4$ to both sides.

$0 = x(x - 5)$ — Remove the common factor x.

$x = 0 \qquad x - 5 = 0$ — Set each factor equal to 0.

$x = 5$

Check

$x = 0$: $2 - 0 + \sqrt{0 + 4} \neq 0$

$2 + \sqrt{4} \neq 0$

$2 + 2 \neq 0$

$4 \neq 0$

It does not check. Thus, $x = 4$ is an extraneous root.

$x = 5$: $2 - 5 + \sqrt{5 + 4} \stackrel{?}{=} 0$

$-3 + \sqrt{9} \stackrel{?}{=} 0$

$-3 + 3 \stackrel{?}{=} 0$

$0 = 0\checkmark$

The only solution is $x = 5$.

Now do the problems in Exercise 9.4.

9.5 Graphing Quadratic Equations

In Chapter 6 we learned how to graph linear equations. The graphs of these equations always yield straight lines. In this section we study the graph of quadratic equations of the form $y = ax^2 + bx + c$, where $a \neq 0$. The graph of these types of equations always yields a curve called a *parabola*. Three examples of the graph of a parabola are illustrated in Figure 9.8.

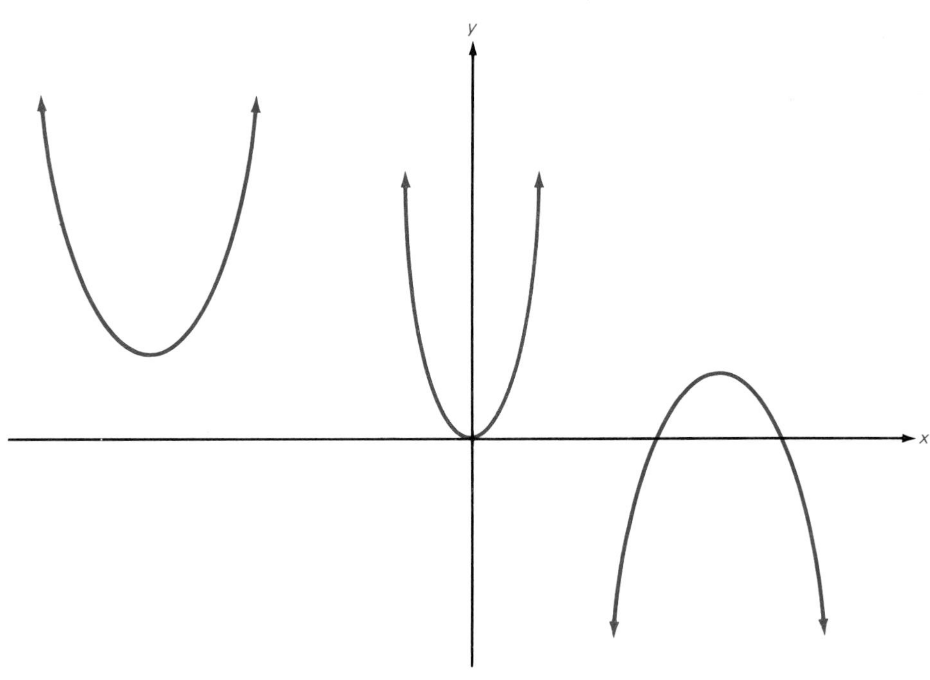

FIGURE 9.8

Since we observe that parabolas are smooth-shaped curves we will find it necessary to plot several points (usually five to eight points) in order to have a reasonably accurate graph. To pick these points we usually pick small, convenient values of x and then solve for the corresponding value for y.

EXAMPLE A Graph: $y = x^2$.

Solution Let us pick some small convenient values for x

$$\text{if } x = -3: \quad y = (-3)^2 = 9$$

$$x = -2 \quad y = (-2)^2 = 4$$

$$x = -1 \quad y = (-1)^2 = 1$$

$$x = 0 \quad y = (0)^2 = 0$$

$$x = 1 \quad y = (1)^2 = 1$$

$$x = 2 \quad y = (2)^2 = 4$$

$$x = 3 \quad y = (3)^2 = 9$$

The data above suggest that we will use the ordered pairs $(-3, 9)$, $(-2, 4)$, $(-1, 1)$, $(0, 0)$, $(1, 1)$, $(2, 4)$, and $(3, 9)$ to graph the parabola. We can also list these data in table form rather than listing the ordered pairs.

x	-3	-2	-1	0	1	2	3
y	9	4	1	0	1	4	9

The graph is shown in Figure 9.9.

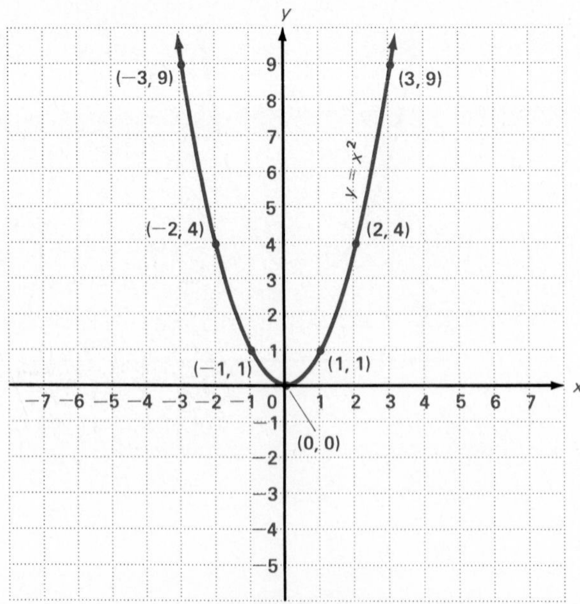

FIGURE 9.9

EXAMPLE B Graph: $y = x^2 - 3x$.

Solution Again we pick small, convenient values for x.

x	$y = x^2 - 3x$	y
-2	$y = (-2)^2 - 3(-2) = 4 + 6 = 10$	10
-1	$y = (-1)^2 - 3(-1) = 1 + 3 = 4$	4
0	$y = (0)^2 - 3(0) = 0 - 0 = 0$	0
1	$y = (1)^2 - 3(1) = 1 - 3 = -2$	-2
2	$y = (2)^2 - 3(2) = 4 - 6 = -2$	-2
3	$y = (3)^2 - 3(3) = 9 - 9 = 0$	0
4	$y = (4)^2 - 3(4) = 16 - 12 = 4$	4
5	$y = (5)^2 - 3(5) = 25 - 15 = 10$	10

We see the sketch of $y = x^2 - 3x$ in Figure 9.10.

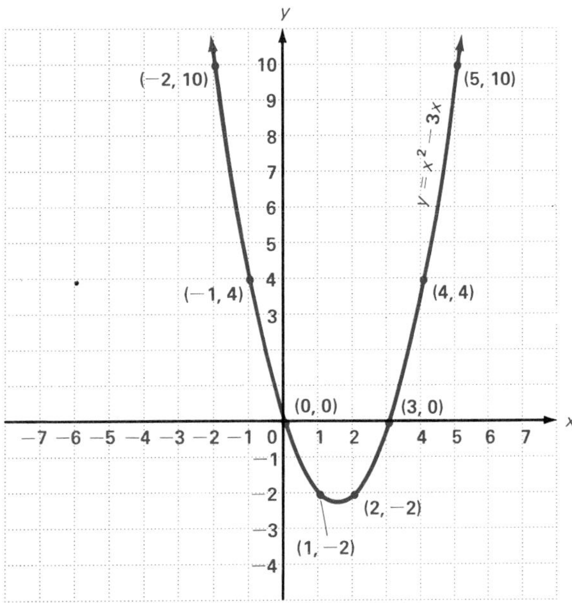

FIGURE 9.10

? How do we know if we have plotted enough points?

Usually we would like to see enough of the curve to see where it crosses the x-axis and where the highest and lowest points are. In many cases, eight points is enough to graph a parabola, but if you see that you do not have enough points to sketch the parabola in a certain region, it is good to pick a value of x in that region and determine y. In Figure 9.10, if we only plotted the points in the table, we would not know how far down to draw the curve. (The lowest point on a parabola is called the *vertex*.) Thus we would probably pick between $x = 1$ and $x = 2$. The most logical choice would be $x = \frac{3}{2}$. Thus if $x = \frac{3}{2}$,

$$y = \left(\frac{3}{2}\right)^2 - 3\left(\frac{3}{2}\right) = \frac{9}{4} - \frac{9}{2}$$

$$= \frac{9}{4} - \frac{18}{4} = -\frac{9}{4}$$

Plotting the point $(\frac{3}{2}, -\frac{9}{4})$ will greatly improve the accuracy of our graph of the parabola.

Do all equations of the form $y = ax^2 + bx + c$ always open upward? No. If the coefficient of x^2 is negative (if $a < 0$), the parabola opens downward. We will see this in Examples C and E. The *vertex* of a parabola opening downward is the highest point of the curve.

> **All equations of the form $ax^2 + bx + c$ are**
>
> **1. parabolas that open *downward* if $a < 0$**
>
> **2. parabolas that open upward if $a > 0$.**

EXAMPLE C Graph: $y = -x^2 + 2x + 3$.

Solution We pick small, convenient values for x.

x	$y = -x^2 + 2x + 3$	y
-2	$y = -(-2)^2 + 2(-2) + 3 = -4 - 4 + 3 = -5$	-5
-1	$y = -(-1)^2 + 2(-1) + 3 = -1 - 2 + 3 = 0$	0
0	$y = -(0)^2 + 2(0) + 3 = 0 + 0 + 3 = 3$	3
1	$y = -(1)^2 + 2(1) + 3 = -1 + 2 + 3 = 4$	4
2	$y = -(2)^2 + 2(2) + 3 = -4 + 4 + 3 = 3$	3
3	$y = -(3)^2 + 2(3) + 3 = -9 + 6 + 3 = 0$	0
4	$y = -(4)^2 + 2(4) + 3 = -16 + 8 + 3 = -5$	-5

The graph of this parabola is shown in Figure 9.11. The vertex of the parabola is the point (1, 4). It is the highest point of the curve.

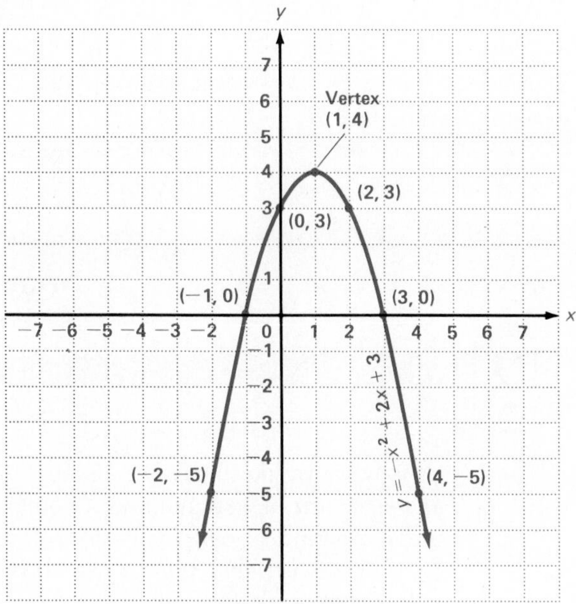

FIGURE 9.11

For all parabolas a *line of symmetry* can be drawn through the vertex (Figure 9.12). If you fold the graph on this line, the two halves of the curve will correspond exactly. When you draw a graph of a parabola you should be able to observe this property of symmetry. You should make it a goal to draw enough of the parabola so that the two symmetrical sides are clearly evident.

What if the two sides do not look very symmetric? First, see if you just need to plot a few more points on one side of the parabola. We will illustrate this point in

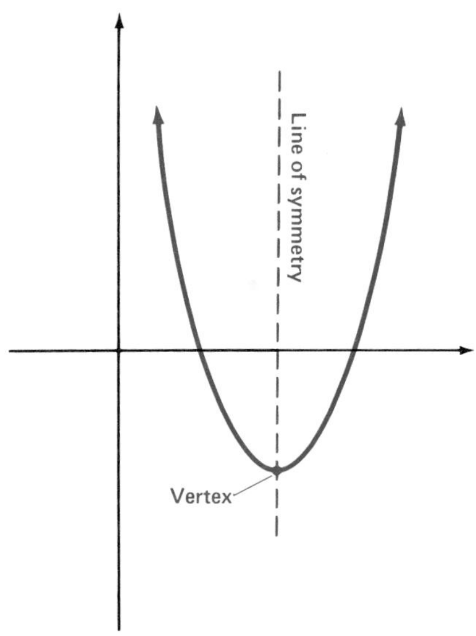

FIGURE 9.12

Example D. If this is not the case, you should double-check the substitution of values for x to obtain y. It is quite easy to make a careless error in evaluating the expression to obtain y.

EXAMPLE D Graph: $y = (x - 3)^2 - 2$.

Solution We determine the following:

x	$y = (x - 3)^2 - 2$	y
-1	$y = (-1 - 3)^2 - 2 = (-4)^2 - 2 = 16 - 2 = 14$	14
0	$y = (0 - 3)^2 - 2 = (-3)^2 - 2 = 9 - 2 = 7$	7
1	$y = (1 - 3)^2 - 2 = (-2)^2 - 2 = 4 - 2 = 2$	2
2	$y = (2 - 3)^2 - 2 = (-1)^2 - 2 = 1 - 2 = -1$	-1
3	$y = (3 - 3)^2 - 2 = (0)^2 - 2 = 0 - 2 = -2$	-2

Our graph is shown in Figure 9.13. The graph does not look complete. Clearly, our figure does

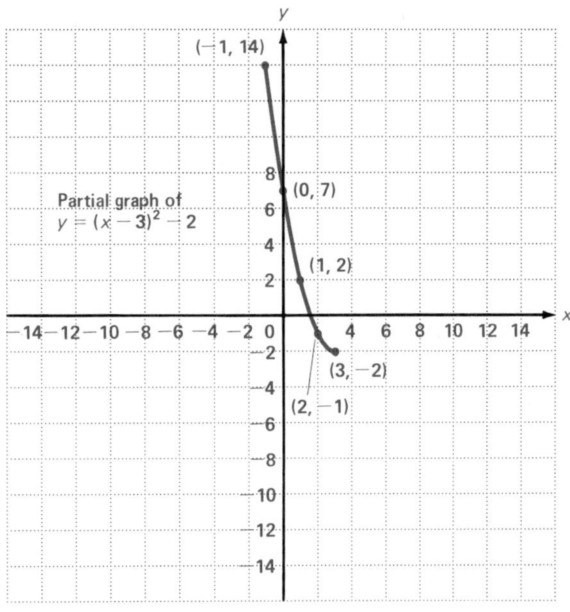

FIGURE 9.13

not look symmetrical! Perhaps we should try larger values of x, such as $x = 4$, $x = 5$, etc., and see if that will improve our sketch.

x	$y = (x - 3)^2 - 2$	y
4	$y = (4 - 3)^2 - 2 = (1)^2 - 2 = 1 - 2 = -1$	-1
5	$y = (5 - 3)^2 - 2 = (2)^2 - 2 = 4 - 2 = 2$	2
6	$y = (6 - 3)^2 - 2 = (3)^2 - 2 = 9 - 2 = 7$	7
7	$y = (7 - 3)^2 - 2 = (4)^2 - 2 = 16 - 2 = 14$	14

Now we have a set of points that will enable us to see the region of the parabola that was needed Our new sketch is shown in Figure 9.14. The symmetry of the parabola can now be observed.

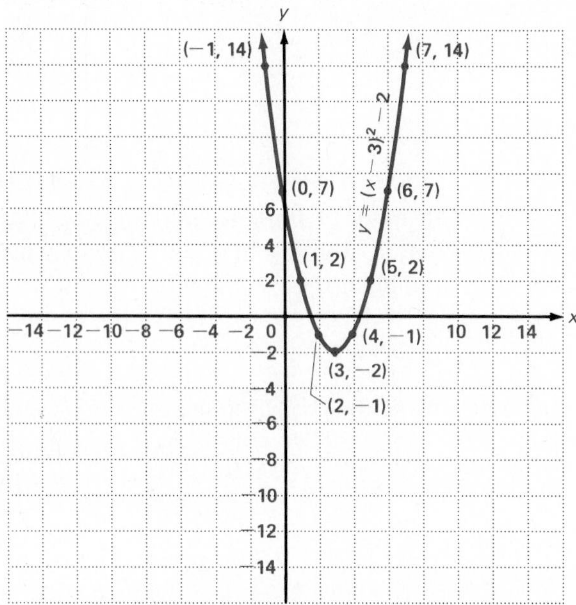

FIGURE 9.14

? Isn't there some quick way to find the vertex of the parabola? It seems like too much guesswork or luck!

There is a convenient formula that is developed in higher-level math courses. Using the methods of calculus, it can be shown that

> **The x coordinate of the vertex of the parabola $y = ax^2 + bx + c$ is equal to $-b/2a$. (The vertex will occur where $x = -b/2a$.)**

(We can also develop this formula $x = -b/2a$ by a geometric argument. This is left as an exercise for the student.) If we start with this value of x and choose a few x values that lie to the right of this on the number line, and a few x values that lie to the left of this on the number line, the process of graphing can be done more conveniently.

EXAMPLE E Graph: $y = -2x^2 + 6x + 4$.

Solution The graph of the parabola $y = ax^2 + bx + c$ has a vertex at $x = -b/2a$. In this case $a = -2$, $b = 6$, and $c = 4$. Thus the parabola will have a vertex at

$$x = \frac{-b}{2a} = \frac{-6}{2(-2)} = \frac{-6}{-4} = \frac{3}{2}$$

We start with that value of x and pick a few convenient values of x to the left and to the right of it on the number line.

x	$y = -2x^2 + 6x + 4$	y
-1	$y = -2(-1)^2 + 6(-1) + 4 = -2 - 6 + 4 = -4$	-4
0	$y = -2(0)^2 + 6(0) + 4 = 0 + 0 + 4 = 4$	4
1	$y = -2(1)^2 + 6(1) + 4 = -2 + 6 + 4 = 8$	8
$\frac{3}{2}$ (or $1\frac{1}{2}$)	$y = -2(\frac{3}{2})^2 + 6(\frac{3}{2}) + 4 = -2(\frac{9}{4}) + 6(\frac{3}{2}) + 4 = -\frac{9}{2} + 9 + 4$ $= -\frac{9}{2} + 13 = -\frac{9}{2} + \frac{26}{2} = \frac{17}{2}$	$\frac{17}{2}$ or $8\frac{1}{2}$
2	$y = -2(2)^2 + 6(2) + 4 = -8 + 12 + 4 = 8$	8
3	$y = -2(3)^2 + 6(3) + 4 = -18 + 18 + 4 = 4$	4
4	$y = -2(4)^2 + 6(4) + 4 = -32 + 24 + 4 = -4$	-4

Since $a = -2$ this parabola opens downward. We plot the points obtained in the table and Figure 9.15 is the result.

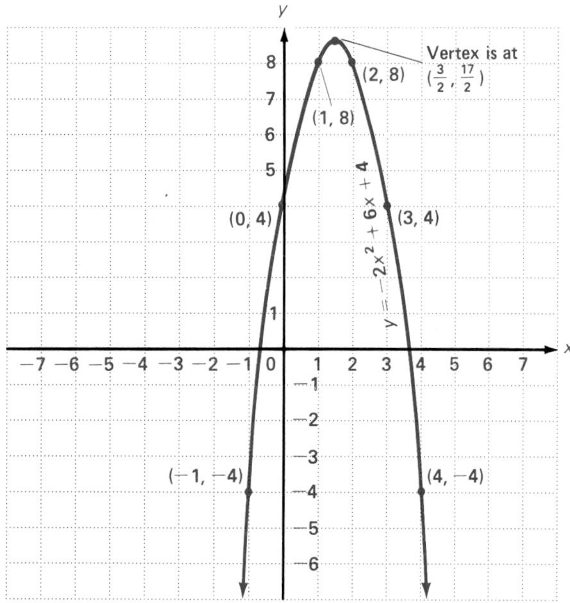

FIGURE 9.15

Take a few minutes to review Examples A to E and then try these practice problems.

Practice Problems for Exercise 9.5

Graph.

1. $y = -2x^2$

2. $y = x^2 - 3$

3. $y = x^2 + 2x - 3$

Now turn to the answers following Exercise 9.5 and check your solutions for these practice problems.

Exercise 9.5

Graph. Plot five to eight points for each graph.

1. $y = 2x^2 - 1$

2. $y = x^2 + 2$

3. $y = -\frac{1}{3}x^2$

4. $y = \frac{1}{2}x^2$

5. $y = x^2 - 2x$

6. $y = x^2 + 3x$

7. $y = x^2 - 4x + 4$

8. $y = 2x^2 + x - 3$

9. $y = 3 - 2x^2$

10. $y = 6x - 3x^2$

11. $y = -x^2 - 2x + 3$

12. $y = -2x^2 - 4x + 1$

13. $y = (2x - 1)^2$

14. $y = (3x + 2)^2$

15. $y = (x - 2)^2 - 1$

16. $y = (x + 1)^2 + 2$

17. $y = -(x - 2)^2$

18. $y = -(x - 1)^2$

19. $y = x^2 - 8x + 16$

20. $y = x^2 + 6x + 9$

21. $y = 3x^2 - 6x - 10$

22. $y = 2x^2 + 3x - 8$

23. $y = 5x^2 + 3x - 2$

24. $y = 5x^2 - x - 6$

Challenge Problems

Graph.

25. $\frac{y + 15}{4} = x^2 + x$

26. $x = \frac{y + 15 - 8x^2}{14}$

27. Show from a graphical point of view that a vertex of the parabola defined by $y = ax^2 + bx + c$ will occur when $x = -\frac{b}{2a}$ by using the quadratic formula in the form $x = -\frac{b}{2a} \pm \frac{\sqrt{b^2 - 4ac}}{2a}$. (Hint: Show that if $b^2 - 4ac > 0$ the two x-intercepts occur at $x = -\frac{b}{2a} \pm \frac{\sqrt{b^2 - 4ac}}{2a}$ and use the property of symmetry.)

Solutions to Practice Problems for Exercise 9.5

1. $y = -2x^2$ Since $a = -2 < 0$ we know the parabola opens downward.

Our table of values is as follows:

x	$y = -2x^2$	y
$x = -2$	$y = -2(-2)^2 = -2(4) = -8$	-8
$x = -1$	$y = -2(-1)^2 = -2(1) = -2$	-2
$x = 0$	$y = -2(0)^2 = -2(0) = 0$	0
$x = 1$	$y = -2(1)^2 = -2(1) = -2$	-2
$x = 2$	$y = -2(-2)^2 = -2(4) = -8$	-8

The graph is shown in Figure 9.16.

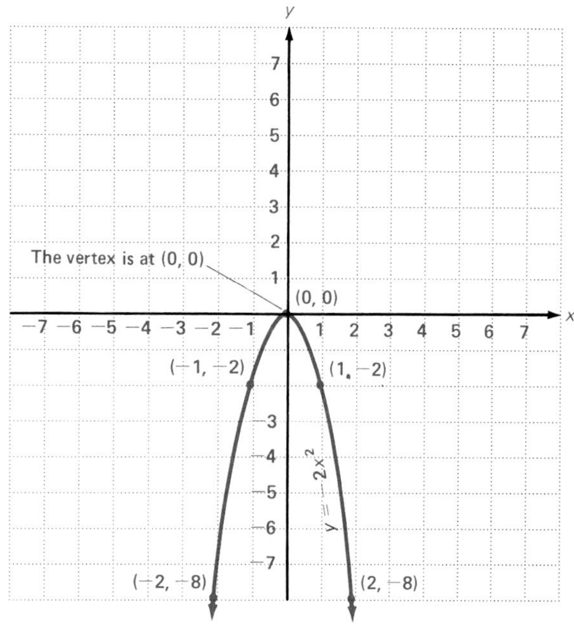

FIGURE 9.16

2. $y = x^2 - 3$ Since $a = 1 > 0$ we know the parabola opens upward.

Our table of values is as follows:

x	$y = x^2 - 3$	y
$x = -3$	$y = (-3)^2 - 3 = 9 - 3 = 6$	6
$x = -2$	$y = (-2)^2 - 3 = 4 - 3 = 1$	1
$x = -1$	$y = (-1)^2 - 3 = 1 - 3 = -2$	-2
$x = 0$	$y = (0)^2 - 3 = 0 - 3 = -3$	-3
$x = 1$	$y = (1)^2 - 3 = 1 - 3 = -2$	-2
$x = 2$	$y = (2)^2 - 3 = 4 - 3 = 1$	1
$x = 3$	$y = (3)^2 - 3 = 9 - 3 = 6$	6

The graph is shown in Figure 9.17.

3. $y = x^2 + 2x - 3$ The vertex occurs when $x = \dfrac{-b}{2a} = \dfrac{-2}{2(1)} = -1$. Since $a = 1 > 0$ the curve opens upward. We plot a few x values smaller than -1 and a few greater than -1.

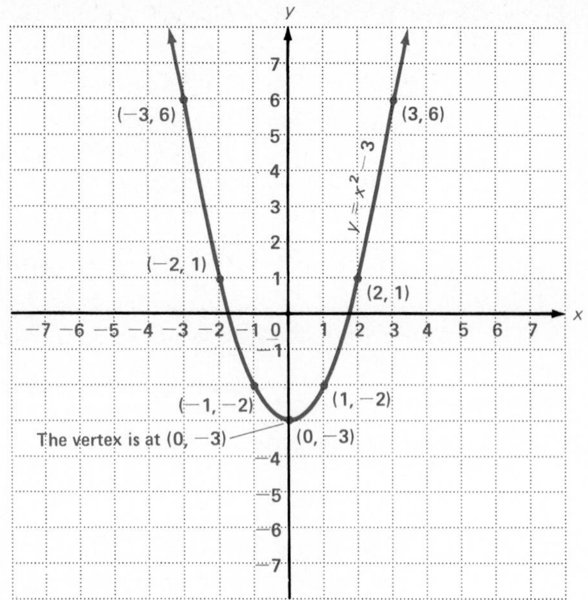

FIGURE 9.17

Our table of values for $y = x^2 + 2x - 3$ is as follows:

x	$y = x^2 + 2x - 3$	y
$x = -4$	$y = (-4)^2 + 2(-4) - 3 = 16 - 8 - 3 = 5$	5
$x = -3$	$y = (-3)^2 + 2(-3) - 3 = 9 - 6 - 3 = 0$	0
$x = -2$	$y = (-2)^2 + 2(-2) - 3 = 4 - 4 - 3 = -3$	-3
$x = -1$	$y = (-1)^2 + 2(-1) - 3 = 1 - 2 - 3 = -4$	-4
$x = 0$	$y = (0)^2 + 2(0) - 3 = 0 + 0 - 3 = -3$	-3
$x = 1$	$y = (1)^2 + 2(1) - 3 = 1 + 2 - 3 = 0$	0
$x = 2$	$y = (2)^2 + 2(2) - 3 = 4 + 4 - 3 = 5$	5

The graph is shown in Figure 9.18.

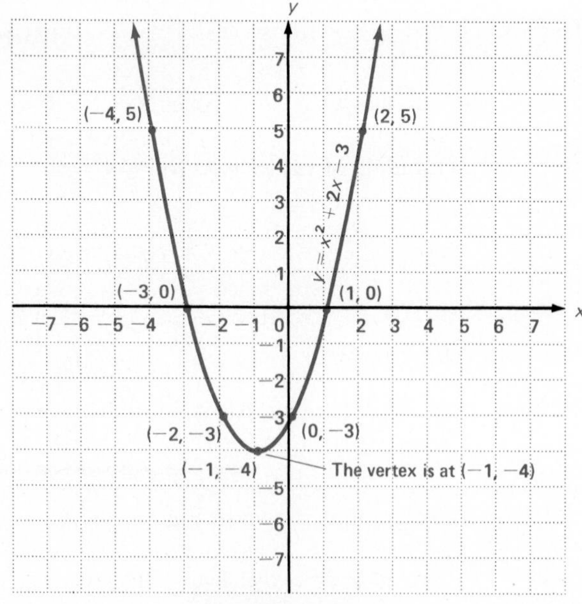

FIGURE 9.18

Now do the problems in Exercise 9.5

9.6 Word Problems Involving Quadratic Equations

(If you feel you are weak in word problems, you may want to review Section 5.1 before you study this section.) Certain types of word problems lead to quadratic equations. This often occurs in problems when one variable quantity is multiplied by another. It is particularly important to check the solutions to the quadratic equation with conditions stated in the word problem. Often a particular solution to the quadratic equation will be eliminated by the conditions of the word problem.

EXAMPLE A The length of a rectangle is 3 meters longer than twice the width. The area of the rectangle is 44 square meters. Find the length and width of the rectangle.

Solution Let x = the width in meters. Then $2x + 3$ = the length in meters (Figure 9.19).

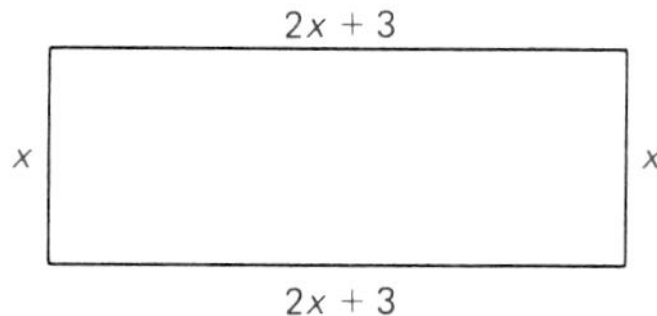

FIGURE 9.19

Area = (width)(length)

$44 = x(2x + 3)$	Substituting x for the width and $(2x + 3)$ for the length, we have.
$44 = 2x^2 + 3x$	Remove parentheses.
$0 = 2x^2 + 3x - 44$	Subtract 44 from both sides.
$0 = (2x + 11)(x - 4)$	Factor.
$2x + 11 = 0 \qquad x - 4 = 0$	Set each factor equal to 0, simplify and solve.
$2x = -11 \qquad x = 4$	
$x = -5\frac{1}{2}$	

(Note that $x = -5\frac{1}{2}$ is not a valid solution. It would not make sense to have a rectangle with a negative number as a width.)

Since $x = 4$, the width of the rectangle is 4 meters. The length is $2x + 3$, so we have $2(4) + 3 = 8 + 3 = 11$. Thus the length of the rectangle is 11 meters.

Check Is the length 3 meters more than twice the width?

$$11 \stackrel{?}{=} 3 + 2(4) \qquad 11 = 3 + 8\checkmark$$

Is the area of the rectangle 44 square meters?

$$4 \times 11 \stackrel{?}{=} 44 \qquad 44 = 44\checkmark$$

EXAMPLE B The hypotenuse of a right triangle is 25 meters in length. One leg is 17 meters longer than the other. Find the length of each leg.

Solution Let x = the length in meters of the shorter leg. Then $x + 17$ = the length of the longer leg (Figure 9.20).

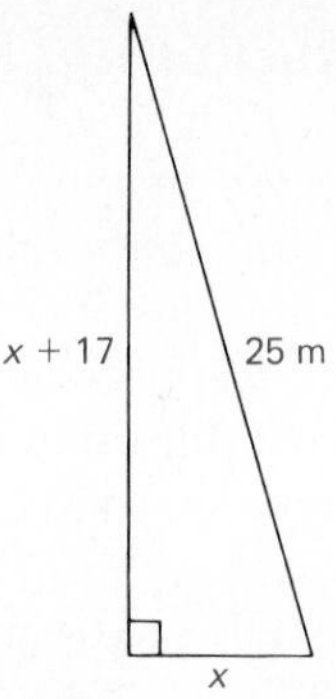

FIGURE 9.20

We will use the Pythagorean theorem, $a^2 + b^2 = c^2$.

$x^2 + (x + 17)^2 = (25)^2$	We substitute x and $x + 17$ for the legs and 25 for the hypotenuse.
$x^2 + x^2 + 34x + 289 = 625$	Remove parentheses.
$2x^2 + 34x + 289 = 625$	Collect like terms.
$2x^2 + 34x - 336 = 0$	Subtract 625 from each side.

Now observe an important procedure here. This equation would be more difficult to solve due to the large numbers. We can obtain an equivalent (and more convenient) equation by dividing each term on both sides of the equation by 2.

$x^2 + 17x - 168 = 0$	The resulting equation.
$(x + 24)(x - 7) = 0$	Factor the quadratic.
$x + 24 = 0 \qquad x - 7 = 0$	Set each factor equal to zero and solve.
$x = -24 \qquad x = 7$	

(Note that $x = -24$ is not a valid solution. It would not make sense to have a triangle with a side of length -24.) Thus one leg is 7 meters in length. $x + 17 = 7 + 17 = 24$ meters. The second leg is 24 meters in length.

Check Do the two legs differ by 17 meters?

$$24 - 7 \stackrel{?}{=} 17 \qquad 17 = 17\checkmark$$

Do the sides of the triangle satisfy the Pythagorean theorem?

$$7^2 + 24^2 \stackrel{?}{=} 25^2 \qquad 49 + 576 \stackrel{?}{=} 625 \qquad 625 = 625\checkmark$$

Sometimes it may help to use two variables in the initial part of the problem. Then one variable can be eliminated by substitution.

EXAMPLE C The Ski Club is renting a bus to travel to Mount Snow. The members agreed to share the cost of \$180 equally. On the day of the ski trip three members were sick with the flu and could not go. This raised by \$10 the share of each person going on the trip. How many people originally planned to attend?

Solution Let

s = the number of students in the ski club

c = the cost for each student in the original group

We observe:

Number of students × cost per student = total cost

$$\textbf{(1)}\quad s \cdot c = 180$$

Now if 3 people are sick, the number of students drops by 3 but the cost for each increases by \$10. The total is still \$180. Therefore,

(2)	$(s - 3)(c + 10) = 180$	
(3)	$sc - 3c + 10s - 30 = 180$	Remove parentheses.
(4)	$sc - 3c + 10s = 210$	Add 30 to both sides.

Now we want to know what s is, so we desire an equation that contains s as the only variable. We will use substitution.

$c = \dfrac{180}{s}$	Solve equation (1) for c.
$s\left(\dfrac{180}{s}\right) - 3\left(\dfrac{180}{s}\right) + 10s = 210$	Substitute that expression for c in equation (4). We now have an equation that contains s as the only variable.
$180 - \dfrac{540}{s} + 10s = 210$	Remove parentheses.
$\dfrac{-540}{s} + 10s = 30$	Subtract 180 from both sides.
$s\left(\dfrac{-540}{s}\right) + s(10s) = s(30)$	Multiply each term by LCD = s.
$-540 + 10s^2 = 30s$	Remove parentheses.
$10s^2 - 30s - 540 = 0$	Add $-30s$ to both sides. We notice this is a quadric equation.
$s^2 - 3s - 54 = 0$	Divide each term on both sides of the equation by 10. This equivalent equation is easier to work with.
$(s - 9)(s + 6) = 0$	Factor.
$s - 9 = 0 \qquad s + 6 = 0$	Set each factor equal to zero and solve.
$s = 9 \qquad s = -6$	

(We do not consider $s = -6$. We cannot have a negative number of students.)

The number of students originally in the ski club was 9.

Check Was the cost increased by \$10 if the number dropped from 9 students to 6 students?

9 people in the club would mean that each would pay \$20. $(180 \div 9 = 20)$

If the number is reduced by 3, there were 6 people who took the trip. Their cost was \$30 $(180 \div 6 = 30)$. The increase is \$10.

$$20 + 10 \stackrel{?}{=} 30 \qquad 30 = 30\checkmark$$

Many examples in the sciences require the use of quadratic equations. You will study these in more detail if you take a course in physics or calculus in college. Often a quadratic equation is given as part of the problem. Consider this example.

EXAMPLE D When an object is thrown upward, its height (S) in meters is given, approximately, by the quadratic equation

$$S = -5t^2 + vt + h$$

The letter h represents the initial height in meters. The letter v represents the initial velocity of the object thrown. The letter t represents the time in seconds starting from the time the object is thrown.

Suppose that the initial height above the ground is 6 meters (Figure 9.21). The ball is thrown upward with an initial velocity of 29 meters/second. At what time t will the ball hit the ground?

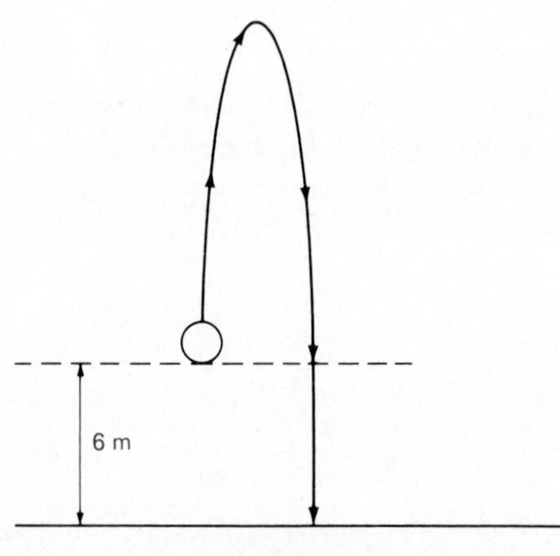

FIGURE 9.21

Solution

$$S = -5t^2 + vt + h$$

In this case $S = 0$ since the ball will hit the ground. The initial upward velocity is $v = 29$ meters/second. The initial height is 6 meters, so $h = 6$.

$0 = -5t^2 + 29t + 6$ — Substituting all values into the equation.

$5t^2 - 29t - 6 = 0$ — Transpose to left side. (Most students can factor more readily if the squared variable is positive.)

$(5t + 1)(t - 6) = 0$ — Factor (or use quadratic formula).

$5t + 1 = 0 \qquad t - 6 = 0$ — Set each factor $= 0$

$5t = -1 \qquad t = 6$ — Solve for t.

$$t = -\frac{1}{5}$$

(Note that since we want a positive time t in seconds we do not consider the value $t = -1/5$.)

Thus the ball will strike the ground 6 seconds after you throw it.

Check Can you verify this solution?

Now try these practice problems.

Practice Problems for Exercise 9.6

Solve each of the following word problems.

1. A number is added to twice its reciprocal. The result is $\frac{19}{3}$. Find the number(s).
2. A triangle has an area of 35 square centimeters. The altitude of the triangle is 3 centimeters shorter than the base. Find the altitude and the base of the triangle.
3. Minette is fencing in a garden that borders the back of a large barn. She has 120 feet of fencing. She would like a rectangular garden that measures 1350 square feet in area. She wants to use the back of the barn, so she only needs to use fencing on three sides. What dimensions should she use for her garden?

Now turn to the answers following Exercise 9.6 and check your solutions for these practice problems.

Exercise 9.6

Solve each of the following word problems.

1. The area of a rectangle is 140 square meters. The width is 3 meters longer than one-half of the length. Find the length and the width of the rectangle.

2. The area of a triangle is 33 square centimeters. The base of the triangle is 1 centimeter less than double the altitude. Find the altitude and the base of the triangle.

3. A number is added to its reciprocal. The result is $\frac{34}{15}$. Find all numbers that satisfy these conditions.

4. A number is added to triple the value of its reciprocal. The result is $\frac{13}{2}$. Find all the numbers that satisfy these conditions.

5. Find two positive numbers whose difference is 10 and whose product is 56.

6. The sum of two numbers is -17. The product of these two numbers is 66. Find the numbers.

7. The diagonal of a rectangle is 13 meters (Figure 9.22). The width of the rectangle is 7 meters shorter than its length. Find the length and the width of the rectangle.

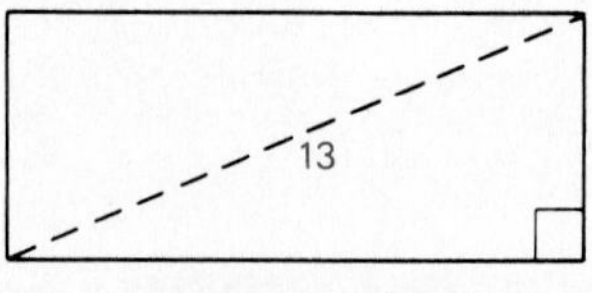

FIGURE 9.22

8. Manuel placed a ladder 20 feet long against the house (Figure 9.23). The distance from the top of the ladder to the bottom of the house is 4 feet greater than the distance from the bottom of the house to the foot of the ladder. How far is the foot of the ladder from the house? How far up on the house does the top of the ladder touch the building?

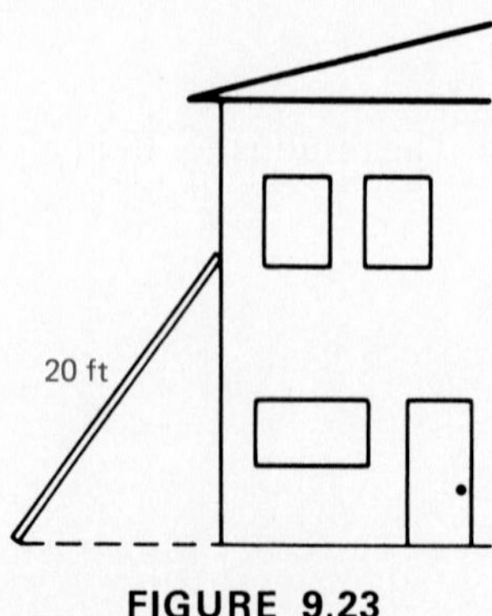

FIGURE 9.23

Use the following information for problems 9 and 10.

When an object is thrown upward, its height (S), in meters, is given (approximately) by the quadratic equation $S = -5t^2 + vt + h$, where

$v =$ the upward initial velocity in meters/second

$t =$ the time of flight in seconds

$h =$ the height above level ground from which the object is thrown

9. You stand on a platform 6 meters high and you throw the ball straight up at a velocity of 13 meters/second. At what time t will the ball hit the ground? How far from the ground is the ball after 2 seconds have elapsed from the time of the throw? (Assume that the ball is 6 meters from the ground when it leaves your hand.)

10. You are standing on the edge of a sheer cliff overlooking the ocean. The place where you stand is 180 meters from the ocean. You drop a pebble into the water (by "dropping" the pebble there is no initial velocity, so $v = 0$). How many seconds will it take to hit the water? How far has the pebble dropped in 3 seconds?

11. The members of the varsity football team decided to give the coach a gift that cost \$80. On the day that they collected money, 10 people were absent. They were in a hurry, so they collected an equal amount from each member and bought the gift. Later the team captain said that if everyone had been present to chip in, each person would have contributed \$4 less. How many people actually contributed? How many people in total were on the football team?

12. One positive number is 2 more than twice another positive number. The difference of the squares of these numbers is 119. Find the numbers that satisfy these conditions.

13. A pilot is testing a new experimental craft. The cruising speed is classified information. In a test the jet traveled 2400 miles. The pilot revealed if he had increased his speed 200 miles/hour, the total trip would take 1 hour less. Can you determine the cruising speed of the jet?

14. A gymnasium floor is being covered by square shock-absorbing tiles. The old gym floor required 864 square tiles. The new tiles are 2 inches larger in both length and width than the old tiles. The new flooring will require only 600 tiles. What is the length of a side of one of the new shock-absorbing tiles? (Hint: Since the area is the same in each case, write an expression for area with old tiles and one with new tiles. Set the two expressions equal to each other.)

15. Overtime production is expensive, but if sufficient extra goods are produced, it is worth putting a shift on overtime. In a certain company, the additional hourly cost in dollars for producing x additional stereo sets is given by the cost equation: $C = 2x^2 - 7x$. If the additional hourly cost is $15, how many additional stereo sets were produced?

16. A boat generator on a fishing boat is required to produce 64 watts of power. The amount of current I measured in amperes needed to produce the power for this generator is given by the equation $P = 40I - 4I^2$. What is the *minimum* number of amperes required to produce the necessary power?

17. An open box is constructed from a sheet of tin. A square of 0.5 centimeter is cut from each corner (Figure 9.24). The box is then constructed by folding up each side. The volume of the open box is 36 cubic centimeters (36 cc). The width of the box is 1 centimeter less than the length. What are the dimensions of the box?

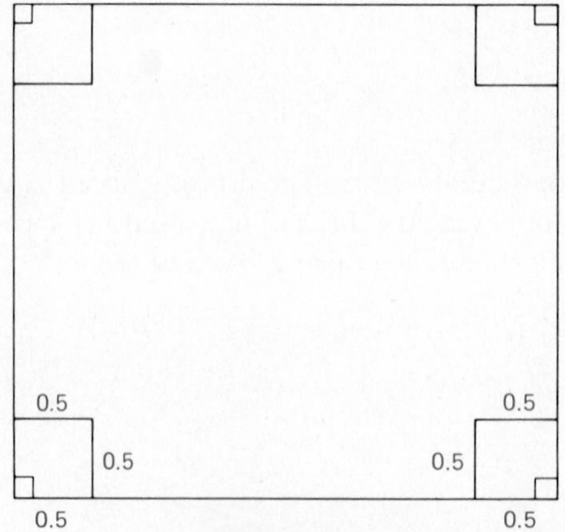

FIGURE 9.24

18. Repeat problem 17 given the following changes: (a) the box is constructed from a *square* piece of tin and (b) the size of the square cut from each corner is 5 centimeters, and (c) the volume of the open box is 320 cc.

Solutions to Practice Problems for Exercise 9.6

1. A number is added to twice its reciprocal. The result is $\frac{19}{3}$. Find the number(s).

Let $x =$ the number; then

$$\frac{1}{x} = \text{the reciprocal of a number}$$

Note: Be sure that you can find a reciprocal of a number. If a number is 4, its reciprocal is $\frac{1}{4}$. If a number is $\frac{3}{5}$, its reciprocal is $\frac{5}{3}$.

$$\left(\text{since } 1 \div \frac{3}{5} = 1 \times \frac{5}{3} = \frac{5}{3}\right)$$

$$x + 2\left(\frac{1}{x}\right) = \frac{19}{3}$$ Write an equation expressing conditions of the problem.

$$x + \frac{2}{x} = \frac{19}{3}$$ Simplify.

$$3x(x) + 3\cancel{x}\left(\frac{2}{\cancel{x}}\right) = \cancel{3}x\left(\frac{19}{\cancel{3}}\right)$$ Multiply by the LCD $= 3x$ to remove fractions.

$$3x^2 + 6 = 19x$$ Simplify.

$$3x^2 - 19x + 6 = 0$$ Subtract $19x$ from both sides.

$$(3x - 1)(x - 6) = 0$$ Factor.

$$3x - 1 = 0 \qquad x - 6 = 0$$ Solve.

$$3x = 1 \qquad x = 6$$

$$x = \frac{1}{3}$$

Check If the number is $\frac{1}{3}$, its reciprocal is 3:

$$\frac{1}{3} + 2(3) = \frac{1}{3} + 6 = \frac{1}{3} + \frac{18}{3} = \frac{19}{3} \checkmark$$

If the number is 6, its reciprocal is $\frac{1}{6}$:

$$6 + 2\left(\frac{1}{6}\right) = 6 + \frac{1}{3} = \frac{18}{3} + \frac{1}{3} = \frac{19}{3}$$

Thus the two answers are possible. The number is either $\frac{1}{3}$ or 6.

2. A triangle has an area of 35 square centimeters. The altitude of the triangle is 3 centimeters shorter than the base. Find the altitude and the base of the triangle.

$$\text{Area of triangle} = \frac{ab}{2}$$

Let

x = the length in centimeters of the base of the triangle

$x - 3$ = the length in centimeters of the altitude of the triangle (Figure 9.25)

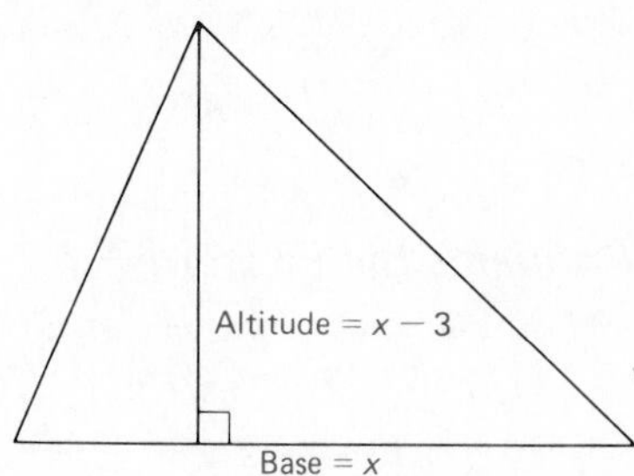

FIGURE 9.25

$$35 = \frac{x(x-3)}{2}$$ Substitution into the area formula.

$$2(35) = \not{2}\left[\frac{x(x-3)}{\not{2}}\right]$$ Multiply both sides by the LCD = 2 to remove fractions.

$$70 = x(x-3) = x^2 - 3x$$ Simplify.

$$0 = x^2 - 3x - 70$$ Subtract 70 from both sides.

$$0 = (x - 10)(x + 7)$$ Factor.

$$x - 10 = 0 \qquad x + 7 = 0$$ Set each factor = 0 and solve for x.

$$x = 10 \qquad x = -7$$

$x = -7$ is meaningless here since our triangle has positive dimensions. Thus the base is 10 centimeters.

$$x - 3 = 10 - 3 = 7$$ The altitude is 7 centimeters.

Check Is the altitude 3 centimeters shorter than the base?

$$10 - 7 \stackrel{?}{=} 3 \qquad 3 = 3\checkmark$$

Is the area of the triangle 35 square centimeters?

$$35 = \frac{(7)(10)}{2} = \frac{70}{2} = 35\checkmark$$

3. Minette is fencing in a garden that borders the back of a large barn. She has 120 feet of fencing. She would like a rectangular garden that measures 1350 square feet in area. She wants to use the back of the barn, so she needs to use fencing only on three sides. What dimensions should she use for her garden?

Draw some type of sketch, as in Figure 9.26, so that you can visualize what is being described.

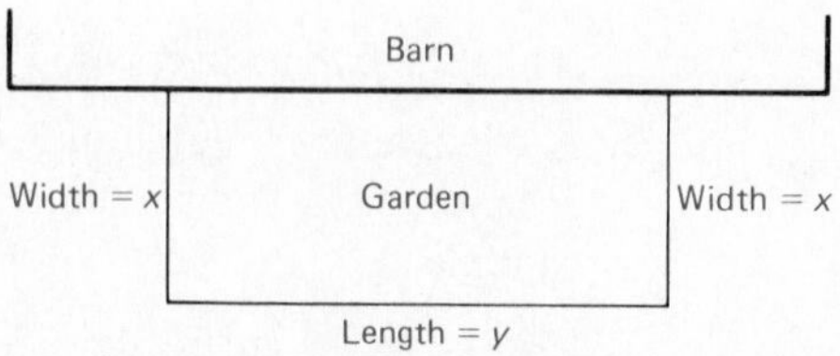

FIGURE 9.26

Let

$$x = \text{the width of the garden in feet}$$

$$y = \text{the length of the garden in feet}$$

Since the 120 feet of fencing will cover the width twice and the length once:

(1) $2x + y = 120$	Equation describing conditions for 120 feet of fencing.

Now the area formula is $A = (\text{width}) \times (\text{length})$:

(2) $1350 = (x)(y)$	Equation for the area of 1350 square feet.
$y = 120 - 2x$	Now if we solve equation (1) for y, we have
(3) $1350 = x(120 - 2x)$	Replace y by this expression in equation (2).
$1350 = 120x - 2x^2$	Simplify.
$2x^2 - 120x + 1350 = 0$	Add $2x^2$ and add $-120x$ to both sides.
$x^2 - 60x + 675 = 0$	Divide each term of the equation by 2.
$(x - 15)(x - 45) = 0$	Factor.
$x = 15 \qquad x = 45$	Set each parentheses $= 0$ and solve for x in each case.

First Solution

If the width $= 15$, then the length is

$$y = 120 - 2x$$

$$y = 120 - 2(15) = 120 - 30$$

$$y = 90$$

Check

(a) Does this provide 120 feet of fencing?

$$\text{Width} + \text{width} + \text{length} \stackrel{?}{=} 120 \text{ feet of fencing}$$

$$15 + 15 + 90 \stackrel{?}{=} 120 \qquad 120 = 120\checkmark$$

(b) Does the area amount to 1350?

$$(\text{length}) \cdot (\text{width}) \stackrel{?}{=} 1350$$

$$(90)(15) \stackrel{?}{=} 1350 \qquad 1350 = 1350\checkmark$$

Second Solution

If the width $= 45$ then the length is

$$y = 120 - 2x$$

$$y = 120 - 2(45) = 120 - 90$$

$$y = 30$$

Check

(a) Width + width + length = 120 feet of fencing

$$45 + 45 + 30 = 120$$

(b) (Length) $\cdot$ (width) $= 1350$

$$(45)(30) = 1350$$

Thus we see that Minette has two choices that satisfy the given requirements (see Figure 9.27). She would probably choose the shape that is the most practical for her. (If her barn is not 90 feet long, she could not use this first solution!)

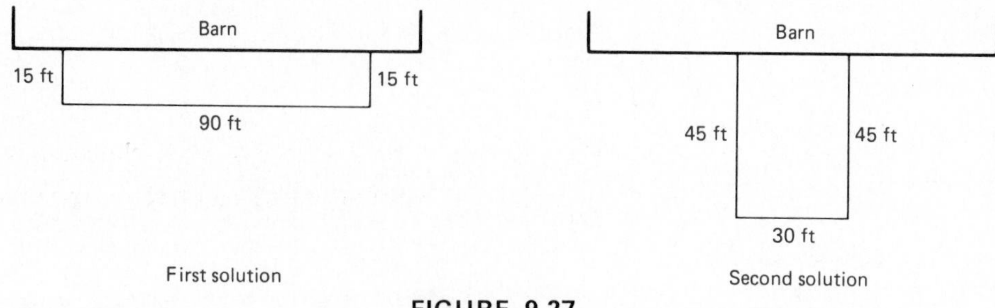

FIGURE 9.27

Remark: Isn't there a shorter way to do this problem? Yes. Use only the variable x to represent the length and width. If you let $x =$ the width and $120 - 2x =$ the length, you can immediately write equation (3), which is

$$1350 = x(120 - 2x)$$

However, many students find that this is difficult to do, and they would prefer to use two variables. Try to work these problems in a way that seems fairly easy to you.

Now do the problems in Exercise 9.6.

An Application for Chapter 9: Maximizing Profit

There are a number of real-life applications of quadratic equations. One of great interest to many people is the area of analysis of profits in a small business. In many instances it has been shown that the profit of a small business can be calculated by or approximated by a quadratic equation. The profit picture can be illustrated by graphing the parabola defined by the quadratic equation that predicts the profit. Consider the following illustration.

Last semester Juan Perez decided to sell custom T-shirts to students at his college. He obtained the T-shirts at $2 each, had them lettered at a local clothing shop with the message requested, and sold them to students for $5 each. He made so much money he decided during the next semester to set up a small business for himself. He rented a store and a machine that will imprint lettering or designs on the shirts, hired two students to run the store, installed a telephone, and started running his business. His profit for the next four months was as follows:

	Number of T-Shirts Ordered and Sold	*Profit at the end of Month After Paying all Expenses*
September	60	–$ 43.00 (loss)
October	120	$188.00 (profit)
November	260	$237.00 (profit)
December	380	–$267.00 (loss)

Initially, Juan realized that as sales increased, so did profits. He hired more help and placed more advertising as sales increased. The profits increased until a very busy December when the loss to the company was $267! Juan asked his roommate, a math major, to help him analyze his company. The math major put these values into a computer that has a program to create polynomial equations as models to relate experimental data. The computer program produced an equation that predicted the monthly profit of Juan's company. The equation predicted the profit in terms of the variable t (the number of T-shirts ordered and sold in a given month):

$$P = -\frac{7}{400}t^2 + 7t - 400$$

Juan had three questions:

1. What is my profit if I make and sell 100 T-shirts each month? 200? 300? 400?
2. How many T-shirts do I need to sell each month to break even?
3. What is the ideal number of T-shirts to sell each month to maximize profits?

A Challenge for You!

Before you see Juan's solution see if you can do this problem by yourself.

Solution Since Juan had taken a course in beginning algebra using this text, he was able to answer all the questions himself.

1. He substituted $t = 100, 200, 300$, and 400 into the profit equation and obtained the corresponding P values. The actual work takes a few minutes because of the large numbers. Can you verify his answers below?

Number of T-shirts (t)	100	200	300	400
Profit (P)	125	300	125	−400

Using these values together with his original values of t and P for September, October, November, and December, he had a total of eight points. He plotted these points and drew a graph of the resulting parabola (Figure 9.28).

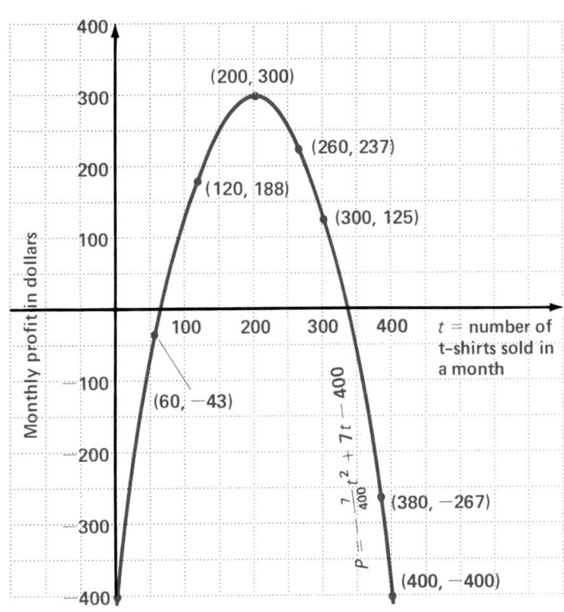

FIGURE 9.28

2. From his graph he estimated that $P = 0$ when $t = 70$ and when $t = 330$. To obtain a more accurate answer, he took the equation

$$P = 0 = -\frac{7}{400}t^2 + 7t - 400$$

and solved it for t using the quadratic formula, where $a = -\frac{7}{400}$, $b = 7$, and $c = -400$. (Remember, it is usually best to have a, b, c as integers but the quadratic formula will work if a, b, c are rational numbers.)

$$t = \frac{-b \pm \sqrt{b^2 - 4ac}}{2a}$$

$$t = \frac{-7 \pm \sqrt{(7)^2 - 4\left(-\frac{7}{400}\right)(-400)}}{2\left(-\frac{7}{400}\right)}$$ Substituting above values of a, b, and c.

$$t = \frac{-7 \pm \sqrt{49 - 28}}{-\frac{7}{200}}$$ Simplifying the expression.

$$t = \frac{-7 \pm \sqrt{21}}{-\frac{7}{200}}$$ Using the property of division of fractions, we can simplify this to the product of two fractions.

$$t = \left(\frac{-7 \pm \sqrt{21}}{1}\right) \div \left(-\frac{7}{200}\right)$$

$$= \left(\frac{-7 \pm \sqrt{21}}{1}\right) \cdot \left(\frac{200}{-7}\right)$$

$$t = +200 \pm \frac{200\sqrt{21}}{7}$$ We will use the table of square roots (in Appendix A. Now since $\sqrt{21} \doteq 4.583$, we can approximate the values of t.

$t \doteq 200 + \frac{200(4.583)}{7}$	$t \doteq 200 - \frac{200(4.583)}{7}$	Substitute.
$t \doteq 200 + 130.943$	$t \doteq 200 - 130.943$	Evaluate $\frac{(200)(4.583)}{7}$.
$t \doteq 330.943$	$t \doteq 69.057$	Now round to the nearest whole number.
$t \doteq 331$	$t \doteq 69$	

Thus his graphical solution was only off by a very slight amount. The correct answer to the nearest whole number of T-shirts is that he will break even if he sells 69 T-shirts or 331 T-shirts. The maximum profit on the graph appears to be when $t = 200$.

3. To verify that the vertex appears to be at $t = 200$, he evaluated P at $t = 199$ and $t = 201$.

If $t = 199$: $P = \$299.98$ to the nearest cent

If $t = 201$: $P = \$299.98$ to the nearest cent

Since the values of P are less if t is greater or less than 200, he concluded that due to the shape of the parabola the vertex was at (200, 300). The maximum profit takes place if he sells 200 T-shirts each month.

A better technique to find the vertex of a parabola would be to use the approach of Example E in Section 9.5.

We can verify that the maximum value is at $t = 200$ without using the graph. The vertex of any parabola of the form $y = ax^2 + bx + c$ will occur when $x = -b/2a$. Thus the vertex of

$$P = \frac{-7}{400}t^2 + 7t - 400$$

will occur when

$$t = \frac{-b}{2a} = \frac{-7}{2(-7/400)} = \frac{-7}{-7/200} = (+7)\cdot\left(\frac{200}{7}\right) = 200$$

Supplementary Problems for Chapter 9

Solve these problems by any desired method. If there is no real number, so state.

1. $x^2 - 3x - 18 = 0$

2. $x^2 + 3x - 10 = 0$

3. $5x^2 = 2x - 7x^2$

4. $8x^2 + 5x = 2x^2 - 6x$

5. $2x^2 + 8x - 5 = 0$

6. $3x^2 - 6x + 2 = 0$

7. $5x^2 - 7x = 8$

8. $5x^2 = 7x + 6$

9. $3x^2 + 1 = 6 - 8x$

10. $4x^2 + 4x = x^2 + 5$

11. $5x^2 + 7x + 1 = 0$

12. $x^2 + 5x = -1 + 2x$

13. $2x^2 + 10x = 2x - 7$

14. $3x^2 + 6x = 2x^2 - 9$

15. $4x^2 + 9x - 9 = 0$

16. $5x^2 - 11x + 2 = 0$

17. $2y - 10 = 10y(y - 2)$

18. $3y(3y + 2) = 3y + 2$

19. $7(y^2 + 2) = 2(2y + 7)$

20. $4(5 - y) = 5(y^2 + 4)$

21. $\frac{3y - 2}{4} = \frac{y^2 - 2}{y}$

22. $\frac{y^2 + 5}{2y} = \frac{2y - 1}{3}$

23. $\frac{(y - 2)^2}{20} + 3 + y = 0$

24. $\frac{(y + 2)^2}{5} + 2y = -9$

Use the Pythagorean theorem to find the missing side of each right triangle.

25.

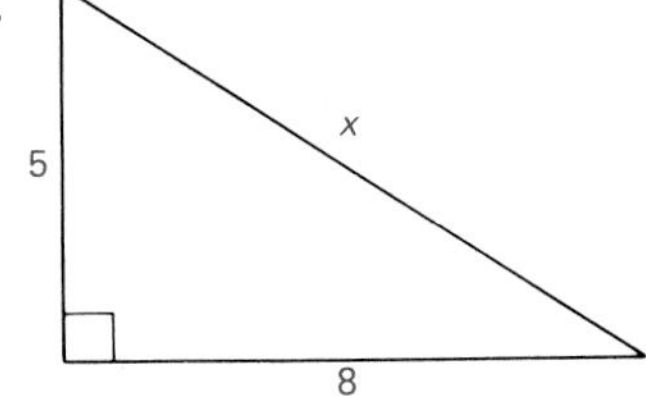

26.

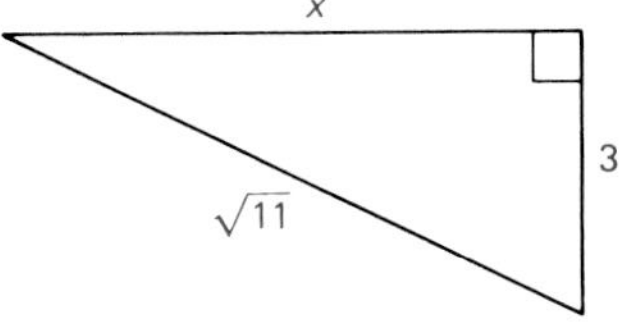

27.

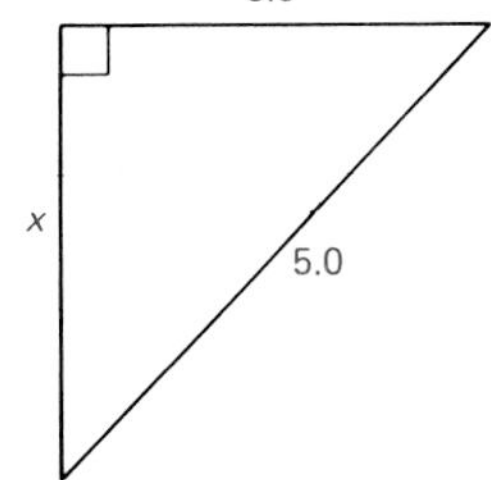

28.

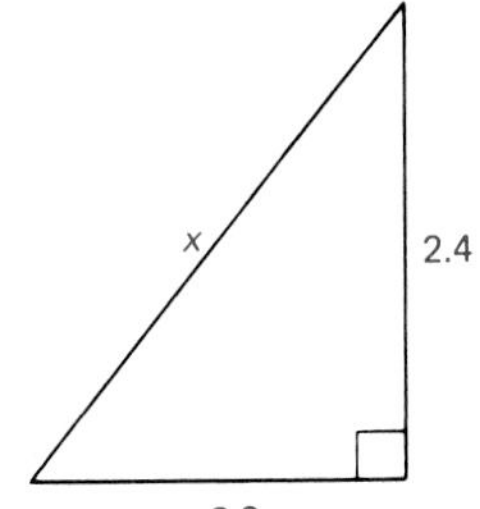

29. A flag pole is 24 meters tall. A man stands 18 meters from the base of the pole. How far is it from the feet of the man to the top of the pole?

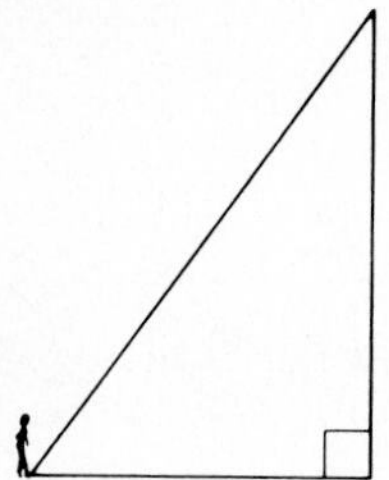

30. A city is 20 miles east of a major airport. A small town is directly south of the city. The town is 50 miles from the airport. How far is it from the town to the city?

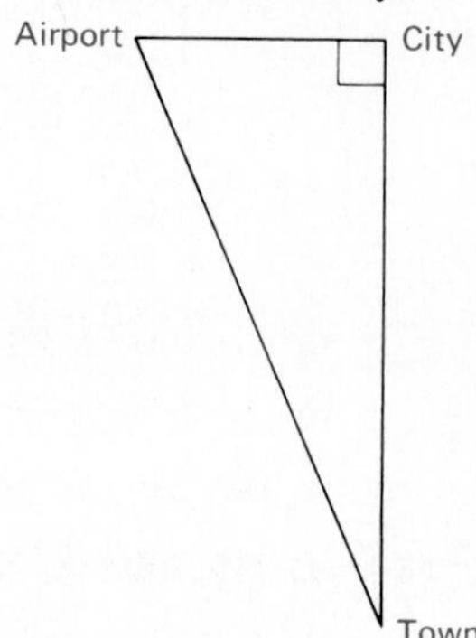

Solve the following radical equations. Be sure to verify your answers.

31. $\sqrt{x} - 6 = -x$

32. $x + \sqrt{2x - 5} = 10$

33. $6 - \sqrt{5x - 1} = x + 1$

34. $1 - 4x + \sqrt{x + 2} = 4 - 4x$

In problems 35 to 44, graph the quadratic equations.

35. $y = x^2 - 3x - 4$

36. $y = \frac{1}{2}x^2 - 2$

37. $y = 3x^2 + 2x$

38. $y = 5 - x^2$

39. $y = -x^2 + 2x - 1$

40. $y = \frac{1}{12}x^2 - \frac{1}{2}x + \frac{1}{4}$

41. $y = (x - 2)^2 + 4$

42. $y = (x + 5)^2 - 3$

43. $y = -2x^2 + 12x - 17$

44. $y = -3x^2 - 2x + 4$

45. The length of a room exceeds its width by 4 feet. A rug covers the floor area except for a border 2 feet wide all around it. The area of the border is 68 square feet. What is the area of the rug?

46. The area of a triangle is 35 square centimeters. The base is 3 centimeters longer than the altitude of the triangle. Find the length of the base and the altitude.

47. Alice drove 90 miles to visit her cousin. Her average speed on the trip home was 15 mph faster than her speed on the trip going. Her total travel time for both going and returning was 3.5 hours. What was her average rate of speed on each trip?

48. A rectangular fieldhouse at Landham University can seat 720 people. The Dean decided to change the original seat arrangement. The number of rows needed to seat the students in the new arrangement would be four less if each row had chairs for 6 more students. How many people are seated in each row in the original seat arrangement? How many rows were in the original seat arrangement?

49. A picture in the Richardson Gallery measures 6 inches by 12 inches. It is surrounded by a frame of uniform width. The area of the frame is exactly twice the area of the picture. Find the width of the frame.

Chapter 9 Summary

Methods to Solve a Quadratic Equation

A quadratic equation in *standard form* is an equation of the form $ax^2 + bx + c = 0$, where a, b, and c are real numbers and $a \neq 0$. The roots of quadratic equations can be obtained in one of three ways.

1. Factoring
   (a) Place the equation in standard form: $ax^2 + bx + c = 0$.
   (b) Factor, if possible.

(c) Set each factor equal to 0.

(d) Solve each of the resulting equations.

2. Completing the square

(a) Perform any necessary steps to transform the equation to the form $ax^2 + bx =$ some constant.

(b) If $a \neq 1$, divide each term of the equation by a.

(c) Square half of the numerical coefficient of the linear term. Add the result to both sides of the equation.

(d) Factor the left side, then take the square root of both sides of the equation.

(e) Solve each resulting equation for x.

3. The Quadratic Formula

(a) Place the equation in standard form: $ax^2 + bx + c = 0$.

(b) Determine the values of a, b, and c.

(c) Substitute into the formula

$$x = \frac{-b \pm \sqrt{b^2 - 4ac}}{2a}$$

(d) Simplify the result to obtain the two values of x.

Number of Solutions to a Quadratic Equation of the Form $ax^2 + bx + c = 0$

1. A quadratic equation may have *two distinct real roots* (when $b^2 - 4ac > 0$).
2. A quadratic equation may have *one real root.* When you solve an equation of this type you obtain the same answer twice. This is often called a double root. (This will occur when $b^2 - 4ac = 0$.)
3. A quadratic equation may have *no real roots.* When you solve an equation of this type you will obtain an expression containing a *square root of a negative number*. (This will occur when $b^2 - 4ac < 0$.)

The Pythagorean Theorem

In any right triangle if c is the length of the hypotenuse and a and b are the lengths of the two legs, then

$$c^2 = a^2 + b^2$$

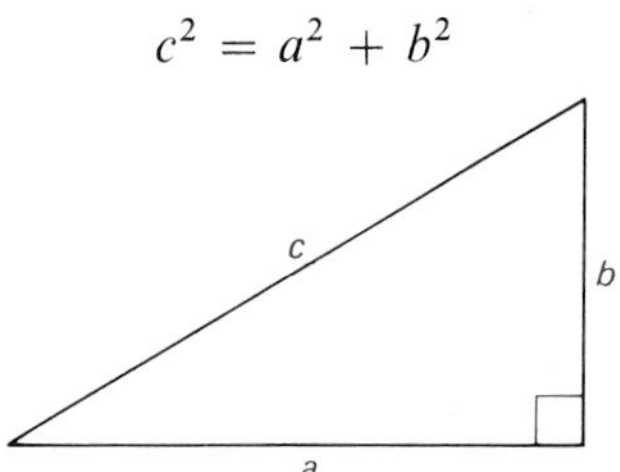

Radical Equations

To solve an equation containing a square root radical:

1. Perform algebraic operations to obtain one radical by itself on one side of the equation.
2. Square each side of the equation.
3. Simplify, if possible.
4. Solve the resulting equation.
5. Check all apparent solutions! Solutions to radical equations *must* be verified.

Graphing Quadratic Equations of the Form $y = ax^2 + bx + c$

1. The graph of these equations yields a smooth curve called a *parabola.*
2. The parabola opens upward if $a > 0$ and downward if $a < 0$.
3. For all parabolas a *line of symmetry* can be drawn (Figure 9.29). If you fold the graph on this line, the two halves of the curve will correspond exactly.

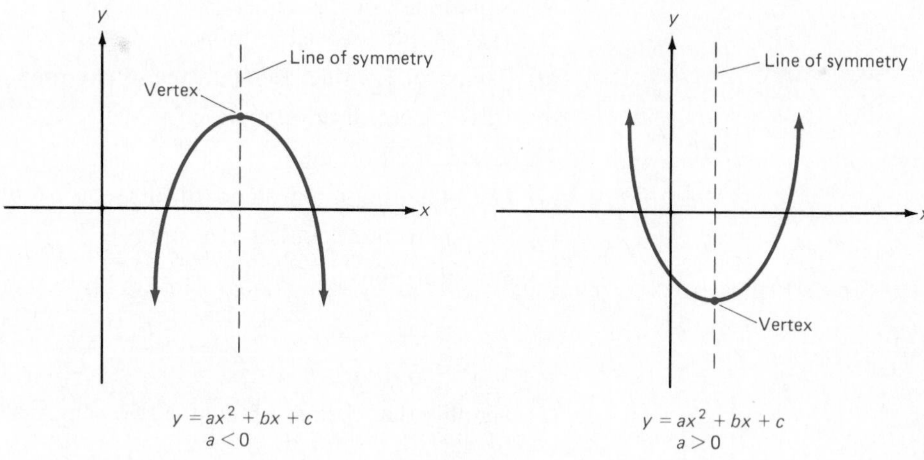

FIGURE 9.29

4. To draw a parabola pick five to eight values of x and find the corresponding y values. Connect the resulting points by a smooth curve.
5. The vertex of any parabola of these types will be at the lowest point of a parabola opening upward (if $a > 0$) or the highest point of a parabola opening downward (if $a < 0$). The x-coordinate of the vertex is given by $x = \frac{-b}{2a}$.

   The line of symmetry passes through the vertex.

Practice Quiz for Chapter 9

Sections 9.1–9.3

Solve by the *factoring method.*

1. $x^2 - 6x - 27 = 0$

2. $7x = x(x - 4)$

Solve by the *completing the square method.*

3. $x^2 - 2x = 3$

4. $2x^2 - 6x + 3 = 0$

Solve by using the *quadratic formula.*

5. $x^2 + 4x - 10 = 0$

6. $2x^2 + 5x - 12 = 0$

Solve each of the following problems, if possible. You may use any method. If there is no real number solution, so state.

7. $2x^2 = -5x + 3$

8. $8 = x(x + 2)$

9. $1 + 3x^2 + 5x = 3$

10. $(x + 1)(x - 2) = 4$

11. $2x^2 = -10x - 11$

12. $10x^2 - 7x = -6$

Sections 9.4-9.6

Use the Pythagorean theorem to find the missing side of each triangle. Be sure to simplify your answer.

1.

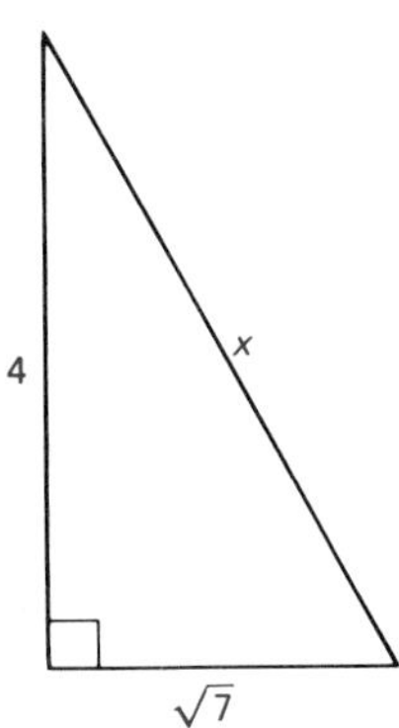

2.

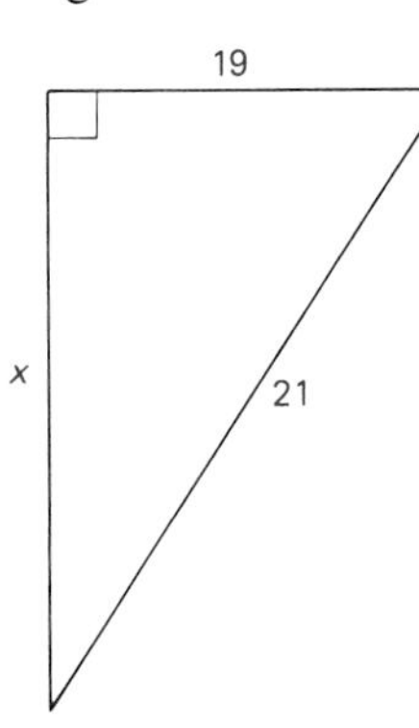

3. An observer on the ground is exactly 500 meters from the top of a radio broadcast antenna (Figure 9.30). She is 450 meters from the center of the base of the antenna. How tall is the antenna? Use the square root table to approximate your answer.

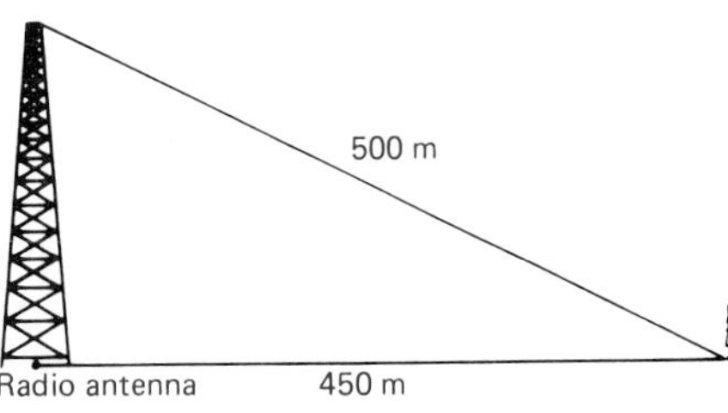

FIGURE 9.30

Solve for the variable. Be sure to verify your answer(s).

4. $\sqrt{2x + 4} = 4$ **5.** $\sqrt{4x + 5} = x$ **6.** $\sqrt{3y - 2} + 2 = y$

In problems 7 to 10, graph the quadratic equations.

7. $y = \frac{1}{2}x^2$ **8.** $y = 3x^2 - 4$

9. $y = 2x^2 + 4x - 5$ **10.** $y = 2x - 3x^2$

11. One number is five greater than another. The sum of the squares of these numbers is 157. Find all the numbers that satisfy these conditions.

12. The number of diagonals in a polygon is

$$\frac{x(x-3)}{2} \quad \text{where } x \text{ is the number of sides}$$

How many sides would be contained in a polygon that has 44 diagonals?

13. An insurance office has a square floor space. The area of the office is 49 square meters. The manager wants to expand the office to a larger square office area. He wants 32 square meters more office space. If he is able to do this, how many meters should be added to each side of the original square office?

Practice Test for Chapter 9

Solve problems 1 to 4 by any desired method. If there is no real number solution, so state.

1. $5x^2 + 7x = 2x(x - 3)$ **2.** $3x^2 + 13x = 10$

3. $2x^2 = 2x - 5$ **4.** $6x^2 - 7x - 5 = 0$

Find the missing side by using the Pythagorean theorem.

5.

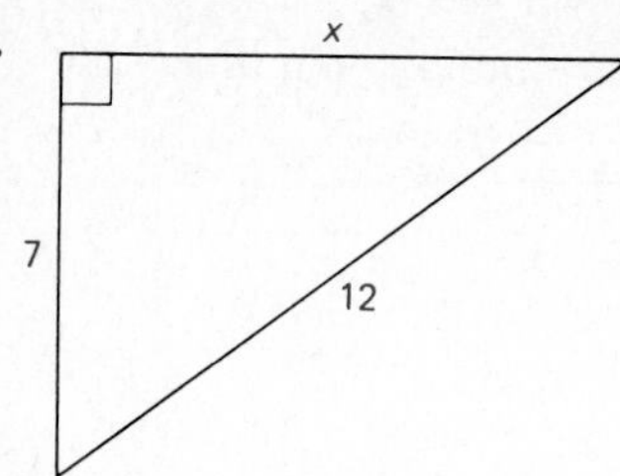

6.

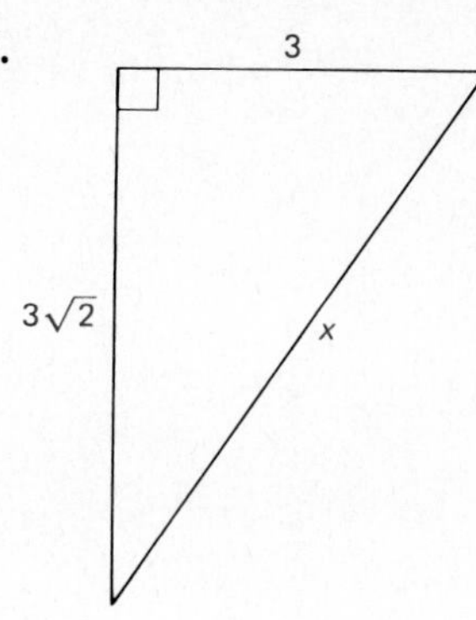

Solve the following radical equations. Verify your solutions.

7. $6 - \sqrt{2x + 1} = 0$

8. $x = 5 + \sqrt{x + 7}$

Graph the following quadratic equations.

9. $y = 3x^2 - 5x$

10. $y = -x^2 + 8x - 12$

11. When an object is thrown upward, its height (S) in meters is given (approximately) by the quadratic equation $S = -5t^2 + vt + h$ where

$v =$ the upward initial velocity in meters per second

$t =$ the time of flight in seconds

$h =$ the height above level ground from which the object is thrown

Suppose that a ball is thrown upward with a velocity of 33 meters/second at a height of 14 meters above the ground. When will it strike the ground?

12. The Blakes took a car trip. The first part of the trip was 120 miles and they traveled at a constant speed. For the next 144 miles they traveled exactly 10 mph slower. They made the trip in 6 hours. How fast did they travel for each part of the trip?

Cumulative Examination

The following problems cover the content of Chapters 1 to 9. Follow the directions for each problem and simplify your answers.

1. Simplify:

$$-2x + 3y\{7 - 2[x - (4x + y)]\}$$

2. Evaluate if $x = -2$ and $y = 3$:

$$2x^2 - 3xy - 4y$$

3. Solve for x:

$$\frac{1}{2}(x + 4) - \frac{2}{3}(x - 7) = 4x$$

4. Solve for x and graph the resulting inequality on a number line:

$$5x + 3 - (4x - 2) \leq 6x - 8$$

5. Factor completely: $4x^2 - 18x - 10$

6. Factor completely: $25x^4 - 25$

7. Combine:

$$\frac{2}{x - 3} - \frac{3}{x^2 - x - 6} + \frac{4}{x + 2}$$

8. Simplify:

$$\frac{\frac{3}{x} + \frac{5}{2x}}{1 + \frac{2}{x + 2}}$$

9. Find the slope of the line and then graph the line:

$$5x - 2y - 3 = 0$$

10. Two quarter-circles are attached to a rectangle. The dimensions are indicated in the figure. Find the area of the region. Use 3.14 as an approximation for π.

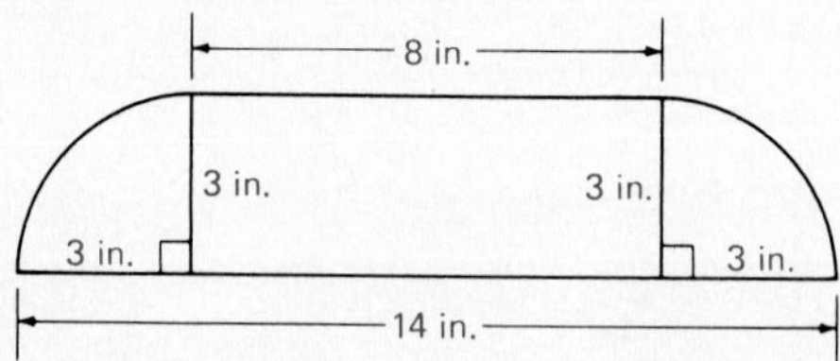

11. Solve for x and y:

$$2x + 3y = -1$$

$$3x + 5y = -4$$

12. Solve for a and b:

$$a - \frac{3}{4}b = \frac{1}{4}$$

$$\frac{3}{2}a + \frac{1}{2}b = -\frac{9}{2}$$

13. Simplify and combine:

$$2x\sqrt{50} + \sqrt{98x^2} - 3x\sqrt{18}$$

14. Simplify:

$$\frac{3 + \sqrt{7}}{\sqrt{5} - \sqrt{7}}$$

15. Solve for x:

$$3x^2 - 2x - 5 = 0$$

16. Solve for y:

$$2y^2 = 6y - 1$$

17. A number is tripled and then increased by 6. The result is 21. What is the original number?

18. A rectangular region has a perimeter of 38 meters. The length of the rectangle is 2 meters shorter than double the width. What are the dimensions of the rectangle?

19. A lady invested \$7000 for 1 year. She placed part of the money in a tax-free bond earning 10% interest. She placed the rest in a money market earning 14% interest. At the end of 1 year she had earned \$860 in interest. How much did she invest at each interest rate?

20. A benefit concert was held on the campus to raise scholarship money. A total of 360 tickets were sold. Admission prices were \$5 for reserved seats and \$3 for general admission. Total receipts were \$1480. How many reserved tickets were sold? How many general admission tickets were sold?

Appendix A

Table of Square Roots

X	$\sqrt{X}$	X	$\sqrt{X}$	X	$\sqrt{X}$	X	$\sqrt{X}$	X	$\sqrt{X}$
1	1.000	41	6.403	81	9.000	121	11.000	161	12.689
2	1.414	42	6.481	82	9.055	122	11.045	162	12.728
3	1.732	43	6.557	83	9.110	123	11.091	163	12.767
4	2.000	44	6.633	84	9.165	124	11.136	164	12.806
5	2.236	45	6.708	85	9.220	125	11.180	165	12.845
6	2.449	46	6.782	86	9.274	126	11.225	166	12.884
7	2.646	47	6.856	87	9.327	127	11.269	167	12.923
8	2.828	48	6.928	88	9.381	128	11.314	168	12.961
9	3.000	49	7.000	89	9.434	129	11.358	169	13.000
10	3.162	50	7.071	90	9.487	130	11.402	170	13.038
11	3.317	51	7.141	91	9.539	131	11.446	171	13.077
12	3.464	52	7.211	92	9.592	132	11.489	172	13.115
13	3.606	53	7.280	93	9.644	133	11.533	173	13.153
14	3.742	54	7.348	94	9.695	134	11.576	174	13.191
15	3.873	55	7.416	95	9.747	135	11.619	175	13.229
16	4.000	56	7.483	96	9.798	136	11.662	176	13.266
17	4.123	57	7.550	97	9.849	137	11.705	177	13.304
18	4.243	58	7.616	98	9.899	138	11.747	178	13.342
19	4.359	59	7.681	99	9.950	139	11.790	179	13.379
20	4.472	60	7.746	100	10.000	140	11.832	180	13.416
21	4.583	61	7.810	101	10.050	141	11.874	181	13.454
22	4.690	62	7.874	102	10.100	142	11.916	182	13.491
23	4.796	63	7.937	103	10.149	143	11.958	183	13.528
24	4.899	64	8.000	104	10.198	144	12.000	184	13.565
25	5.000	65	8.062	105	10.247	145	12.042	185	13.601
26	5.099	66	8.124	106	10.296	146	12.083	186	13.638
27	5.196	67	8.185	107	10.344	147	12.124	187	13.675
28	5.292	68	8.246	108	10.392	148	12.166	188	13.711
29	5.385	69	8.307	109	10.440	149	12.207	189	13.748
30	5.477	70	8.367	110	10.488	150	12.247	190	13.784
31	5.568	71	8.426	111	10.536	151	12.288	191	13.820
32	5.657	72	8.485	112	10.583	152	12.329	192	13.856
33	5.745	73	8.544	113	10.630	153	12.369	193	13.892
34	5.831	74	8.602	114	10.677	154	12.410	194	13.928
35	5.916	75	8.660	115	10.724	155	12.450	195	13.964
36	6.000	76	8.718	116	10.770	156	12.490	196	14.000
37	6.083	77	8.775	117	10.817	157	12.530	197	14.036
38	6.164	78	8.832	118	10.863	158	12.570	198	14.071
39	6.245	79	8.888	119	10.909	159	12.610	199	14.107
40	6.325	80	8.944	120	10.954	160	12.649	200	14.142

Appendix B

Ratio, Proportion, and Metric Measurement

The *ratio* of one number a to another number b is the quotient $a \div b$ or a/b. For example, the ratio of 7 to 9 can be written as $\frac{7}{9}$.

A *proportion* is an equation stating that two ratios are equal. The concept of equal ratios is very useful in converting from one size, unit, or amount to another. Suppose that we have two equal ratios in the proportion

$$\frac{a}{b} = \frac{c}{d}$$

By using the techniques of solving fractional equations we know that the LCD $= bd$.

$(\not{b}d)\dfrac{a}{\not{b}} = (b\not{d})\dfrac{c}{\not{d}}$	We multiply each side by the LCD.
$da = bc$	We cancel common factors.
$ad = bc$	Since multiplication is commutative, $da = ad$.

Thus we have proved the following:

If

$$\frac{a}{b} = \frac{c}{d} \qquad \textbf{then } ad = bc$$

for all real numbers a, b, c, and d where $b \neq 0$, $d \neq 0$.

This formula is sometimes called "cross-multiplying." It can be applied only if you have *one* fraction and nothing else on each side of the equation.

We can use a proportion to solve a variety of applied problems involving two ratios.

EXAMPLE A Michael took 5 hours to drive 245 miles on the turnpike. If he continues at the same rate, how many hours will it take him to drive a distance of 392 miles?

Solution Let x = the number of hours it will take to drive 392 miles. If 5 hours are needed to drive 245 miles, then x hours are needed to drive 392 miles. We will write this as a proportion.

$$\begin{matrix}\text{Initial time} \rightarrow \\ \text{Initial distance} \rightarrow\end{matrix} \frac{5}{245} = \frac{x}{392} \begin{matrix}\leftarrow \text{Second time} \\ \leftarrow \text{Second distance}\end{matrix}$$

$5(392) = 245x$ — If $\frac{a}{b} = \frac{c}{d}$, then $ad = bc$. (Cross-multiplying.)

$1960 = 245x$ — Multiply the two numbers on left side of the equation.

$\frac{1960}{245} = x$ — Divide both sides by 245.

$8 = x$ — Simplify.

So it would take Michael 8 hours to drive 392 miles.

EXAMPLE B If $\frac{3}{4}$ inch on a map represents an actual distance of 20 miles, how long of a distance is represented by $4\frac{1}{8}$ inches on the same map?

Solution Let x = the distance represented by $4\frac{1}{8}$ inches.

$$\begin{matrix}\text{Initial measurement on map} \rightarrow \\ \text{Initial distance} \rightarrow\end{matrix} \frac{\frac{3}{4}}{20} = \frac{4\frac{1}{8}}{x} \begin{matrix}\leftarrow \text{Second measurement on the map} \\ \leftarrow \text{Second distance}\end{matrix}$$

$\left(\frac{3}{4}\right)(x) = (20)\left(4\frac{1}{8}\right)$ — Cross-multiplying.

$\left(\frac{3}{4}\right)(x) = (\overset{5}{\cancel{20}})\left(\frac{33}{\underset{2}{\cancel{8}}}\right)$ — Transform $4\frac{1}{8}$ to equivalent improper fraction of $\frac{33}{8}$ and simplify.

$\frac{3x}{4} = \frac{165}{2}$ — Multiplication of fractions.

$\cancel{4}\left(\frac{3x}{\cancel{4}}\right) = \overset{2}{\cancel{4}}\left(\frac{165}{\cancel{2}}\right)$ — Multiply each side by 4.

$3x = 330$ — Divide both sides by three.

$x = 110$

$4\frac{1}{8}$ inches on the map represents an actual distance of 110 miles.

Metric measurements are becoming more common every day in the United States. The basic unit of length in the metric system is the meter. The meter is divided into 100 centimeters, for making smaller measurements. One kilometer is equivalent to 1000 meters. The following table illustrates the equivalent values of various metric length measurements.

Metric Length

*1 kilometer (km)	=	1000 meters
1 hectometer (hm)	=	100 meters
1 dekameter (dam)	=	10 meters
*1 meter (m)	=	1 meter
1 decimeter (dm)	=	0.1 meter
*1 centimeter (cm)	=	0.01 meter
*1 millimeter (mm)	=	0.001 meter

The four most common units of metric measurement are indicated by an asterisk.

The following are three commonly used metric units of measurement of weight:

Metric Weight

1 kilogram (kg)	=	1000 grams
1 gram (g)	=	1 gram
1 milligram (mg)	=	0.001 gram

The following are three commonly used metric units of measurement of liquid capacity:

Metric Volume

1 kiloliter (kl)	=	1000 liters
1 liter (L)	=	1 liter
1 milliliter (ml)	=	0.001 liter

The eight most common relationships between the two systems are listed in the following table. These values are approximate.

TABLE OF METRIC CONVERSION RATIOS

Length:	1 inch = 2.54 centimeters
	39.37 inches = 1 meter
	1 mile = 1.61 kilometers
Weight:	1 pound = 454 grams
	2.20 pounds = 1 kilogram
Liquid capacity:	1 quart = 946 milliliters
	1.06 quarts = 1 liter
	1 gallon = 3.785 liters

We will illustrate how these metric conversion ratios can be used in Examples C, D, and E.

EXAMPLE C A box weighs 170 grams. How many pounds is that?

Solution Let x = the number of pounds in 170 grams. If 1 pound is 454 grams, x pounds will be 170 grams.

$$\begin{array}{r} \text{Initial number of pounds} \rightarrow \\ \text{Initial number of grams} \rightarrow \end{array} \frac{1}{454} = \frac{x}{170} \begin{array}{l} \leftarrow \text{Second number of pounds} \\ \leftarrow \text{Second number of grams} \end{array}$$

$170 = 454x$ — If $\frac{a}{b} = \frac{c}{d}$, then $ad = bc$. (Cross multiplying.)

$\frac{170}{454} = x$ — Divide both sides by 454.

$0.374 = x$ — Carry out the division to three decimal places.

The box weighs approximately 0.374 pound.

EXAMPLE D A sign on the highway reads "Smithville 20 kilometers." How many miles away is Smithville?

Solution Let $x =$ the number of miles in 20 kilometers. If 1 mile is 1.61 kilometers, x miles will be 20 kilometers.

$$\begin{matrix}\text{Initial number of miles} \rightarrow \\ \text{Initial number of kilometers} \rightarrow\end{matrix} \frac{1}{1.61} = \frac{x}{20} \begin{matrix}\leftarrow \text{Second number of miles} \\ \leftarrow \text{Second number of kilometers}\end{matrix}$$

$20 = 1.61x$ — Cross multiplying.

$\frac{20}{1.61} = x$ — Divide each side by 1.61.

$12.4 = x$ — Carry out the division to one decimal place.

Thus Smithville is about 12.4 miles away from the location indicated by the sign.

EXAMPLE E Anita purchased 42 gallons of gasoline for her car. How many liters did she buy?

Solution Let $x =$ the number of liters in 42 gallons. If 1 gallon is 3.785 liters, then 42 gallons will be x liters.

$$\begin{matrix}\text{Initial number of gallons} \rightarrow \\ \text{Initial number of liters} \rightarrow\end{matrix} \frac{1}{3.785} = \frac{42}{x} \begin{matrix}\leftarrow \text{Second number of gallons} \\ \leftarrow \text{Second number of liters}\end{matrix}$$

$x = 42(3.785)$ — Cross multiplying.

$x = 158.97$ — Simplify.

If we round this to the nearest whole liter, we have approximately 159 liters.

Exercises for Appendix B

Solve the following proportions.

1. $\frac{7}{5} = \frac{21}{x}$

2. $\frac{3}{8} = \frac{x}{14}$

3. $\frac{x}{17} = \frac{12}{5}$

4. $\frac{16}{x} = \frac{3}{4}$

Answer the following questions.

5. If 190 cups of coffee can be made from 5 pounds of coffee, how many cups can be made from 3 pounds of coffee?

6. It took 6 hours to register 380 students. How long will it take (at the same rate) for 500 students to be registered?

7. Ben's engine is using a mixture of 4 pints of oil for every 7 gallons of gas. He has 11 pints of oil on hand to prepare a mixture. How many gallons of gas should he use?

8. A family punch uses 5 quarts of ginger ale for every 8 quarts of fruit drink. How many quarts of ginger ale should be mixed with 31 quarts of fruit drink?

Use the Table of Metric Conversion Ratios to do problems 9 to 18.

9. How many centimeters are in 9 inches?

10. How many kilometers are in 13 miles?

11. How many meters are in 110 inches?

12. How many inches are in 7.5 centimeters?

13. A box weighing 2.4 pounds would weigh how many grams?

14. A man weighs 78 kilograms. How many pounds does he weigh?

15. A container holds 26 liters. How many quarts does it hold?

16. A heating unit holds 18 gallons of water. How many liters of water does it hold?

17. A 400-milliliter container of acid was used in an experiment. How many quarts of acid were used? How many pints of acid were used?

18. A mechanical sprayer holds $1\frac{5}{8}$ quarts of paint. How many milliliters of paint will it hold? How many liters of paint will it hold?

Appendix C

Negative Exponents and Scientific Notation

If n is a positive integer and $x \neq 0$, then x^{-n} is defined by the rule

$$x^{-n} = \frac{1}{x^n}$$

EXAMPLE A Write with positive exponents.

1. y^{-3}
2. z^{-6}
3. w^{-1}
4. x^{-12}

Solution

1. $y^{-3} = \frac{1}{y^3}$
2. $z^{-6} = \frac{1}{z^6}$
3. $w^{-1} = \frac{1}{w^1} = \frac{1}{w}$
4. $x^{-12} = \frac{1}{x^{12}}$

To evaluate a numerical expression with a negative exponent, first write the expression with a positive exponent.

EXAMPLE B Evaluate.

1. 2^{-5}
2. 3^{-4}

Solution

1. $2^{-5} = \frac{1}{2^5} = \frac{1}{32}$
2. $3^{-4} = \frac{1}{3^4} = \frac{1}{81}$

All the previously studied laws of exponents in Section 1.3 can be applied when the exponent is any integer (positive, negative, or zero). These laws are the following:

$$(x^a)(x^b) = x^{a+b} \qquad (x^a)^b = x^{ab}$$

$$\frac{x^a}{x^b} = x^{a-b} \quad \text{if } a > b \qquad (xy)^a = x^a y^a$$

$$\frac{x^a}{x^b} = x^0 = 1 \quad \text{if } a = b \qquad \left(\frac{x}{y}\right)^a = \frac{x^a}{y^a}$$

$$\frac{x^a}{x^b} = \frac{1}{x^{b-a}} \quad \text{if } a < b$$

By using the definition of a negative exponent and the properties of fractions we can derive two more helpful properties of exponents.

$$\frac{1}{x^{-n}} = x^n \qquad \frac{x^{-m}}{y^{-n}} = \frac{y^n}{x^m}$$

EXAMPLE C Simplify the following. Write the expression so that no negative exponents appear.

1. $\frac{1}{x^{-6}}$ **2.** $\frac{x^{-3}y^{-2}}{z^{-4}}$ **3.** $x^{-2}y^3$

Solution **1.** $\frac{1}{x^{-6}} = x^6$ **2.** $\frac{x^{-3}y^{-2}}{z^{-4}} = \frac{z^4}{x^3y^2}$ **3.** $x^{-2}y^3 = \frac{y^3}{x^2}$

EXAMPLE D Simplify the following. Write each expression so that no negative exponents appear.

1. $(3x^{-4}y^2)^{-3}$ **2.** $\frac{x^2y^{-4}}{x^{-5}y^3}$

Solution **1.** $(3x^{-4}y^2)^{-3} = 3^{-3}x^{12}y^{-6} = \frac{x^{12}}{3^3y^6} = \frac{x^{12}}{27y^6}$

2. $\frac{x^2y^{-4}}{x^{-5}y^3} = \frac{x^2x^5}{y^4y^3} = \frac{x^7}{y^7}$

First Method for part 2: First rewrite the expression so that only positive exponents appear. Then simplify using the product rule.

Alternative Method for part 2: The same answer may be obtained by first using the quotient rule to subtract exponents and then removing negative exponents.

The rule

$$\frac{x^a}{x^b} = x^{a-b}$$

holds for all integers a and b. Thus

$$\frac{x^2y^{-4}}{x^{-5}y^3} = x^{2-(-5)}y^{-4-3} = x^{2+5}y^{-7} = \frac{x^7}{y^7}$$

One common use of negative exponents is in writing numbers in scientific notation. Scientific notation is most useful in expressing very large and very small numbers.

Scientific Notation

A positive number is written in scientific notation if it is in the form $a \times 10^n$, where $1 \leq a < 10$ and n is an integer.

EXAMPLE E Write in scientific notation.

1. 3567 **2.** 157,000,000

Solution **1.** $3567 = 3.567 \times 1000 = 3.567 \times 10^3$

2. $157{,}000{,}000 = 1.57 \times 100{,}000{,}000 = 1.57 \times 10^8$

Numbers that are smaller than 1 will have a negative power of 10 if they are written in scientific notation.

EXAMPLE F Write in scientific notation.

1. 0.061 2. 0.0034 3. 0.000052

Solution

1. $0.061 = 6.1 \times \frac{1}{100} = 6.1 \times \frac{1}{10^2} = 6.1 \times 10^{-2}$

2. $0.0034 = 3.4 \times \frac{1}{1000} = 3.4 \times \frac{1}{10^3} = 3.4 \times 10^{-3}$

3. $0.000052 = 5.2 \times 10^{-5}$ (Do you see that we moved the decimal point 5 places and that the exponent of 10 is -5?)

The reverse procedure is used to write a number in ordinary decimal notation.

EXAMPLE G Write in ordinary decimal notation.

1. 1.568×10^2
2. 7.432×10^{-3}
3. 1.8×10^{-7}
4. 2.986×10^5

Solution

1. $1.568 \times 10^2 = 156.8$ (decimal shifts 2 places)
2. $7.432 \times 10^{-3} = 0.007432$ (decimal shifts 3 places)
3. $1.8 \times 10^{-7} = 0.00000018$ (decimal shifts 7 places)
4. $2.986 \times 10^5 = 298{,}600$ (decimal shifts 5 places)

To perform a calculation involving very large or very small numbers it is usually helpful to express the numbers in scientific notation and to use the laws of exponents.

EXAMPLE H Use scientific notation and the laws of exponents to find these answers. Leave your answer in scientific notation.

1. (32,000,000)(1,500,000,000,000)
2. $\frac{0.00063}{0.021}$

Solution

1. (32,000,000)(1,500,000,000,000)

$= (3.2 \times 10^7)(1.5 \times 10^{12})$ Change to scientific notation.

$= 3.2 \times 1.5 \times 10^7 \times 10^{12}$ Rearrange the order. Multiplication is commutative.

$= 4.8 \times 10^{19}$ Multiply 3.2×1.5 and use $x^a \cdot x^b = x^{a+b}$.

2. $\frac{0.00063}{0.021} = \frac{6.3 \times 10^{-4}}{2.1 \times 10^{-2}}$ Change to scientific notation.

$= \frac{6.3}{2.1} \times \frac{10^{-4}}{10^{-2}}$ Rearrange the order. We are actually using the definition of multiplication of fractions.

$= 3.0 \times 10^{-4-(-2)}$ Calculate $6.3 \div 2.1$ and then use properties of exponents.

$= 3.0 \times 10^{-2}$ Simplify (since $-4 - (-2) = -4 + 2 = -2$).

Exercises for Appendix C

Simplify each of the following and express your answer with positive exponents. Assume that all variables are nonzero.

1. $3x^{-2}$

2. $4xy^{-4}$

3. $(2xy)^{-1}$

4. $(3x^2y^3)^{-2}$

5. $\dfrac{3xy^{-2}}{z^{-3}}$

6. $\dfrac{4x^{-2}y^{-3}z^0}{y^4}$

7. $\dfrac{(3x)^{-2}}{(3x)^{-3}}$

8. $\dfrac{(2ab^2)^{-3}}{(2ab^2)^{-4}}$

9. $x^6y^{-2}z^{-3}w^{-10}$

10. $2x^2y^{-3}z^{-4}w^5$

11. $(4^{-3})(2^0)$

12. $(6^{-2})(3^2)$

13. $\left(\dfrac{3xy^2}{z^4}\right)^{-2}$

14. $\left(\dfrac{2a^3b^0}{c^2}\right)^{-3}$

15. $\dfrac{x^{-2}y^{-3}}{x^4y^{-2}}$

16. $\dfrac{a^{-6}b^3}{a^{-2}b^{-5}}$

17. $(-2x^3y^{-2})^{-3}$

18. $(-3x^{-4}y^2)^{-4}$

Write in scientific notation.

19. 123,780

20. 0.063

21. 0.000742

22. 889,610,000,000

23. 7,652,000,000

24. 0.00000001963

Write in decimal notation.

25. 5.63×10^4

26. 1.776×10^8

27. 3.3×10^{-5}

28. 1.99×10^{-1}

29. 9.83×10^5

30. 3.5×10^{-8}

Evaluate by using scientific notation and the laws of exponents. Leave your answer in scientific notation.

31. $\dfrac{(5{,}000{,}000)(16{,}000)}{8{,}000{,}000{,}000}$

32. $(0.0075)(0.0000002)(0.001)$

33. $(0.0002)^5$

34. $\dfrac{0.000000006}{0.000012}$

35. $(150{,}000{,}000)(0.00005)(0.002)$

36. $(30{,}000{,}000)^4$

Glossary

Abscissa (6.1) The first component in an ordered pair of numbers. In the ordered pair $(-3, 6)$ the abscissa is -3.

Absolute Value of a Number (2.5) The absolute value of a number x is the distance between 0 and the number x on the number line. It is written as $|x|$. $|x| = x$ if $x \geq 0$, but $|x| = -x$ if $x < 0$.

Altitude of a Geometric Figure (6.5) The height of the geometric figure. In the three figures shown the altitude is labeled by the distance a.

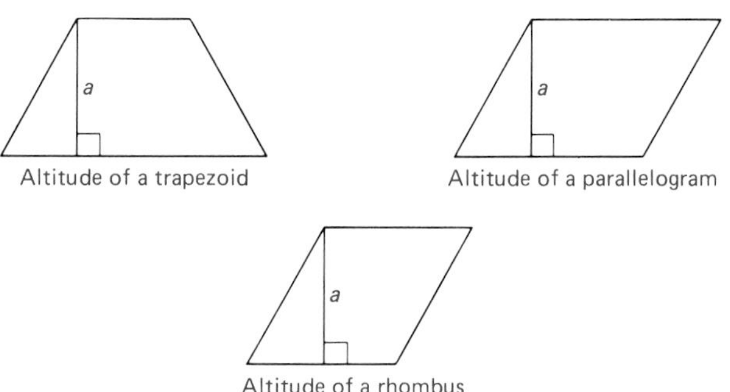

Altitude of a trapezoid

Altitude of a parallelogram

Altitude of a rhombus

Altitude of a Triangle (1.6) and (6.5) The height of any given triangle. In the three triangles shown the altitude is labeled by the distance a.

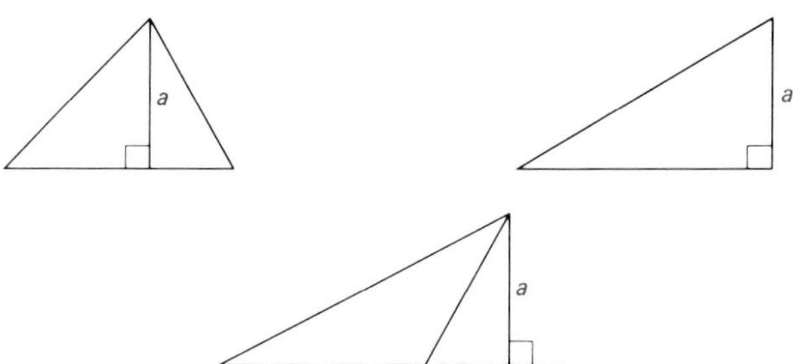

Associative Property of Addition (1.1) If a, b, and c are real numbers, then

$$a + (b + c) = (a + b) + c$$

This property states if three numbers are added, it does not matter *which two numbers* are added first, the result will be the same.

Associative Property of Multiplication (1.2) If a, b, and c are real numbers, then

$$a \times (b \times c) = (a \times b) \times c$$

This property states if three numbers are multiplied, it does not matter *which two numbers* are multiplied first, the result will be the same.

Base (1.3) The number or variable that is raised to a power. In the expression 2^6 the number 2 is the base.

Binomial (1.5) A polynomial of two terms. The expressions $a + 2b$, $6x^3 + 1$, and $5a^3b^2 + 6ab$ are all binomials.

Circumference of a Circle (6.5) The distance around a circle. The circumference of a circle is given by the formula $C = \pi d$ or $C = 2\pi r$, where d is the diameter of the circle and r is the radius of the circle.

Coefficient (1.3) A coefficient is a factor or a group of factors in a product. In the term $4xy$ the coefficient of y is $4x$ but the coefficient of xy is 4. In the term $-5x^3y$ the coefficient of x^3y is -5.

Commutative Property for Addition (1.1) If a and b are any real numbers, then $a + b = b + a$.

Commutative Property for Multiplication (1.2) If a and b are any real numbers, then $ab = ba$.

Complex Fraction (4.4) A fraction that contains a fraction in the numerator or in the denominator or both. These three fractions are complex fractions:

$$\frac{7 + \frac{1}{x}}{x^2 + 2} \qquad \frac{1 + \frac{1}{5}}{2 - \frac{1}{7}} \qquad \text{and} \qquad \frac{1}{\frac{3}{4}}$$

Constant (2.3) Symbol or letter that is used to represent exactly one single number during a particular problem or discussion.

Coordinates of a Point (6.1) An ordered pair of numbers (x, y) that specify the location of a point on a rectangular coordinate system.

Degree of a Polynomial (3.1) The degree of the highest degree term of a polynomial. The degree of $5x^3 + 2x^2 - 6x + 8$ is 3. The degree of $5x^2y^2 + 3xy + 8$ is 4.

Degree of a Term of a Polynomial (3.1) The sum of the exponents of the variables in the term. The degree of $3x^3$ is 3. The degree of $4x^5y^2$ is 7.

Denominator (0.1) and (4.1) The bottom number or algebraic expression in a fraction. The denominator of

$$\frac{3x - 2}{x + 4}$$

is $x + 4$. The denominator of $\frac{3}{7}$ is 7. The denominator of a fraction may not be zero.

Dependent Equations (7.3) Two equations are dependent if every value that satisfies one equation satisfies the other. A system of two dependent equations in two variables will not have a unique solution.

Diagonal of a Four-Sided Figure (6.5) A line connecting two nonadjacent corners of the figure. In each of the figures shown line AC is a diagonal.

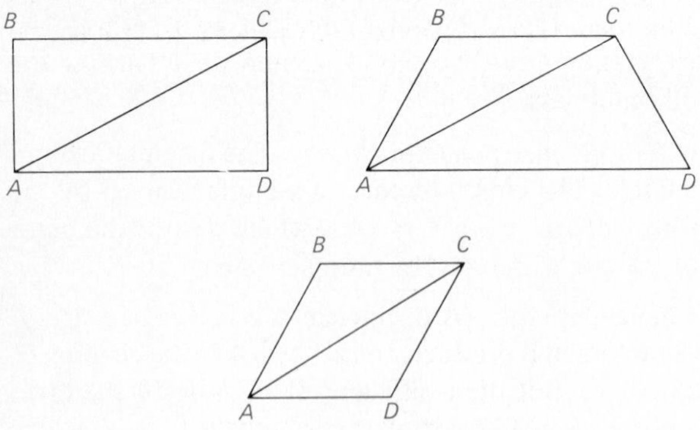

Difference (5.1) The result of subtracting one number or expression from another. The mathematical expression $x - 6$ can be written in words as the difference between x and 6.

Difference of Two Squares Polynomial (3.4) A polynomial of the form $a^2 - b^2$ which may be factored by using the formula

$$a^2 - b^2 = (a + b)(a - b).$$

Distributive Property (1.5) For all real numbers a, b, and c it is true that $a(b + c) = ab + ac$.

Dividend (0.4) The number that is to be divided by another. In the problem $30 \div 5 = 6$ the three parts are as follows:

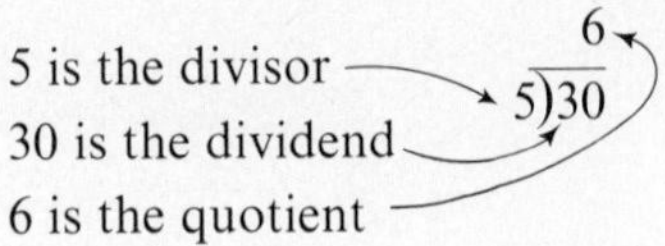

Divisor (0.4) The number you divide into another.

Equilateral Triangle (6.5) A triangle with three sides equal in length and three angles which measure 60°. Triangle ABC is an equilateral triangle.

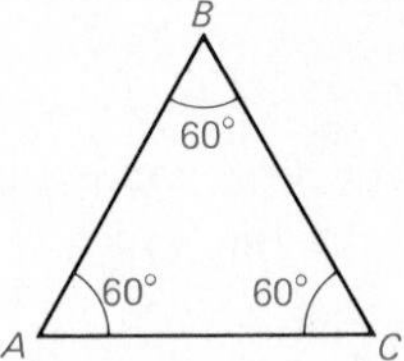

Even Integers (1.2) Integers that are exactly divisible by 2, such as . . . , −4, −2, 0, 2, 4, 6, ...

Exponent (1.3) The number that indicates the power of a base. If the number is a positive integer it indicates how many times the base is multiplied. In the expression 2^6 the exponent is 6.

Extraneous Solution (4.5) and (9.4) An obtained solution to an equation, which, when substituted back into the original equation, does *not* yield an identity. $x = 2$ is an extraneous solution to the equation

$$\frac{x}{x - 2} - 4 = \frac{2}{x - 2}.$$

An extraneous solution is also called an extraneous root.

Factor (0.1) and (3.2) When two or more numbers, variables, or algebraic expressions are multiplied, each one of them is called a factor. If we write $3 \cdot 5 \cdot 2$, the factors are 3, 5, and 2. If we write $2xy$, the factors are 2, x, and y. In the expression $(x - 6)(x + 2)$ the factors are $(x - 6)$ and $(x + 2)$.

Fractions

Algebraic Fractions (4.1) The indicated quotient of two algebraic expressions.

$$\frac{x^2 + 3x + 2}{x - 4} \qquad \text{and} \qquad \frac{y - 6}{y + 8}$$

are algebraic fractions. In these fractions the value of the denominator cannot be zero.

Numerical Fractions (0.1) A set of numbers used to describe parts of whole quantities. A numerical fraction can be represented by the quotient of two integers where the denominator is not zero. The numbers $\frac{1}{5}$, $-\frac{2}{3}$, $\frac{8}{2}$, $-\frac{4}{31}$, $\frac{8}{1}$, and $-\frac{12}{1}$ are all numerical fractions. The set of rational numbers can be represented by numerical fractions.

Hypotenuse of a Right Triangle (9.4) The side opposite the right angle in any right triangle. The hypotenuse is always the longest side of a triangle. Side AB is the hypotenuse of triangle ABC.

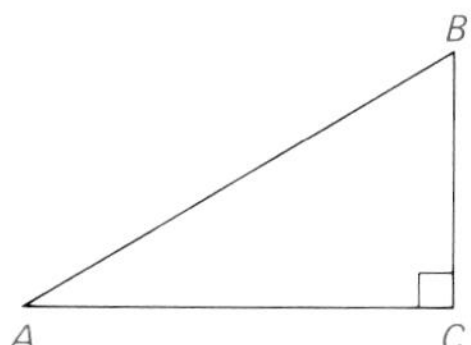

Imaginary Number (8.1) A number that is not a real number. Imaginary numbers can be created by taking the square root of a negative real number. The numbers $\sqrt{-9}$, $\sqrt{-7}$, and $\sqrt{-12}$ are all imaginary numbers.

Improper Fraction (0.1) A numerical fraction whose numerator is larger than or equal to its denominator. $\frac{8}{3}$, $\frac{5}{2}$, and $\frac{7}{7}$ are improper fractions.

Identity (2.1) A statement that is always true. The equations $5 = 5$, $7 + 4 = 7 + 4$, and $x + 8 = x + 8$ are examples of identities.

Inconsistent System of Equations (7.3) A system of equations that does not have a common solution.

Independent Equations (7.3) Two equations that are not dependent are said to be independent.

Index of a Radical (8.6) The number indicating the order of a root. The index number for $\sqrt[3]{x}$ is 3. The index number of $\sqrt[4]{5}$ is 4. The index number for a square root is not written but understood to be 2. The index number for $\sqrt{x}$ is 2.

Inequality (2.4) and (6.4) A mathematical relationship between quantities that are not equal. $x \leq -3$, $w > 5$, and $x < 2y + 1$ are mathematical inequalities.

Integers (1.1) The set of numbers $\ldots, -5, -4, -3, -2, -1, 0, 1, 2, 3, 4, 5, \ldots$.

Intercepts of an Equation (6.2) The point or points where the graph of the equation crosses the x-axis or the y-axis or both. (See X-intercept.)

Irrational Numbers (8.6) A real number that cannot be expressed in the form $\frac{a}{b}$, where a and b are integers and $b \neq 0$. $\sqrt{2}$, $\sqrt[3]{3}$, π, $5 + 3\sqrt{2}$, and $-4\sqrt{7}$ are irrational numbers.

Isosceles Triangle (6.5) A triangle with two equal sides and two equal angles. Triangle ABC is an isosceles triangle. Angle BAC is equal to angle ACB. Side AB is equal in length to side BC.

Least Common Denominator of Numerical Fractions (0.2) The smallest whole number that is exactly divisible by all denominators of a group of fractions. The least common denominator (LCD) of $\frac{1}{6}$, $\frac{2}{3}$, and $\frac{3}{5}$ is 30. The least common denominator is also called the lowest common denominator.

Leg of a Right Triangle (9.4) One of the two shortest sides of a right triangle. Side AC and side BC are legs of triangle ABC.

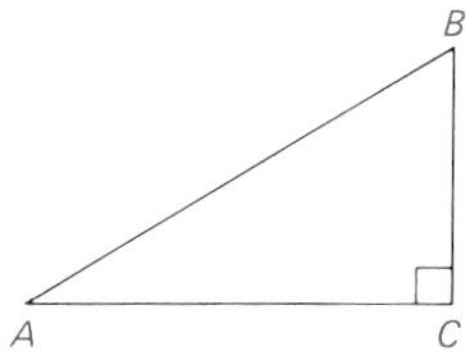

Like Terms (1.4) Terms that have identical variables and exponents. In the expression $5x^3 + 2xy^2 + 6x^2 - 3xy^2$ the term $2xy^2$ and the term $-3xy^2$ are like terms.

Line of Symmetry of a Parabola (9.5) A line that can be drawn through a parabola such that if the graph were folded on this line, the two halves of the curve would correspond exactly. The line of symmetry through the parabola formed by $y = ax^2 + bx + c$ is given by $x = \frac{-b}{2a}$.

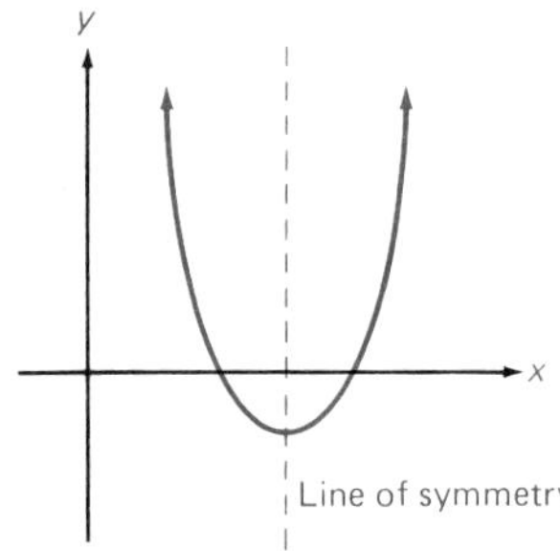

Linear Equation in Two Variables (6.2) An equation of the form $Ax + By = C$, where A, B, and C are real numbers. The graph of a linear equation in two variables is a straight line.

Mixed Fraction (0.1) A fraction that consists of an integer written next to a proper fraction. $2\frac{1}{3}$, $4\frac{6}{7}$, and $3\frac{3}{8}$ are all mixed fractions. Mixed fractions are sometimes called mixed numbers or mixed numerals.

Monomial (1.5) A polynomial of one term. The expressions $3xy$, $5a^2b^3cd$, and -6 are all monomials.

Natural Numbers (0.1) The set of numbers 1, 2, 3, 4, 5, ... This set is also called the set of counting numbers.

Numeral (0.1) The symbol used to describe a number.

Numerator (0.1) The top number or algebraic expression in a fraction. The numerator of

$$\frac{x + 3}{5x - 2}$$

is $x + 3$. The numerator of $\frac{12}{13}$ is 12.

Numerical Coefficient (1.3) The number that is multiplied by a variable or a group of variables. The numerical coefficient of $5x^3y^2$ is 5. The numerical coefficient of $-6abc$ is -6. The numerical coefficient of x^2y is 1. A numerical coefficient of 1 is not usually written.

Odd Integers (1.2) Integers that are not exactly divisible by 2, such as $-3, -1, 1, 3, 5, 7, 9, \ldots$.

Opposite of a Number (1.1) Two numbers that are the same distance from zero on the number line but lie on different sides of it are considered opposites. The opposite of -6 is 6. The opposite of $\frac{22}{7}$ is $-\frac{22}{7}$.

Ordered Pair (6.1) A pair of numbers represented in a specified order. An ordered pair is often used to specify a location on a graph. Every point on a rectangular coordinate system can be represented by an ordered pair (x, y).

Ordinate (6.1) The second component in an ordered pair of numbers. In the ordered pair (7, 5) the ordinate is 5.

Origin (6.1) The point (0, 0) in the rectangular coordinate system.

Parabola (9.5) A curve created by graphing a quadratic equation. The curves shown are all parabolas. The graph of the equation $y = ax^2 + bx + c$, where $a \neq 0$, will always yield a parabola.

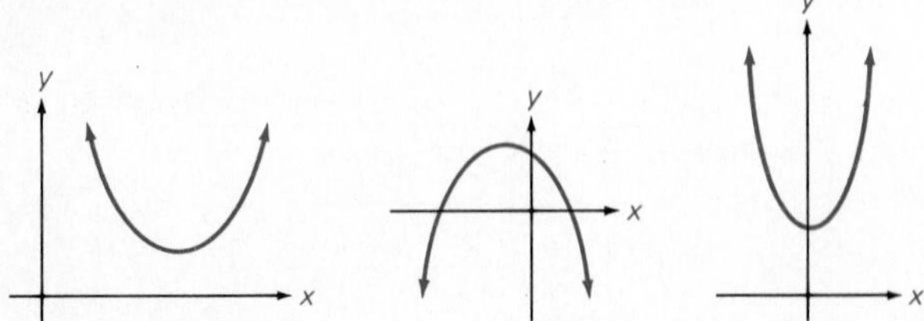

Parallel Lines (7.3) Two straight lines that never intersect. The graph of an inconsistent system of two equations in two variables will result in parallel lines.

Parallelogram (6.5) A four-sided figure with opposite sides parallel. Figure $ABCD$ is a parallelogram.

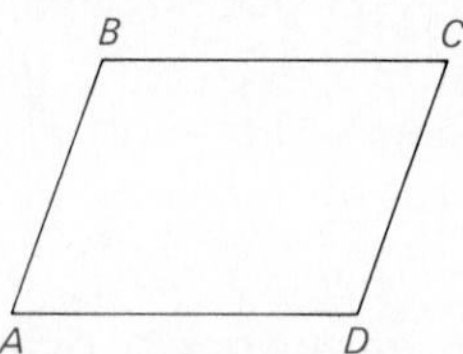

Percent (0.5) Hundredths or "per one hundred"; indicated by the % symbol. Thirty-seven hundredths $(\frac{37}{100}) = 37\%$ (thirty-seven percent).

Perfect Square Number (3.4) A number that is the square of an integer. The numbers 1, 4, 9, 16, 25, 36, 49, 64, 81, 100, 121, 144, . . . are considered perfect square numbers.

Perfect Square Trinomial (3.4) A polynomial of the form $a^2 + 2ab + b^2$ or $a^2 - 2ab + b^2$ which may be factored using one of the following formulas:

$$a^2 + 2ab + b^2 = (a + b)^2$$

or

$$a^2 - 2ab + b^2 = (a - b)^2$$

Perimeter (6.5) The distance around any closed geometric figure. The perimeter of this triangle is 13. The perimeter of this rectangle is 20.

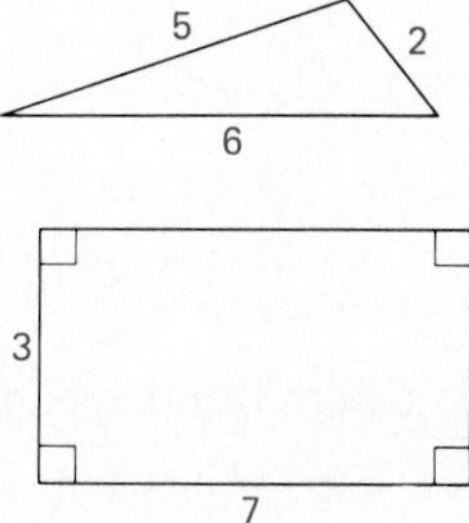

Pi (6.5) An irrational number, denoted by the symbol π, which is approximately equal to 3.141592654. In most cases 3.14 can be used as a sufficiently accurate approximation for π.

Polynomial (1.5) Expressions that contain terms with nonnegative integer exponents. The expressions $5ab + 6$, $x^3 + 6x^2 + 3$, -12, and $x + 3y - 2$ *are all polynomials.* The expressions $x^{-2} + 2x^{-1}$, $2\sqrt{x} + 6$, and $\frac{5}{x} + 2x^2$ *are not polynomials.*

Polynomial in x (3.1) An expression of the form

$$ax^n + bx^{n-1} + cx^{n-2} + \cdots + p$$

where n is a nonnegative integer and the constants $a, b, c, \ldots, p$ are real numbers.

Prime Numbers (0.1) All natural numbers greater than 1 whose only natural number factors are 1 and itself. The first eight prime numbers are 2, 3, 5, 7, 11, 13, 17, and 19.

Prime Polynomial (3.5) A prime polynomial is a polynomial that cannot be factored by the methods of elementary algebra. $x^2 + x + 1$ is a prime polynomial.

Principal (5.4) In monetary problems, the principal is the original amount of money invested or borrowed.

Principal Square Root (8.1) For any given nonnegative number N the principal square root of N (written $\sqrt{N}$) is the nonnegative number a if and only if $a^2 = N$. The principal square root of 25 ($\sqrt{25}$) is 5.

Proper Fraction (0.1) A numerical fraction whose numerator is less than its denominator. $\frac{3}{7}$, $\frac{2}{5}$, and $\frac{8}{9}$ are proper fractions.

Proportion (Appendix B) A proportion is an equation stating that two ratios are equal.

$$\frac{a}{b} = \frac{c}{d}$$

is a proportion.

Pythagorean Theorem (9.4) In any right triangle, if c is the length of the hypotenuse and a and b are the lengths of the two legs, then $c^2 = a^2 + b^2$.

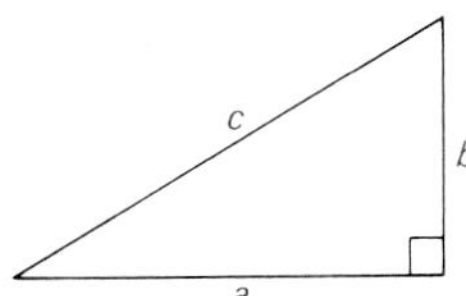

Quadratic Equation (9.1) A quadratic equation is a polynomial equation of degree 2. It is an equation with one variable that contains at least one term with the variable squared but no term with the variable raised to a higher power. $5x^2 + 6x - 3 = 0$, $x^2 = 7$, and $5x^2 = 2x$ are all quadratic equations.

Quadratic Formula (9.3) If $ax^2 + bx + c = 0$ and $a \neq 0$, then

$$x = \frac{-b \pm \sqrt{b^2 - 4ac}}{2a}$$

Quotient (0.4) and (5.1) The result of dividing one number or expression by another. In the problem $12 \div 4 = 3$, the quotient is 3.

Radical (8.1) An expression composed of a radical sign and a radicand. The expressions $\sqrt{5x}$, $\sqrt{\frac{3}{5}}$, $\sqrt{5x + b}$, and $\sqrt[3]{10}$ are called radicals.

Radical Equation (9.4) An equation that contains one or more radicals. $\sqrt{x + 4} = 12$, $\sqrt{x} = 3$, and $\sqrt{3x + 4} = x + 2$ are all radical equations.

Radical Sign (8.1) The symbol $\sqrt{\ }$ used to indicate the root of a number.

Radicand (8.1) The expression beneath the radical sign. The radicand of $\sqrt{5ab}$ is $5ab$.

Ratio (Appendix B) The ratio of one number a to another number b is the quotient $a \div b$ or $\frac{a}{b}$.

Rational Numbers (1.1) and (8.6) A number that can be expressed in the form $\frac{a}{b}$, where a and b are integers and $b \neq 0$. $\frac{7}{3}$, $-\frac{2}{5}$, $7/-8$, $\frac{5}{1}$, 1.62 and 2.7156 are rational numbers.

Rationalizing

Rationalizing the Denominator (8.5) The process of transforming a fraction that contains a radical in the denominator to an equivalent fraction that does not contain one.

Rationalizing the Expression (8.5) The process of transforming a radical that contains a fraction to an equivalent expression that does not contain a fraction inside a radical.

Original Fraction	*Equivalent Fraction That Has Been Rationalized*
$\frac{3}{\sqrt{2}}$	$\frac{3\sqrt{2}}{2}$

Original Radical	*Equivalent Expression after It Has Been Rationalized*
$\sqrt{\frac{3}{5}}$	$\frac{\sqrt{15}}{5}$

Real Number (8.1) Any number that is rational or irrational. 2, 7, $\sqrt{5}$, $\frac{3}{8}$, π, $-\frac{7}{5}$, and $-3\sqrt{5}$ are all real numbers.

Rectangle (6.5) A four-sided figure with opposite sides parallel and all interior angles measuring 90°. The opposite sides of a rectangle are equal.

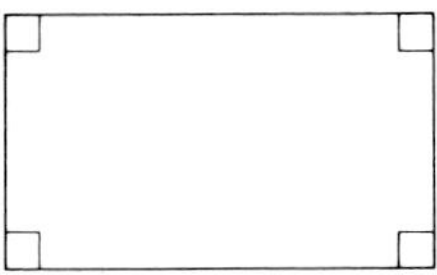

Rhombus (6.5) A parallelogram with four equal sides. Figure $ABCD$ is a rhombus.

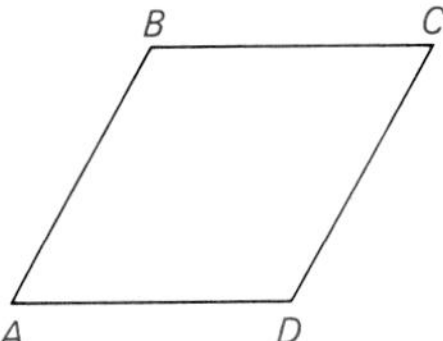

Right Angle (6.5) An angle that measures 90°. Right angles are usually labeled in a sketch by using a small square to indicate that it is a right angle. Here angle ABC is a right angle.

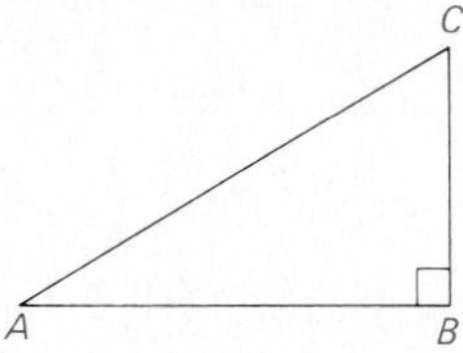

Right Triangle (6.5) A triangle that contains a right angle.

Root of an Equation (2.1) and (9.1) A value of the variable that makes an equation into a true statement. The root of an equation is also called the solution of an equation.

Solution of an Equation (2.1) A number which, when substituted into a given equation, yields an identity. The solution of an equation is also called the root of an equation.

Solution of a Linear Inequality (2.4) The possible values that make a linear inequality true.

Solution of an Inequality in Two Variables (6.4) The set of all possible ordered pairs that when substituted into the inequality will yield an identity.

Solution to a System of Two Equations in Two Variables (7.1) An ordered pair that can be substituted into each equation and an identity will be obtained in each case.

Scientific Notation (Appendix C) A positive number is written in scientific notation if it is in the form $a \times 10^n$, where $1 \le a < 10$ and n is an integer.

Slope-Intercept Form (6.3) The slope-intercept form of the equation of a line that has slope m and the y intercept at $(0, b)$ is given by $y = mx + b$.

Slope of a Line (6.3) The ratio of change in y over the change in x for any two different points on a nonvertical line. The slope (m) is determined by

$$m = \frac{y_2 - y_1}{x_2 - x_1}$$

where $x_2 \neq x_1$ for any two points (x_1, y_1) and (x_2, y_2) on a nonvertical line.

Square (6.5) A rectangle with four equal sides.

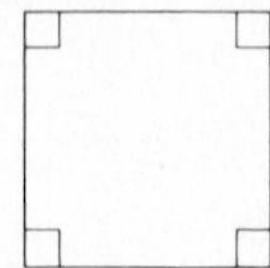

Square Root (8.1) For any given nonnegative number N the square root of N is the number a if $a^2 = N$. One square root of 16 is 4 since $(4)^2 = 16$. Another square root of 16 is -4 since $(-4)^2 = 16$.

Standard Form of a Quadratic Equation (9.1) A quadratic equation placed in the form $ax^2 + bx + c = 0$.

Subscript of a Variable (5.3) A small number or letter written slightly below and to the right of a variable. In the expressions $5 = 2(x - x_0)$ the subscript of x is 0. In the expression $t_f = 5(t_a - b)$ the subscript of the first t is f. The subscript of the second t is a. A subscript is used to indicate a different value of the variable.

System of Equations (7.1) A set of two or more equations that must be considered together.

Term (1.4) A number, a variable, or a product of numbers and variables. In the expression $a^3 - 3a^2b + 4ab^2 + 6b^3 + 8$ there are five terms. They are a^3, $-3a^2b$, $4ab^2$, $6b^3$, and 8. The terms of a polynomial are separated by plus and minus signs.

Trapezoid (6.5) A four-sided figure with two sides parallel. The parallel sides are called the bases of the trapezoid. Figure $ABCD$ is a trapezoid.

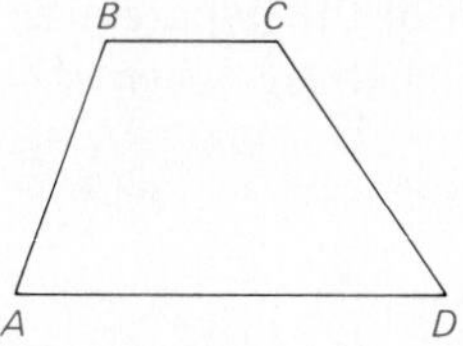

Trinomial (1.5) A polynomial of three terms. The expressions $x^2 + 6x - 8$ and $a + 2b - 3c$ are trinomials.

Variable (1.3) A letter that is used to represent a number or a set of numbers.

Vertex of a Parabola (9.5) The lowest point on a parabola opening upward or the highest point on a parabola opening downward. The x-coordinate of the vertex of the parabola formed by $y = ax^2 + bx + c$ is given by $x = \dfrac{-b}{2a}$.

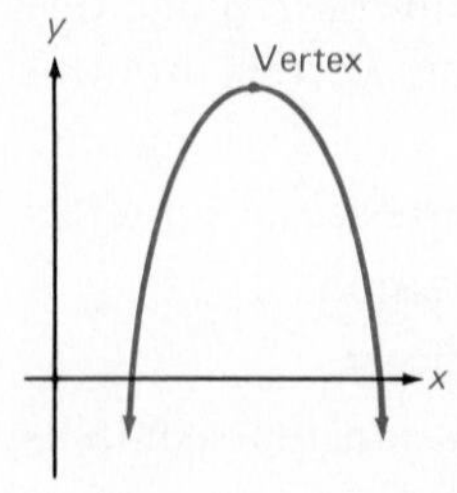

Parabola opening downward

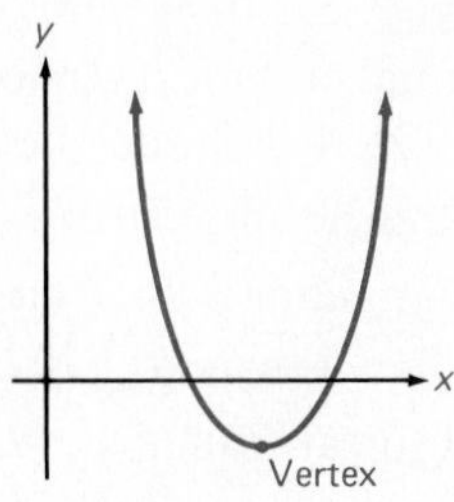

Parabola opening upward

Whole Numbers (0.1) The set of numbers 0, 1, 2, 3, 4, 5, ...

X-Intercept (6.2) The number a is the x-intercept of a line if the line crosses the x-axis at $(a, 0)$. The x-intercept of line l on the graph shown is 4.

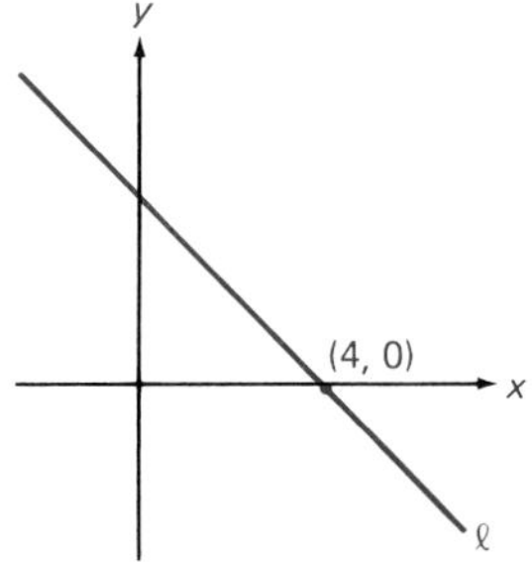

Y-Intercept (6.2) The number b is the y-intercept of a line if the line crosses the y-axis at $(0, b)$. The y-intercept of line p on the accompanying graph is 3.

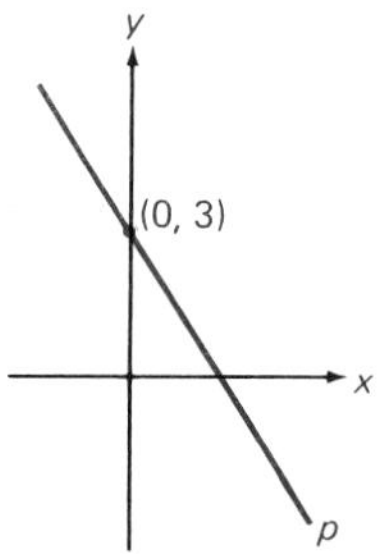

CHAPTER 0

Pretest Chapter 0

1. $\frac{7}{9}$ **2.** $3\frac{3}{4}$ **3.** 18 **4.** $\frac{65}{63}$ or $1\frac{2}{63}$ **5.** $\frac{73}{40}$ or $1\frac{33}{40}$ **6.** $\frac{10}{9}$ or $1\frac{1}{9}$ **7.** $\frac{28}{39}$ **8.** 10.651 **9.** 2.0664 **10.** 0.45
11. 0.625 **12.** 0.32 **13.** 391.2 **14.** 7.2% **15.** 900

Exercise 0.1

1. $\frac{3}{4}$ **2.** $\frac{5}{7}$ **3.** $\frac{2}{5}$ **4.** $\frac{16}{5}$ **5.** $\frac{20}{7}$ **6.** $\frac{17}{3}$ **7.** $1\frac{5}{8}$ **8.** $5\frac{4}{7}$ **9.** $2\frac{1}{3}$ **10.** 12 **11.** 20 **12.** 21 **13.** 6
14. 45 **15.** 39 **16.** $\frac{171}{310}$ **17.** He will need to work correctly 112 *more* problems.

Exercise 0.2

1. $\frac{9}{11}$ **2.** $\frac{3}{17}$ **3.** $\frac{41}{30}$ or $1\frac{11}{30}$ **4.** $\frac{3}{28}$ **5.** $\frac{19}{18}$ or $1\frac{1}{18}$ **6.** $\frac{17}{40}$ **7.** $\frac{41}{24}$ or $1\frac{17}{24}$ **8.** $\frac{13}{48}$ **9.** $\frac{88}{15}$ or $5\frac{13}{15}$
10. $\frac{107}{72}$ or $1\frac{35}{72}$ **11.** $\frac{59}{12}$ or $4\frac{11}{12}$ **12.** $\frac{63}{10}$ or $6\frac{3}{10}$ **13.** $\frac{139}{24}$ or $5\frac{19}{24}$ **14.** $\frac{11}{4}$ or $2\frac{3}{4}$ **15.** $\frac{247}{24} = 10\frac{7}{24}$ miles
16. $\frac{49}{12} = 4\frac{1}{12}$ hours

Exercise 0.3

1. $\frac{6}{55}$ **2.** 10 **3.** $\frac{10}{3}$ or $3\frac{1}{3}$ **4.** $\frac{5}{3}$ or $1\frac{2}{3}$ **5.** $\frac{24}{7}$ or $3\frac{3}{7}$ **6.** $\frac{1}{7}$ **7.** $\frac{2}{3}$ **8.** $\frac{33}{20}$ or $1\frac{13}{20}$ **9.** $\frac{9}{35}$ **10.** $\frac{10}{3}$ or $3\frac{1}{3}$
11. 20 **12.** $\frac{5}{3}$ or $1\frac{2}{3}$ **13.** $71\frac{1}{2}$ yards **14.** $2\frac{4}{5}$ miles **15.** $7\frac{1}{5}$ cubic feet of water

Exercise 0.4

1. 0.625 **2.** 0.15 **3.** 0.032 **4.** $\frac{3}{25}$ **5.** $\frac{9}{8}$ or $1\frac{1}{8}$ **6.** $\frac{29}{50}$ **7.** 9.38 **8.** 0.00153 **9.** 159.651 **10.** 118.56
11. 2.834 **12.** 0.00789 **13.** 0.0001116 **14.** 77.28 **15.** 0.35 **16.** 0.34 **17.** 261.5 **18.** 35.8 **19.** 1360
20. 243 **21.** 0.00352 **22.** 16.432 **23.** 666.9 miles **24.** 17.9 miles per gallon (rounded to nearest tenth)
25. $301.60

Exercise 0.5

1. 62.4% **2.** 0.3% **3.** 156% **4.** 0.06% **5.** 0.04 **6.** 0.0024 **7.** 2.5 **8.** 0.0358 **9.** 358 **10.** 0.162
11. 296.8 **12.** 0.176 **13.** 5% **14.** 5% **15.** 150% **16.** 0.4% **17.** 2220 **18.** 3.16 **19.** 156
20. 170 **21.** 14,100 people **22.** $36.00 **23.** 65% **24.** $2,500,000
25. 53.89% (rounded to nearest hundredth) **26.** $4.54 per hour (rounded to nearest cent)

Practice Test Chapter 0

1. $\frac{3}{4}$ **2.** $\frac{3}{10}$ **3.** $\frac{1}{3}$ **4.** $\frac{3}{5}$ **5.** $\frac{19}{7}$ **6.** $6\frac{4}{5}$ **7.** $6\frac{3}{4}$ **8.** 15 **9.** 15 **10.** 12 **11.** 22 **12.** $\frac{11}{12}$ **13.** $\frac{11}{24}$
14. $\frac{4}{15}$ **15.** $\frac{17}{48}$ **16.** $5\frac{23}{30}$ **17.** $3\frac{19}{20}$ **18.** $\frac{8}{15}$ **19.** $1\frac{11}{12}$ **20.** $\frac{3}{5}$ **21.** $\frac{2}{7}$ **22.** $2\frac{8}{11}$ **23.** $10\frac{1}{2}$ **24.** $2\frac{3}{8}$

25. $2\frac{6}{7}$ **26.** 2 **27.** 4 **28.** $\frac{3}{10}$ **29.** $\frac{2}{3}$ **30.** $\frac{3}{20}$ **31.** 6 **32.** 4.462 **33.** 7.737 **34.** 11.427 **35.** 4.814 **36.** 13.7 **37.** 111.1121 **38.** 0.00862 **39.** 362,341 **40.** 0.07956 **41.** 10.368 **42.** 0.00186 **43.** 0.07132 **44.** 0.002 **45.** 90 **46.** 0.07 **47.** 0.5 **48.** 37.5% **49.** $\frac{6}{25}$ **50.** 0.014 **51.** 0.361 **52.** 0.0002 **53.** 1.253 **54.** 260 **55.** 90 **56.** 200 **57.** 18 **58.** 12.5% **59.** 60% **60.** 40 **61.** 250 **62.** 97,200,000 **63.** 75% **64.** 24,000 **65.** $291 **66.** $6\frac{1}{4}$ feet

CHAPTER 1

Pretest Chapter 1

1. 3 **2.** -4 **3.** -42 **4.** $\frac{8}{3}$ or $2\frac{2}{3}$ **5.** $9a^3b^2$ **6.** 64 **7.** $6x^5y^5$ **8.** $-\dfrac{3y^2}{2x^2}$ **9.** $\dfrac{16x^{12}y^4}{z^8}$

10. $5x^2 - 11xy - 6x^2y$ **11.** $-6x^2 + 4x^2y - 2xz$ **12.** -17 **13.** 18 **14.** $-x^2 + 6xy$

Exercise 1.1.1

1. -4 **2.** -5 **3.** -13 **4.** -18 **5.** $\frac{1}{3}$ **6.** $-\frac{4}{5}$ **7.** 0.4 **8.** -0.3 **9.** -25 **10.** -38 **11.** $\frac{1}{12}$ **12.** $-\frac{1}{2}$ **13.** -11 **14.** -9 **15.** 3 **16.** 2 **17.** -2 **18.** -1 **19.** 13 **20.** -16

Exercise 1.1.2

1. 3 **2.** 6 **3.** 10 **4.** -4 **5.** -8 **6.** -11 **7.** -5 **8.** -6 **9.** $\frac{11}{15}$ **10.** $\frac{27}{20}$ or $1\frac{7}{20}$ **11.** -0.9 **12.** -1.4 **13.** 7 **14.** 48 **15.** -4 **16.** -19 **17.** $-\frac{11}{12}$ **18.** $-\frac{23}{40}$ **19.** 4.47 **20.** 0.16 **21.** -41 **22.** 24 **23.** $-5\frac{4}{5}$ or $-\frac{29}{5}$ **24.** $3\frac{2}{7}$ or $\frac{23}{7}$

Exercise 1.2

1. -36 **2.** -20 **3.** 30 **4.** -35 **5.** -60 **6.** 18 **7.** $-\frac{9}{5}$ or $-1\frac{4}{5}$ **8.** -24 **9.** -24 **10.** -55 **11.** $\frac{4}{3}$ **12.** $\frac{5}{16}$ **13.** 0 **14.** -3 **15.** 4 **16.** 26 **17.** $-\frac{10}{21}$ **18.** $-\frac{3}{10}$ **19.** -30 **20.** $\frac{1}{14}$ **21.** $\frac{9}{16}$ **22.** $-\frac{2}{3}$ **23.** -2.0 **24.** -40 **25.** 0.264 **26.** 0.054 **27.** 144 **28.** -24 **29.** -24 **30.** 36 **31.** 0 **32.** -16

Exercise 1.3.1

1. 8 **2.** 16 **3.** 81 **4.** 27 **5.** 343 **6.** 625 **7.** -27 **8.** -8 **9.** 64 **10.** 81 **11.** -216 **12.** -343 **13.** $\frac{1}{16}$ **14.** $\frac{1}{8}$ **15.** $\frac{8}{125}$ **16.** $\frac{16}{81}$ **17.** $4^3 = 64$ **18.** $7^2 = 49$ **19.** $(-5)^4 = 625$ **20.** $(-4)^4 = 256$ **21.** $(-7)^3 = -343$ **22.** $(-6)^3 = -216$

Exercise 1.3.2

1. $3x^3y^2$ **2.** $4a^4b^3$ **3.** $-3a^2b^2c^3$ **4.** $-7x^2y^2z$ **5.** Coefficient $= -8$; bases $= x, y$; exponents $= 5, 2$ **6.** Coefficient $= 6$; bases $= x, y$; exponents $= 11, 1$ **7.** 3^{15} **8.** 2^{13} **9.** $(5^3)(2^6)$ Rule does not apply. **10.** $(3^5)(8^2)$ Rule does not apply. **11.** $-72x^5$ **12.** $-60x^3$ **13.** $50x^3$ **14.** $26x^3$ **15.** $-6a^3b^4$ **16.** $-10x^4y^3$ **17.** $-30x^4y^2$ **18.** 0 **19.** $80x^4y^5$ **20.** $72x^6y^{11}$ **21.** $-60x^5y^8$ **22.** $-28a^5b^6$ **23.** $-5a^4b^3$ **24.** $-3x^3y$ **25.** 0 **26.** $12a^8b^6$ **27.** $-10x^3yw^4z$ or $-10w^4x^3yz$ **28.** $-24w^5z^6xy$ or $-24w^5xyz^6$ **29.** $-30a^3b^3c^4$ **30.** $-48a^4b^4c^2$ **31.** $-72x^5y^3$ **32.** $180x^3y^5$

Exercise 1.3.3

1. $\dfrac{1}{x^2}$ **2.** a^2 **3.** 2^{13} **4.** $\dfrac{1}{4^8}$ **5.** $\dfrac{a^8}{4}$ **6.** $4b^3$

7. $\dfrac{x^7}{y^9}$ This fraction cannot be simplified. The bases are not the same.

8. $\dfrac{2^{20}}{5^6}$ Unless you multiply out these large numbers this fraction cannot be simplified. The bases are not the same.

9. $\dfrac{1}{2b}$ **10.** $\dfrac{2x^2}{y^2}$ **11.** $-4x^4y^5$ **12.** $-\dfrac{5}{a^2}$ **13.** $-\dfrac{1}{2a^2b}$ **14.** 1 **15.** -1 **16.** $-\dfrac{1}{3x^5}$ **17.** $-\dfrac{3x^2}{y^3z}$

18. $\dfrac{10m^5}{n^3}$ **19.** $\dfrac{y^5}{10x^4}$ **20.** $-3x^3z^5$ **21.** $6x^2$ **22.** $9x^2$ **23.** $\dfrac{y^2}{16x^3}$ **24.** $\dfrac{9y^5}{x^2}$ **25.** $\dfrac{3a}{4}$ **26.** $\dfrac{3a^4}{4}$

27. $\frac{5x^6}{7y^8}$ 28. $-13x^5y^2$ 29. $-\frac{xyz}{2}$ 30. $\frac{3y^5}{2x^3}$ 31. $2x$ 32. $6a^8$ 33. $-\frac{2}{3a^3}$ 34. $3x^{10}y^4$

Exercise 1.3.4

1. x^{12} 2. w^{40} 3. x^7y^{14} 4. $a^{24}b^{16}c^8$ 5. $16a^4b^8c^{28}$ 6. $3^8x^4y^8$ *or* $6561x^4y^8$ 7. $\frac{7^8}{w^{16}}$ 8. $\frac{12^5x^5}{y^{10}}$
9. $\frac{81a^4}{16b^4}$ 10. $\frac{36x^2}{25y^6}$ 11. $81a^8b^{12}$ 12. $-32a^{25}b^{10}c^5$ 13. $\frac{1}{2x}$ 14. $\frac{a}{4b^4}$ 15. $27a^4b^7$ 16. $-8a^7b^{11}$
17. $\frac{8x^6y^9}{27}$ 18. $\frac{16y^8}{9z^6}$

Exercise 1.4

1. $10x$ 2. $-6a$ 3. $-28x^3$ 4. $-4a^2$ 5. $18x^4 + 7x^2$ 6. $3xy - 2y + 6x$ These terms cannot be combined.
7. $5a + 2b - 7a^2$ These terms cannot be combined. 8. $8a^3 - 6a^2$ 9. $-4ab - 7$ 10. -7 11. $7.1x - 3.5y$
12. $2.3ab + 5.1b$ 13. $\frac{3}{4}x^2 - \frac{10}{3}y$ 14. $-\frac{3}{10}a^2 - 5b$ 15. $-5x - 7y - 20$ 16. $-8a + b + 7$
17. $5x^2y - 10xy^6 - xy^2$ 18. $-7bcd - 7cd$ 19. $5bc - 6ac$ 20. $5x^2y - 8x^2$ 21. $3x^2 - 2x + 1$
22. $-2x^2 - 7x + 7$ 23. $-10y^2 - 16y + 12$ 24. $-11y^2 - 5y - 7$

Exercise 1.5

1. $a^2 - 2ab$ 2. $2x^2 - xy$ 3. $-8a - 24b$ 4. $-6x^2 + 4$ 5. $-2x^3 + 2x^2y$ 6. $-15a^3 - 10ab^2$
7. $-a - 2b + c^2$ 8. $-3x^2 - 6xy + 3x$ 9. $8x - 32y - 76$ 10. $-5x^2y + 2xy^2 + xy$ 11. $-81x + 45y - 72$
12. $-15x - 45 + 35y$ 13. $x^2 + \frac{2}{3}x - \frac{1}{3}$ 14. $\frac{1}{4}x^2 + \frac{1}{2}x - 2$ 15. $\frac{1}{5}x^3 + 2x^2 - \frac{4}{5}x$ 16. $y^3 - \frac{4}{3}y^2 - 2y$
17. $5x^2 + 10xy + 5xz - 5x$ 18. $6a^2 + 3ab - 3ac - 12a$ 19. $-x^4 + 3x^3 - 2x^2 + x$
20. $-a^4b - 4a^3b + 2a^2b + ab$ 21. $6x^3y + 9x^2y^3 - 3x^2y^2$ 22. $-8x^3y^2 + 4x^2y^3 - 4xy^4$
23. $0.15xy - 0.5x^2y - 0.1x$ 24. $-0.12a^2 + 0.08ab - 0.008ab^2$ 25. $-a + 10b + 12ab$ 26. $14x - 16y$
27. $15x^2 - 12xy + 3y^2$ 28. $-17xy + 9y^2 + 2x^2$

Exercise 1.6.1

1. 24 2. 4 3. 7 4. 6 5. 6 6. 13 7. −29 8. −2 9. 14 10. 21 11. 42 12. 14
13. $\frac{9}{4}$ or $2\frac{1}{4}$ 14. $\frac{11}{18}$ 15. 0.848 16. 1.458

Exercise 1.6.2

1. −3 2. 9 3. 33 4. 39 5. −2 6. −2 7. −56 8. 34 9. 26 10. 20
11. (a) 27 (b) 3 (c) 3 12. (a) −10 (b) 6 (c) 5 13. 49 14. 42 15. 122 16. 29 17. 77°F
18. 14°F 19. \$8580 20. \$1080

Exercise 1.7

1. $-3x - 4y$ 2. $-2y - 3z$ 3. $5a + 3b$ 4. $-2x - 6y$ 5. $-x^4 - x^2y^2$ 6. $-7a^2b^2 + 10ab^2$
7. $25x - 245$ 8. $-4x + 20y$ 9. $4x^2 + 6$ 10. $8x^3 - 4x^2 + 12x$ 11. $-3y^2 + 2y + 8$
12. $-12y^3 - 8y^2 + 32y$ 13. $4x - 6y - 3$ 14. $5a - 5ab$ 15. $-x^3 + 2x^2 - 3x - 12$ 16. $5x^3 + 38x^2 - 2x + 4$
17. $15x - 10y$ 18. $8b^2 - 27b$ 19. $4a^2 - 2b$ 20. $2x^3 + 2x^2 - 3x$

Practice Quiz

Sections 1.1–1.2 Practice Quiz

1. 4 2. 7 3. 5 4. −7 5. −1 6. 2 7. $-\frac{7}{12}$ 8. −15 9. −18 10. $-\frac{15}{8}$ or $-1\frac{7}{8}$ 11. 8
12. 48 13. −2.2 14. 1.2

Sections 1.3–1.5 Practice Quiz

1. $(-5)^3x^4y^2$ 2. 81 3. 125 4. 2^{17} 5. $-30x^2y^7$ 6. $12x^2b^6c^4$ 7. $\frac{1}{2x^3}$ 8. $\frac{3xy^2}{5}$
9. $-3xy + 6yz - 3xz$ 10. $-5x^3 + 3x^2 - x + 9$ 11. $-12x^2 + 6x - 3$ 12. $6x^2y - 10x^3y^2 - 14x^4y^4$
13. $27x^{12}y^3$ 14. $\frac{a^6}{64c^{24}}$

Sections 1.6–1.7 Practice Quiz

1. 5 **2.** 5 **3.** 9 **4.** 88 **5.** 20 **6.** 24 **7.** $5x^2 - 16xy - 9y$ **8.** $7a + 3b$ **9.** $-2x + 12y - 4xy - 4xy^2$
10. $12x - 28y - 12$

Practice Test Chapter 1

1. 4 **2.** $\frac{1}{3}$ **3.** 0.2 **4.** 96 **5.** -70 **6.** 4 **7.** -3 **8.** $-\frac{10}{3}$ or $-3\frac{1}{3}$ **9.** -64 **10.** $-6x^7y^7$
11. $\dfrac{-7x^3}{5}$ **12.** $-6xy - 5x^2y - 4xy^2$ **13.** $6a^2b^2 + 4ab^3 - 14a^2b^3$ **14.** 8 **15.** 122 **16.** $14x - 16y$
17. $-21x + 22y - 27$ **18.** $-3a - 9ab + 3b^2 - 3ab^2$ **19.** $125x^3y^{18}$ **20.** $\dfrac{49a^{14}b^4}{9}$

CHAPTER 2

Pretest Chapter 2

1. $-\frac{7}{2}$ **2.** $-\frac{1}{7}$ **3.** -2 **4.** $\frac{17}{10}$ or 1.7 **5.** $w = \dfrac{P - 2L}{2}$ or $\dfrac{P}{2} - L$ **6.** $h = \dfrac{2A - 2b^2}{b}$
7. $x = \dfrac{2y - 13}{-2a}$ *or* $x = \dfrac{-2y + 13}{2a}$ **8.** $x > -4$

-4 -3 -2 -1 0 1 x

9. $x \geq -\frac{4}{5}$ **10.** 4 **11.** $x = -1$ and -5 **12.** $x = \frac{1}{2}$ and $\frac{5}{6}$

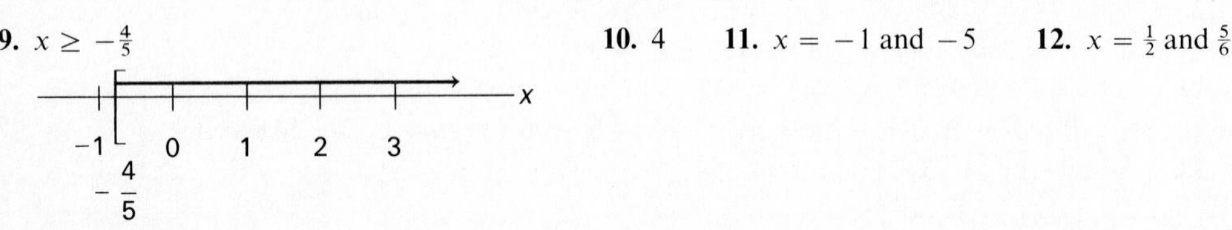

Exercise 2.1.1

1. 4 **2.** 3 **3.** -2 **4.** -1 **5.** 21 **6.** 20 **7.** -5 **8.** 7 **9.** 15 **10.** 1 **11.** 0 **12.** 0
13. $\frac{1}{12}$ **14.** $\frac{4}{5}$ **15.** 5 **16.** 5 **17.** -2.9 **18.** 3.5 **19.** -2.3 **20.** -1.2

Exercise 2.1.2

1. 4 **2.** 4 **3.** $-\frac{8}{3}$ **4.** $-\frac{5}{3}$ **5.** 50 **6.** 0.18 **7.** 15 **8.** -32 **9.** -7 **10.** 8 **11.** 0.4
12. 30 **13.** -0.8 **14.** -0.36 **15.** $\frac{9}{4}$ **16.** $-\frac{9}{4}$

Exercise 2.1.3

1. $\frac{24}{7}$ or $3\frac{3}{7}$ **2.** 15 **3.** 11 **4.** 5 **5.** 10 **6.** $-\frac{5}{6}$ **7.** 7 **8.** 5 **9.** $-\frac{2}{7}$ **10.** -1 **11.** 6 **12.** $-\frac{5}{16}$
13. 5 **14.** $-\frac{6}{5}$ **15.** -7 **16.** 2 **17.** 2 **18.** $-\frac{5}{2}$ **19.** $-\frac{1}{4}$ **20.** 1 **21.** -0.25 **22.** 1
23. $-\frac{14}{3}$ or -4.7 (to nearest tenth) **24.** $\frac{10}{7}$ or 1.4 (to nearest tenth)

Exercise 2.2.1

1. -11 **2.** 4 **3.** $-\frac{1}{3}$ **4.** $\frac{1}{2}$ **5.** $\frac{5}{2}$ **6.** 1 **7.** 3 **8.** 4 **9.** 5 **10.** -33 **11.** 2 **12.** 2 **13.** 8
14. $\frac{1}{2}$ **15.** -6 **16.** -5 **17.** 0 **18.** 4

Exercise 2.2.2

1. 1 **2.** 20 **3.** 24 **4.** 15 **5.** 20 **6.** $\frac{3}{8}$ **7.** 7 **8.** 4 **9.** 1 **10.** $\frac{5}{8}$ **11.** 4 **12.** -13 **13.** -22
14. -3 **15.** 3 **16.** 3 **17.** -12 **18.** 2

Exercise 2.3

1. $b = \dfrac{2A}{h}$ **2.** $\pi = \dfrac{A}{r^2}$ **3.** $p = \dfrac{I}{rt}$ **4.** $t^2 = \dfrac{2S}{g}$ **5.** $m = \dfrac{y - b}{x}$ **6.** $t = \dfrac{A - P}{Pr}$ **7.** $\pi = \dfrac{3V}{4r^3}$ **8.** $S = \dfrac{360A}{\pi r^2}$
9. $R = \dfrac{E}{I}$ **10.** $d = \dfrac{M}{Afj}$ **11.** $e = \dfrac{L - 2d - \pi r}{\pi}$ or $e = \dfrac{L - 2d}{\pi} - r$ **12.** $n = \dfrac{L - a + d}{d}$ or $n = \dfrac{L - a}{d} + 1$

13. $y = \dfrac{5x + 8}{6}$ or $y = \dfrac{5x}{6} + \dfrac{4}{3}$ **14.** $x = \dfrac{-5y + 12}{4}$ or $y = 3 - \dfrac{5y}{4}$ **15.** $x = \dfrac{3y + 8}{2}$ or $x = 4 + \dfrac{3y}{2}$

16. $y = \dfrac{-5x + 6}{3}$ or $y = 2 - \dfrac{5x}{3}$ **17.** $x = \dfrac{y + 8}{a}$ **18.** $y = \dfrac{-2 - x}{2b}$ **19.** $t = \dfrac{x + 18}{2b}$ **20.** $x = \dfrac{-3a - 40}{7}$

Exercise 2.4.1

1. $>$ **2.** $<$ **3.** $<$ **4.** $>$ **5.** $<$ **6.** $<$ **7.** $<$ **8.** $<$ **9.** $<$ **10.** $>$ **11.** $>$ **12.** $<$
13. $x \geq -1$ **14.** $x > -\frac{3}{2}$ **15.** $x < -20$ **16.** $x \leq -35.5$

17. 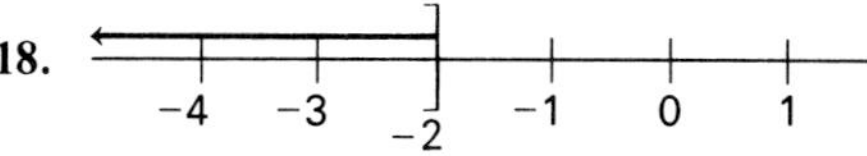 **18.** 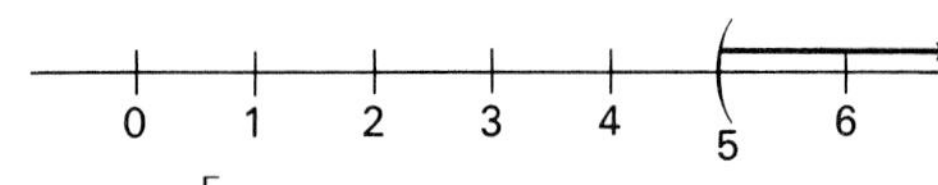

19. 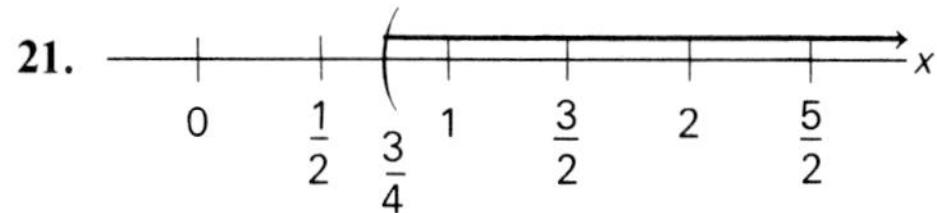  **20.** 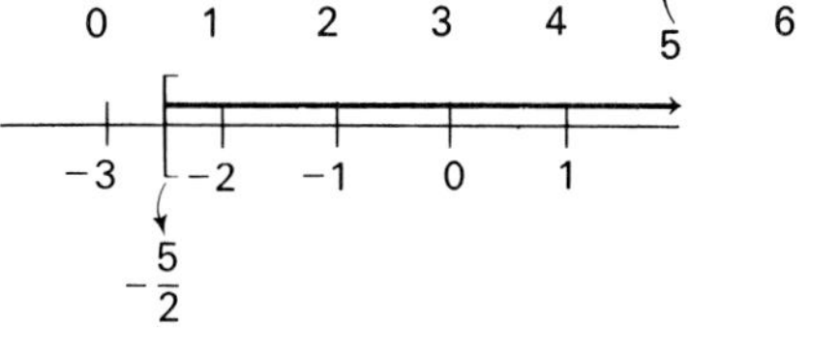

21. 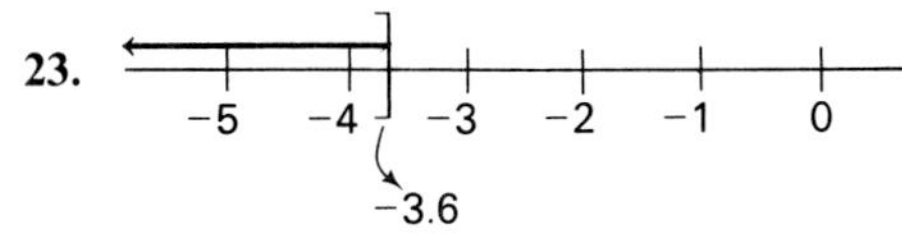 **22.** 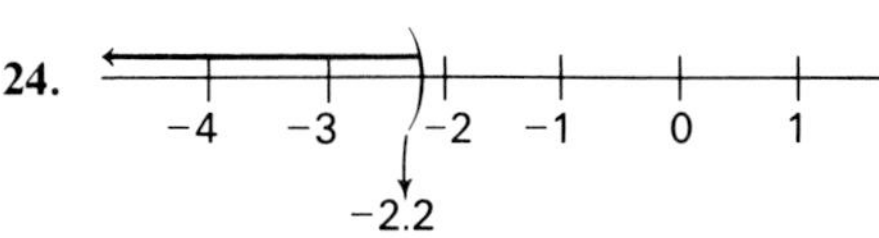

23.  **24.** 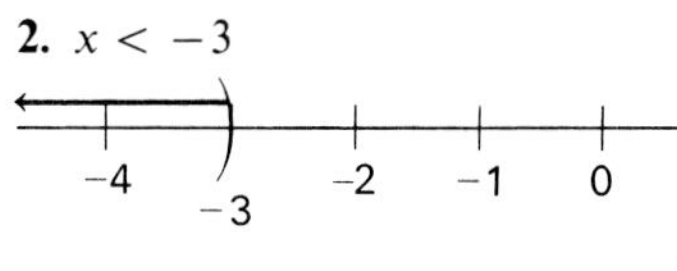 **25.** $c < 56$

26. $V > 580$ **27.** $h \geq 37$ **28.** $n \leq 6$

Exercise 2.4.2

1. $x < \frac{7}{2}$ **2.** $x < -3$

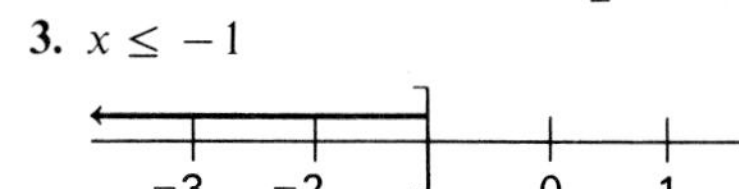 

3. $x \leq -1$ **4.** $x \geq 6$

 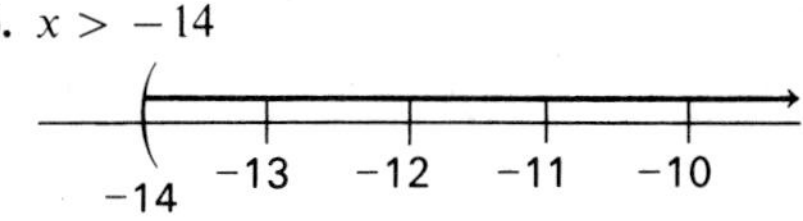

5. $x > -6$ **6.** $x > -14$

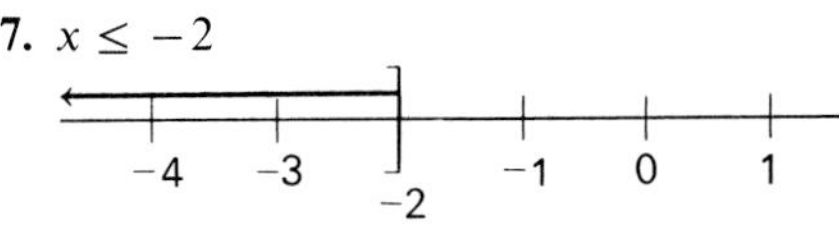 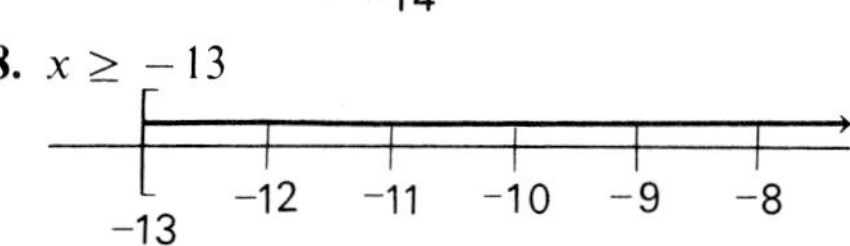

7. $x \leq -2$ **8.** $x \geq -13$

9. $x < -1$ **10.** $x < 3$ **11.** $x \geq -6$ **12.** $x \geq -1$ **13.** $x \geq -6$ **14.** $x \leq \frac{29}{2}$ **15.** $x > 0$ **16.** $x \geq 2$

Exercise 2.5

1. 4 **2.** 0 **3.** $\frac{6}{7}$ **4.** 5 **5.** -2 **6.** 4 **7.** $\frac{1}{35}$ **8.** $\frac{5}{24}$ **9.** $x = 4$ and -5 **10.** $x = 5$ and $-\frac{11}{3}$
11. $x = \frac{14}{3}$ and $-\frac{14}{3}$ **12.** $x = 18$ and -18 **13.** $x = 6$ and -10 **14.** $x = 19$ and -11 **15.** $x = 9$ and -15
16. $x = \frac{11}{4}$ and $-\frac{9}{4}$ **17.** $x = \frac{9}{10}$ and $\frac{3}{2}$ **18.** $x = \frac{7}{30}$ and $\frac{11}{10}$ **19.** $x = 6$ and -18 **20.** $x = \frac{17}{6}$ and $-\frac{5}{2}$
21. $x = -\frac{14}{3}$ and $\frac{34}{3}$ **22.** $x = \frac{7}{5}$ and $\frac{47}{5}$ **23.** $x = \frac{25}{12}$ and $-\frac{3}{4}$ **24.** $x = -\frac{1}{10}$ and $-\frac{11}{10}$

Practice Quizzes

Sections 2.1–2.3 Practice Quiz

1. -3 **2.** $-\frac{4}{3}$ **3.** 3 **4.** 15 **5.** $\frac{5}{8}$ **6.** $g = \dfrac{2s}{t^2}$ **7.** $c = \frac{5}{9}(F - 32)$ or $c = \dfrac{5F - 160}{9}$

8. $x = \dfrac{3y - 8}{6}$ or $x = \dfrac{y}{2} - \dfrac{4}{3}$

Sections 2.4–2.5 Practice Quiz

1. $x < -1$ **2.** $x \le -3$

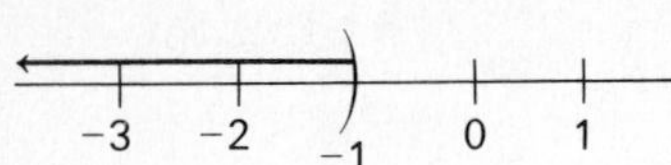

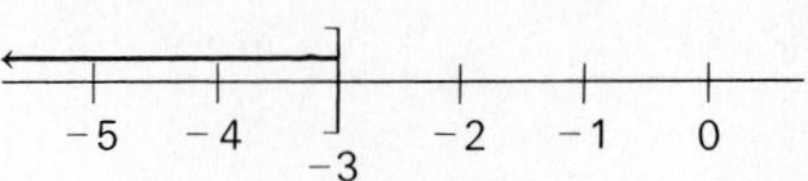

3. $x \le -\frac{8}{3}$ **4.** -3 **5.** $x = -\frac{3}{2}$ and 1 **6.** $x = \frac{4}{3}$ and 0

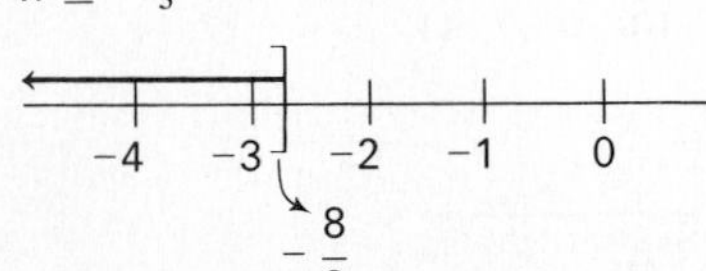

Practice Test Chapter 2

1. 2 **2.** -1.2 **3.** 7 **4.** $w = \dfrac{-3x + 6}{4}$ or $w = \dfrac{3}{2} - \dfrac{3x}{4}$ **5.** $a = \dfrac{2A - hb}{h}$ or $a = \dfrac{2A}{h} - b$

6. $y = \dfrac{10ax - 5}{8ax}$ or $y = \dfrac{5}{4} - \dfrac{5}{8ax}$ **7.** $x \le -\frac{1}{2}$ **8.** $x > -\frac{5}{4}$

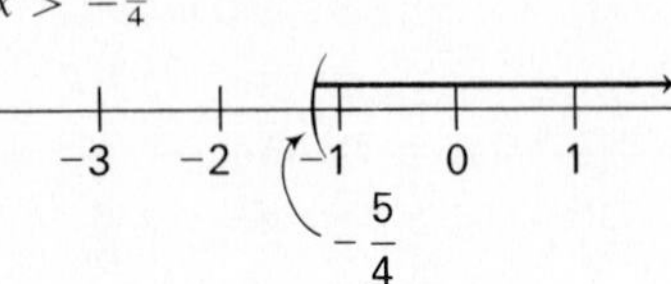

9. $x = -9$ and 27 **10.** $x = \frac{20}{3}$ and $-\frac{16}{3}$

CHAPTER 3

Pretest Chapter 3

1. $12x^2 + xy - 6y^2$ **2.** $4x^3 - 13x^2 + 11x - 2$ **3.** $25a^2b^2 - 60ab + 36$ **4.** $64x^2 - 121y^2$
5. $3xy^2(4x + 2y - 3)$ **6.** $(a - 2b)(5 - 3x)$ **7.** $(x - 24)(x + 2)$ **8.** $(5x - 2)(3x - 2)$
9. $(9x^2 + 4)(3x + 2)(3x - 2)$ **10.** $(7x - 2y)^2$ **11.** $3x(2x - 1)(x + 3)$ **12.** $2y^2(4x - 3)^2$

Exercise 3.1.1

1. $6ab^2 - 4a^2b^3 - 10ab^5$ **2.** $-2x^4y + 16x^3y - 4x^2y^3$ **3.** $-15x^4y^2 + 6x^3y^2 - 18x^2y^2$ **4.** $15b^5 - 30b^4 + 40ab^4$
5. $x^2 + 7x + 12$ **6.** $x^2 + 8x + 12$ **7.** $x^2 - 6x - 16$ **8.** $x^2 - 3x - 18$ **9.** $x^2 - 9x + 20$ **10.** $x^2 - 11x + 30$
11. $-3x^2 - 13x - 4$ **12.** $-5x^2 - 16x - 3$ **13.** $7x^2 + 14xy - 4x - 8y$ **14.** $3x^2 + 9xy - 5x - 15y$
15. $10y^2 + 11y - 6$ **16.** $12y^2 - y - 6$ **17.** $15y^2 - 38y + 7$ **18.** $12y^2 - 16y + 5$ **19.** $30x^2 + 43xy - 8y^2$
20. $15x^2 + 34xy - 16y^2$ **21.** $35x^2 + 21xz - 10xy - 6yz$ **22.** $4a^2 + 12af - 5ab - 15bf$
23. $4ac - 3ab - 32bc + 24b^2$ **24.** $xz - 8xy - 3yz + 24y^2$ **25.** $-28x^2 + 2xy + 6y^2$ **26.** $-40x^2 - 21xy + 27y^2$
27. $15a^2 - 28ab^2 + 12b^4$ **28.** $16a^2 - 46ab^2 + 15b^4$ **29.** $4x^4 - 25y^4$ **30.** $64x^4 - 48x^2y^2 + 9y^4$
31. $64x^2 - 32x + 4$ **32.** $25x^2 - 30x + 9$ **33.** $25x^4 + 20x^2y^2 + 4y^4$ **34.** $9x^4 - 42x^2y^2 + 49y^4$
35. $15t^2 + 8tp - 63p^2$ **36.** $72w^2 + 25wz - 63z^2$ **37.** $6a^2b^2 - 31abd + 35d^2$ **38.** $30x^2y^2 + 14xyz - 8z^2$

Exercise 3.1.2

1. $x^2 - 16$ **2.** $y^2 - 64$ **3.** $4x^2 - 25y^2$ **4.** $81w^2 - 16z^2$ **5.** $100p^2 - 49q^2$ **6.** $121x^2y^2 - 1$
7. $144x^4 - 49$ **8.** $1 - 64a^2b^4$ **9.** $9y^2 + 6y + 1$ **10.** $16x^2 - 8x + 1$ **11.** $9 - 24y^2 + 16y^4$
12. $64 + 48x + 9x^2$ **13.** $16a^2 + 40ab + 25b^2$ **14.** $36w^2 + 60wz + 25z^2$ **15.** $25x^2y^2 - 60xyz + 36z^2$
16. $81a^2b^2 - 198abc + 121c^2$ **17.** $49x^2 - 9y^2$ **18.** $144a^2 - 25b^2$ **19.** $16x^4 - 56x^2y + 49y^2$
20. $9a^2 - 60ab^2 + 100b^4$ **21.** $121a^2 + 110ab + 25b^2$ **22.** $81x^2 - 49y^4$ **23.** $x^3 - 4x^2 + 5x - 2$
24. $x^3 + 6x^2 + 8x - 3$ **25.** $9y^3 + 3y^2z - 8yz^2 - 4z^3$ **26.** $4w^3 + 6w^2z - 2z^3$ **27.** $4x^4 - 7x^3 + 2x^2 - 3x - 1$
28. $3x^4 + 2x^3 - 13x^2 - 2x + 2$ **29.** $2x^3 + 3x^2 - 17x - 30$ **30.** $3x^3 + 13x^2 - 28x + 12$
31. $x^4 + x^3 - 4x^2 - 7x - 3$ **32.** $y^4 - 5y^3 + 11y^2 - 14y + 4$
33. Square the first term, add to double the product of the first and second terms, and finally add to the square of the second term.
34. $4x + 7$ **35.** Any problems of the type $(a + b)(a - b) = a^2 - b^2$.

Exercise 3.2.1

1. $6x(x-1)$ **2.** $5b(a-1)$ **3.** $7ab(3c-2b)$ **4.** $9wz(2-3w)$ **5.** $6(3abc+2d-ab)$
6. $r(9pq-12p^2q-14x)$ **7.** $x(ax-2bx-cx+d)$ **8.** $a(a^3y-a^2y^2+ay^3-2)$ **9.** $5x(12x^2-10x+5)$
10. $2x^5(3x^4-4x^2+2)$ **11.** $2\pi r(h+r)$ **12.** $9ab(ab-4)$ **13.** $7x(2xy-5y-9)$ **14.** $4b(5z^2-ab+3a+2a^2b)$
15. $(a+b)(2+3x)$ **16.** $(x-2y)(5-z)$ **17.** $(x-3y)(7a-4)$ **18.** $(c+2d)(5b-2a)$
19. $(x^2+1)(3+2y+w)$ **20.** $(bc-1)(5a+b+c)$ **21.** $(x+2y)(4y+1)$ **22.** $(x-2y)(3x^2-1)$

Exercise 3.2.2

In this exercise and throughout the chapter you should remember that the product of the two binomials can be written in either order. Since multiplication is commutative it does not matter which quantity is written first in your answer.

1. $(x-2)(3+y)$ **2.** $(x-5)(2+y)$ **3.** $(x+y)(a-2b)$ **4.** $(x+2y)(4-3w)$ **5.** $(5+6c)(a+2b)$
6. $(2+5z)(x+3y)$ **7.** $(a-b)(5-2x)$ **8.** $(y-4)(x-3)$ **9.** $(y-2)(y-3)$ **10.** $(x+4)(x-5)$
11. $(x+y)(3a-1)$ **12.** $(x-1)(a+2b)$ **13.** $(y+3x)(2a-1)$ **14.** $(2b+c)(2a-1)$ **15.** $(7x+2y^2)(4x+3w)$
16. $(4w+5x)(2x+7y^2)$ **17.** $(4a-3b)(2a+5e)$ **18.** $(7a-2b)(3a-4e)$

Exercise 3.3.1

1. $(x+8)(x+1)$ **2.** $(x+5)(x+3)$ **3.** $(x+6)(x+2)$ **4.** $(x+20)(x+1)$ **5.** $(x-3)(x-1)$
6. $(x-4)(x-2)$ **7.** $(x-11)(x-2)$ **8.** $(x-14)(x-1)$ **9.** $(x+4)(x-3)$ **10.** $(x+4)(x-2)$
11. $(x-14)(x+1)$ **12.** $(x-8)(x+2)$ **13.** $(x+5)(x-4)$ **14.** $(x+9)(x-2)$ **15.** $(x-8)(x+3)$
16. $(x-13)(x+2)$ **17.** $(x-5)(x-2)$ **18.** $(x+6)(x-3)$ **19.** $(x-2)(x+1)$ **20.** $(x+7)(x+3)$
21. $(x+7)(x-2)$ **22.** $(x-5)(x+3)$ **23.** $(x-5)(x-4)$ **24.** $(x-6)(x-2)$ **25.** $(x+15)(x+2)$
26. $(x-7)(x+4)$ **27.** $(x-5)(x+1)$ **28.** $(x-7)(x-1)$ **29.** $(x+8)(x-2)$ **30.** $(x-10)(x-3)$
31. $(x-8)(x-4)$ **32.** $(x-9)(x+3)$ **33.** $(x+7)(x-3)$ **34.** $(x-9)(x-2)$ **35.** $(x+8)(x+5)$
36. $(x+10)(x+5)$ **37.** $(x-22)(x+1)$ **38.** $(x+14)(x-2)$ **39.** $(x+12)(x-3)$ **40.** $(x-9)(x-4)$
41. $(x-7)(x+6)$ **42.** $(x-8)(x+7)$ **43.** $(x-9)(x-7)$ **44.** $(x+16)(x+3)$

Exercise 3.3.2

1. $(3x+1)(x+2)$ **2.** $(2x+1)(x+3)$ **3.** $(2x-1)(x-2)$ **4.** $(3x-2)(x-2)$ **5.** $(2x+5)(x-1)$
6. $(3x-2)(x+1)$ **7.** $(2x+1)(x-3)$ **8.** $(2x+3)(x-2)$ **9.** $(5x-2)(x+1)$ **10.** $(3x+2)(2x-1)$
11. $(3x-2)(2x-3)$ **12.** $(3x-2)(2x-1)$ **13.** $(2x-5)(x+4)$ **14.** $(3x-2)(2x+5)$ **15.** $(3x+1)(3x+2)$
16. $(4x+3)(x+2)$ **17.** $(3x+2)(2x-3)$ **18.** $(3x+2)(x-5)$ **19.** $(4x-1)(2x-1)$ **20.** $(2x-1)(2x-3)$
21. $(2x-1)(2x+9)$ **22.** $(8x+9)(x-1)$ **23.** $(9x-4)(x-1)$ **24.** $(5x-1)(x-2)$ **25.** $(5x+2)(x-3)$
26. $(3x+2)(x-4)$ **27.** $(6x-1)(2x-3)$ **28.** $(9x-4)(x+1)$ **29.** $(5x-2)(3x+2)$ **30.** $(8x-3)(x-1)$
31. $(5x+3)(2x+3)$ **32.** $(4x+1)(3x+2)$ **33.** $(8x+1)(2x-1)$ **34.** $(3x-7)(x+2)$

Exercise 3.4

1. $(3x+4)(3x-4)$ **2.** $(6x+1)(6x-1)$ **3.** $(1+10s)(1-10s)$ **4.** $(12x+5)(12x-5)$ **5.** $(9x+7y)(9x-7y)$
6. $(4x^2+1)(2x+1)(2x-1)$ **7.** $(5+11x)(5-11x)$ **8.** $(3x+7)(3x-7)$ **9.** $(9x+10y^2)(9x-10y^2)$
10. $(5x^2+4y^2)(5x^2-4y^2)$ **11.** $(7x+1)^2$ **12.** $(6x+1)^2$ **13.** $(y-3)^2$ **14.** $(y-4)^2$ **15.** $(3x-4)^2$
16. $(2x+5)^2$ **17.** $(9x+2y)^2$ **18.** $(6x+5y)^2$ **19.** $(3x^2-y)^2$ **20.** $(y^2-11)^2$ **21.** $(7x+9)(7x+1)$
22. $(5x-8)(5x-2)$ **23.** $(x^3+5)(x^3-5)$ **24.** $(x^4+1)(x^2+1)(x+1)(x-1)$ **25.** $(x^5+6y^5)(x^5-6y^5)$
26. $(x^2+7y^3)(x^2-7y^3)$ **27.** $(2x^4+3)^2$ **28.** $(2x^3+1)^2$
29. You cannot factor $9x^2+1$. No two binomials when multiplied will yield $9x^2+1$. For example,
$(3x+1)(3x+1) = 9x^2 - 1 \neq 9x^2 + 1$
$(3x+1)^2 = 9x^2 + 6x + 1 \neq 9x^2 + 1$

Exercise 3.5

Note: If your answer does not seem to be the same as the following, stop to check if you forgot to remove a common factor in the first step. Students find this is a common error.

1. $3(x^2+2)(x^2-2)$ **2.** $(y+8)^2$ **3.** $2(x-11)(x+6)$ **4.** $(4x+9y)(4x-9y)$ **5.** $(2x-3)(x-4)$
6. $2x(y+5)(y-5)$ **7.** $(x-10y)(x+7y)$ **8.** $(x^2+2)(2x-7)$ **9.** $(x-5)(a-4)$ **10.** $3(x^2-3x+5)$
11. $5x(x+3)(x-3)$ **12.** $3(3y+2)(2y-1)$ **13.** $5xy^3(x-1)^2$ **14.** $3(2x-3)^2$ **15.** $3xy(3z+2)(3z-2)$
16. $2x(3x-1)^2$ **17.** $3(x+7)(x-5)$ **18.** $4(x-9)(x+2)$ **19.** $5(x-4)(x-2)$ **20.** Prime
21. $(2x^2+1)(x+2)(x-2)$ **22.** $(2x^2+1)(x^2-5)$ **23.** Prime **24.** $4x(x+5)(x-3)$ **25.** $4(2x-1)(x+4)$
26. $6(2x-1)(x-2)$ **27.** $a(x+2y)(a-3)$ **28.** $x(x+y)(2a-5)$

Practice Quizzes

Section 3.1–3.2 Practice Quiz

1. $a^2 - ab - 6b^2$ **2.** $5x^2 + 19xy - 4y^2$ **3.** $x^3 + 3x^2 - 10x - 24$ **4.** $3x^3 + 17x^2 - 12x + 2$
5. $25p^2 + 40pr + 16r^2$ **6.** $49x^4 - y^2$ **7.** $5(a - 2b + 5ab)$ **8.** $x^2(3 - 7x + 24y)$ **9.** $(2x + y)(x - 2)$
10. $(x + 2y)(3 - y)$ **11.** $(a - c)(a - 2b)$ **12.** $(x + 4)(ab - 2)$

Section 3.3–3.5 Practice Quiz

1. $(5x - 3)^2$ **2.** $(x - 7)(x - 6)$ **3.** $2(2x + 5)(2x + 1)$ **4.** $y^2(4x + 1)(4x - 1)$ **5.** $2(x - 5)(x + 3)$
6. $y^2(3x + 2y)^2$ **7.** Prime **8.** $(2x - 7)(x + 4)$ **9.** $3(x - 5)(x - 2)$ **10.** $c(x - 2)(a + b)$
11. $3y^2(x^2 + 4)(x + 2)(x - 2)$ **12.** $4a^2(a + 3b)(a - 2b)$

Practice Test Chapter 3

1. $10a^2 + 7ab - 12b^2$ **2.** $6x^3 - 11x^2 - 19x - 6$ **3.** $49x^4 + 28x^2y^2 + 4y^4$ **4.** $81x^2 - 4y^2$ **5.** $(x + 14)(x - 2)$
6. $(3x + 10y)(3x - 10y)$ **7.** $(5x + 1)(2x + 5)$ **8.** $(3a - 5b)^2$ **9.** $x(7 - 9x + 14y)$ **10.** $(x + 2y)(3x - 2w)$
11. $2x(3x - 4)(x - 2)$ **12.** $c(5a - b)(a - 2b)$ **13.** $4(5x^2 + 2y^2)(5x^2 - 2y^2)$ **14.** $(3x - 4y)(3x - y)$

CHAPTER 4

Pretest Chapter 4

1. $\dfrac{b}{2a - b}$ **2.** $\dfrac{3x - 2}{4x + 3}$ **3.** $\dfrac{2a - b}{6}$ **4.** $\dfrac{x}{x + 5}$ **5.** $\dfrac{1}{x + 3}$ **6.** $\dfrac{y}{x - 1}$ **7.** $\dfrac{3y^2 + 18y + 4}{2y(y + 2)}$ **8.** $\dfrac{-2y - 7}{(2y + 3)(y - 1)}$

9. $\dfrac{2a - 3}{a(5a + 1)}$ or $\dfrac{2a - 3}{5a^2 + a}$ **10.** $\dfrac{a^2 - 2a - 2}{3a^2(a + 1)}$ or $\dfrac{a^2 - 2a - 2}{3a^3 + 3a^2}$ **11.** $x = -5$

12. No solution; although you get $x = 5$, checking your solution gives you divison by zero, which is not allowed.

Exercise 4.1

1. 3 **2.** $\frac{1}{2}$ **3.** $\dfrac{1}{3 - x}$ **4.** $\dfrac{x}{3}$ **5.** $\dfrac{2}{x - 4}$ **6.** $\dfrac{-(2x + 1)}{2x - 1}$ or $\dfrac{2x + 1}{1 - 2x}$ **7.** $x - 3y$ **8.** $\dfrac{x + y^2}{xy}$ **9.** $\dfrac{3x}{x - 3y}$

10. $\dfrac{x + 4}{x + 1}$ **11.** $\dfrac{x + 2}{x}$ **12.** $\dfrac{x + 2}{x - 2}$ **13.** $\dfrac{x + 3}{4x - 1}$ **14.** $\dfrac{4x - 6}{2x + 3}$ or $\dfrac{2(2x - 3)}{2x + 3}$ **15.** $\dfrac{3x - 5}{4x - 1}$ **16.** $\dfrac{y + 3}{y + 4}$

17. $\dfrac{x - 4}{x - 2}$ **18.** The fraction can be factored into $\dfrac{2(x + 3)(x - 2)}{(x + 4)(x - 1)}$. However, this fraction cannot be reduced.

19. $\dfrac{x + 5}{x - 3}$ **20.** $\dfrac{2x + 3}{2x + 1}$ **21.** $-\frac{3}{2}$ **22.** $-\dfrac{1}{x^2}$ **23.** $\dfrac{-(2x + 3)}{x + 5}$ **24.** $\dfrac{-(x + 7)}{2x + 5}$ **25.** $\dfrac{4x + 5}{2x - 1}$ **26.** $\dfrac{2x + 1}{3x - 8}$

27. $\dfrac{-(2y + 3)}{y + 2}$ **28.** $\dfrac{-(5y + 2)}{2y + 1}$ **29.** $\dfrac{a - b}{2a - b}$ **30.** $\dfrac{a + 5b}{3a - b}$ **31.** $\dfrac{4x + 5y}{x + 2y}$ **32.** $\dfrac{2x - y}{3x + 4y}$

Exercise 4.2

1. $\dfrac{x + 5}{x + 7}$ **2.** $\dfrac{x + 2}{x - 2}$ **3.** $\dfrac{x^2}{2x - 4}$ or $\dfrac{x^2}{2(x - 2)}$ **4.** $\dfrac{2}{3x}$ **5.** $\dfrac{x + 4}{x - 4}$ **6.** $\dfrac{x + 2}{x + 3}$ **7.** $\dfrac{(x + 8)(x + 7)}{7x(x - 3)}$

8. $\dfrac{3(x + 2)}{x - 4}$ or $\dfrac{3x + 6}{x - 4}$ **9.** $\dfrac{x + 1}{(x + 3)(x + 2)}$ **10.** 2 **11.** $\dfrac{4(x + y)}{3(x - y)}$ or $\dfrac{4x + 4y}{3x - 3y}$ **12.** $\dfrac{x(x + 3)}{3}$ or $\dfrac{x^2 + 3x}{3}$

13. $\dfrac{y(x + y)}{(x + 1)^2}$ **14.** $\dfrac{3(2a + b)}{2(a + 2b)}$ or $\dfrac{6a + 3b}{2a + 4b}$ **15.** $\dfrac{3(2 - x)}{x + 5}$ or $\dfrac{-3x + 6}{x + 5}$ **16.** $\dfrac{-3}{x - 2}$ or $\dfrac{3}{2 - x}$ **17.** $\dfrac{(x + 4)(x - 5)}{2x + 3}$

18. $\dfrac{3x + 2}{(4x + 5)(x - 4)}$ **19.** $\dfrac{(2x + 1)(4x - 1)}{(x + 1)(x - 6)}$ **20.** $\dfrac{a + 3}{a + 1}$

Exercise 4.3

1. $\dfrac{5y + 3}{y^3}$, LCD = y^3 **2.** $\dfrac{8y + 2}{xy^2}$, LCD = xy^2 **3.** $\dfrac{2x - 1}{(x + 2)(x - 2)}$, LCD = $(x + 2)(x - 2)$

4. $\frac{2x+9}{(x+3)(x+2)}$, LCD = $(x+3)(x+2)$ 5. $\frac{4y^2-4y}{(y+2)(y-2)}$, LCD = $(y+2)(y-2)$

6. $\frac{4y}{(y-1)(y+1)}$, LCD = $(y-1)(y+1)$ 7. $\frac{14a+6}{3a(a+3)}$, LCD = $3a(a+3)$

8. $\frac{10a+5b+2ab}{2ab(2a+b)}$, LCD = $2ab(2a+b)$ 9. $\frac{4z+x}{6xyz}$, LCD = $6xyz$ 10. $\frac{x^2-3x+24}{4x^2}$, LCD = $4x^2$

11. $\frac{2x+5}{(x+2)(x+1)(x+3)}$, LCD = $(x+2)(x+1)(x+3)$ 12. $\frac{6x+20}{(x+5)^2(x-5)}$, LCD = $(x+5)^2(x-5)$

13. $\frac{a+10b}{12}$, LCD = 12 14. $\frac{a+5}{6}$, LCD = 6 15. $\frac{-7a-13}{4a}$, LCD = $4a$ 16. $\frac{7x-12}{12x}$, LCD = $12x$

17. $\frac{-4x+34}{(2x-3)(x+2)}$, LCD = $(2x-3)(x+2)$ 18. $\frac{9x+2}{(3x-4)(4x-3)}$, LCD = $(3x-4)(4x-3)$

19. $\frac{-7x}{(x+3)(x-1)(x-4)}$, LCD = $(x+3)(x-1)(x-4)$ 20. $\frac{-4x-2}{x(x-2)^2}$, LCD = $x(x-2)^2$

21. $\frac{-17y^2-10y}{(4y-1)(2y+1)(y-5)}$, LCD = $(4y-1)(2y+1)(y-5)$ 22. $\frac{x^2-5x-2}{(x+3)(x+2)(x-1)}$, LCD = $(x+3)(x+2)(x-1)$

23. $\frac{5x^2+28x-140}{5x(x-5)}$, LCD = $5x(x-5)$ 24. $\frac{3a-3b+2ab+a^3-a^2b}{ab(a-b)}$, LCD = $ab(a-b)$

25. $\frac{-24y}{(y+6)(y-6)}$, LCD = $(y+6)(y-6)$ 26. $\frac{3x}{x-5}$, LCD = $x-5$ 27. $\frac{6}{y+3}$, LCD = $(y+1)(y+3)$

28. $\frac{3}{y+4}$, LCD = $(y+4)(y-5)$ 29. $\frac{2x^2+2}{(x+1)(x-1)}$, LCD = $(x+1)(x-1)$

30. $\frac{3x^2+3x+2}{(x+3)(x-3)}$, LCD = $(x+3)(x-3)$ 31. $\frac{-x+25}{-2(x+5)(x-2)}$ or $\frac{x-25}{2(x+5)(x-2)}$, LCD = $2(x+5)(x-2)$

32. $\frac{-4y^2+3y+1}{-2y(3y+1)(y-3)}$ or $\frac{4y^2-3y-1}{2y(3y+1)(y-3)}$, LCD = $2y(3y+1)(y-3)$ 33. $\frac{7}{6(x-2)}$, LCD = $12(x-2)$

34. $\frac{7}{2(5x-1)}$, LCD = $6(5x-1)$ 35. $\frac{2x^2-2x-10}{(x+3)^2(x+1)}$, LCD = $(x+3)^2(x+1)$

36. $\frac{-2x^2-x+8}{(x+2)^2(x+3)}$, LCD = $(x+2)^2(x+3)$ 37. $\frac{2x-6y}{(x-3y)(x-4y)}=\frac{2}{x-4y}$, LCD = $(x-3y)(x-4y)$

38. $\frac{42x+6xy}{(y+7)(y-7)}=\frac{6x}{y-7}$, LCD = $(y+7)(y-7)$

Exercise 4.4

1. $\frac{3y}{2x}$ 2. $\frac{3}{b^2}$ 3. $y+x$, LCD = xy 4. $\frac{1+x}{x^2}$, LCD = x in numerator 5. $\frac{1}{xy(x-y)}$, LCD = xy in numerator

6. $\frac{1}{xy}$, LCD = xy in numerator 7. $\frac{x-3}{x}$; LCD = x^2 in numerator, x in denominator

8. $\frac{x}{x-4}$; LCD = x in numerator, x^2 in denominator 9. $\frac{-2x}{3x+3}$, LCD = $x+1$ in numerator

10. $\frac{xy-3}{xy(x+2y)}$ or $\frac{xy-3}{x^2y+2xy^2}$, LCD = xy in numerator 11. $\frac{3a^2+9}{a^3+2a}$, LCD = a in numerator, LCD = 3 in denominator

12. $\frac{a^2+1}{3-a^2}$, LCD = a 13. $\frac{x+2y}{x-y}$; LCD = $(x-y)$ in numerator, LCD = $(x+2y)$ in denominator

14. $\frac{2y^3+y^2-8y-4}{2y^2+2y}$ or $\frac{(2y+1)(y+2)(y-2)}{2y(y+1)}$; LCD = y in numerator, LCD = $(2y+1)$ in denominator

15. $\frac{x^2-2x}{x^2+4}$; LCD = 4 in numerator, LCD = $4x$ in denominator 16. $\frac{1}{x+1}$, LCD = $2x$

17. $\frac{2x-5}{3x+9}$, LCD = $(x+3)(x-3)$ in numerator 18. $\frac{5x-20}{x+2}$, LCD = $(x+4)(x-4)$ in denominator

Exercise 4.5

1. $x = 3$, LCD $= 6x$ **2.** $x = -1$, LCD $= x(2x - 1)$ **3.** $x = 2$, LCD $= x(4 - x)$ **4.** $x = 6$, LCD $= 12x$
5. $y = \frac{17}{2}$, LCD $= 5(3y + 2)$ **6.** $y = -5$, LCD $= y(y + 3)$ **7.** $x = -2$, LCD $= x(x + 1)$
8. $x = -4$, LCD $= 2(x + 6)$ **9.** $x = 4$, LCD $= 2x$ **10.** $x = 8$, LCD $= 5x$ **11.** $x = -\frac{14}{3}$, LCD $= 6x$
12. $x = -11$, LCD $= 9x$
13. An apparent solution is $x = 2$ for LCD $= (x - 2)$. However there is no solution to this problem. **14.** An apparent solution is $x = -1$ for LCD $= (x + 1)$ but this leads to division by zero. Therefore there is no solution to this problem.
15. $x = -3$, LCD $= (x + 1)(x - 1)$ **16.** $x = 0$, LCD $= (2x + 1)(2x - 1)$ **17.** $y = -5$, LCD $= (y + 3)(y - 1)$
18. $y = 1$, LCD $= (3y - 2)(y + 1)$ **19.** $x = 2$, LCD $= (x + 3)(x - 1)$ **20.** $x = 5$, LCD $= 6(x + 2)$
21. An apparent solution is $x = -4$ for LCD $= (x + 4)(x - 4)$ but this leads to division by zero, so there is *no solution.*
22. An apparent solution is $x = -3$ for LCD $= (x + 3)(x - 3)$ but this leads to division by zero, so there is *no solution.*

Practice Quizzes

Section 4.1–4.3 Practice Quiz

1. $\dfrac{5}{x}$ **2.** $\dfrac{a - 1}{3a + 4}$ **3.** $\dfrac{7(7 - a)}{(a - 8)(3a + 1)}$ **4.** $\dfrac{-x(x - 4)}{x + 2}$ **5.** $\dfrac{x - 6}{2x + 3}$ **6.** $\dfrac{a - 3b}{a(a^2 + 2)}$ **7.** $\dfrac{x + 5}{4x}$
8. $\dfrac{3 + 5x^2}{x(x - 3)}$, LCD $= x(x - 3)$ **9.** $\dfrac{5}{4(y - 2)}$, LCD $= 20(y - 2)$ **10.** $\dfrac{-2x^2 - 13x - 6}{(x + 2)^2(x + 1)}$, LCD $= (x + 2)^2(x + 1)$
11. $\dfrac{x - 5y}{3x + y}$ **12.** $\dfrac{x + 2y}{2x - 3y}$

Section 4.4–4.5 Practice Quiz

1. $\dfrac{1}{x(x - 3)}$ **2.** $\dfrac{-1}{y}$, LCD $= xy$ in numerator, LCD $= x$ in denominator
3. $\dfrac{6y - 1}{3y}$, LCD $= 9y^2$ in numerator, LCD $= 3y$ in denominator **4.** $\dfrac{-2}{5}$, LCD $= (y + 1)(y - 1)$
5. An apparent solution is $x = 2$ for LCD $= (x - 2)$. But this leads to division by zero. So there is *no solution.*
6. $x = 2$, LCD $= 3x(x + 4)$ **7.** $x = 6$, LCD $= (x + 2)(x - 2)$ **8.** $x = 3$, LCD $= (x + 2)(x + 1)$

Practice Test Chapter 4

1. $\dfrac{2}{3a}$ **2.** $\dfrac{-2x^2(y - 2)}{y + 2}$ or $\dfrac{2x^2(2 - y)}{y + 2}$ **3.** $\frac{5}{12}$ **4.** $\dfrac{1}{3b^2}$ **5.** $\dfrac{(4a + 7)(a - 2)}{(a + 2)^2}$
6. $\dfrac{23x - 64}{2(x - 2)(x - 3)}$, LCD $= 2(x - 2)(x - 3)$ **7.** $\dfrac{x - a}{ax}$, LCD $= axy$
8. $\dfrac{-x^2 + 4x + 12}{(x - 6)(x + 3)} = -\dfrac{x + 2}{x + 3}$, LCD $= (x - 6)(x + 3)$ **9.** $\dfrac{x}{4}$, LCD $= 6y$ in numerator, LCD $= 3xy$ in denominator
10. $\dfrac{x + 1}{x(x + 3)^2}$, LCD $= x(x + 3)$ in numerator **11.** $\dfrac{2x - 3y}{4x + y}$ **12.** $x = 3$, LCD $= x(x + 2)$
13. $x = 4$, LCD $= (x + 3)(x - 2)$

CHAPTER 5

Pretest Chapter 5

1. $2x - \dfrac{x}{3}$ or $2x - \frac{1}{3}x$ **2.** $0.40x + 150$ **3.** x = number
$7x + 9 = 44$
$x = 5$

4. $12 = 3x$ = first number
$4 = x$ = second number
$7 = \dfrac{3x}{2} + 1$ = third number
$x + (3x) + \left(\dfrac{3x}{2} + 1\right) = 23$

5. 17 m $= 2x - 3$ = first side
10 m $= x$ = second side
14 m $= 2x - 6$ = third side
$(2x - 3) + x + (2x - 6) = 41$

6. $x + 3\frac{1}{2}$ = weight of second package $= 7\frac{1}{2}$ lb
x = weight of first package = 4 lb
$x + 3\frac{1}{2} - 2 = x + 1\frac{1}{2}$ = weight of third package $= 5\frac{1}{2}$ lb
$(x + 3\frac{1}{2}) + x + (x + 1\frac{1}{2}) = 17$

7. \$400 $= x =$ one investment at 12% interest
\$600 $= 1000 - x =$ second investment at 9% interest
$0.12x + 0.09(1000 - x) = 102$ at 9% interest

8. $x + 4 =$ number of dimes $= 7$
$x =$ number of nickels $= 3$
$x + 7 =$ number of quarters $= 10$
$0.05x + 0.10(x + 4) + 0.25(x + 7) = 3.35$

Exercise 5.1

1. $12x$
2. $\frac{x}{8}$
3. $x - 5$
4. $x + 15$
5. $\frac{2x}{7} - 3$ or $\frac{2}{7}x - 3$
6. $2x + 4$
7. $2x - 15$
8. $2x + \frac{x}{3}$ or $2x + \frac{1}{3}x$
9. $4x - \frac{x}{2}$ or $4x - \frac{1}{2}x$
10. $x - \frac{x}{6}$ or $x - \frac{1}{6}x$
11. $2x + 5$
12. $0.12x + 50$
13. $x + 3x + \frac{x}{2}$
14. $2(x - 12)$
15. $0.18\left(\frac{x}{2}\right)$ or $0.09x$
16. $\frac{5}{9}(x - 32)$
17. $\frac{9x + 160}{5}$
18. $\frac{3x}{4} - 5$
19. $x + 100 + 0.20x$
20. $3x - 400 - 0.50x$

Exercise 5.2

1. $60 = x =$ number
$\frac{x}{4} - 5 = 10$

2. $26 = x =$ original number
$7\left(\frac{x}{2}\right) = 91$

3. $9 = x =$ original number
$12x + 9 = 117$

4. $13 = x =$ original number
$7x + 3x = 130$

5. $-10 = x =$ number
$4x - 30 = 7x$

6. $-10 = x =$ number
$5x + 30 = 2x$

7. $8 = x =$ smaller number
$26 = 3x + 2 =$ larger number
$x + (3x + 2) = 34$

8. $11 = x =$ smaller number
$38 = 4x - 6 =$ larger number
$x + (4x - 6) = 49$

9. $14 = x =$ second number
$7 = \frac{x}{2} =$ first number
$42 = 3x =$ third number
$x + \frac{x}{2} + (3x) = 63$

10. $6 = x =$ second number
$24 = 4x =$ first number
$4 = \frac{2x}{3} =$ third number
$x + (4x) + \left(\frac{2x}{3}\right) = 34$

11. $80 = x =$ original number
$\frac{x}{4} + \frac{x}{8} + \frac{x}{5} = 46$

12. $96 = x =$ original number
$\frac{x}{8} + \frac{x}{3} + \frac{x}{6} = 60$

Exercise 5.3

1. 13 cm $= 2x - 3 =$ length
8 cm $= x =$ width
$2(2x - 3) + 2x = 42$

2. 360 ft $= x =$ width
520 ft $= 2x - 200 =$ length
$2x + 2(2x - 200) = 1760$

3. 16 m $= x =$ first side
8 m $= \frac{x}{2} =$ second side
13 m $= \frac{x}{2} + 5 =$ third side
$x + \frac{x}{2} + \left(\frac{x}{2} + 5\right) = 37$

4. 34 cm $= x + 10 =$ longest side
24 cm $= x =$ second longest side
20 cm $= x - 4 =$ shortest side
$(x + 10) + x + (x - 4) = 78$

5. 19 mi $= x =$ width
27 mi $= 3x - 30 =$ length
$2x + 2(3x - 30) = 92$

6. $20{,}600 = x =$ Monday
$20{,}250 = x - 350 =$ Tuesday
$23{,}700 = x + 3100 =$ Wednesday
$x + (x - 350) + (x + 3100) = 64{,}550$

7. \$10,392 $= x + 2220 =$ Alice
\$9,024 $= x + 852 =$ Michael
\$8,172 $= x =$ Bettina
$(x + 2220) + (x + 852) + x = 27{,}588$

8. 44 lb $= \frac{2x}{3} =$ Rita
66 lb $= x =$ Margaret
46 lb $= \frac{2x}{3} + 2 =$ Tony
$\left(\frac{2x}{3}\right) + x + \left(\frac{2x}{3} + 2\right) = 156$

9. 2.4 ppm $= 3x =$ first pollutant
0.8 ppm $= x =$ second pollutant
9.6 ppm $= 12x =$ third pollutant
$(3x) + x + (12x) = 12.8$

10. 0.0045A $= \frac{x}{2} =$ first transistor
0.009A $= x =$ second transistor
0.0135A $= \frac{3x}{2} =$ third transistor
$\left(\frac{x}{2}\right) + x + \left(\frac{3x}{2}\right) = 0.027$

11. 0.0085A $= \frac{x}{4} =$ first transistor
0.034A $= x =$ second transistor
0.102A $= 12\left(\frac{x}{4}\right) = 3x =$ third transistor
$\left(\frac{x}{4}\right) + x + (3x) = 0.1445$

12. Time for Martha's trip $t = \dfrac{154 \text{ mi}}{55 \text{ mph}}$

Time for trip by ambulance $t = \dfrac{154 \text{ mi}}{70 \text{ mph}}$

Martha took 36 minutes longer.

13. Time for trip to airport $t = \dfrac{38 \text{ mi}}{50 \text{ mph}}$

Time for return trip to house $t = \dfrac{38 \text{ mi}}{47.5 \text{ mph}}$

His return trip was 2.4 minutes longer.

14. Time for trip to Maine $v = \dfrac{88 \text{ mi}}{4 \text{ hr}}$

Time for trip coming home $v = \dfrac{114 \text{ mi}}{3.8 \text{ hr}}$

He averaged 8 mi/hr faster going home.

15. 38 mpg = smallest car
21 mpg = second car
9 mpg = x = "old beast"
$\dfrac{38 + 21 + x}{3} = 22\frac{2}{3} = \frac{68}{3}$

16. \$350 = Greg
\$700 = Rich
\$420 = Hector
\$1950 = x = Constanos
$\dfrac{\$350 + \$700 + \$420 + x}{4} = \$855$

Exercise 5.4

1. 360 mi = x = number of miles driven
$0.08x + 4(18) = 100.80$

2. 158 mi = x = number of miles driven
$0.20x + 2(38) = 107.60$

3. $17 = x$ = number of dictionaries
$8x + 5(10) = 186$

4. $37 = x$ = number of tables
$3x + 2(20) = 151$

5. $13 = x$ = number of quarters
$9 = x - 4$ = number of nickels
$0.25x + 0.05(x - 4) = 3.70$

6. $11 = x$ = number of dimes
$17 = x + 6$ = number of quarters
$0.10x + 0.25(x + 6) = 5.35$

7. $6 = x$ = number of dimes
$9 = x + 3$ = number of quarters
$18 = 2(x + 3)$ = number of nickels
$0.10x + 0.25(x + 3) + 0.05(2x + 6) = 3.75$

8. $5 = x$ = number of nickels
$8 = x + 3$ = number of dimes
$16 = 2(x + 3)$ = number of quarters
$0.05x + 0.10(x + 3) + 0.25(2x + 6) = 5.05$

9. $7 = x$ = number of nickels
$5 = x - 2$ = number of quarters
$15 = 2x + 1$ = number of half-dollars
$0.05x + 0.25(x - 2) + 0.50(2x + 1) = 9.10$

10. $5 = x$ = number of dimes
$15 = 3x$ = number of nickels
$13 = x + 8$ = number of quarters
$0.10x + 0.05(3x) + 0.25(x + 8) = 4.50$

11. x = one investment at 12% interest
$4000 - x$ = second investment at 14% interest
\$2600 at 12%
\$1400 at 14%
$0.12x + 0.14(4000 - x) = 508$

12. x = one amount at 8% interest
$7200 - x$ = second amount at 10% interest
\$3200 at 8%
\$4000 at 10%
$0.08x + 0.10(7200 - x) = 656$

13. \$5000 = x = original amount of principal
$0.14x$ = interest on the principal
$x + 0.14x = 5700$

14. \$14,000 = x = last year's salary
$0.13x$ = cost of living increase
$x + 0.13x = 15{,}820$

Challenge Problems

15. $70° = 2x$ = first angle
$35° = x$ = second angle
$75° = 2x + 5$ = third angle
$(2x) + x + (2x + 5) = 180$

16. Total gallons pumped is 18,000. Leakage is 360 gallons
$6x = 360$
$x = 60$ gal/hr

17. $225 = x$ = number of miles driven where costs are the same.
$0.24x + 3(35) = 0.16x + 3(41)$
The Peerless Agency would be a better buy if he drives more than 225 miles.

18. 7.38 cm = x = original length
$12 = x + (1.32)(3.5)$

Practice Quizzes

Section 5.1–5.2 Practice Quiz

1. $\dfrac{x}{8} + 5$

2. $\frac{2}{3}(x - 6)$

3. $-10 = x$ = number
$5x + 3 = 4x - 7$

4. $\frac{29}{5} = x$ = smaller number
$\frac{91}{5} = 4x - 5$ = one number
$x + 4x - 5 = 24$

Section 5.3–5.4 Practice Quiz

1. $16 \text{ m} = 2x + 4 = \text{length}$
$6 \text{ m} = x = \text{width}$
$2(2x + 4) + 2x = 44$

2. $\$310 = x + 110 = \text{first}$
$\$200 = x = \text{second}$
$\$346 = x + 146 = \text{third}$
$x + 110 + x + x + 146 = 856$

3. $300 = x = \text{miles}$
$0.24x + 3(19) = 129$

4. $9 = x = \text{number of dimes}$

$6 = \dfrac{2x}{3} = \text{number of nickels}$

$11 = x + 2 = \text{number of quarters}$

$0.10x + 0.05\left(\dfrac{2x}{3}\right) + 0.25(x + 2) = 3.95$

Practice Test Chapter 5

1. $36 = x = \text{original number}$

$\dfrac{x}{2} + \dfrac{x}{9} + \dfrac{x}{12} = 25$

2. $-12 = x = \text{original number}$
$3x + 6 = 2(x - 3)$

3. $x = \text{one investment at } 14\% \text{ interest}$
$4000 - x = \text{second investment at } 11\% \text{ interest}$
$0.14x + 0.11(4000 - x) = 482$
$1400 at 14%
$2600 at 11%

4. $8 \text{ ppm} = 2x = \text{first pollutant}$
$4 \text{ ppm} = x = \text{second pollutant}$

$3 \text{ ppm} = 0.75x \text{ or } \dfrac{3x}{4} = \text{third pollutant}$

$2x + x + 0.75x = 15$

5. $2x = \text{number of nickels} = 16$
$x = \text{number of quarters} = 8$
$x - 1 = \text{number of dimes} = 7$

$0.05(2x) + 0.25(x) + 0.10(x - 1) = 3.50$

6. $\frac{2}{3}x = \text{first side} = 20 \text{ m}$
$x = \text{second side} = 30 \text{ m}$
$\frac{2}{3}x - 4 = \text{third side} = 16 \text{ m}$
$(\frac{2}{3}x) + x + (\frac{2}{3}x - 4) = 66$

CHAPTER 6

Pretest Chapter 6

1.

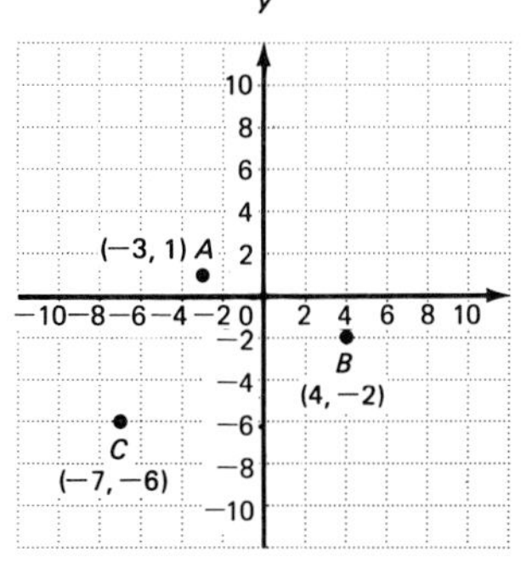

2.

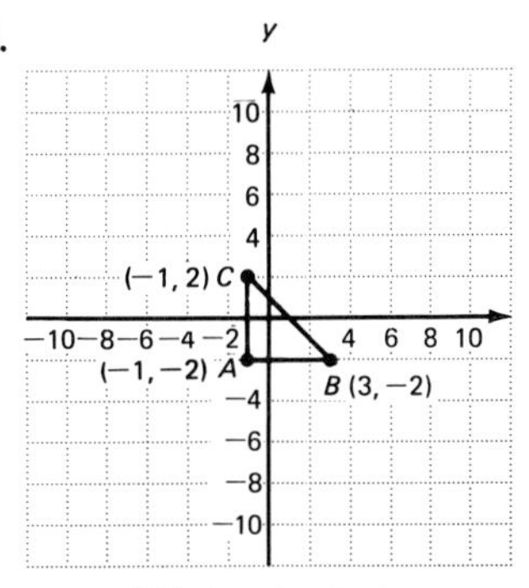

Right isosceles triangle

3.

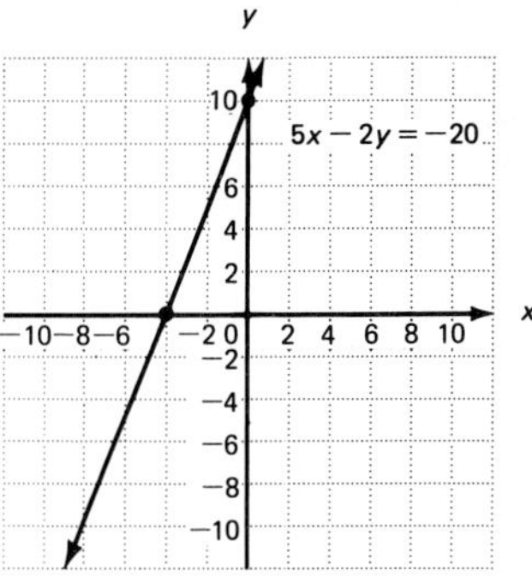

4.

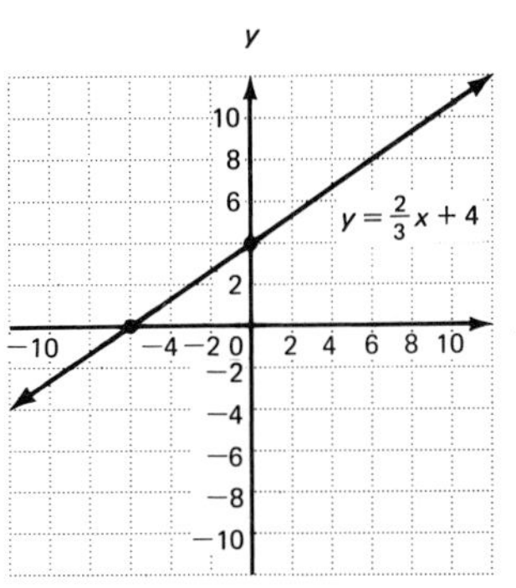

5. $m = \frac{2}{7}$ **6.** $m = \frac{4}{3}$ **7.** $y = 4x - 5$

8.

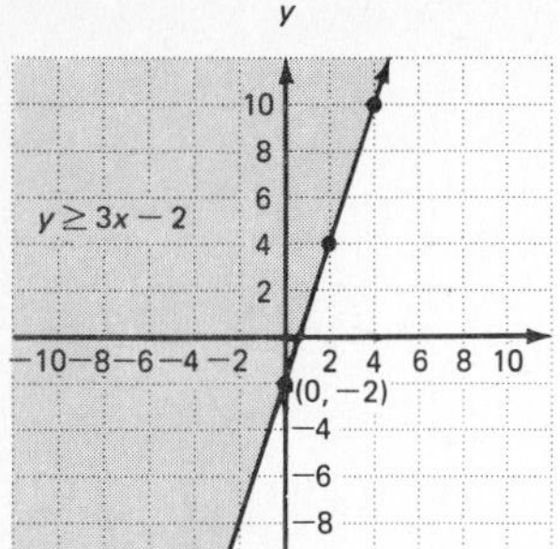

9.

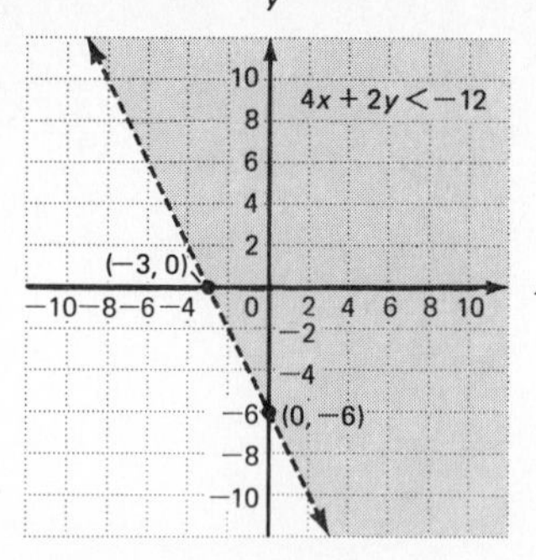

10. 34 square meters

11. 37.5 inches each 12. $A = \frac{5}{2}(b + 14)$ or $A = \frac{5b}{2} + 35$

Exercise 6.1

1–2.

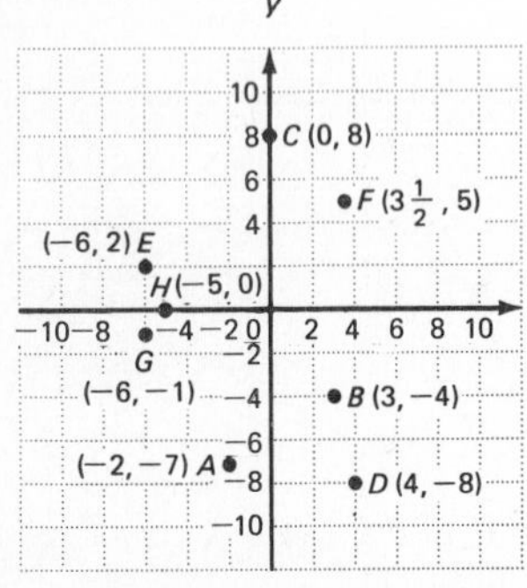

3. Abscissa of point $I = -6$, ordinate of point $J = -3$

4. Abscissa of point $K = 0$, ordinate of point $L = 6$ 5. y-coordinate of point $M = 3$, x-coordinate of point $N = -5.5$

6. y-coordinate of point $P = -6$, x-coordinate of point $Q = -5$

7. $R = (-3, -5)$, $S = (-4.5, 0)$, $X = (3, -5)$, $Y = (2.5, 6)$ 8. $T = (-6, 4)$, $V = (3.5, -2)$, $W = (6.5, 2)$ $Z = (0, -3.5)$

9. Zero 10. Zero 11. $(-3, 4), (-3, 3), (-2, 2), (-2, 1), (-1, 1), (0, 1), (1, 1), (1, 2), (2, 3), (2, 4)$ 12. $(-5, -3), (-4, -4), (-4, -5), (-3, -3), (-2, -2), (-2, -1), (-1, -3), (0, -4), (0, -5), (1, -3)$

Exercise 6.2

1.

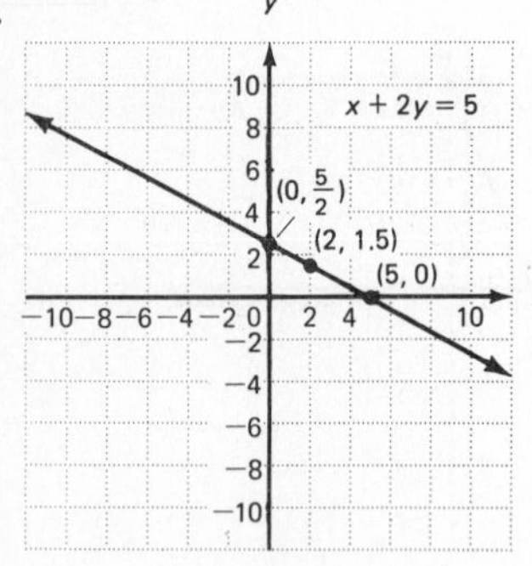

2.

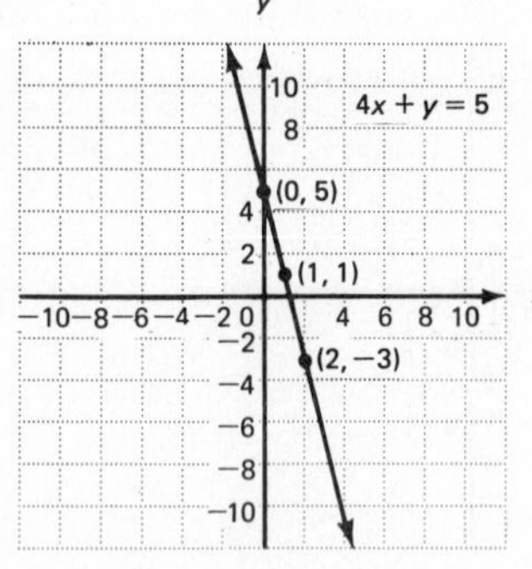

3.

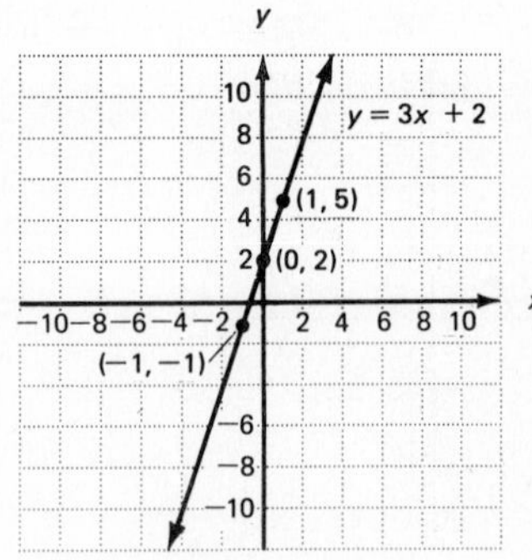

4.

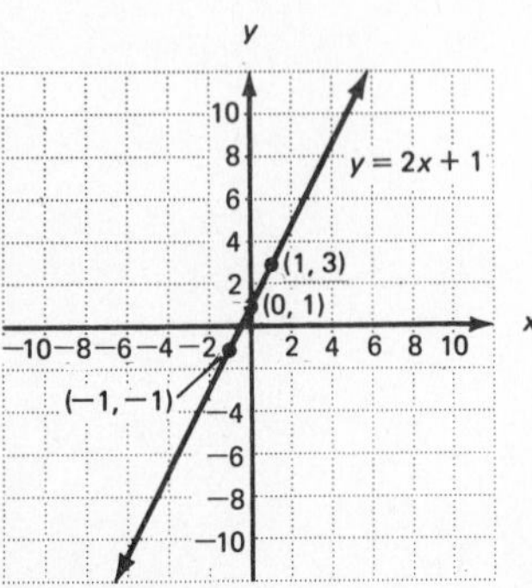

5.

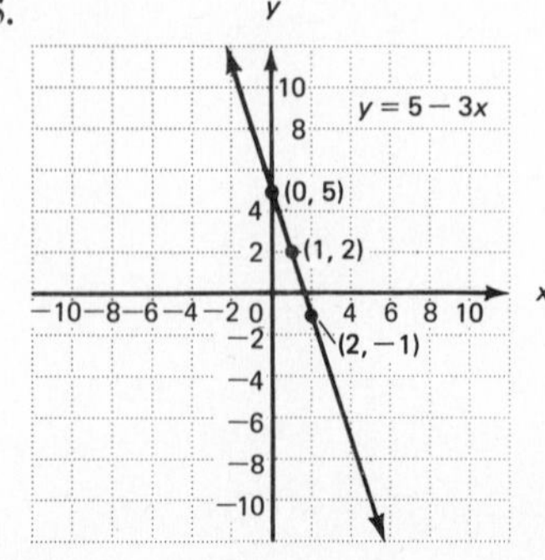

6.

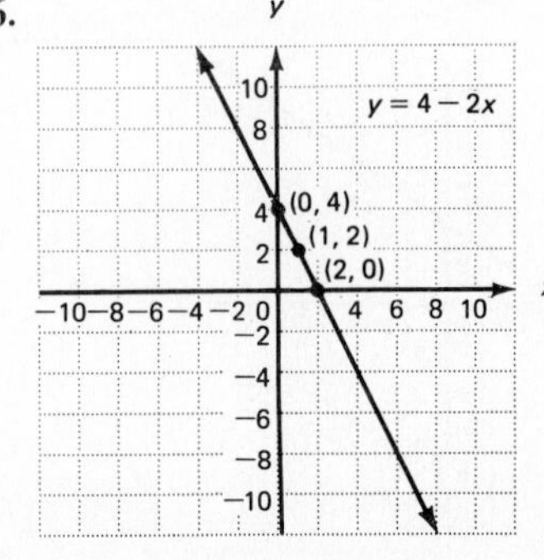

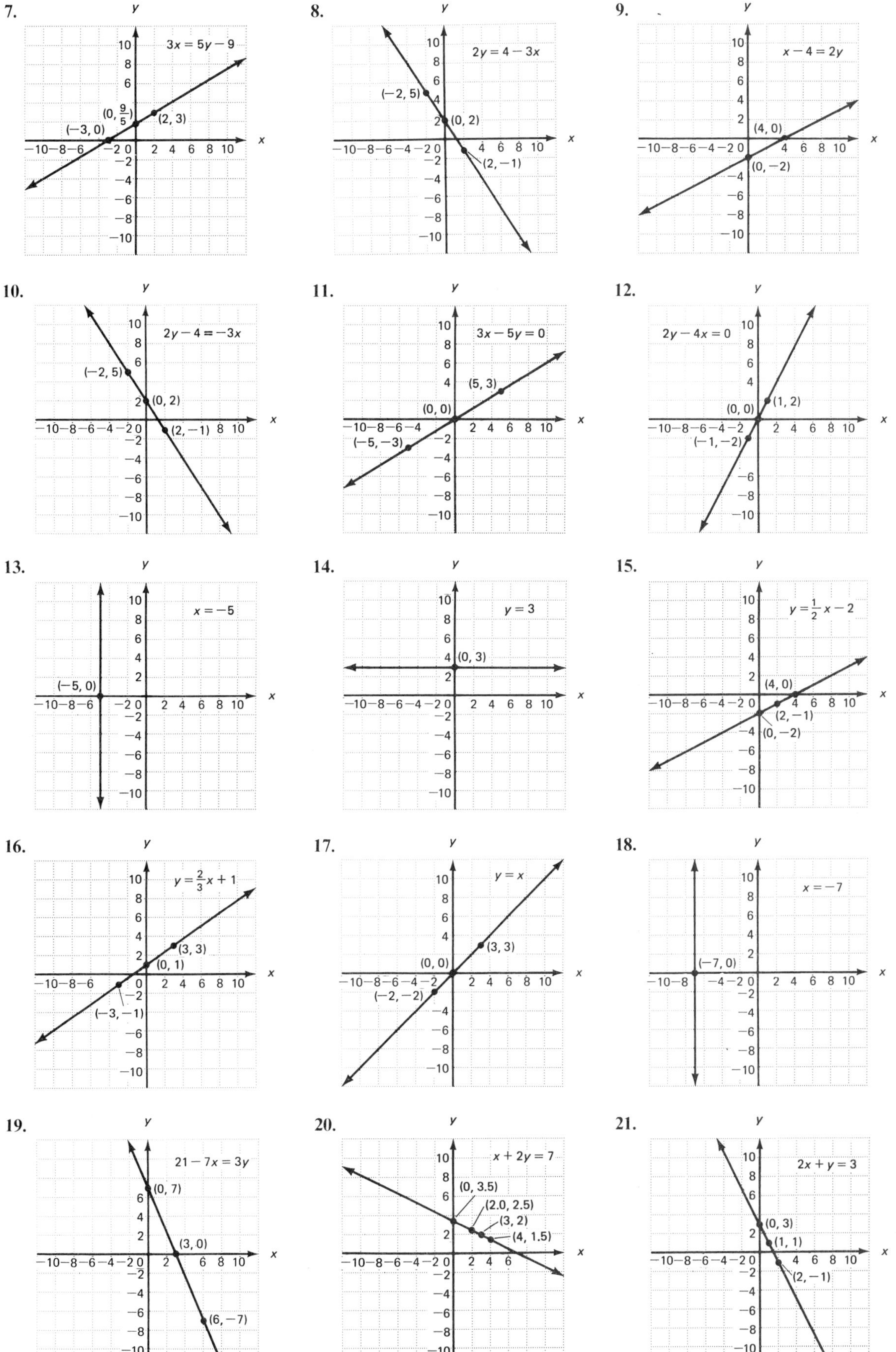
7.
3x = 5y − 9
(0, 9/5)
(2, 3)
(−3, 0)
8.
2y = 4 − 3x
(−2, 5)
(0, 2)
(2, −1)
9.
x − 4 = 2y
(4, 0)
(0, −2)
10.
2y − 4 = −3x
(−2, 5)
(0, 2)
(2, −1)
11.
3x − 5y = 0
(5, 3)
(0, 0)
(−5, −3)
12.
2y − 4x = 0
(1, 2)
(0, 0)
(−1, −2)
13.
x = −5
(−5, 0)
14.
y = 3
(0, 3)
15.
y = 1/2 x − 2
(4, 0)
(2, −1)
(0, −2)
16.
y = 2/3 x + 1
(3, 3)
(0, 1)
(−3, −1)
17.
y = x
(3, 3)
(0, 0)
(−2, −2)
18.
x = −7
(−7, 0)
19.
21 − 7x = 3y
(0, 7)
(3, 0)
(6, −7)
20.
x + 2y = 7
(0, 3.5)
(2.0, 2.5)
(3, 2)
(4, 1.5)
21.
2x + y = 3
(0, 3)
(1, 1)
(2, −1)

22.

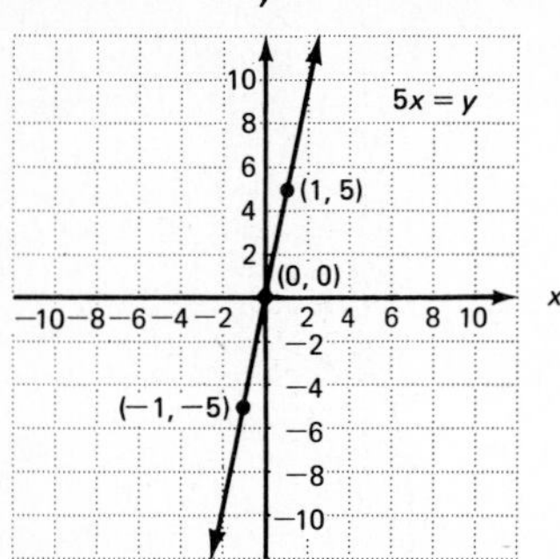

Exercise 6.3

1. $-\frac{1}{6}$ **2.** 1 **3.** $\frac{3}{5}$ **4.** $\frac{5}{8}$ **5.** 2 **6.** $-\frac{1}{8}$ **7.** 0 **8.** 0 **9.** -9 **10.** $-\frac{16}{3}$ **11.** $-\frac{10}{9}$ **12.** $\frac{5}{3}$
13. No, $x_1 = x_2$. We cannot find the slope of a vertical line. **14.** $y = -\frac{3}{4}x + 6$ **15.** $y = -7x + \frac{2}{5}$
16. $m = -3$, y-intercept $= -2$ **17.** $m = 4$, y-intercept $= 5$ **18.** $m = -\frac{5}{2}$, y-intercept $= \frac{3}{2}$
19. $m = -\frac{7}{3}$, y-intercept $= \frac{4}{3}$ **20.** $m = \frac{4}{3}$, y-intercept $= -4$ **21.** $m = 6$, y-intercept $= -2$ **22.** $m = \frac{7}{2}$, y-intercept $= \frac{3}{2}$
23. $m = \frac{4}{5}$, y-intercept $= -\frac{2}{5}$ **24.** $m = -6$, y-intercept $= -12$ **25.** $m = -6$, y-intercept $= 12$ **26.** $y = 6x + 20$
27. $y = 5x - 21$ **28.** $y = -2x + 11$ **29.** $y = -4x + 27$ **30.** $y = \frac{2}{3}x$ **31.** $y = \frac{2}{5}x - \frac{2}{5}$ **32.** $y = -3x + \frac{7}{2}$
33. $y = -2x + \frac{19}{3}$ **34.** $y = -4$ (horizontal line has slope of 0) **35.** $x = -3$ (vertical line has no slope)

Your answers should be *approximately* close to the following depending on how you estimate your points and slopes.

36. $y = \frac{4}{5}x + \frac{4}{5}$ **37.** $y = \frac{1}{4}x + 5$ **38.** $y = -\frac{2}{3}x + 3$ **39.** $y = -\frac{1}{2}x - 4$ **40.** $y = 2x + 3$ **41.** $y = 4x - 1$
42. $y = -3x + 1$ **43.** $y = -5x + 3$

Exercise 6.4

1.

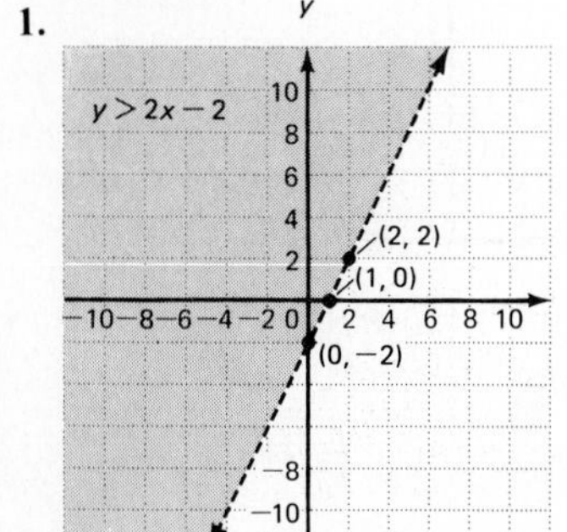

2.

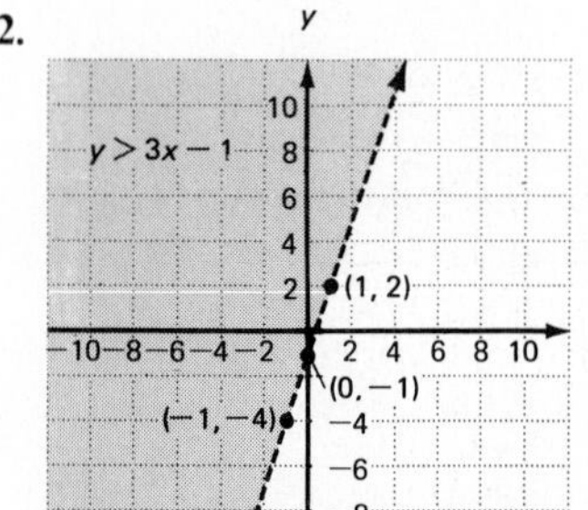

3.

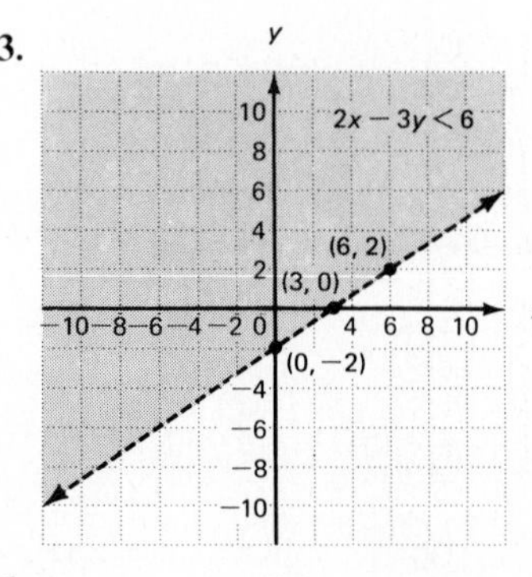

4.

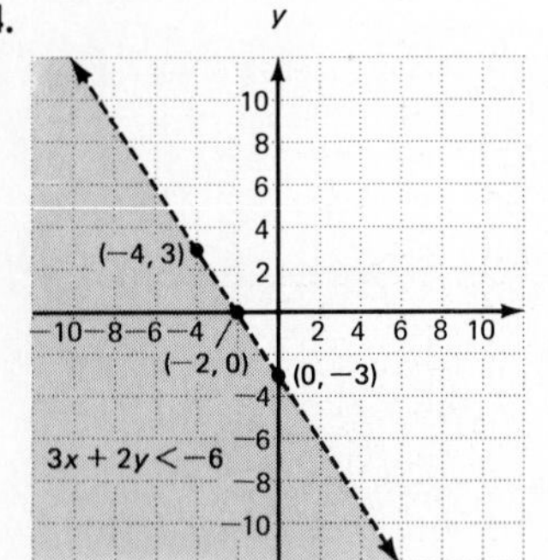

5.

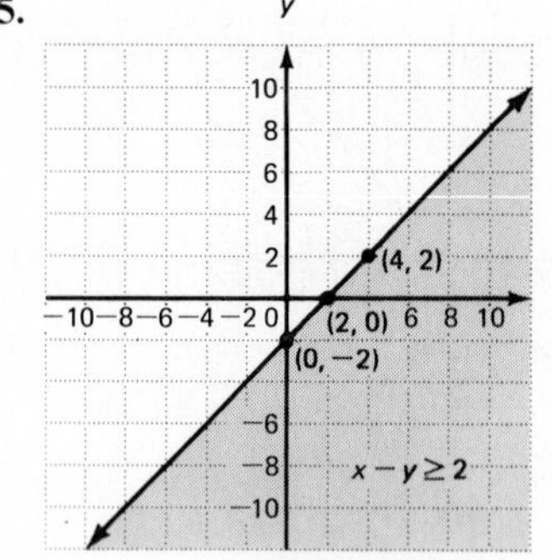

6.

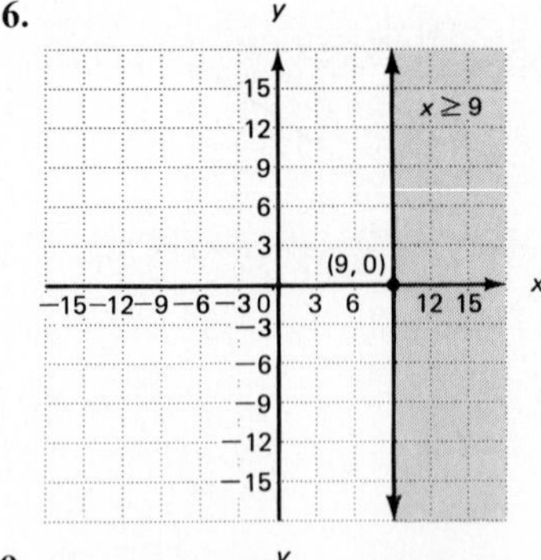

7.

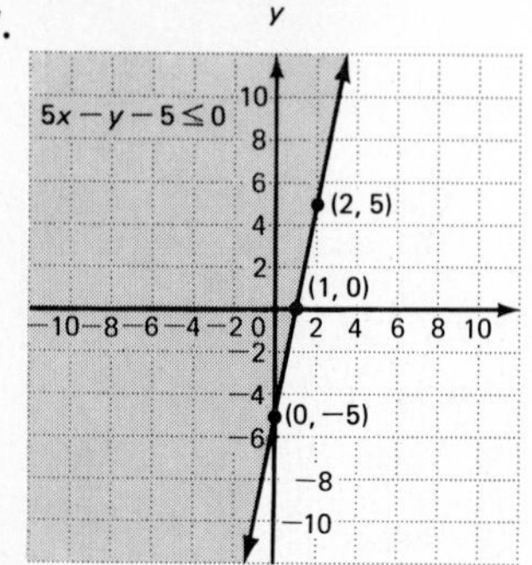

8.

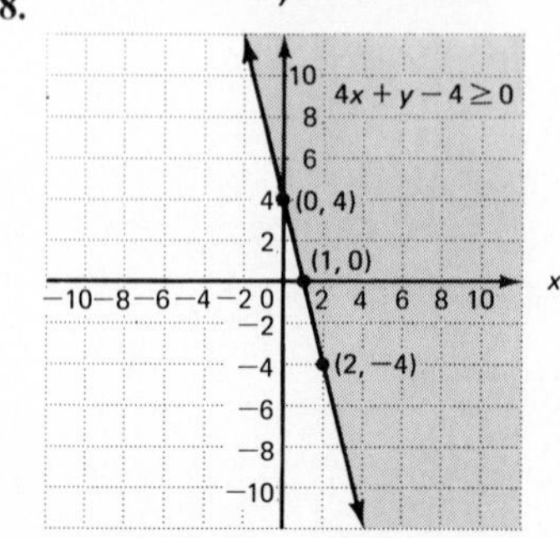

9.

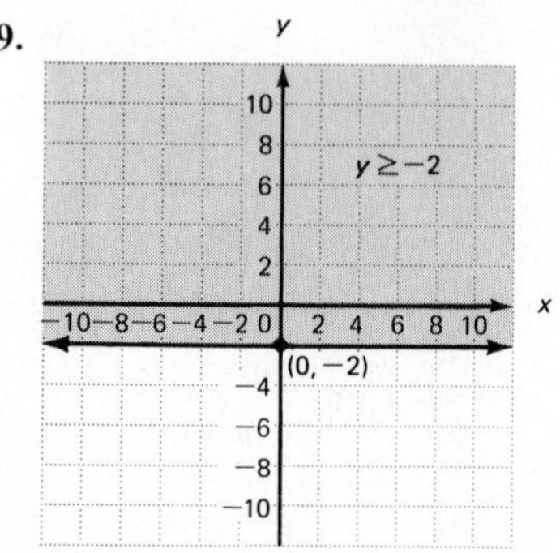

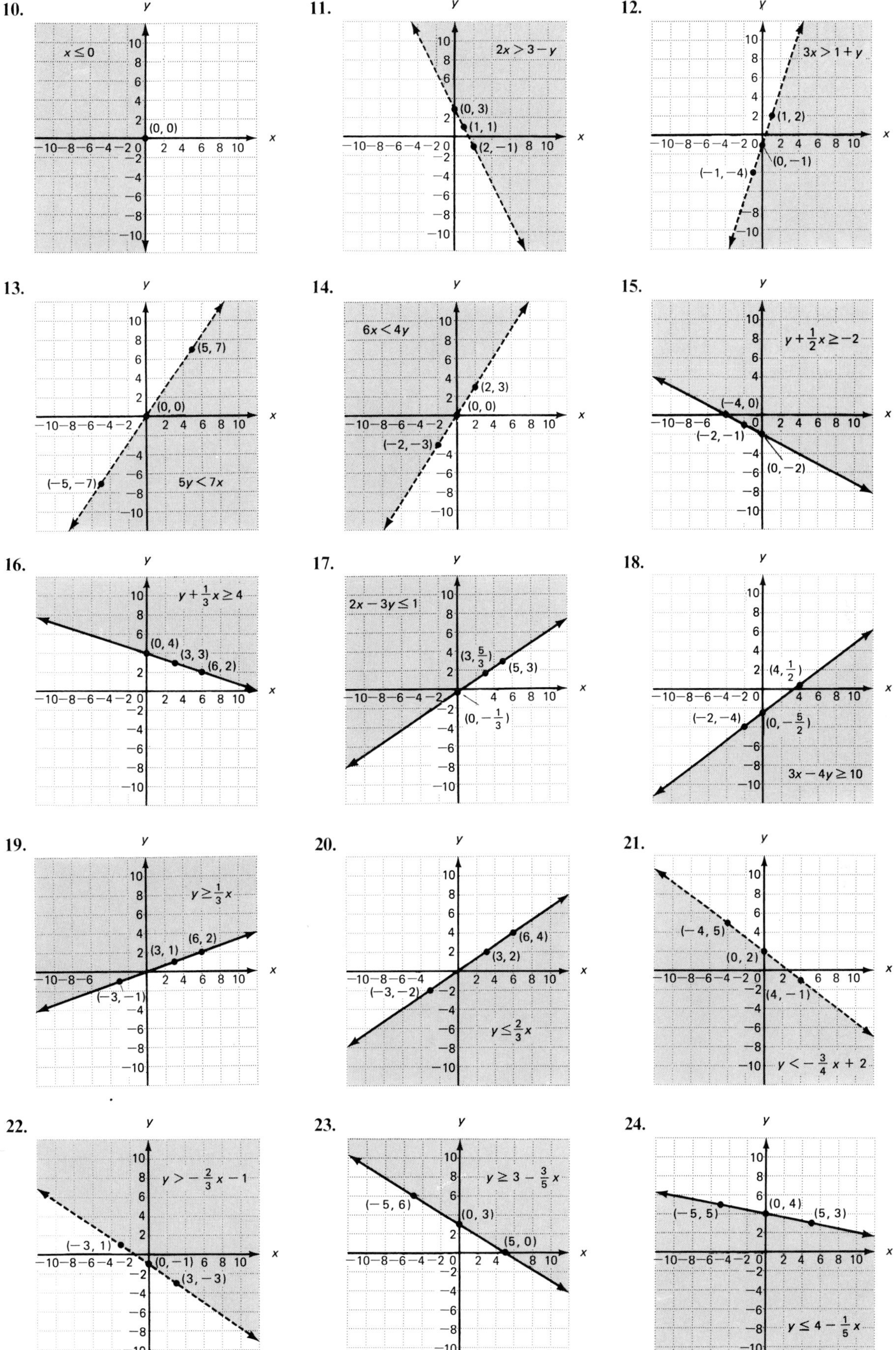
10.
$x \le 0$
(0, 0)
11.
$2x > 3 - y$
(0, 3)
(1, 1)
(2, −1)
12.
$3x > 1 + y$
(1, 2)
(0, −1)
(−1, −4)
13.
(5, 7)
(0, 0)
(−5, −7)
$5y < 7x$
14.
$6x < 4y$
(2, 3)
(0, 0)
(−2, −3)
15.
$y + \frac{1}{2}x \ge -2$
(−4, 0)
(−2, −1)
(0, −2)
16.
$y + \frac{1}{3}x \ge 4$
(0, 4)
(3, 3)
(6, 2)
17.
$2x - 3y \le 1$
$(3, \frac{5}{3})$
(5, 3)
$(0, -\frac{1}{3})$
18.
$(4, \frac{1}{2})$
(−2, −4)
$(0, -\frac{5}{2})$
$3x - 4y \ge 10$
19.
$y \ge \frac{1}{3}x$
(3, 1)
(6, 2)
(−3, −1)
20.
(6, 4)
(3, 2)
(−3, −2)
$y \le \frac{2}{3}x$
21.
(−4, 5)
(0, 2)
(4, −1)
$y < -\frac{3}{4}x + 2$
22.
$y > -\frac{2}{3}x - 1$
(−3, 1)
(0, −1)
(3, −3)
23.
$y \ge 3 - \frac{3}{5}x$
(−5, 6)
(0, 3)
(5, 0)
24.
(−5, 5)
(0, 4)
(5, 3)
$y \le 4 - \frac{1}{5}x$

Exercise 6.5

1. 61 ft **2.** 70 ft **3.** 32 m **4.** 62.8 m **5.** 42 sq. in. **6.** 70 sq. in. **7.** Area = $10a$ **8.** $V = 60l$
9. $S = \frac{11}{2}$ km or $5\frac{1}{2}$ km **10.** Length = $\frac{11}{2}$ km or $5\frac{1}{2}$ km **11.** $\angle A = 65°$ **12.** $P = 7.8$ m
13. No; there are two possible situations; perimeter I = 15.5 ft, perimeter II = 19 ft **14.** 0.56 ft shorter
15. Area of square = 16 sq. ft. Area of circles = 25.12 sq. ft. Total area = 41.12 sq. ft
16. Area rectangle = 8 sq. in. Area semicircle = 1.57 sq. in. Area quarter-circle = 3.14 sq. in. Total area = 12.71 sq. in.
17. Area triangle = 40 sq. yd. Area quarter-circle = 19.625 sq. yd. Total area = 59.625 sq. yd. Cost = $23.85
18. Area small semicircle = 56.52 sq. ft. Area large semicircle = 100.48 sq. ft. Difference = 43.96 sq. ft. Cost = $8.79 (to nearest cent)

Practice Quizzes

Section 6.1–6.3 Practice Quiz

1. $A = (-2, 2)$, $B = (4, 1)$, $C = (6, 7)$, $D = (2, -3)$, $E = (1, 5)$, $F = (-5, 5)$, $G = (-4, -2)$, $H = (0, -5)$

2.

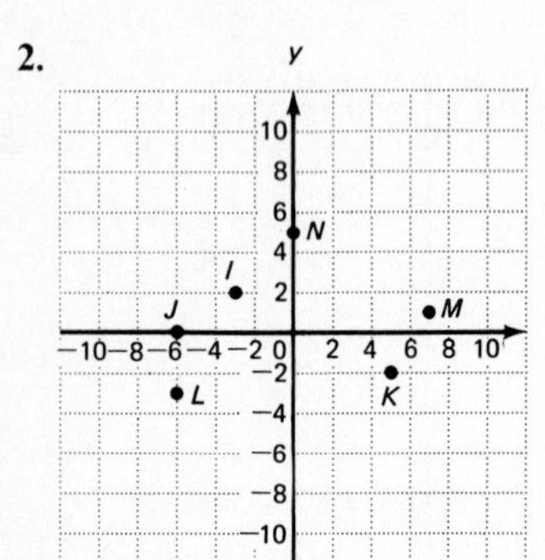

3.

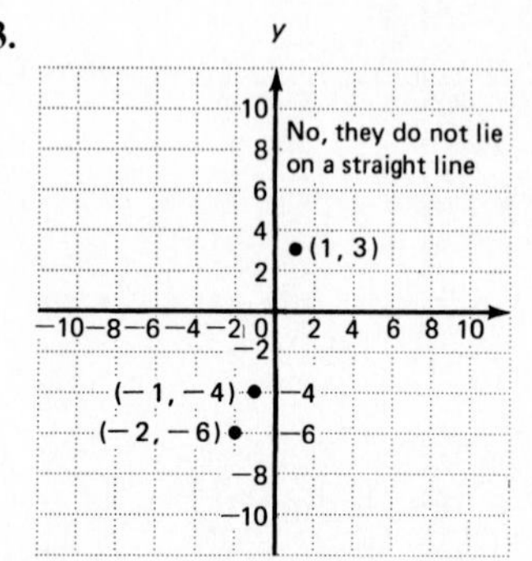

4.

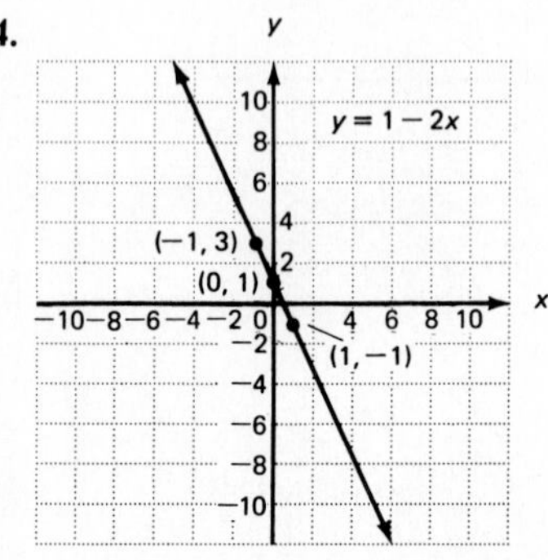

5.

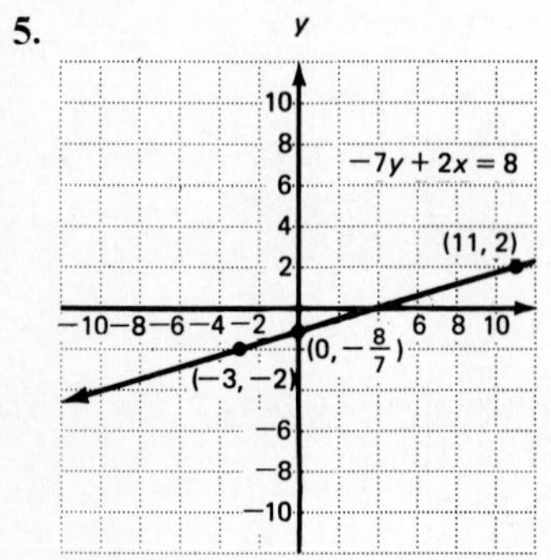

6.

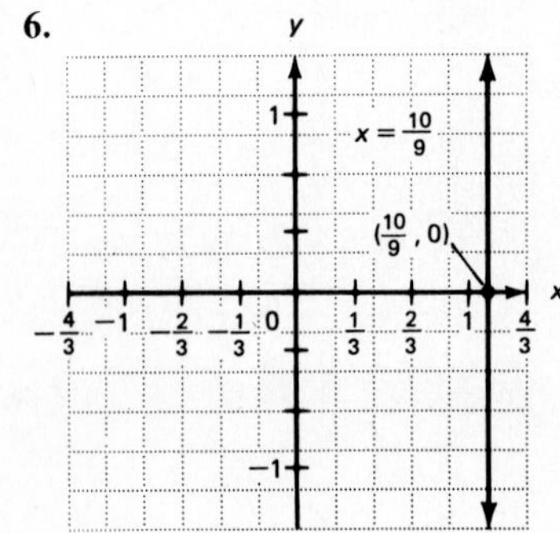

7.

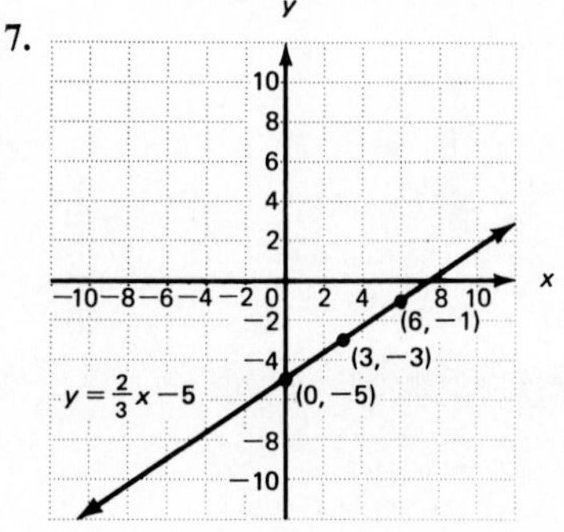

8. Equation can be simplified to $y = -\frac{7}{2}$ or $y = -3\frac{1}{2}$

9. $m = -5$

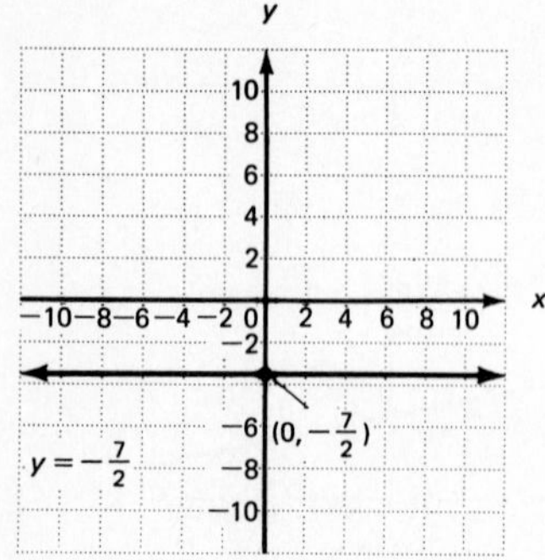

10. $m = \frac{1}{3}$, y-intercept = 2 **11.** $y = -\frac{2}{3}x - \frac{13}{3}$ **12.** $y = \frac{1}{3}x + \frac{11}{3}$ **13.** $y = -\frac{1}{15}x + \frac{2}{3}$
14. This is a vertical line, $x = -4$.

Section 6.4–6.5 Practice Quiz

1.

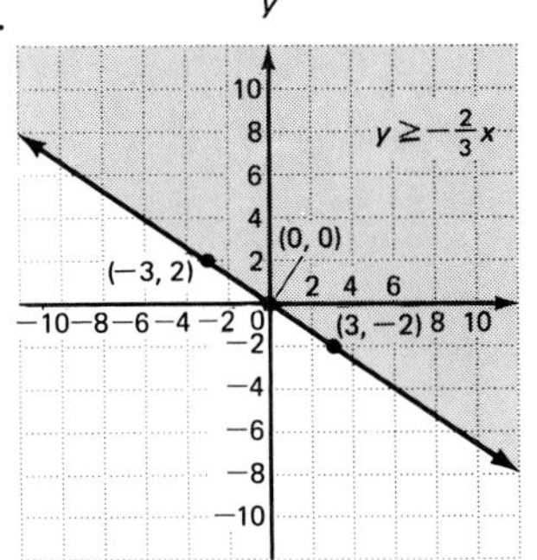

2.

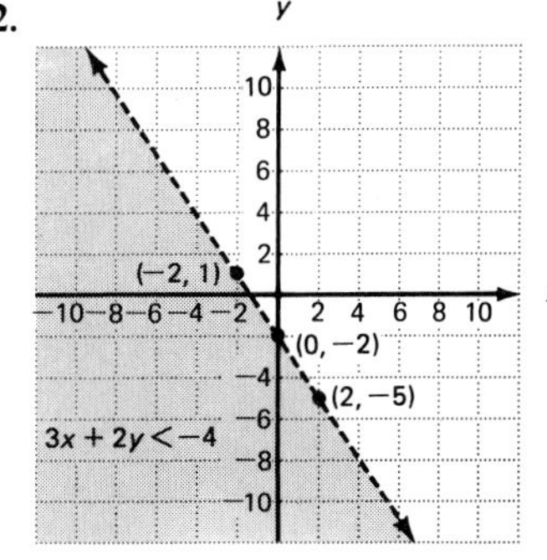

3.

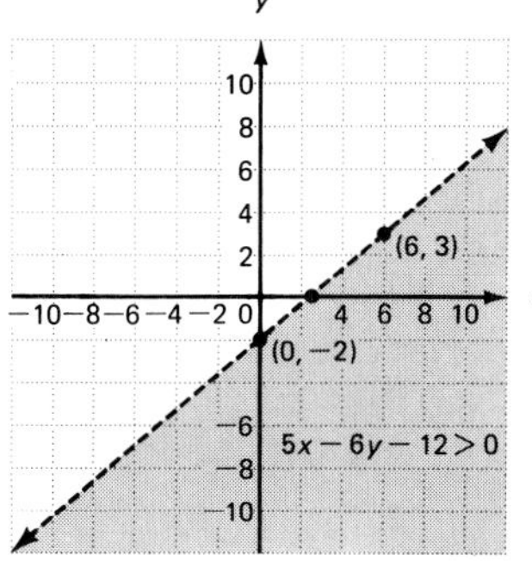

4.

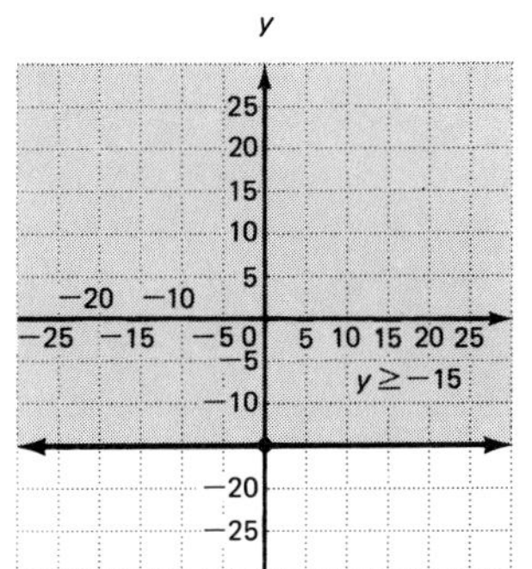

5. $c = 94.2$ in. 6. $r = \frac{29}{4}$ m or $7\frac{1}{4}$ m 7. $A = 32b$ 8. $c = \frac{74}{5}$ m or $14\frac{4}{5}$ m

9. Perimeter of the rug = 64 ft. Cost = \$11.52

10. Area square = 36 cm^2. Area circle = 28.26 cm^2. Shaded area = 7.74 cm^2.

11. Since we do not know the height, we designate the height by h. $V = 0.000168h$

Practice Test Chapter 6

1.

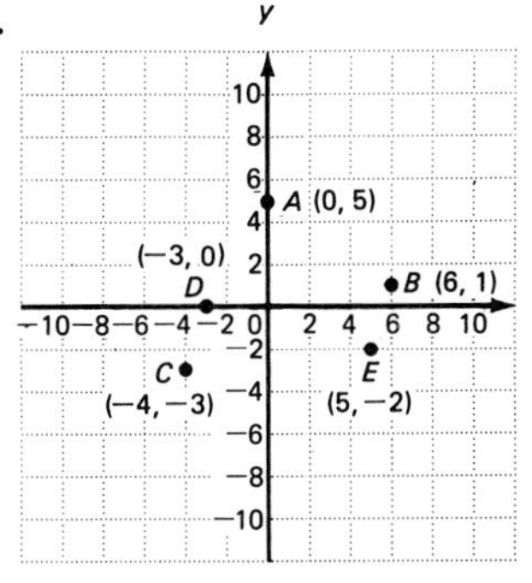

2.

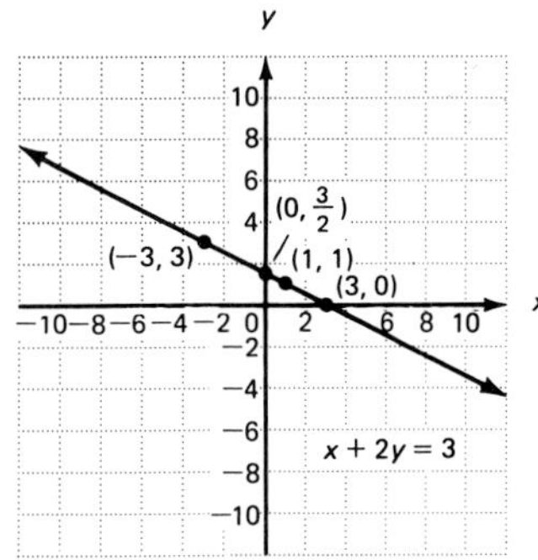

The equation $6x - 3 = 5x - 2y$ can be simplified to $x + 2y = 3$.

3.

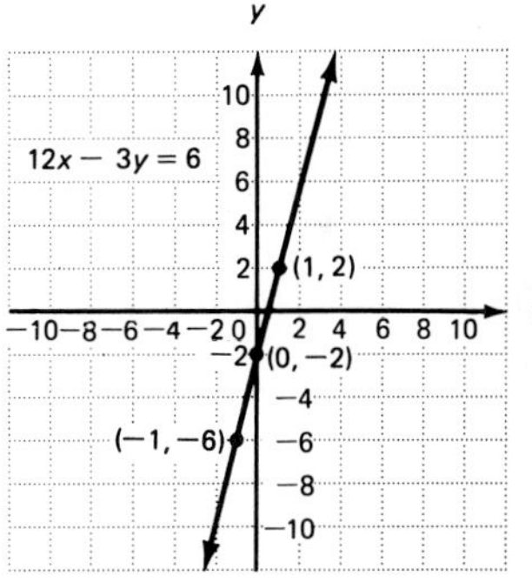

4. $m = -\frac{3}{2}$, y-intercept $= \frac{5}{2}$ 5. $y = \frac{1}{2}x - 4$ 6. $m = \frac{1}{2}$

7.

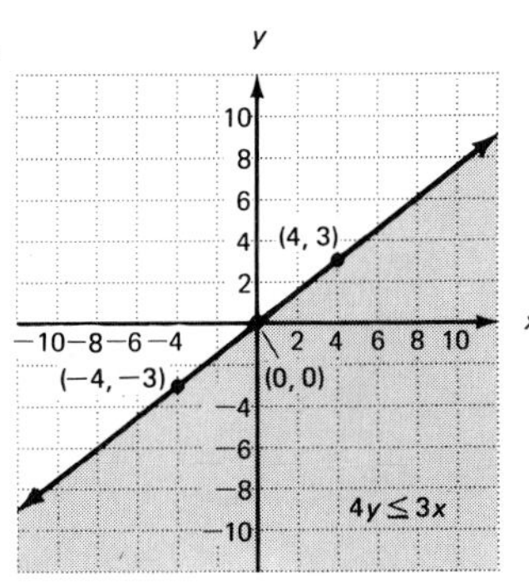

8.

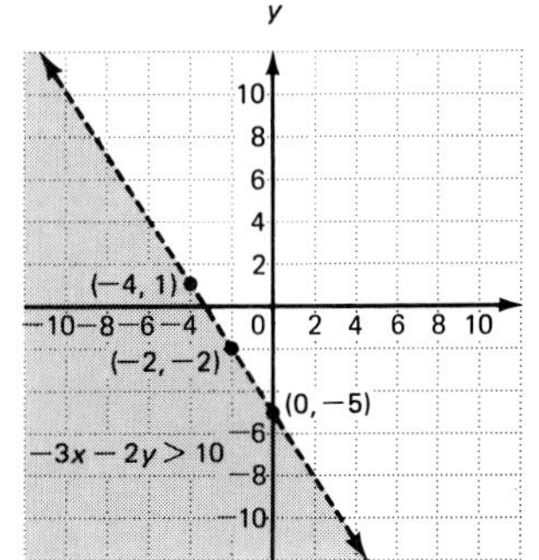

9. $c = 138.16$ in. 10. Area $= 12a$

11. Area triangle = 1.5 sq. yd. Area of 9×2 rectangle = 18 sq. yd. Area of 1.5×12 rectangle = 18 sq. yd. Total area = 37.5 sq. yd. Cost = \$450.00 **12.** $V = 4187$ cubic inches.

CHAPTER 7

Pretest Chapter 7

1. $x = -3, y = -2$ **2.** $x = -1, y = 4$ **3.** $x = 2, y = 0$ **4.** $x = -3, y = -1$

5.

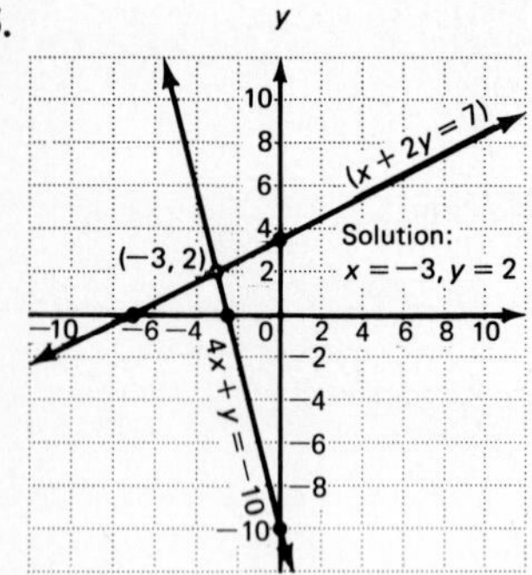

6.

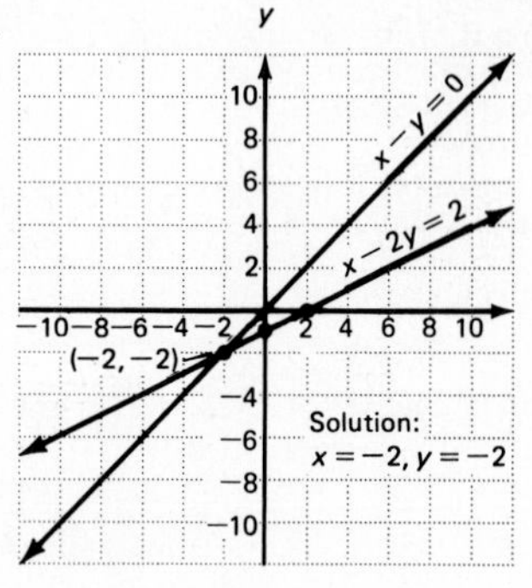

7. Inconsistent system of equations

8. Dependent equations **9.** $x = 3, y = \frac{1}{2}$ **10.** $x = 0, y = -5$

11. x = number of letters sorted per hour by James = 290
y = number of letters sorted per hour by Michael = 690
$3x + 2y = 2250$
$2x + 3y = 2650$

12. $170 = x$ = number of childrens' tickets sold
$210 = y$ = number of adults' tickets sold
$x + y = 380$
$2x + 3y = 970$

Exercise 7.1

1. $x = 1, y = 1$ **2.** $x = 3, y = -\frac{3}{2}$ **3.** $x = 3, y = -1$ **4.** $x = 0, y = 1$ **5.** $x = 1, y = 2$ **6.** $x = -1, y = 1$
7. $x = \frac{8}{5}, y = \frac{16}{5}$ **8.** $x = 9, y = 3$ **9.** $x = 5, y = -1$ **10.** $x = 2, y = -2$ **11.** $x = 3, y = 0$ **12.** $x = 1, y = 0$
13. $a = 9, b = 5$ **14.** $a = 1, b = -4$ **15.** $x = 2, y = 0$ **16.** $x = 2, y = -1$ **17.** $c = 3, d = -2$
18. $s = 0, t = -3$ **19.** $x = 2, y = -3$ **20.** $x = \frac{3}{4}, y = \frac{1}{2}$ **21.** $x = -3, y = 6$ **22.** $x = 7, y = 3$

Exercise 7.2

1. $x = -2, y = 3$ **2.** $x = -1, y = 4$ **3.** $x = 2, y = -1$ **4.** $x = 1, y = 2$ **5.** $x = -3, y = 1$ **6.** $x = 0, y = 5$
7. $x = -2, y = 1$ **8.** $x = -3, y = 4$ **9.** $x = -4, y = 3$ **10.** $x = 8, y = -1$ **11.** $a = -10, b = -5$
12. $a = -6, b = 2$ **13.** $x = -\frac{1}{2}, y = \frac{1}{3}$ **14.** $x = -\frac{9}{4}, y = 1$ **15.** $x = 1, y = 1$ **16.** $x = 2, y = -1$
17. $x = -2, y = \frac{4}{3}$ **18.** $x = \frac{5}{2}, y = -\frac{4}{3}$ **19.** $x = 11, y = 6$ **20.** $x = 1, y = 0$ **21.** $x = 3, y = -5$
22. $x = -4, y = -7$ **23.** $x = -2, y = 2$ **24.** $x = 3, y = -2$

Exercise 7.3

1.

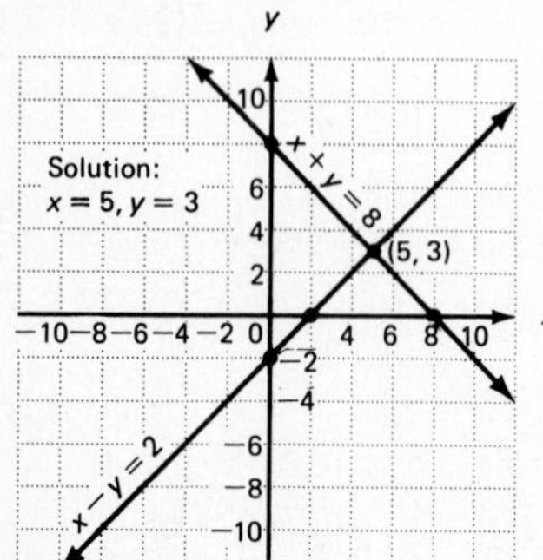

2.

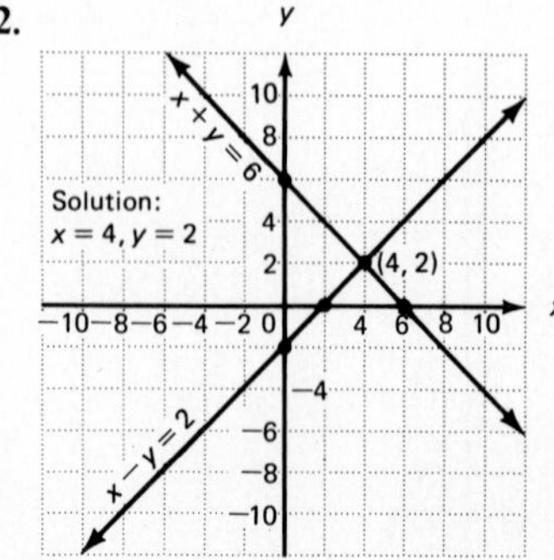

3.

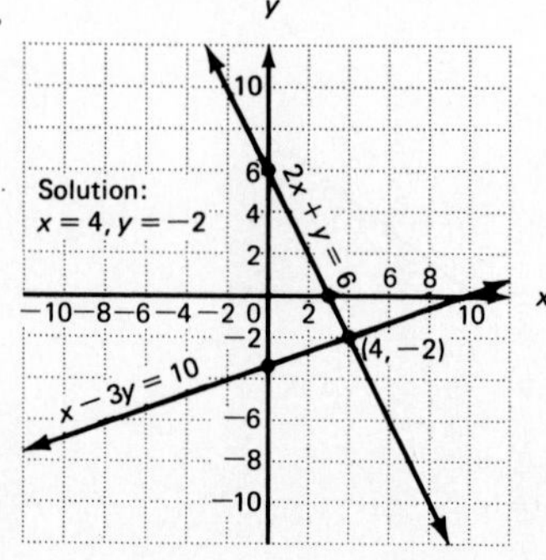

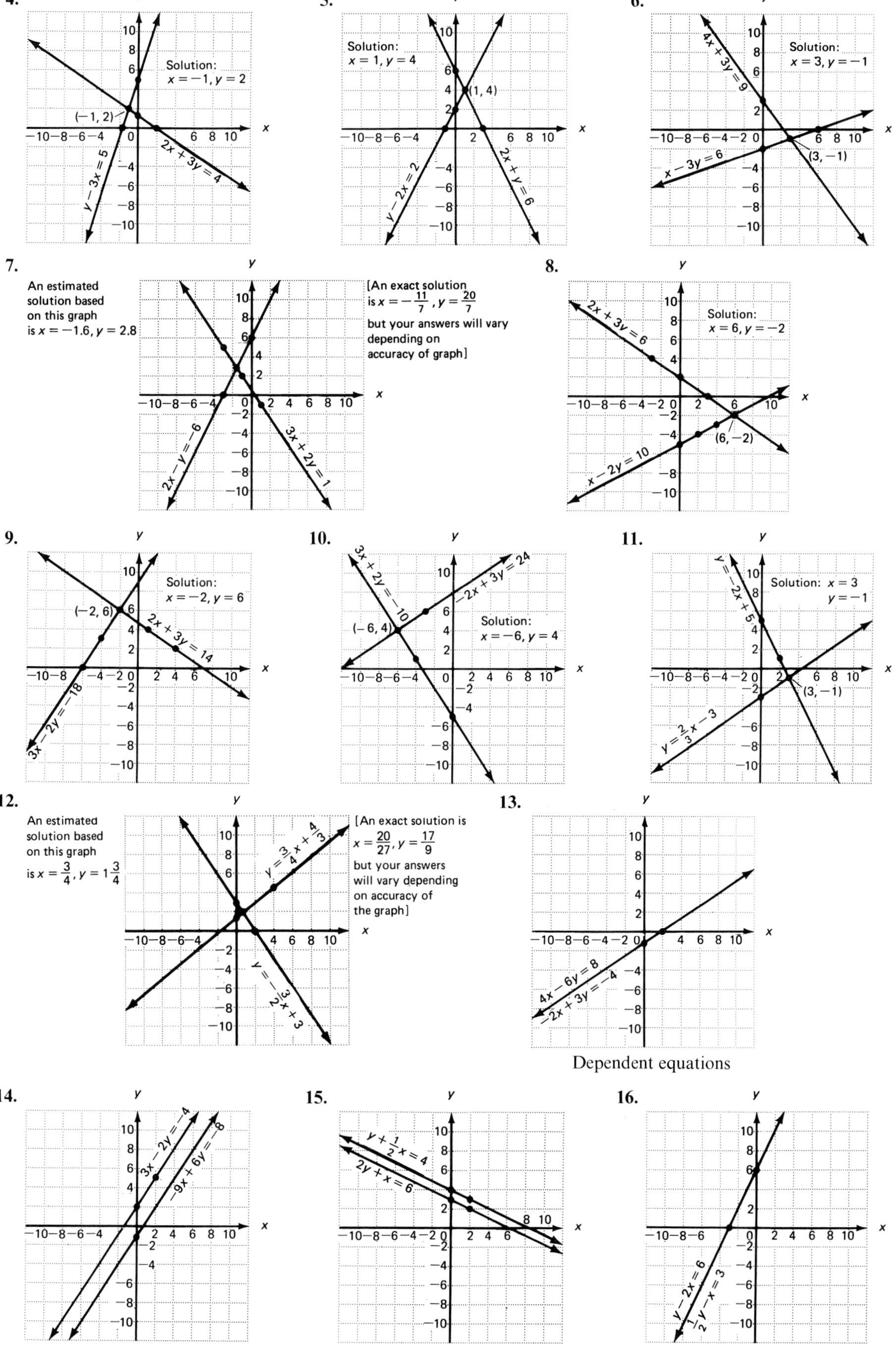
4.
Solution:
$x = -1, y = 2$
(−1, 2)
$2x + 3y = 4$
$y - 3x = 5$
5.
Solution:
$x = 1, y = 4$
(1, 4)
$y - 2x = 2$
$2x + y = 6$
6.
Solution:
$x = 3, y = -1$
$4x + 3y = 9$
$x - 3y = 6$
(3, −1)
7.
An estimated solution based on this graph is $x = -1.6, y = 2.8$
[An exact solution is $x = -\frac{11}{7}, y = \frac{20}{7}$ but your answers will vary depending on accuracy of graph]
$2x - y = -6$
$3x + 2y = 1$
8.
Solution:
$x = 6, y = -2$
$2x + 3y = 6$
$x - 2y = 10$
(6, −2)
9.
Solution:
$x = -2, y = 6$
(−2, 6)
$2x + 3y = 14$
$3x - 2y = -18$
10.
Solution:
$x = -6, y = 4$
(−6, 4)
$3x + 2y = -10$
$-2x + 3y = 24$
11.
Solution: $x = 3$
$y = -1$
$y = -2x + 5$
$y = \frac{2}{3}x - 3$
(3, −1)
12.
An estimated solution based on this graph is $x = \frac{3}{4}, y = 1\frac{3}{4}$
[An exact solution is $x = \frac{20}{27}, y = \frac{17}{9}$ but your answers will vary depending on accuracy of the graph]
$y = \frac{3}{4}x + \frac{4}{3}$
$y = -\frac{3}{2}x + 3$
13.
$4x - 6y = 8$
$-2x + 3y = -4$
Dependent equations
14.
$3x - 2y = -4$
$-9x + 6y = -8$
Inconsistent system of equations
15.
$y + \frac{1}{2}x = 4$
$2y + x = 6$
Inconsistent system of equations
16.
$y - 2x = 6$
$\frac{1}{2}y - x = 3$
Dependent equations

17. $x = -14, y = -5$ **18.** $x = 4, y = \frac{15}{2}$ **19.** Dependent equations **20.** $x = -2, y = 2$ **21.** $x = 3, y = -7$
22. Inconsistent system of equations **23.** Inconsistent system of equations **24.** Dependent equations
25. $x = 3, y = 1$ **26.** $x = -1, y = -3$

Exercise 7.4

1. $13 = x =$ one number
$-17 = y =$ second number
$x + y = -4$
$x - y = 30$

2. $19 = x =$ one number
$5 = y =$ second number
$x + y = 24$
$x - y = 14$

3. $42 = H =$ highway EPA rating
$30 = C =$ city EPA rating
$H - C = 12$
$\frac{H + C}{2} = 36$

4. $59° = x =$ one angle
$31° = y =$ second angle
$x + y = 90$
$x - y = 28$

5. $32 = A =$ number of managers who have attended college
$14 = N =$ number of managers who have not attended college
$A - N = 18$
$A + N = 46$

6. $50 = x =$ number of managers
$46 = y =$ number of sales persons
$5.2x + 3.5y = 421$
$x - y = 4$

7. $170 = x =$ number of mezzanine tickets
$250 = y =$ number of orchestra tickets
$x + y = 420$
$8x + 13y = 4610$

8. $32 = x =$ number of secretaries
$14 = y =$ number of receptionists
$x + y = 46$
$35x + 25y = 1470$

9. $11 = W =$ width in feet
$13 = L =$ length in feet
$2L + 2W = 48$
$4L + 6W = 118$

10. $9 = c =$ number of cars sold
$5 = t =$ number of trucks sold
$c + t = 14$
$4c + 3t = 51$

11. The 3-hour trip was completed at an average speed of 210 miles per hour. The 3.5-hour trip was completed at an average speed of 180 miles per hour.

$x =$ speed of plane in still air in mph $= 195$
$y =$ wind speed in mph $= 15$

$$x + y = 210$$
$$x - y = 180$$

12. The 6-hour trip was completed at an average speed of 500 km/hour. The 5-hour trip was completed at an average speed of 600 km/hour.

$x =$ speed of plane in still air in km/hour $= 550$ km/hour
$y =$ wind speed in km/hour $= 50$

$$x - y = 500$$
$$x + y = 600$$

13. The 4-hour trip was completed at $\frac{45}{4}$ or $11\frac{1}{4}$ miles per hour. The 3-hour trip was completed at 15 miles per hour.

$x =$ speed of boat in miles per hour $= 13\frac{1}{8}$ or 13.125
$y =$ speed of the current in miles per hour $= 1\frac{7}{8}$ or 1.875

$$x + y = 15$$
$$x - y = \frac{45}{4}$$

14. $8 = x =$ original amount measured in qt
$8 = y =$ amount added measured in qt
$x + y = 16$
$0.5x + 0.8y = 10.4$

15. 15 kg $= x =$ number of kg of 50% alloy
5 kg $= y =$ number of kg of 30% alloy
$x + y = 20$
$0.5x + 0.3y = 9$

16. $30 = x =$ number of lb of nuts
$20 = y =$ number of lb of raisins
$x + y = 50$
$2x + 1.5y = 90$

17. $x =$ number of lb of \$0.85 per pound flour $= 4$ lb
$y =$ number of lb of \$0.90 per pound flour $= 6$ lb
$x + y = 10$
$0.85x + 0.90y = (10)(8.8)$

Practice Quizzes

Section 7.1–7.2 Practice Quiz

1. $x = 2, y = -1$ **2.** $x = -2, y = 4$ **3.** $x = 1, y = 3$ **4.** $x = 2, y = -1$ **5.** $x = 2, y = -1$
6. $x = 3, y = -1$ **7.** $x = -4, y = -2$

Section 7.3–7.4 Practice Quiz

1.

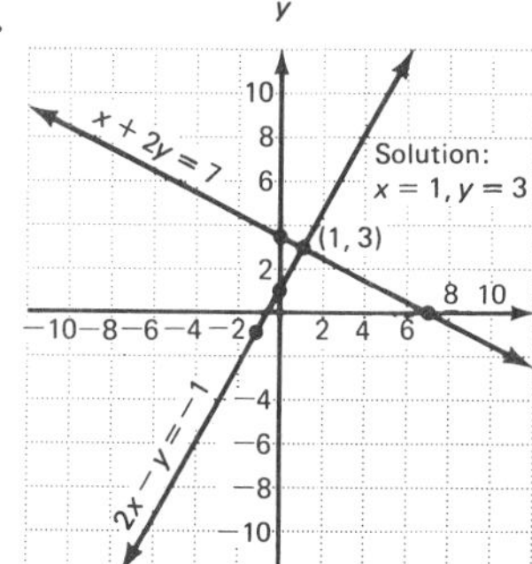

2.

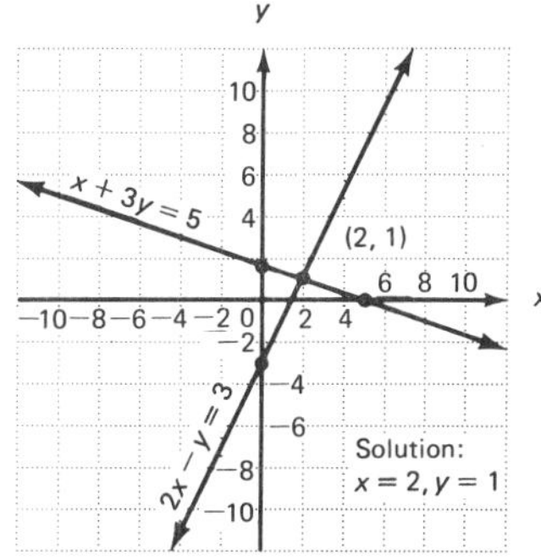

3. $x = 4, y = -1$ **4.** $x = 3, y = -2$ **5.** Dependent equations **6.** Inconsistent system of equations

7. $x = 2, y = 2$ **8.** $44 = x$ = number of cars that were regular size
$42 = y$ = number of cars that were compact size
$x + y = 86$
$2x + 1.5y = 151$

9. $8 = x$ = number of tons of 15% shipment
$16 = y$ = number of tons of 30% shipment
$x + y = 24$
$0.15x + 0.3y = (0.25)(24) = 6$

10. $19 = x$ = boat speed in km/hr
$4 = y$ = current's speed in km/hr
$x + y = 23$
$x - y = 15$

11. $\$3 = x$ = cost of 1 gram of copper
$\$4 = y$ = cost of 1 hour of labor
$5x + 3y = 27$
$4x + 5y = 32$

Practice Test Chapter 7

1. $x = -3, y = -4$ **2.** $x = -3, y = 4$ **3.**

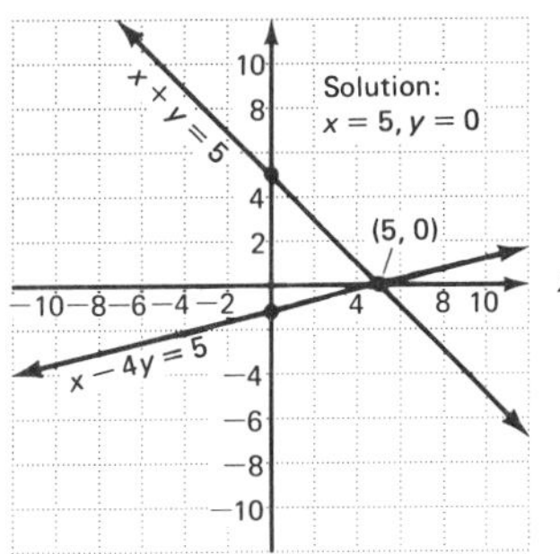

4. $x = -6, y = -2$

5. $x = 4, y = 3$ **6.** $x = 1, y = 2$ **7.** $x = 3, y = 0$ **8.** Inconsistent system of equations

9. Dependent equations **10.** $x = -2, y = 3$

11. x = number of envelopes addressed per hour by the first machine = 500
y = number of envelopes addressed per hour by the second machine = 400

$$8x + 3y = 5200$$
$$4x + 6y = 4400$$

12. The 5-hour trip was completed at an average speed of 400 km/hr. The 4-hour trip was completed at an average speed of 500 km/hr

x = speed of plane in still air in km/hour = 450
y = speed of wind in km/hour = 50

$$x + y = 500$$
$$x - y = 400$$

CHAPTER 8

Pretest Chapter 8

1. 11 **2.** $\frac{5}{8}$ **3.** -10 **4.** No real solution **5.** $2xy^2\sqrt{3xy}$ **6.** $6x^3y\sqrt{y}$ **7.** $4\sqrt{2}$ **8.** $13\sqrt{7} - 7$

9. $30 - 10\sqrt{3}$ **10.** $8 + 6\sqrt{3}$ **11.** $\dfrac{5\sqrt{3}}{3}$ **12.** $\dfrac{3\sqrt{5} - 6\sqrt{3}}{-7}$ or $\dfrac{6\sqrt{3} - 3\sqrt{5}}{7}$

13. 0.775 (Rounded to 3 decimal places) **14.** $5x^2y^3$ **15.** $2x^2y$

16. $R = kv^2$. Evaluate k to obtain $R = \frac{1}{3}v^2$. If $v = 50$, $R = \frac{2500}{3}$ or $R = 833\frac{1}{3}$ pounds.

Exercise 8.1

1. 5 **2.** 6 **3.** 0 **4.** 1 **5.** 7 **6.** 4 **7.** 10 **8.** 3 **9.** −6 **10.** −5 **11.** 0.9 **12.** 0.8 **13.** 0.1 **14.** 0.4 **15.** $\frac{4}{5}$ **16.** $\frac{3}{8}$ **17.** $\frac{7}{8}$ **18.** $\frac{7}{10}$ **19.** Not a real number **20.** Not a real number **21.** Not a real number **22.** Not a real number **23.** 13 **24.** 15 **25.** $-\frac{1}{8}$ **26.** $\frac{9}{10}$ **27.** $\frac{5}{7}$ **28.** $-\frac{3}{8}$ **29.** 0.6 **30.** 0.5 **31.** Not a real number **32.** $\frac{1}{14}$ **33.** $\frac{7}{10}$ **34.** Not a real number **35.** 0.08 **36.** 0.009 **37.** 24 **38.** −19 **39.** −29 **40.** 23 **41.** $\frac{17}{30}$ **42.** $\frac{20}{27}$

Exercise 8.2

1. $3\sqrt{3}$ **2.** $3\sqrt{2}$ **3.** $4\sqrt{2}$ **4.** $5\sqrt{2}$ **5.** $4\sqrt{3}$ **6.** $6\sqrt{2}$ **7.** 64 or 8^2 **8.** 5^5 or 3125 **9.** 56^3 **10.** 39^4 **11.** w^6 **12.** y^3 **13.** xy^3 **14.** x^2y^6 **15.** $5x^4y^2$ **16.** $4xy^{10}$ **17.** $2x\sqrt{2x}$ **18.** $2y^2\sqrt{3y}$ **19.** $3w^2\sqrt{3w}$ **20.** $3y^6\sqrt{2y}$ **21.** $4x^3\sqrt{x}$ **22.** $7x^2\sqrt{x}$ **23.** $5xy^2\sqrt{3xy}$ **24.** $5x^3\sqrt{2xy}$ **25.** $4y\sqrt{3yw}$ **26.** $2xy\sqrt{3y}$ **27.** $5xy\sqrt{3y}$ **28.** $3a^4y^3\sqrt{3y}$ **29.** $3bc^2\sqrt{5ab}$ **30.** $4ac^2\sqrt{2ab}$ **31.** $6ab^2c^4d^6$ **32.** $3a^2b^3c\sqrt{7c}$ **33.** $2ab^2c^4\sqrt{7b}$ **34.** $9x^6y^5w^2\sqrt{yw}$ **35.** $8x^4y^4w^5\sqrt{yw}$ **36.** $3x^2y^3\sqrt{15xy}$ **37.** $2x^3y^5\sqrt{35xy}$

Exercise 8.3

1. $-2\sqrt{3}$ **2.** $-2\sqrt{6}$ **3.** $3\sqrt{3}+5\sqrt{2}$ **4.** $4\sqrt{5}-3\sqrt{6}$ **5.** $-4\sqrt{2x}+5\sqrt{x}$
6. Cannot be simplified, the answer remains as $\sqrt{ab}-2\sqrt{b}+3\sqrt{a}$ **7.** $\sqrt{2}-2\sqrt{3}$ **8.** $\sqrt{5}-2\sqrt{5}=-\sqrt{5}$
9. $5\sqrt{2}+12\sqrt{2}=17\sqrt{2}$ **10.** $4\sqrt{3}+4\sqrt{3}=8\sqrt{3}$ **11.** $4\sqrt{2}-3\sqrt{2}=\sqrt{2}$ **12.** $6\sqrt{3}-4\sqrt{3}=2\sqrt{3}$
13. $5\sqrt{3}+\sqrt{3}-6\sqrt{3}=0$ **14.** $2\sqrt{7}-3\sqrt{7}+6\sqrt{7}=5\sqrt{7}$ **15.** $4\sqrt{3}+2\sqrt{5}+6$
16. $2\sqrt{6}-6\sqrt{6}+6\sqrt{2}=-4\sqrt{6}+6\sqrt{2}$ **17.** $2\sqrt{5}-4\sqrt{5}+12\sqrt{3}=-2\sqrt{5}+12\sqrt{3}$ **18.** $5\sqrt{2}+2\sqrt{7}-3$
19. $3\sqrt{3x}+2\sqrt{3x}=5\sqrt{3x}$ **20.** $5\sqrt{2y}-3\sqrt{2y}=2\sqrt{2y}$ **21.** $6\sqrt{2xy}-5\sqrt{2xy}=\sqrt{2xy}$
22. $6\sqrt{5ab}-2\sqrt{5ab}=4\sqrt{5ab}$ **23.** $1.2\sqrt{3x}-1.0\sqrt{3x}=0.2\sqrt{3x}$ **24.** $-1.5\sqrt{2a}+0.4\sqrt{2a}=-1.1\sqrt{2a}$
25. $2\sqrt{5y}+6\sqrt{5y}-\sqrt{5y}=7\sqrt{5y}$ **26.** $6\sqrt{2w}-3\sqrt{2w}-5\sqrt{2w}=-2\sqrt{2w}$ **27.** $3y\sqrt{y}+2y\sqrt{y}=5y\sqrt{y}$
28. $4x\sqrt{x}-4x\sqrt{x}=0$ **29.** $4x\sqrt{2x}+15\sqrt{2x}$ **30.** $-6a\sqrt{3a}+5\sqrt{3a}$ **31.** $10x\sqrt{2x}-15x\sqrt{2x}=-5x\sqrt{2x}$
32. $-6y\sqrt{3y}+2y\sqrt{3y}=-4y\sqrt{3y}$ **33.** $9x\sqrt{3}-8x\sqrt{3}=x\sqrt{3}$ **34.** $-10y\sqrt{2}+8y\sqrt{2}=-2y\sqrt{2}$
35. $2y\sqrt{6y}-6y\sqrt{6}$ **36.** $6x^2\sqrt{6}-6x^2\sqrt{6}=0$ **37.** $-30x^3\sqrt{2}-10x^3\sqrt{2}=-40x^3\sqrt{2}$ **38.** $6x\sqrt{5x}-4x^2\sqrt{5}$

Exercise 8.4

1. $\sqrt{35}$ **2.** $\sqrt{33}$ **3.** $2\sqrt{11}$ **4.** $5\sqrt{3}$ **5.** $a\sqrt{15}$ **6.** $3\sqrt{2x}$ **7.** $5\sqrt{2x}$ **8.** $2b\sqrt{6}$ **9.** $6\sqrt{30}$
10. $-15a^2\sqrt{6a}$ **11.** $6x^3\sqrt{5}$ **12.** $48\sqrt{2}$ **13.** $-6b\sqrt{a}$ **14.** $10x\sqrt{y}$ **15.** $\sqrt{6}+2\sqrt{15}$ **16.** $\sqrt{10}-4$
17. $2\sqrt{5}-6\sqrt{3}$ **18.** $5\sqrt{3}+15$ **19.** $2x-16\sqrt{5x}$ **20.** $-6\sqrt{ab}-9b$ **21.** $2\sqrt{3}-18+4\sqrt{15}$
22. $5\sqrt{2}-30+10\sqrt{5}$ **23.** $6\sqrt{ab}+2a\sqrt{b}-4a$ **24.** $3\sqrt{xy}+6x\sqrt{y}-12x$ **25.** $16+3\sqrt{14}$ **26.** $18+4\sqrt{15}$
27. $22+18\sqrt{3}$ **28.** $32+11\sqrt{2}$ **29.** $5-\sqrt{21}$ **30.** $24+28\sqrt{3}$ **31.** $-6+9\sqrt{2}$ **32.** $-26+10\sqrt{5}$
33. $34+2\sqrt{14}$ **34.** $19-2\sqrt{15}$ **35.** $29-12\sqrt{5}$ **36.** $43+30\sqrt{2}$ **37.** $53+10\sqrt{6}$ **38.** $29-4\sqrt{30}$
39. $18-12\sqrt{2}$ **40.** $137+20\sqrt{15}$ **41.** $a-2a\sqrt{b}+ab$ **42.** $xy-4x\sqrt{y}+4x$ **43.** $5a-\sqrt{3a}+2a\sqrt{3a}$
44. $5xy-10x\sqrt{2y}+\sqrt{2y}-4$ **45.** $5x\sqrt{5x}-5x^2-10\sqrt{x}+2x\sqrt{5}$ **46.** $4a^2-24a\sqrt{3a}-2a\sqrt{3}+36\sqrt{a}$

Exercise 8.5

1. 4 **2.** $\frac{1}{2}$ **3.** $\frac{1}{2}$ **4.** $\frac{\sqrt{39}}{3}$ **5.** 3 **6.** $\sqrt{3}$ **7.** $\frac{\sqrt{6}}{x^2}$ **8.** $\frac{2\sqrt{3}}{x}$ **9.** $\frac{5\sqrt{6}}{6}$ **10.** $\frac{3\sqrt{5}}{5}$ **11.** $\frac{x\sqrt{3}}{3}$

12. $\frac{\sqrt{6y}}{6}$ **13.** $\frac{2\sqrt{2x}}{x}$ **14.** $\frac{3\sqrt{2y}}{y}$ **15.** $\frac{\sqrt{3}}{2}$ **16.** $\frac{2\sqrt{5}}{5}$ **17.** $\frac{7\sqrt{a}}{a^2}$ **18.** $\frac{5\sqrt{b}}{b^3}$ **19.** $\frac{\sqrt{2x}}{2x^2}$ **20.** $\frac{y\sqrt{5x}}{5x^2}$

21. $\frac{\sqrt{x}}{2x^4}$ **22.** $\frac{\sqrt{10x}}{6x^3}$ **23.** $\frac{\sqrt{14}}{7}$ **24.** $\frac{\sqrt{30}}{10}$ **25.** $\frac{\sqrt{3a}}{ab}$ **26.** $\frac{\sqrt{10b}}{2ab}$ **27.** $\frac{9\sqrt{2x}}{8x}$ **28.** $\frac{3\sqrt{2x}}{10x}$ **29.** $3\sqrt{2}+3$

30. $\frac{2(\sqrt{3}-1)}{2}=\sqrt{3}-1$ **31.** $\frac{\sqrt{2}-\sqrt{5}}{-1}=-\sqrt{2}+\sqrt{5}$ **32.** $\frac{4(\sqrt{7}+\sqrt{2})}{5}$

33. $-x(\sqrt{7}-2\sqrt{2})$ or $-x\sqrt{7}+2x\sqrt{2}$ **34.** $\frac{x(\sqrt{6}-3\sqrt{5})}{-39}$ or $\frac{-x\sqrt{6}+3x\sqrt{5}}{39}$ **35.** $\frac{x(\sqrt{3}+2x)}{3-4x^2}$ or $\frac{\sqrt{3}x+2x^2}{3-4x^2}$

36. $\frac{y(\sqrt{5}+2y)}{5-4y^2}$ or $\frac{\sqrt{5}y+2y^2}{5-4y^2}$ **37.** $2\sqrt{14}-7$ **38.** $-5+\sqrt{30}$ **39.** $\frac{7-2\sqrt{10}}{3}$ **40.** $\frac{2(5+\sqrt{21})}{4}=\frac{5+\sqrt{21}}{2}$

41. $-3\sqrt{35} - 6\sqrt{10} - 6\sqrt{7} - 12\sqrt{2}$ **42.** $\dfrac{\sqrt{30} + 4\sqrt{42} + 2\sqrt{10} + 8\sqrt{14}}{-107}$ **43.** $\dfrac{(x-4)(\sqrt{x}-2)}{x-4} = \sqrt{x} - 2$

44. $\dfrac{(x-9)(\sqrt{x}+3)}{x-9} = \sqrt{x} + 3$

Exercise 8.6

(These answers for problems 1 to 20 are based on the table of square roots in Appendix A. If you use a calculator, you may obtain a slightly different answer.)

1. 7.141 **2.** 8.062 **3.** 9.592 **4.** 6.708 **5.** 9.317 **6.** 5.742 **7.** 13.638 **8.** 12.247 **9.** 5.745 **10.** 2.427 **11.** 18.654 **12.** 20.784 **13.** 10.712 **14.** -4.704 **15.** 0.674 **16.** 0.480 **17.** 1.225 **18.** 0.845 **19.** 0.449 **20.** -1.838 **21.** 3 **22.** 2 **23.** 2 **24.** 3 **25.** 3 **26.** 2 **27.** 1 **28.** 0 **29.** -5 **30.** -3 **31.** Not a real root **32.** 4 **33.** -2 **34.** Not a real root **35.** 3 **36.** -2 **37.** w^2 **38.** y^3 **39.** x^3 **40.** x^4 **41.** x^3 **42.** y^7 **43.** y^2 **44.** x^3 **45.** $2x^2$ **46.** $3y$ **47.** $4x^4y^3$ **48.** $5y^3w^5$ **49.** $6x$ **50.** $10x^2$ **51.** $4w^3$ **52.** $-18x^3\sqrt[3]{3}$ **53.** $\dfrac{2}{x}$ **54.** $\dfrac{5}{3x^3}$

Exercise 8.7

1. $y = \frac{7}{4}x, y = \frac{63}{4}$ **2.** $y = \frac{3}{4}x, y = \frac{15}{4}$ **3.** $y = \frac{3}{2}x^3, y = \frac{1029}{2}$ **4.** $y = \frac{7}{2}\sqrt{x}, y = \frac{35}{2}$

5. $p = \frac{13}{30}d$, $p = 169$ lb/sq. in. **6.** $w_{moon} = 0.16w_{earth}$, $w_{moon} = 144$ kilograms **7.** $t = \frac{7}{8}s^3$, $t = 56$ minutes

8. $d = 16t^2$, $d = 784$ feet **9.** $y = \frac{48}{x}, y = \frac{48}{7}$ **10.** $y = \frac{117}{x}, y = \frac{117}{11}$ **11.** $y = \frac{2500}{x^2}, y = \frac{2500}{9}$ **12.** $y = \frac{324}{x^3}, y = \frac{3}{2}$

13. $C = \frac{210}{R}, C = \frac{105}{2}$ ampere **14.** $H = \frac{18{,}000}{t}, H = \frac{9000}{7}$ BTU/hr **15.** $w = \frac{1.6 \times 10^{10}}{r^2}, w = \frac{4000}{9}$ pounds

16. $p = \frac{0.48}{s^3}$, $p = 7.5$ lb/sq. in.

Practice Quizzes

Section 8.1–8.3 Practice Quiz

1. 12 **2.** No real solution **3.** $\frac{5}{6}$ **4.** 0.7 **5.** $-\frac{8}{9}$ **6.** $\frac{1}{13}$ **7.** $4x^3\sqrt{3x}$ **8.** $7x^3y\sqrt{2y}$ **9.** $9ab^4\sqrt{a}$ **10.** $2^{10}x^2y^3z^3\sqrt{xz}$ **11.** $5\sqrt{5} - 5$ **12.** $6\sqrt{2}$ **13.** $3\sqrt{3}$ **14.** $3\sqrt{6x}$ **15.** $4a\sqrt{2a}$

Section 8.4–8.7 Practice Quiz

1. $4x\sqrt{6}$ **2.** $x\sqrt{10} - 5\sqrt{2x}$ **3.** $11 - 4\sqrt{6}$ **4.** $18 - 2\sqrt{30} + 3\sqrt{15} - 5\sqrt{2}$

5. $2a\sqrt{15} - 4a\sqrt{3a} + \sqrt{5b} - 2\sqrt{ab}$ **6.** $\dfrac{x\sqrt{y}}{y}$ **7.** $\dfrac{5\sqrt{3}}{6}$

8. $\dfrac{2\sqrt{5} + 4\sqrt{3}}{-7}$ or $\dfrac{-2\sqrt{5} - 4\sqrt{3}}{7}$ **9.** $\dfrac{6 - \sqrt{6} + 6\sqrt{2} - 2\sqrt{3}}{10}$ **10.** 5.449 **11.** $-2x^2$ **12.** $5x^3y$

13. No real solution **14.** $\dfrac{2x}{y^2}$ **15.** $A = kr^2$, $A = 314r^2$, $A = 78.5$ square meters

Practice Test Chapter 8

1. 9 **2.** $\frac{3}{10}$ **3.** $4xy^3\sqrt{3y}$ **4.** $10xz^2\sqrt{xy}$ **5.** $\sqrt{5}$ **6.** $8\sqrt{a} + 5\sqrt{2a}$ **7.** $12ab$ **8.** $3\sqrt{2} - \sqrt{6} + 45$

9. $10 - 4\sqrt{6}$ **10.** $19 + \sqrt{10}$ **11.** $\dfrac{\sqrt{5x}}{5}$ **12.** $\dfrac{\sqrt{3}}{2}$ **13.** $\dfrac{2\sqrt{15} - \sqrt{6} - 2\sqrt{10} + 2}{18}$

14. $\dfrac{3a(\sqrt{5} - \sqrt{2})}{3} = a\sqrt{5} - a\sqrt{2}$ **15.** 1.549 (rounded to nearest thousandth) **16.** $3x^2y^3$ **17.** $-2x^2$

18. $I = \dfrac{k}{d^2}, I_{old} = \dfrac{k}{(8)^2}, I_{new} = \frac{1}{4}I_{old}$, $d = 16$ inches

CHAPTER 9

Pretest Chapter 9

1. $x = 14, -2$ **2.** $x = 0, \frac{4}{3}$ **3.** $x = \frac{2}{5}, 4$ **4.** $x = -4 + \sqrt{11}, -4 - \sqrt{11}$ **5.** $x = \dfrac{-3 + \sqrt{65}}{4}, \dfrac{-3 - \sqrt{65}}{4}$

6. $x = \frac{5}{3}, -1$ **7.** $x = \frac{1}{4}, -2$ **8.** $x = \sqrt{52} = 2\sqrt{13}$ **9.** $x = \sqrt{11}$ **10.** $x = 7$

11.

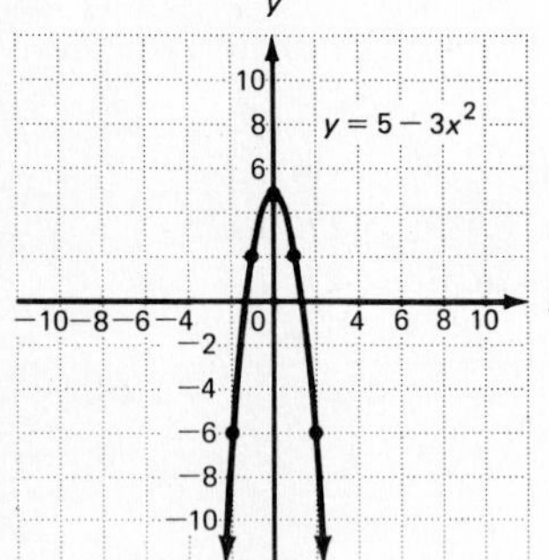

12.

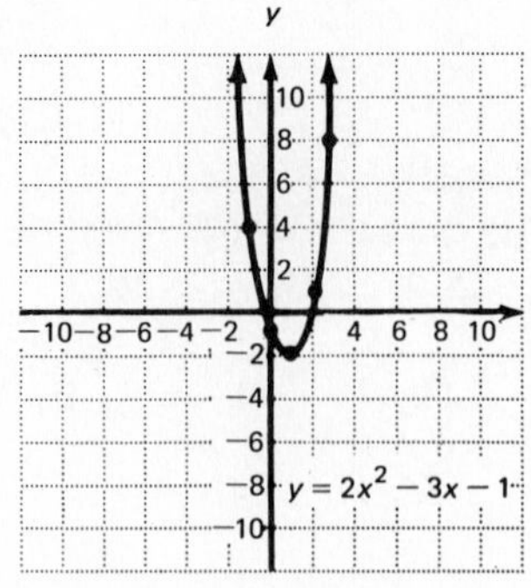

13. $6 = x = \text{base}$
$13 = 2x + 1 = \text{altitude}$
$\frac{1}{2}x(2x + 1) = 39$

14. $\frac{4}{3}, \frac{2}{3} = x = \text{number}$
$\frac{3}{4}, \frac{3}{2} = \dfrac{1}{x} = \text{reciprocal}$
$x + \dfrac{2}{x} = \dfrac{17}{6}$

Exercise 9.1

1. 5, −2 **2.** −6, 4 **3.** −5, −8 **4.** −7, −9 **5.** $-\frac{1}{2}, 3$ **6.** $-\frac{1}{3}, 2$ **7.** $-\frac{1}{4}, 3$ **8.** $\frac{1}{2}, -8$ **9.** 8, −2
10. 7, −3 **11.** 5, 2 **12.** 6, 1 **13.** $\frac{2}{3}, \frac{3}{2}$ **14.** $\frac{3}{5}, -\frac{5}{2}$ **15.** 0, 8 **16.** $0, \frac{1}{5}$ **17.** $0, \frac{2}{3}$ **18.** $0, \frac{4}{3}$ **19.** 0, 1
20. $0, \frac{7}{6}$ **21.** −5, 4 **22.** 8, 3 **23.** 6 **24.** 7 **25.** $-\frac{1}{2}$ **26.** $\frac{2}{3}$ **27.** 6, 2 **28.** 5, 3 **29.** −4, −2
30. −3, −2 **31.** 9, −6 **32.** 3, 6 **33.** 4, −1 **34.** 1 **35.** −6, 5 **36.** 7, −1 **37.** 0, −2 **38.** 0, 6

Exercise 9.2

1. $x = \pm 7$ **2.** $x = \pm 10$ **3.** $x = \pm 5\sqrt{3}$ **4.** $x = \pm 2\sqrt{10}$ **5.** $x = \pm 5$ **6.** $x = \pm 4$ **7.** $x = \pm\sqrt{3}$
8. $x = \pm\sqrt{3}$ **9.** $x = \pm 6$ **10.** $x = \pm 5$ **11.** $x = 3 \pm \sqrt{5}$ **12.** $x = -2 \pm \sqrt{3}$ **13.** $x = \dfrac{-5 \pm \sqrt{2}}{2}$

14. $x = \dfrac{4 \pm \sqrt{6}}{3}$ **15.** $(x + 7)^2 = 64$, $x = 1, -15$ **16.** $(x - 1)^2 = 6$, $x = 1 - \sqrt{6}, 1 + \sqrt{6}$ **17.** $(x - 2)^2 = 15$, $x = 2 + \sqrt{15}$, $x = 2 - \sqrt{15}$ **18.** $(x + 2)^2 = 16$, $x = 2, -6$

19. $(x + 3)^2 = 2$, $x = -3 + \sqrt{2}, -3 - \sqrt{2}$ **20.** $(x - 6)^2 = 40$, $x = 6 + 2\sqrt{10}$, $x = 6 - 2\sqrt{10}$ **21.** $(x - 5)^2 = 29$, $x = 5 + \sqrt{29}$, $x = 5 - \sqrt{29}$ **22.** $(x + 10)^2 = 28$, $x = -10 + 2\sqrt{7}$, $x = -10 - 2\sqrt{7}$

23. $(x + \frac{3}{2})^2 = \frac{9}{4}$, $x = 0, -3$ **24.** $(x + \frac{5}{2})^2 = \frac{37}{4}$, $x = \dfrac{-5 + \sqrt{37}}{2}$, $x = \dfrac{-5 - \sqrt{37}}{2}$ **25.** $(x - \frac{1}{2})^2 = \frac{25}{4}$, $x = 3, -2$ **26.** $(x + \frac{1}{2})^2 = \frac{81}{4}$, $x = 4, -5$ **27.** $(x - \frac{3}{4})^2 = \frac{81}{16}$, $x = 3, -\frac{3}{2}$

28. $(x + \frac{5}{2})^2 = \frac{3}{4}$, $x = \dfrac{-5 + \sqrt{3}}{2}$, $x = \dfrac{-5 - \sqrt{3}}{2}$ **29.** $(x - 1)^2 = \frac{1}{4}$, $x = \frac{3}{2}, \frac{1}{2}$ **30.** $(x - 1)^2 = \frac{5}{3}$, $x = \dfrac{3 + \sqrt{15}}{3}$, $x = \dfrac{3 - \sqrt{15}}{3}$ **31.** $(y + 1)^2 = \frac{9}{2}$, $y = \dfrac{-2 + 3\sqrt{2}}{2}$, $y = \dfrac{-2 - 3\sqrt{2}}{2}$ **32.** $(y - \frac{7}{4})^2 = \frac{121}{16}$, $y = \frac{9}{2}, y = -1$

33. $(y + \frac{5}{6})^2 = \frac{49}{36}$, $y = \frac{1}{3}, -2$ **34.** $(y - \frac{4}{3})^2 = \frac{13}{9}$, $y = \dfrac{4 + \sqrt{13}}{3}$, $y = \dfrac{4 - \sqrt{13}}{3}$

Exercise 9.3

1. $x = \frac{-4 \pm \sqrt{12}}{2} = -2 \pm \sqrt{3}$ **2.** $x = \frac{2 \pm \sqrt{20}}{2} = 1 \pm \sqrt{5}$ **3.** $x = \frac{5 \pm \sqrt{33}}{2}$ **4.** $x = \frac{-5 \pm \sqrt{13}}{2}$

5. $x = \frac{9}{2}, -1$ **6.** $x = \frac{6 \pm \sqrt{28}}{4} = \frac{3 \pm \sqrt{7}}{2}$ **7.** $x = \frac{-1 \pm \sqrt{21}}{10}$ **8.** $x = \frac{-12 \pm \sqrt{184}}{4} = \frac{-6 \pm \sqrt{46}}{2}$

9. $x = \frac{5 \pm \sqrt{17}}{4}$ **10.** $x = \frac{2}{5}, -1$ **11.** $x = \frac{3 \pm \sqrt{33}}{12}$ **12.** $x = \frac{3}{4}, 1$ **13.** $x = \frac{-5 \pm \sqrt{65}}{4}$

14. $x = \frac{-12 \pm \sqrt{240}}{16} = \frac{-3 \pm \sqrt{15}}{4}$ **15.** $x = \frac{6 \pm \sqrt{108}}{18} = \frac{1 \pm \sqrt{3}}{3}$ **16.** $x = -\frac{1}{3}, \frac{5}{2}$ **17.** No real solution

18. $x = 4, 1$ **19.** $x = 6, 1$ **20.** No real solution **21.** $y = \frac{-7 \pm \sqrt{109}}{10}$ **22.** $y = \frac{-5 \pm \sqrt{73}}{4}$ **23.** $m = -\frac{3}{2}, 3$

24. $m = -\frac{1}{3}, 2$ **25.** $t = \frac{2}{3}, \frac{3}{2}$ **26.** $t = -\frac{1}{3}, \frac{3}{2}$ **27.** $s = \frac{1}{3}, 3$ **28.** $p = -5, 3$ **29.** $p = \frac{1 \pm \sqrt{21}}{2}$

30. $s = \frac{1 \pm \sqrt{245}}{2} = \frac{1 \pm 7\sqrt{5}}{2}$ **31.** $y = \frac{3 \pm \sqrt{233}}{14}$ **32.** $\frac{-9 \pm \sqrt{93}}{2}$ **33.** $x = \pm\sqrt{\frac{7}{3}} = \pm\frac{\sqrt{21}}{3}$ **34.** $x = 0, -\frac{5}{7}$

Exercise 9.4

1. $c = 5$ **2.** $a = \sqrt{32} = 4\sqrt{2}$ **3.** $b = 5$ **4.** $c = 17$ **5.** $c = \sqrt{11}$ **6.** $c = \sqrt{32} = 4\sqrt{2}$ **7.** $a = 6$
8. $b = 12$ **9.** $a = \sqrt{18} = 3\sqrt{2}$ **10.** $a = \sqrt{2}$ **11.** 9.9 feet **12.** 98 feet **13.** 127.3 feet **14.** 22.4 feet
15. $x = 121$ **16.** $x = 12.25$ **17.** $x = 17$ **18.** $x = 44$ **19.** $x = -\frac{2}{3}$ **20.** $x = 8$ **21.** $x = 7$ **22.** $x = 6$
23. $x = \frac{81}{2}$ **24.** $x = 3$ **25.** $x = 5$ **26.** $x = 3, 2$ **27.** $y = 0, 3$ **28.** $y = 0$ **29.** $y = 0, 1$ **30.** $y = 3, 4$
31. $y = 5$ **32.** $y = \frac{1}{3}$ **33.** No solution **34.** $x = \frac{3}{2}$

Exercise 9.5

1.

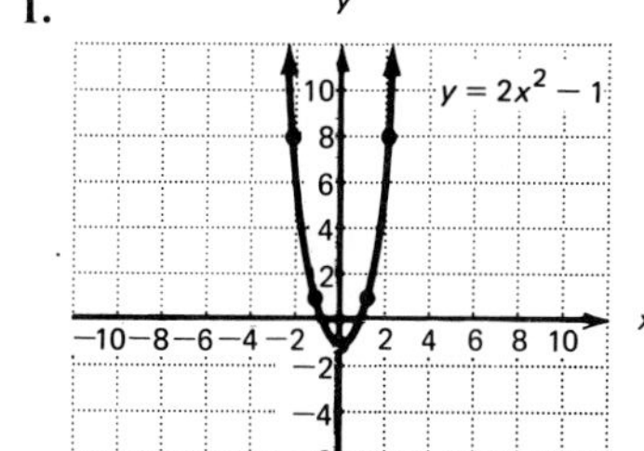

2.

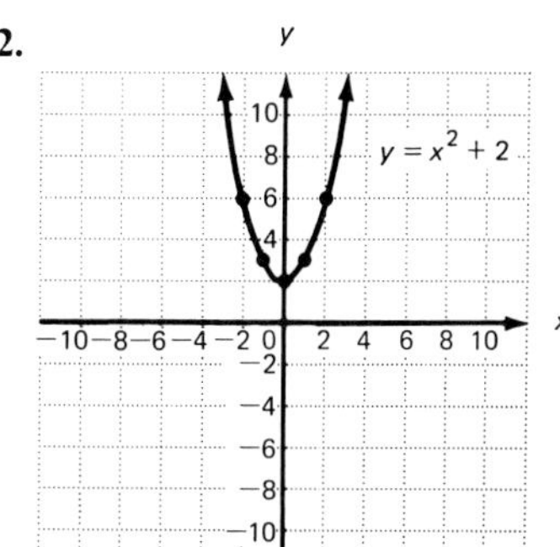

3.

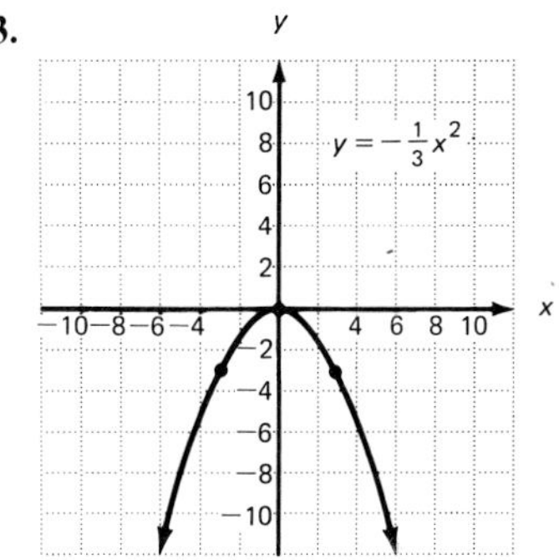

4.

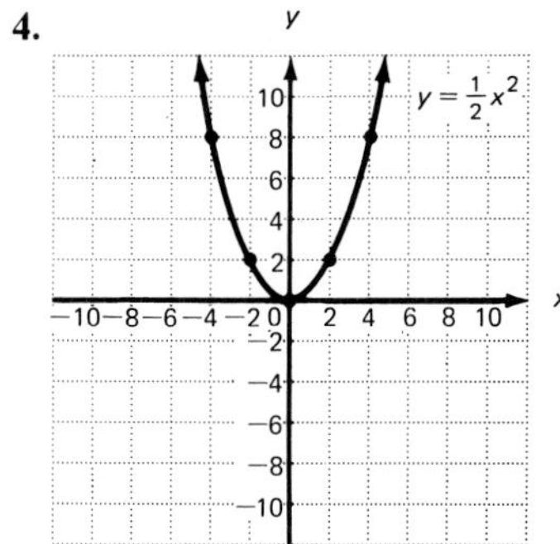

5.

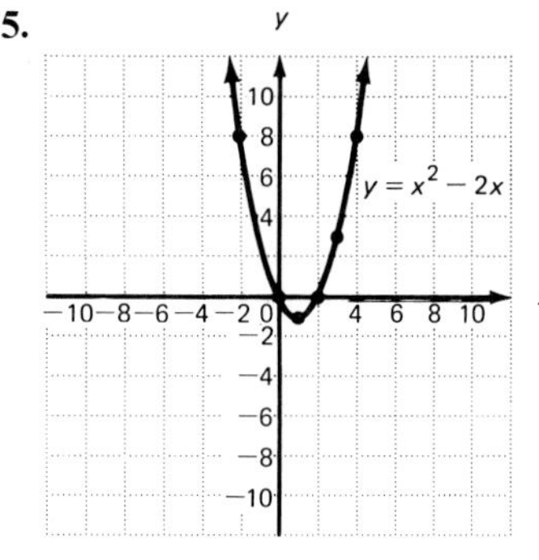

6.

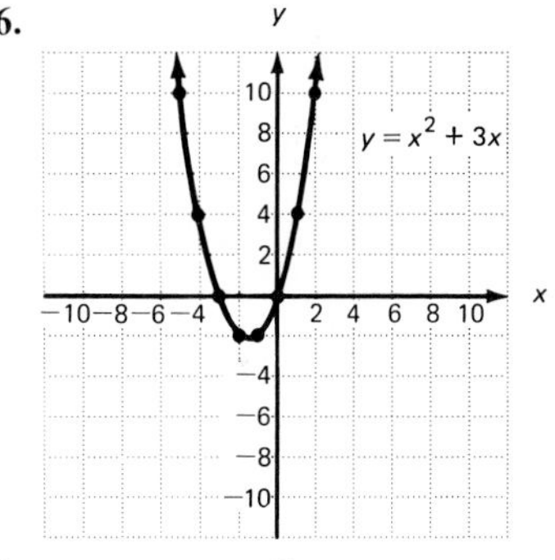

7.

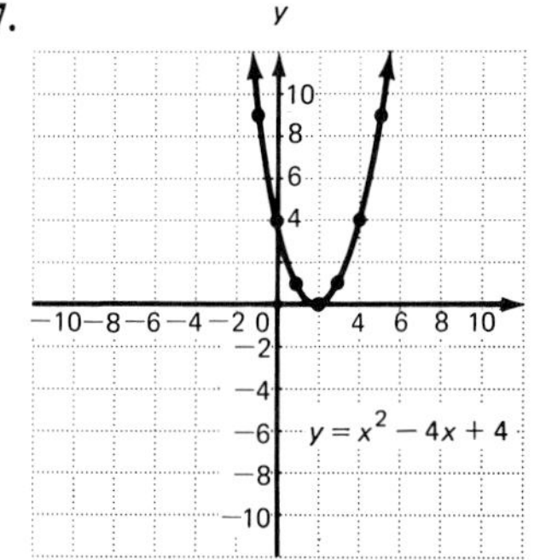

8.

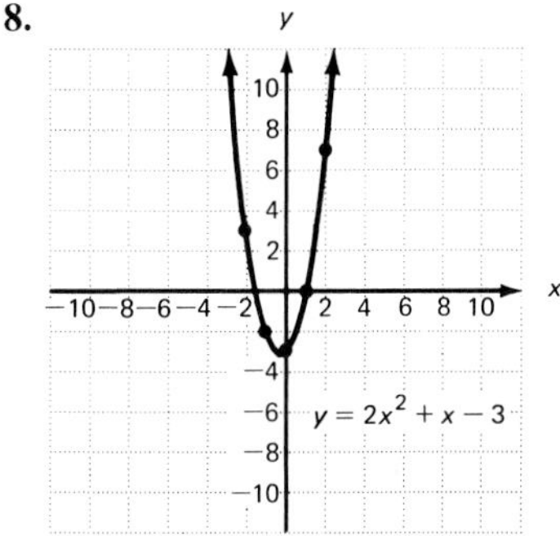

9.

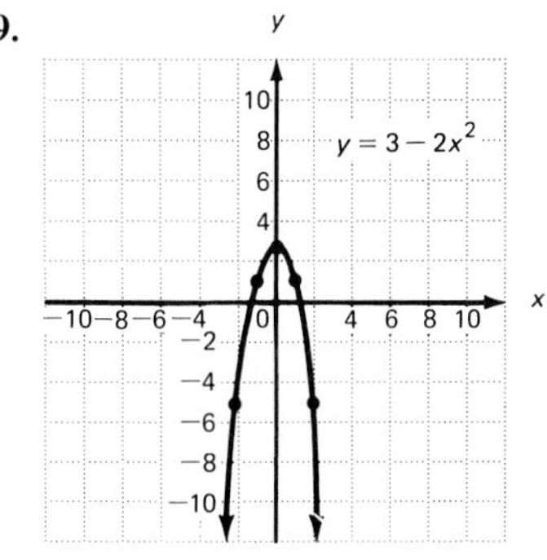

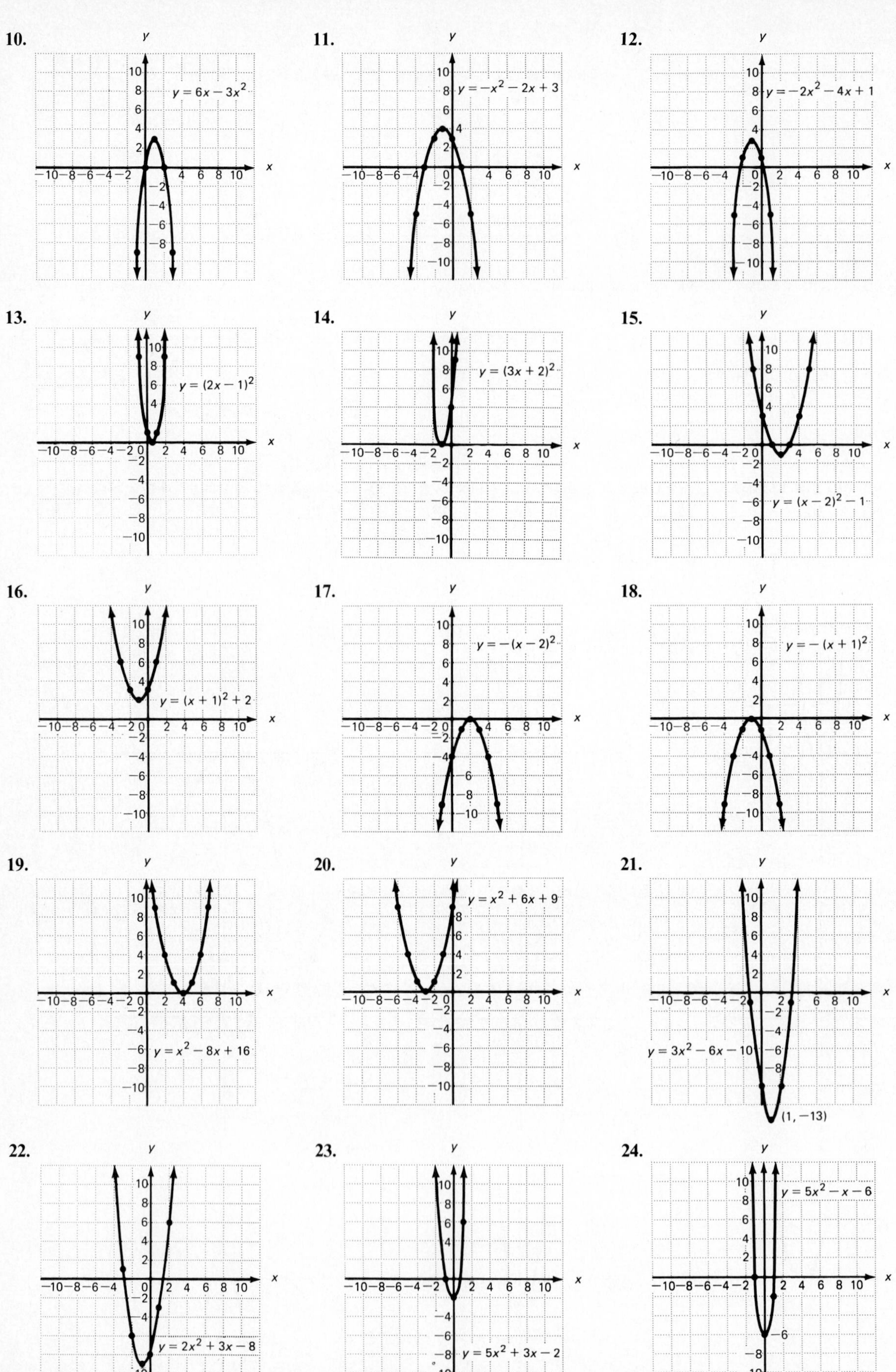
10.
$y = 6x - 3x^2$
11.
$y = -x^2 - 2x + 3$
12.
$y = -2x^2 - 4x + 1$
13.
$y = (2x - 1)^2$
14.
$y = (3x + 2)^2$
15.
$y = (x - 2)^2 - 1$
16.
$y = (x + 1)^2 + 2$
17.
$y = -(x - 2)^2$
18.
$y = -(x + 1)^2$
19.
$y = x^2 - 8x + 16$
20.
$y = x^2 + 6x + 9$
21.
$y = 3x^2 - 6x - 10$
(1, −13)
22.
$y = 2x^2 + 3x - 8$
23.
$y = 5x^2 + 3x - 2$
24.
$y = 5x^2 - x - 6$

Exercise 9.6

1. 14 m $= x =$ length
10 m $= \frac{x}{2} + 3 =$ width
$x\left(\frac{x}{2} + 3\right) = 140$

2. 11 cm $= 2x - 1 =$ base
6 cm $= x =$ altitude
$\frac{1}{2}(x)(2x - 1) = 33$

3. $\frac{5}{3}, \frac{3}{5} = x =$ number
$\frac{3}{5}, \frac{5}{3} = \frac{1}{x} =$ reciprocal
$x + \frac{1}{x} = \frac{34}{15}$

4. $\frac{1}{2}, 6 = x =$ number
$2, \frac{1}{6} = \frac{1}{x} =$ reciprocal
$x + \frac{3}{x} = \frac{13}{2}$

5. The numbers are
$14 = x =$ one number
$4 = y =$ second number
$x - y = 10$
$xy = 56$
$\frac{56}{y} - y = 10$
$y^2 + 10y - 56 = 0$

6. $-11 = x =$ one number
$-6 = y =$ second number
$x + y = -17$
$xy = 66$
$\frac{66}{y} + y = -17$
$y^2 + 17y + 66 = 0$

7. 12 m $= x =$ length
5 m $= x - 7 =$ width
$x^2 + (x - 7)^2 = 169$

8. 12 feet $= x =$ house to ladder (bottom)
16 feet $= x + 4 =$ distance up the house
$x^2 + (x + 4)^2 = 400$

9. $t = 3$ seconds
$-5t^2 + 13t + 6 = 0$
at $t = 2$ seconds the ball is 12 meters above the ground

10. $t = 6$ seconds
$-5t^2 + 180 = 0$
at $t = 3$ seconds, the pebble has dropped 135 meters

11. 20 people on team
10 people contributed
(student)(cost/student) = 80 therefore $(s)(c) = 80$
$(s - 10)(c + 4) = 80$
$4s^2 - 40s - 800 = 0$

12. $5 = x =$ another number
$12 = 2x + 2 =$ one number
$(2x + 2)^2 - x^2 = 119$

13. $vt = 2400$, $v = 600$ mph
$(v + 200)(t - 1) = 2400$
$v^2 + 200v - 480{,}000 = 0$

14. $10 = x =$ length in inches of the side of the old tile
$12 = x + 2 =$ length in inches of the side of the new tile

15. 5 were produced
$2x^2 - 7x = 15$

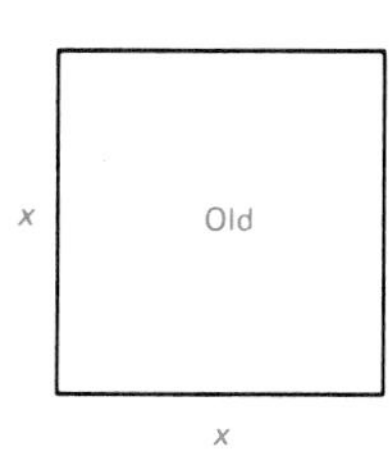

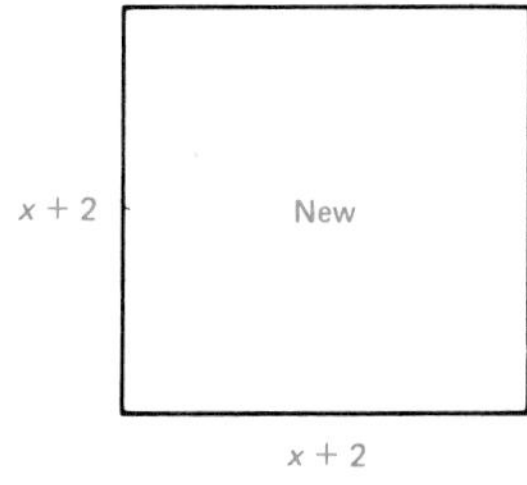

$864x^2 =$ area $\quad 600(x + 2)^2 =$ same area
$864x^2 = 600(x + 2)^2$
This equation can be simplified to $11x^2 - 100x - 100 = 0$

16. Either 8 amperes or 2 amperes will produce 64 watts of power; thus the *minimum amount* of amperes is 2.
$-4I^2 + 40I = 64$

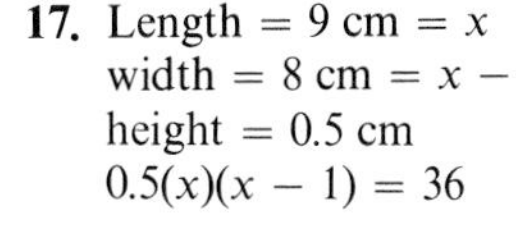

17. Length = 9 cm $= x$
width = 8 cm $= x - 1$
height = 0.5 cm
$0.5(x)(x - 1) = 36$

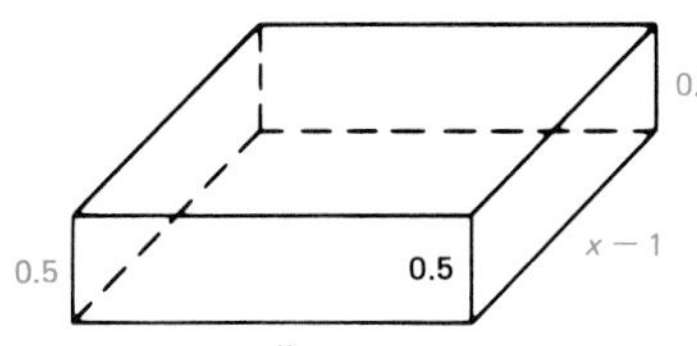

18. length = 8 cm $= x$
width = 8 cm $= x$
height = 5 cm
$5(x)(x) = 320$

Practice Quizzes

Section 9.1–9.3 Practice Quiz

1. $x = 9, -3$ **2.** $x = 0, 11$ **3.** $x = 3, -1$ **4.** $x = \frac{3 + \sqrt{3}}{2}, \frac{3 - \sqrt{3}}{2}$ **5.** $x = -2 + \sqrt{14}, -2 - \sqrt{14}$

6. $x = \frac{3}{2}, -4$ **7.** $x = \frac{1}{2}, -3$ **8.** $x = -4, 2$ **9.** $x = \frac{1}{3}, -2$ **10.** $x = 3, -2$ **11.** $x = \frac{-5 + \sqrt{3}}{2}, \frac{-5 - \sqrt{3}}{2}$

12. No real number

Section 9.4–9.6 Practice Quiz

1. $x = \sqrt{23}$ **2.** $x = 4\sqrt{5}$ **3.** $x = 217.9$ m **4.** $x = 6$ **5.** $x = 5$ **6.** $y = 6$

7.

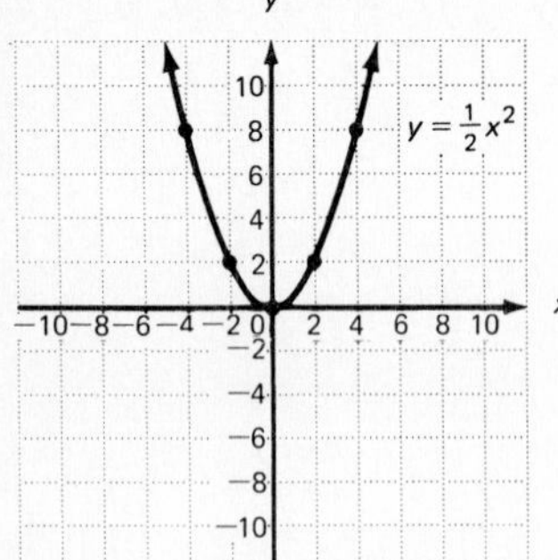

8.

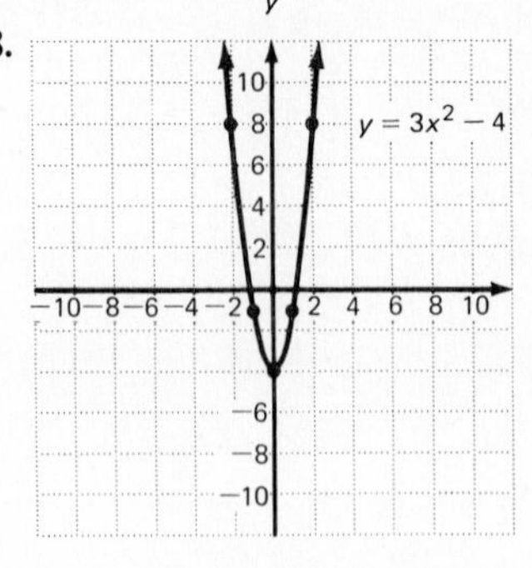

9.

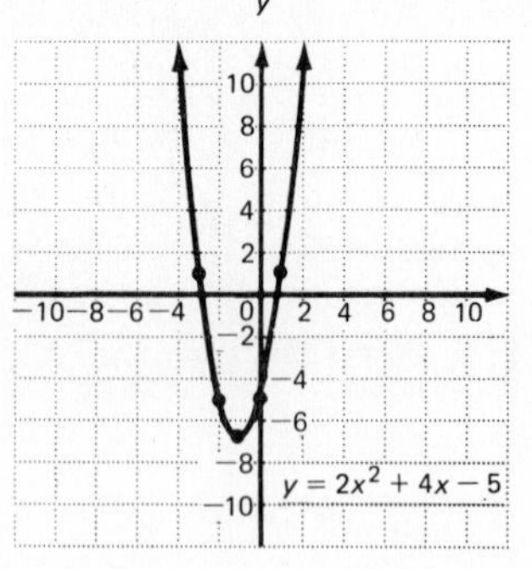

10.

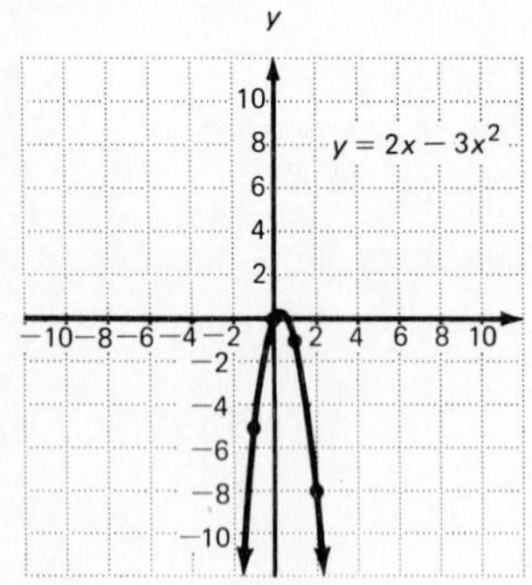

11.

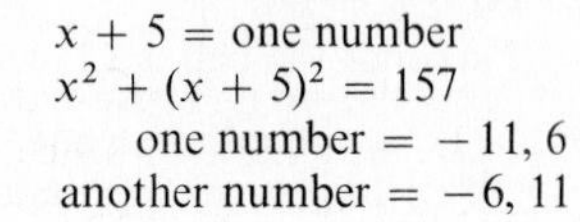

x = another number
$x + 5$ = one number
$x^2 + (x + 5)^2 = 157$
one number $= -11, 6$
another number $= -6, 11$

12. let x = number of sides $= 11$

$$\frac{x(x - 3)}{2} = 44$$

13. Old New
$s^2 = 49$ $s^2 = 81$
$s = 7$ $s = 9$
2 meters added

Practice Test Chapter 9

1. $x = 0, -\frac{13}{3}$ **2.** $x = \frac{2}{3}, -5$ **3.** No real number **4.** $x = \frac{5}{3}, -\frac{1}{2}$ **5.** $x = \sqrt{95}$ **6.** $x = 3\sqrt{3}$ **7.** $x = \frac{35}{2}$

8. $x = 9$ **9.**

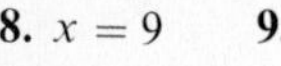

10.

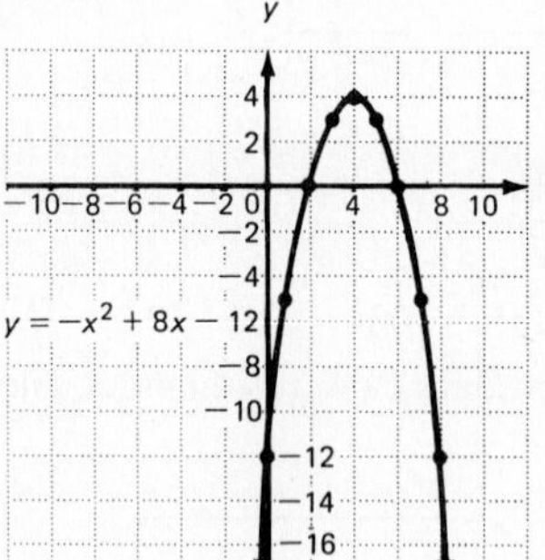

11. $t = 7$ seconds later
$-5t^2 + 33t + 14 = 0$

12. $vt = 120$
$(v - 10)(6 - t) = 1421$
$v^2 - 54v + 200 = 0$
$V = 50$, so 50 mph for first part of trip. They went 50 mph for 2.4 hours.
$V - 10 = 40$, so 40 mph for second part of trip. They went 40 mph for 3.6 hours.

CUMULATIVE EXAMINATION: CHAPTERS 1–9

1. $-2x + 21y + 18xy + 6y^2$ **2.** 14 **3.** $x = \frac{8}{5}$ **4.** $x \geq \frac{13}{5}$

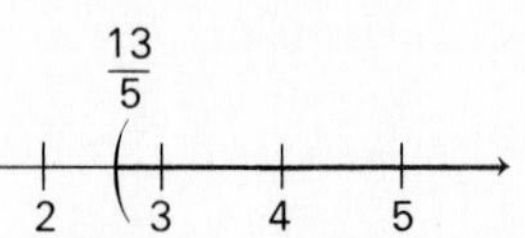

5. $2(2x + 1)(x - 5)$ **6.** $25(x^2 + 1)(x + 1)(x - 1)$ **7.** $\frac{6x - 11}{(x + 2)(x - 3)}$ or $\frac{6x - 11}{x^2 - x - 6}$ **8.** $\frac{11(x + 2)}{2x(x + 4)}$ or $\frac{11x + 22}{2x^2 + 8x}$

9. Slope $= m = \frac{5}{2}$

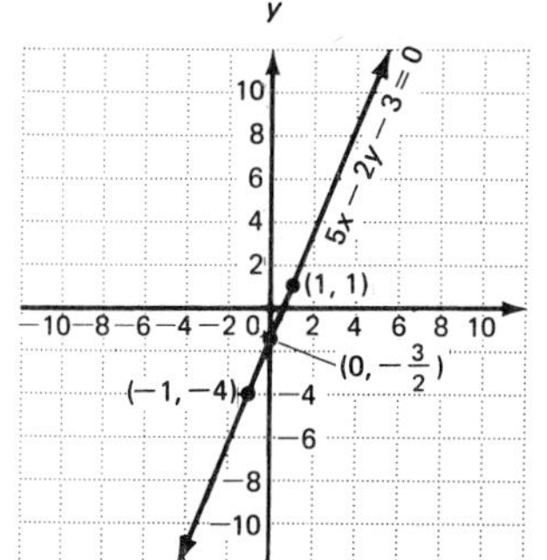

10.

Area of rectangle	= 24 sq. in.
Area of 2 of the quarter circle regions	= 14.13 sq. in.
Total area	= 38.13 sq. in.

11. $x = 7, y = -2$ **12.** $a = -2, b = -3$ **13.** $8x\sqrt{2}$ **14.** $\frac{3\sqrt{5} + \sqrt{35} + 3\sqrt{7} + 7}{-2}$ or $\frac{-3\sqrt{5} - \sqrt{35} - 3\sqrt{7} - 7}{2}$

15. $x = \frac{5}{3}, x = -1$ **16.** $x = \frac{6 \pm \sqrt{28}}{4} = \frac{3 \pm \sqrt{7}}{2}$ **17.** $x =$ the number $= 5$
$3x + 6 = 21$

18. $x =$ the width in meters $= 7$ m
$2x - 2 =$ the length in meters $= 12$ m
$2(x) + 2(2x - 2) = 38$

19. $x =$ the amount invested at 10% interest = \$3000
$7000 - x =$ the amount invested at 14% interest = \$4000
$0.10x + 0.14(7000 - x) = 860$

20. $x =$ number of reserved tickets = 200 tickets
$y =$ number of general admission tickets = 160 tickets

$x + y = 360$
$5x + 3y = 1480$

APPENDIX B

1. $x = 15$ **2.** $x = 5.25$ or $5\frac{1}{4}$ **3.** $x = 40.8$ or $40\frac{4}{5}$ **4.** $x = 21.\bar{3}$ or $21\frac{1}{3}$ **5.** 114 cups
6. Approximately 7.89 hour **7.** Approximately 19.25 cm or $19\frac{1}{4}$ cm **8.** Approximately 19.375 gal. or $19\frac{3}{8}$ gal
9. Approximately 22.86 cm **10.** Approximately 20.93 km **11.** Approximately 2.79 m **12.** Approximately 2.95 inches
13. Approximately 1089.6 grams **14.** Approximately 171.6 pounds **15.** Approximately 27.56 quarts
16. Approximately 68.13 liters **17.** Approximately 0.423 quarts or 0.846 pints
18. Approximately 1537 milliliters or 1.537 liters

APPENDIX C

1. $\frac{3}{x^2}$ **2.** $\frac{4x}{y^4}$ **3.** $\frac{1}{2xy}$ **4.** $\frac{1}{9x^4y^6}$ **5.** $\frac{3xz^3}{y^2}$ **6.** $\frac{4}{x^2y^7}$ **7.** $3x$ **8.** $2ab^2$ **9.** $\frac{x^6}{y^2z^3w^{10}}$ **10.** $\frac{2x^2w^5}{y^3z^4}$

11. $\frac{1}{64}$ **12.** $\frac{9}{36} = \frac{1}{4}$ **13.** $\frac{z^8}{9x^2y^4}$ **14.** $\frac{c^6}{8a^9}$ **15.** $\frac{1}{x^6y}$ **16.** $\frac{b^8}{a^4}$ **17.** $\frac{y^6}{-8x^9}$ or $-\frac{y^6}{8x^9}$ **18.** $\frac{x^{16}}{81y^8}$

19. 1.2378×10^5 **20.** 6.3×10^{-2} **21.** 7.42×10^{-4} **22.** 8.8961×10^{11} **23.** 7.652×10^9 **24.** 1.963×10^{-8}
25. 56,300 **26.** 177,600,000 **27.** 0.000033 **28.** 0.199 **29.** 983,000 **30.** 0.000000035 **31.** 1.0×10^1
32. 1.5×10^{-12} **33.** 3.2×10^{-19} **34.** 5.0×10^{-4} **35.** 1.5×10^1 **36.** 8.1×10^{29}

Index

A

B

C

D

E

F

G

H

I

K

L

M

N

O

P

Q

R

S

T

U

V

W

X

Y

Z

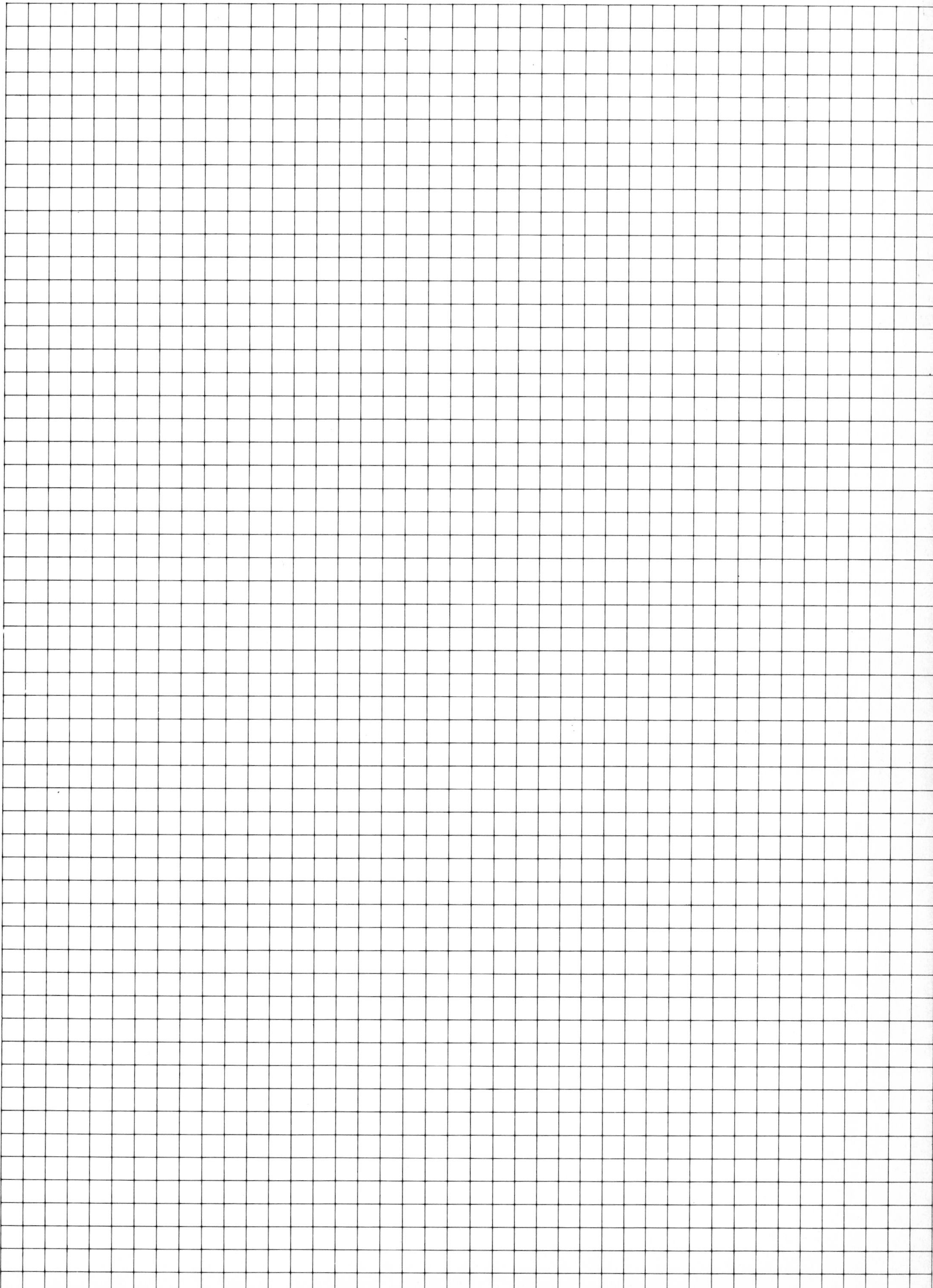

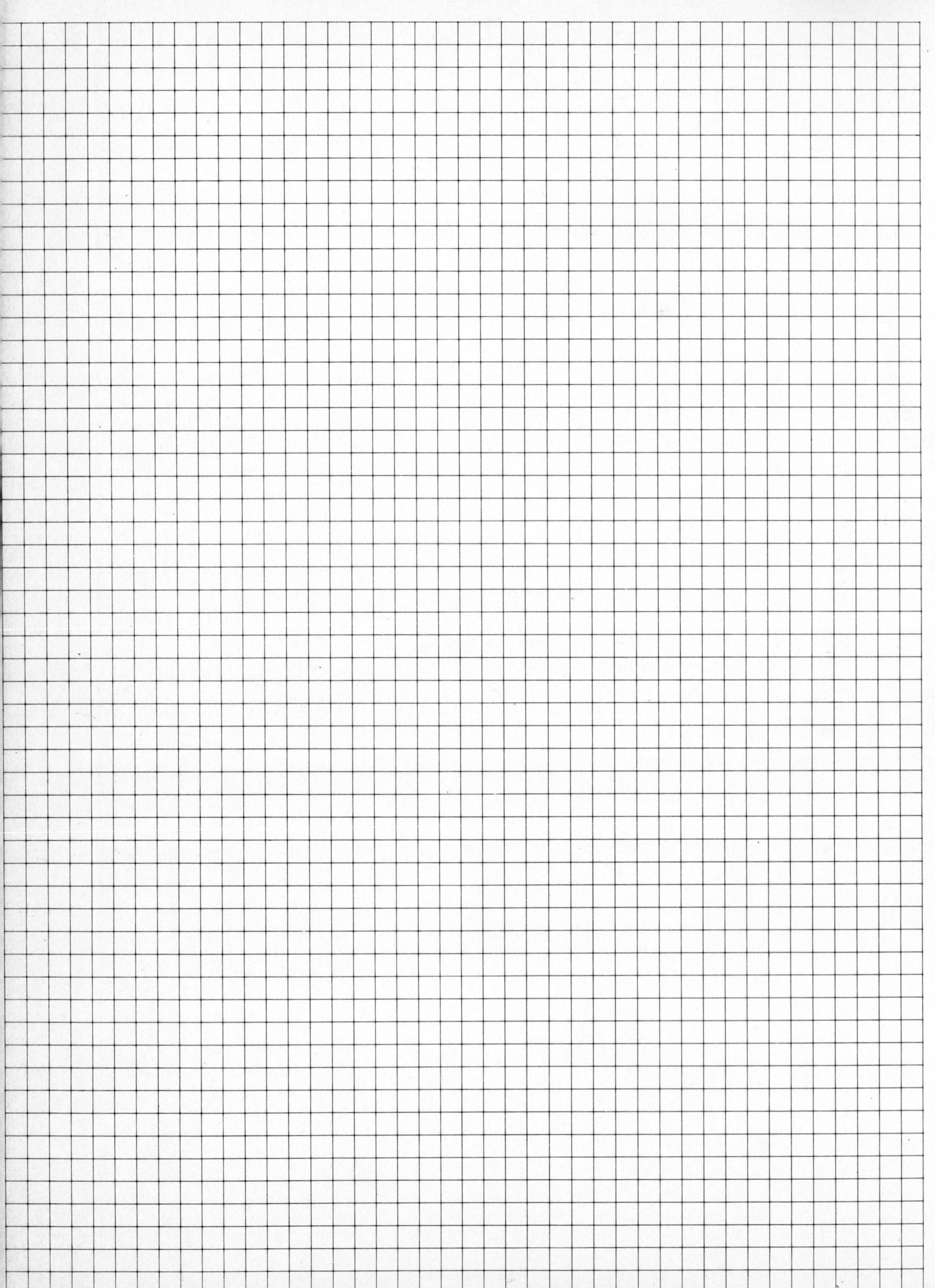